ADVANCES IN CHEMICAL PHYSICS

VOLUME XXIV

Advances in

CHEMICAL PHYSICS

EDITED BY

I. PRIGOGINE

University of Brussels,
Brussels, Belgium

AND

STUART A. RICE

Department of Chemistry
and
The James Franck Institute
The University of Chicago
Chicago, Illinois

VOLUME XXIV

AN INTERSCIENCE® PUBLICATION

JOHN WILEY AND SONS

NEW YORK · LONDON · SYDNEY · TORONTO

AN INTERSCIENCE® PUBLICATION

Library of Congress Catalog Card Number: 58-9935

ISBN 0-471-69929-2

Printed in the United States of America.

10 9 8 7 6 5 4 3 2 1

INTRODUCTION

In the last decades chemical physics has attracted an ever-increasing amount of interest. The variety of problems, such as those of chemical kinetics, molecular physics, molecular spectroscopy, transport processes, thermodynamics, the study of the state of matter, and the variety of experimental methods used, makes the great development of this field understandable. But the consequence of this breadth of subject matter has been the scattering of the relevant literature in a great number of publications.

Despite this variety and the implicit difficulty of exactly defining the topic of chemical physics, there are a certain number of basic problems that concern the properties of individual molecules and atoms as well as the behavior of statistical ensembles of molecules and atoms. This new series is devoted to this group of problems which are characteristic of modern chemical physics.

As a consequence of the enormous growth in the amount of information to be transmitted, the original papers, as published in the leading scientific journals, have of necessity been made as short as is compatible with a minimum of scientific clarity. They have, therefore, become increasingly difficult to follow for anyone who is not an expert in this specific field. In order to alleviate this situation, numerous publications have recently appeared which are devoted to review articles and which contain a more or less critical survey of the literature in a specific field.

An alternative way to improve the situation, however, is to ask an expert to write a comprehensive article in which he explains his view on a subject freely and without limitation of space. The emphasis in this case would be on the personal ideas of the author. This is the approach that has been attempted in this new series. We hope that as a consequence of this approach, the series may become especially stimulating for new research.

Finally, we hope that the style of this series will develop into something more personal and less academic than what has become the standard scientific style. Such a hope, however, is not likely to be completely realized until a certain degree of maturity has been attained—a process which normally requires a few years.

At present, we intend to publish one volume a year, and occasionally several volumes, but this schedule may be revised in the future.

In order to proceed to a more effective coverage of the different aspects of chemical physics, it has seemed appropriate to form an editorial board. I want to express to them my thanks for their cooperation.

I. Prigogine

CONTRIBUTORS TO VOLUME XXIV

J. P. Boon, Universite Libre de Bruxelles, Bruxelles, Belgium

Bernard D. Coleman, Center for Special Studies, Mellon Institute of Science, Carnegie-Mellon University, Pittsburgh, Pennsylvania

H. Ted Davis, Departments of Chemical Engineering and Chemistry, University of Minnesota, Minneapolis, Minnesota

P. A. Fleury, Bell Laboratories, Murray Hill, New Jersey

Joseph Ford, School of Physics, Georgia Institute of Technology, Atlanta, Georgia

M. D. Girardeau, Department of Physics and Institute of Theoretical Science, University of Oregon, Eugene, Oregon

R. M. Mazo, Department of Chemistry and Institute of Theoretical Science, University of Oregon, Eugene, Oregon

CONTENTS

Laser Light Scattering in Fluid Systems 1
By. P. A. Fleury and J. P. Boon

Thermodynamics of Discrete Mechanical Systems with Memory 95
By Bernard D. Coleman

The Transition from Analytic Dynamics to Statistical Mechanics 155
By Joseph Ford

Variational Methods in Statistical Mechanics 187
By M. D. Girardeau and R. M. Mazo

Kinetic Theory of Dense Fluids and Liquids Revisited 257
By H. Ted Davis

Author Index 345

Subject Index 355

LASER LIGHT SCATTERING IN FLUID SYSTEMS

P. A. FLEURY

Bell Laboratories,
Murray Hill, New Jersey

J. P. BOON

Universite Libre De Bruxelles,
Bruxelles, Belgium

CONTENTS

I. Introduction . . . 2

II. Theoretical Considerations A. General Relations . . . 5
B. Generalized Hydrodynamics for $S(\mathbf{k}, \omega)$ . . . 9
C. Internal Degrees of Freedom and Relaxation Processes . . . 14
D. Transport Function and High-Frequency Effects . . . 16
E. Light Scattering and Ultrasonics . . . 20
F. Solutions and Noncritical Mixtures . . . 22

III. Recent Advances in Experimental Technique and Apparatus . . . 25
A. High-Frequency Shifts . . . 25
B. Intermediate-Frequency Shifts . . . 26
C. Low-Frequency Shifts . . . 31

IV. Experimental Results . . . 37
A. Liquid Gas Critical Point . . . 38
B. Critical Mixing . . . 40
C. Brillouin Scattering . . . 40
1. Simple Fluids . . . 40
2. Complex Fluids . . . 41
D. Depolarized Scattering . . . 43
1. Dynamic Shear Waves . . . 44
2. Intermolecular Light Scattering . . . 50
E. Surface Scattering . . . 55
F. Solutions and Macromolecules . . . 57
Literature Search and Bibliography . . . 59

Abstract

Recent theoretical and experimental advances in the field of inelastic light scattering from fluid systems are reviewed. Three basic frequency domains are distinguished by the instrumentation used in each: (1) grating spectrometers, (2) optical interferometers, and (3) electronic spectrometers. Recent advances in each domain are discussed. A generalized hydrodynamic theory is outlined which incorporates both second-order and finite-frequency transport coefficient effects. Scattering experiments on density, anisotropy, and concentration fluctuations are discussed. Studies of surface waves, liquid crystals, intermolecular collisions, and fluid shear waves typify the kinds of experimental advances represented. Critical phenomena associated with various phase transitions in fluids and mixtures are brought up to date, and a discussion of light scattering as currently applied to macromolecular solutions is presented. The review concludes with a cross-referenced bibliography consisting of more than 500 titled entries on the subject of inelastic light scattering in fluid systems.

I. INTRODUCTION

Following the invention of the laser light source there has been an impressive increase in both the quantity and quality of optical scattering experiments performed on fluid systems of physical, chemical, and biological interest. Many of these experiments, particularly in the area of obtaining dynamical information, have been made possible by simultaneous advances in spectroscopic technique. Despite the relative newness of light scattering spectroscopy, several review articles on various aspects of the theory and experimental technique have already appeared. The article by McIntyre and Sengers[10] presents an introduction to the basic concepts of light scattering in fluids and discusses at length the information that can be and has been obtained from such experiments regarding critical phenomena in both pure fluids and fluid mixtures. Mountain[12] has presented formal hydrodynamic theory pertaining to the aspects of dynamic liquid structure inferable from light scattering experiments. An article by Boon and Deguent[32] has extended by techniques of generalized hydrodynamics the theory of light scattering in simple dense fluids. A thorough discussion of the techniques of electronic spectroscopy which utilize the coherence properties of the scattered field has been given by Cummins and Swinney.[3]

Of particular and growing interest to both chemists and biologists are the possible applications of laser scattering in fluids to probe macromolecular dynamics and chemical reaction kinetics. The theory of such experiments has been discussed in detail in two papers by Pecora.[13,14] The experimental as well as theoretical aspects of light scattering in the study of chemical reactions have been reviewed by Yeh and Keeler.[346,347] More wide ranging reviews of light scattering have been presented by Fleury,[9] Chu,[2] and Benedek.[1,217]

The purpose of this article is to review those recent developments not discussed in the above-mentioned works, and to place them in perspective relative to both the more familiar results and the areas showing promise for future exploitation by light scattering spectroscopy. Detailed derivations of theoretical expressions are often omitted here. We concentrate rather on results, both theoretical and experimental. Further, some rather sweeping exclusions must be made at the outset regarding subject matter. We exclude the topics of ordinary Raman effect in fluids, of light scattering from solids and from plasmas, and of stimulated scattering of any kind. Each of these represents such a vast literature and so active a research field in its own right that a proper discussion of them in addition to our main subject would be impossible in a review article of reasonable length.

We begin in Section II with a discussion of theoretical generalities applicable to all types of scattering experiments. Results of the generalized hydrodynamic approach to the calculation of the density correlation functions, and their extension to multicomponent systems and to systems under conditions of nonequilibrium are presented. In addition orientational and relaxation effects on the spectra are examined in Section IV.

In Section III we emphasize recent improvements in experimental techniques and apparatus. These include multiple instruments (the tandem Fabry-Perot and other compound interferometers) on the one hand, and resonance reabsorption of the incident laser frequency, on the other hand. Basic considerations of electronic or "light beating" spectroscopy are reviewed, and recent advances in the theory and construction of digital correlators for collecting and processing data are discussed briefly as well.

Section IV concentrates on new types of experimental results. These include the study of phase transitions in liquid crystals, of intermolecular light scattering in simple fluids, of capillary waves at liquid interfaces, particularly in the vicinity of critical points, and of recent developments in both the theory and experimental observations on critical phenomena. Results in fluid mixtures and macromolecular solutions of interest to chemists and biologists are also discussed.

In the last section, a rather complete bibliography of the literature since the introduction of the laser to light scattering spectroscopy in the early 1960s is included. This bibliography divides the more than 500 citations (with titles included) into a dozen categories for hopefully easy access. To our knowledge it is the most complete compilation to date of the literature in this field, and we hope it will prove useful to both the neophyte and the seasoned research worker. With the exception of stimulated scattering, the subject exclusions mentioned above apply as well to the bibliography. The stimulated scattering citations revealed in our literature search are included here primarily because of their possible interest to readers.

The footnote and reference numbering procedure we adopt here is explained at the beginning of the section, and was dictated by the desire to include a complete bibliography as well as textual references with minimum redundancy and space expenditure.

As we see in the following section, when the kinematics of the scattering process are discussed more fully, the laser's monochromaticity and high degree of directional collimation have made the laser scattering technique valuable for the study of an incredible range of dynamic phenomena in fluids. Using a variety of spectroscopic techniques, fluctuations in the fluid occurring on time scales between 1 cps ranging in a continuous fashion up to 10^{15} cps (approximately the laser frequency itself) can be studied in a scattering experiment. By varying the scattering angle the spatial wavelength of the fluctuations probed can be selected over several orders of magnitude as well, ranging from a few thousand angstroms (for back scattering) to a few centimeters (for forward scattering). The practical upper limit is dictated by diffraction effects which exclude scattering angles smaller than the diffraction angle of the laser beam, $\sim\lambda/d$, where d is the laser beam diameter. This combination of dynamic range in both time and space, which makes light scattering such a powerful tool for the study of fluids, is shown in Fig. 1, where it is compared with those space–time regimes accessible to some other commonly used experimental techniques.

The light scattering domain is divided into three parts corresponding to generally different methods of spectrally analyzing the scattered light. The short-time or high-frequency phenomena occurring on time scales as short as 10^{-14} sec have been probed by means of grating spectrometers. This is the range of times involved, for example, in the intermolecular collision process, a very difficult region to probe by any experimental technique. In the intermediate range such familiar phenomena as Brillouin scattering from thermally excited acoustic waves and Rayleigh scattering from entropy fluctuations have received considerable attention. Similarly, the normal modes of vibration in complicated systems such as liquid crystals fall in this intermediate spectral range. These diverse phenomena have been studied conveniently with optical interferometers, having resolving powers as high as 10^9.

The lowest energy region, corresponding to the highest resolution ($\sim10^{15}$) has only recently emerged and owes its existence to the development of light beating and intensity correlation spectroscopy. Processes of interest in this region include the critical slowing down of density fluctuations near the liquid–gas phase transition, concentration fluctuations in critical mixtures, the diffusive motion of biological macromolecules in solution, and hydrodynamic instabilities in the vicinity of their incipient development.

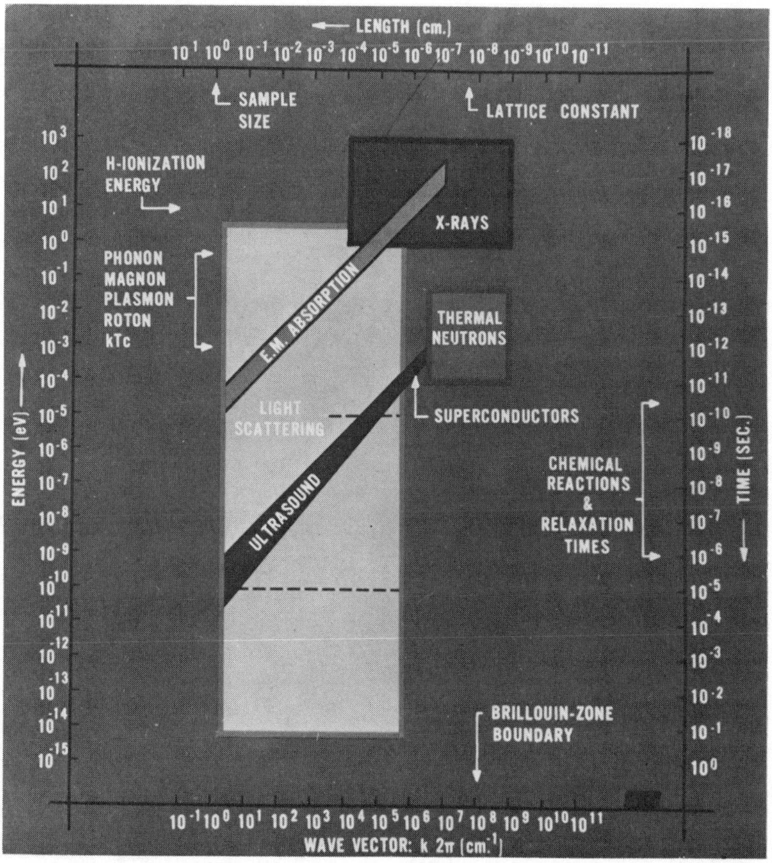

Fig. 1. Space–time or energy–momentum regimes accessible to various experimental techniques. Scattering techniques span rectangular areas by virtue of the additional kinematic range afforded by a pair of probe particles (incident and scattered). Note the logarithmic scales which permit display of large dynamic ranges.

The other scattering techniques depicted in Fig. 1 can probe shorter-wavelength fluctuations, but at present have not nearly the dynamic range in frequency available to light scattering.

II. THEORETICAL CONSIDERATIONS

A. General Relations

The scattering experiment is depicted schematically in Fig. 2, where an incident plane-polarized optical field $\mathbf{E}_1$, of frequency ω_1, wave vector k_1,

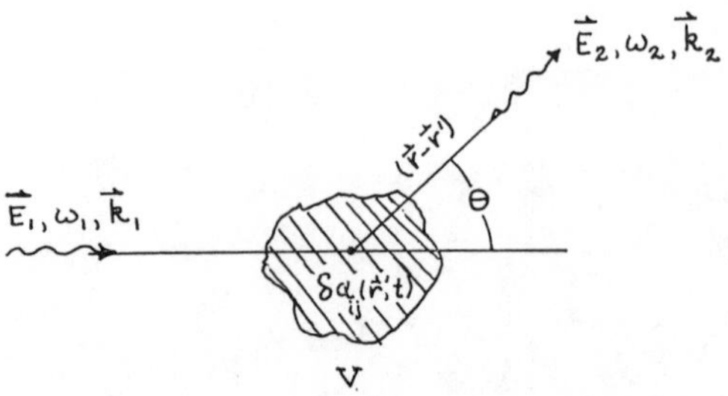

Fig. 2. Schematic diagram of scattering experiment.

and polarization $\hat{\varepsilon}_1$ traverses a medium supporting time- and space-dependent polarizability tensor, $\delta\alpha_{ij}(\mathbf{r}', t)$. For a perfectly homogeneous polarizability, $\delta\alpha_{ij}(\mathbf{r}', t) = \text{Const.} \times \delta\alpha_{ij}(t)$, the reradiated (scattered) field would sum to zero for all but the exact forward direction. For a perfectly static polarizability, $\delta\alpha_{ij}(\mathbf{r}', t) = \text{Const.} \times \delta\alpha_{ij}(\mathbf{r}')$, the scattered light field would have the same frequency (time dependence) as incident field. The experiment generally consists of measuring the strength, polarization, frequency, and wave vector of the scattered field. These quantities carry information on the strength, symmetry, and time and spatial behavior of $\delta\alpha_{ij}$. The magnitude of $\delta\alpha_{ij}$ measures the scattering efficiency or cross section. Under the assumptions that (1) the scattering volume is much larger than λ_1^3, (2) that the scattering volume dimensions are small compared to the distance between the scattering volume and the point of observation ($\mathbf{r}$) of the scattered field, and (3) that $\delta\alpha$ is sufficiently small that only a small fraction of the incident light is scattered, the scattered field may be written[9]:

$$E_2^{(i)}(\mathbf{r}, t) = \frac{-\omega_1^{\,2} \sin \varphi}{|\mathbf{r} - \mathbf{r}'|\, c^2} \eta_1^{\,2} \exp{(i\mathbf{k}' \cdot \mathbf{r} - \omega_1 t)} \times \int_V dV' \exp{(-i\mathbf{k} \cdot \mathbf{r}')} E_1^{(j)} \delta\alpha_{ij}(\mathbf{r}', t) \quad (1)$$

where $(\eta_1\omega_1/c)\, |\mathbf{r} - \mathbf{r}'| = \mathbf{k}' \cdot (\mathbf{r} - \mathbf{r}')$ defines $\mathbf{k}'$ and $\mathbf{k} \equiv \mathbf{k}' - \mathbf{k}_1$ and the superscript indexes the ith Cartesian component of $\mathbf{E}_2$. φ is the angle between the incident polarization direction and the observation direction. Since the scattered intensity spectrum is simply related to the Fourier transform of the time correlation function of the scattered field, we have

$$I(\omega, \mathbf{r}) = \frac{c}{8\pi}\left(\frac{\varepsilon}{\mu}\right)^{1/2} \langle \mathbf{E}_2(\mathbf{r}, \omega_2) \cdot \mathbf{E}_2{}^*(\mathbf{r}, \omega_2) \rangle \quad (2)$$

where the angular brackets indicate an appropriate average over the states of the system. We see from (1) and (2) that the information about the medium revealed by the scattering experiment is contained in the space–time correlation function of the polarizability tensor elements:

$$\langle \delta\alpha_{ij}(\mathbf{r}, t)\delta\alpha_{kl}^*(\mathbf{r} - \mathbf{r}', t - t')\rangle \tag{3}$$

Under assumption 1, above, the volume integral in (1) may be extended to infinity with the result that the scattered field at **r** is strictly proportional to the **q**th Fourier component of $\delta\alpha_{ij}$. The time Fourier transform of $\mathbf{E}_2$ is just

$$\mathbf{E}_2(\omega_2) = \frac{1}{2\pi}\int_{-\infty}^{\infty} \mathbf{E}_2(t)e^{i\omega_2 t}\, dt \tag{4}$$

and is proportional to $\delta\alpha_{ij}(\mathbf{k}, \Omega)e^{i(\omega_1 \pm \Omega)t}$. This reveals that (a) the Ωth Fourier component of $\delta\alpha_{ij}$ is observed and that (b) the scattered field frequency ω_2 is just $\omega_1 \pm \Omega$. Finally, the various polarization components of the scattered intensity are expressed by combining the above results as

$$I(\mathbf{r}, \omega_2) = \frac{V}{8\pi}\left(\frac{\varepsilon}{\mu}\right)^{1/2} \frac{k_1{}^4 \sin{}^2\varphi}{r^2 c} E_1{}^{(j)}E_1{}^{(l)} \langle \delta\alpha_{ij}(\mathbf{k}, \Omega)\, \delta\alpha_{il}^*(\mathbf{k}, \Omega)\rangle \tag{5}$$

where summation on repeated indices is implied.

Implicit in this expression are the kinematics of the scattering process, which correspond to overall momentum and energy conservation in the scattering process.

$$\hbar\mathbf{k}_2 = \hbar\mathbf{k}_1 \pm \hbar\mathbf{k}$$

and (6)

$$\hbar\omega_2 = \hbar\omega_1 \pm \hbar\Omega$$

The (−) and (+) signs correspond to gain and loss, respectively, of a quantum of energy $\hbar\Omega$ by the medium, and are called the Stokes and anti-Stokes components of the spectrum. For a system in thermal equilibrium the spectrum is symmetric about ω_1 when $\hbar\Omega \ll kT$, but more generally exhibits the relative demagnification by $e^{-\hbar\Omega/kT}$ on the anti-Stokes side required by the principle of detailed balance. The wave vectors in (6) form a triangle which for $\Omega \ll \omega_1, \omega_2$ yields the useful relation between scattering angle θ and momentum transfer $\hbar k$:

$$\hbar k = 2\hbar k_1 \sin\frac{\theta}{2} \tag{7}$$

Thus variation of θ permits probing of fluctuations of wavelength $\lambda_1/2 < \lambda < d$, where λ_1 is the laser wavelength ($\sim 10^{-5}\ \text{cm}^{-1}$) and d is the laser beam diameter, thereby defining the diffraction limit.

Equations (5) and (6) represent the fundamental relations for light scattering experiments. However, their very generality is indicative of a lack of detailed physical content. The physical, chemical, and even biological processes which determine the structure and strength of $\delta\alpha_{ij}(\mathbf{r}, t)$ are so incredibly diverse that to write down $\delta\alpha_{ij}$ for a general fluid or fluid mixture would require a detailed microscopic theory of the dynamical aspects of the system affecting the polarizability. Because both single-particle and many-particle polarizabilities are sensitive to locations or configurations of the particles and groups, this means virtually *all* the dynamic aspects. Fortunately different processes contributing to $\delta\alpha_{ij}$ often occur on greatly different time or length scales and may vary differently with parameters like temperature, pressure, concentration, or molecular complexity. Their contributions can then often be separated and the dynamics of a particular process can be *directly* measured by the light scattering experiment.

For example, in a pure fluid of monatomic molecules, the dominant contribution to $\delta\alpha_{ij}$ comes from density fluctuations. These contribute only to diagonal elements of $\overset{\approx}{\delta\alpha}$ so that the scattered light has the same polarization as the incident light. The observed correlation function $\langle\delta\rho(-\mathbf{k}, 0)\, \delta\rho^*(\mathbf{k}, \omega)\rangle$ is simply the familiar dynamic structure factor $S(\mathbf{k}, \omega)$. Various models have been used to calculate S in various approximations: (a) the kinetic regime, single particles[72,76]; (b) the hydrodynamic regime[72,76,133]; and (c) the generalized hydrodynamic regime.[32] When $k\Lambda < 1$ (Λ = particle mean free path), $S(\mathbf{k}, \omega)$ consists of a triplet of peaks centered at $\omega = 0$ and at $\omega = \pm qV_s$ arising from the nonpropagating (Rayleigh) and propagating (Brillouin) components of the density fluctuation at k.[57] These peaks are nearly Lorentzian in shape and of width $\Delta\omega_R = (\kappa/\rho C_p)k^2$ and $\Delta\omega_B = \alpha_s v_s/\pi$; where $\kappa/\rho C_p$ is the thermal diffusivity and α_s, v_s are the sound attenuation and velocity, respectively (see Fig. 3). Studies of the Rayleigh and/or Brillouin spectra were among the first applications of laser spectroscopy, and continue to be of interest because (a) the precise manner in which $\Delta\omega_R$ approaches zero near the liquid–gas critical point has important consequences for the understanding of critical phenomena, and (b) measurements of the Brillouin splitting, width, and shape reveal behavior of sound waves in a frequency range far above that of conventional ultrasonics (10^9–10^{10} Hz), and thus may reveal departures from classical hydrodynamics. The $S(\mathbf{k}, \omega)$ is even more interesting in fluids composed of polyatomic molecules, not only because vibrational relaxation frequencies often lie in the 10^9–10^{10} Hz range, but also because the structure of $S(\mathbf{k}, \omega)$ itself in a relaxing fluid is often qualitatively different from that in a simple fluid. In particular, a new central peak appears whose strength and shape are directly related to the strength and relaxation time of the energy exchange between density fluctuations and the internal degrees of freedom. Further interesting

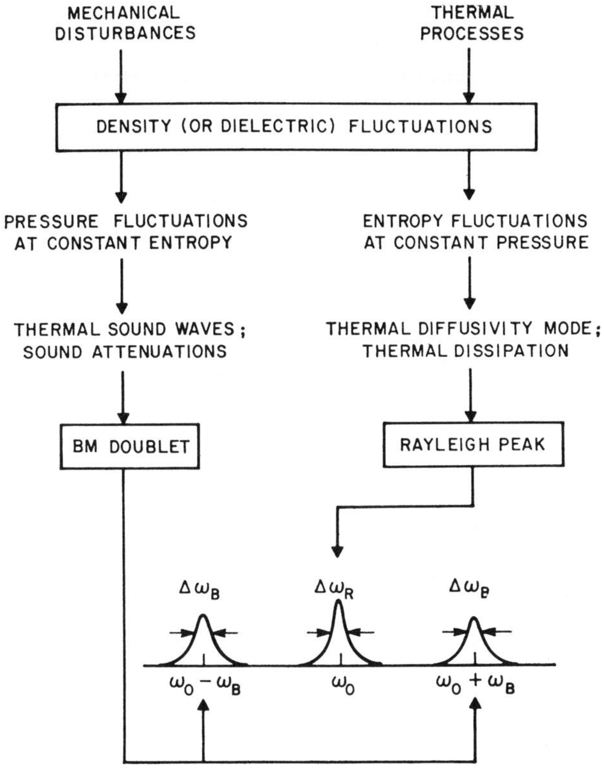

Fig. 3. Representation of the origin of the Rayleigh-Brillouin spectrum for a simple dense fluid.

effects on $S(\mathbf{k}, \omega)$ are expected in the generalized hydrodynamic regime ($\omega\tau_c \sim 1$, and $ka \sim 1$, with τ_c the collision time and a the characteristic intermolecular distance), where the ω and k dependence of thermodynamic and transport coefficients is taken into account. We consider here briefly the results of a description given by a generalized hydrodynamic description[32] which unifies these contributions to $S(k, \omega)$.

B. Generalized Hydrodynamics for $S(\mathbf{k}, \omega)$

In this section, we first present a generalized hydrodynamics analysis as an extension of the hydrodynamic approach and then treat two particular examples as special cases, namely, the case of hypersound dispersion in simple liquids and the case of thermal relaxation resulting in the appearance of a new mode in liquids with internal degrees of freedom. Both cases correspond to actual experimental situations which have been investigated in the laboratory and require the replacement of the usual transport coefficients by the more general concept of *transport functions.*

The generalized hydrodynamic theory presented here is based on the method developed by Ailawadi, Rahman, and Zwanzig[511] for the analysis of current correlation functions. As pointed out by these authors, such a treatment does not constitute a rigorous theory, but presents the advantage of analytic simplicity—quite useful for the interpretation of physical situations encountered in experimental work—leading to the same generalized hydrodynamic equations that have been derived previously by more elaborate statistical mechanical methods.

The basic equations are straightforward generalizations of the Fourier-Laplace transformed versions of the (1) continuity, (2) momentum, and (3) energy equations:

$$s\tilde{\rho}_k(s) + i\mathbf{k}\cdot\tilde{\mathbf{j}}_k(s) = \rho_k(0) \tag{8}$$

$$[s + k^2\tilde{\varphi}_k(s)]\tilde{\mathbf{j}}_k(s) + i\mathbf{k}\frac{k_BT}{S(k)}[\tilde{\rho}_k(s) + \beta\tilde{g}_k(s)] = \mathbf{j}_k(0) \tag{9a}$$

$$[s + q^2\tilde{\varphi}_q(s)]\tilde{\mathbf{j}}_q^{(1)}(s) = \mathbf{j}_q^{(1)}(0) \tag{9b}$$

$$[s + q^2\tilde{\varphi}_q(s)]\tilde{\mathbf{j}}_q^{(2)}(s) = \mathbf{j}_q^{(2)}(0) \tag{9c}$$

$$[s + k^2\tilde{\Phi}_k(s)]\tilde{g}_k(s) + i\mathbf{k}\frac{\gamma - 1}{\beta}\cdot\tilde{\mathbf{j}}_k(s) = g_k(0) \tag{10}$$

where the quantities $\tilde{X}_k(s)$ are the Laplace-Fourier transforms of the fluctuations $\delta X(\mathbf{r}, t)$ defined as

$$\tilde{\rho}_k(s) = \int_0^\infty dt\, e^{-st}\int_V d\mathbf{r}\, e^{i\mathbf{k}\cdot\mathbf{r}}\,\delta\rho(\mathbf{r}, t) \tag{11}$$

with $s = \varepsilon + i\omega$, the complex Laplace variable. The longitudinal current $\mathbf{j}_k$ and the transverse currents $\mathbf{j}_q^{(i)}$ $(i = 1, 2)$ (with the wave vector $\mathbf{q}$ in the plane perpendicular to the $\mathbf{k}$ direction) have been defined as

$$\mathbf{j}_k(t) = \rho_0\int_V d\mathbf{r}\,\mathbf{v}(t)e^{i\mathbf{k}\cdot\mathbf{r}} \tag{12}$$

$$\mathbf{j}_q(t) = \rho_0\int_V d\mathbf{r}\,\mathbf{v}(t)e^{i\mathbf{q}\cdot\mathbf{r}} \tag{13}$$

Similarly, $g(\mathbf{r}, t)$ denotes the heat current density defined as

$$g(\mathbf{r}, t) = \rho_0\,\delta T(\mathbf{r}, t) \tag{14}$$

In (8)–(10) we have employed the generalized coefficients defined as

$$\omega_0^2(k) = k^2 \frac{k_B T}{S(k)} = \frac{k^2}{\rho \chi_T(k)} \tag{15}$$

where $\chi_T(k)$ is the generalized isothermal compressibility

$$k^2 \tilde{\varphi}_k(s) = [\omega_l^2(k) - \omega_0^2(k)] \tilde{\psi}_k^l(s) \tag{16}$$

with

$$\omega_l^2(k) = \frac{\langle [\dot{j}_k(0)]^2 \rangle}{\langle [j_k(0)]^2 \rangle} \tag{17}$$

and where the index l denotes the longitudinal current;

$$k^2 \tilde{\varphi}_q(s) = \omega_t^2(k) \tilde{\psi}_q^L(s) \tag{18}$$

with

$$\omega_t^2(q) = \frac{\langle [\dot{j}_q(0)]^2 \rangle}{\langle [j_q(0)]^2 \rangle} \tag{19}$$

and where the index t denotes the transverse current;

$$k^2 \tilde{\Phi}_k(s) = \omega_\lambda^2(k) \tilde{\psi}_k^\lambda(s) \tag{20}$$

with

$$\omega_\lambda^2(k) = \frac{\langle [\dot{g}_k(0)]^2 \rangle}{\langle [g_k(0)]^2 \rangle} \tag{21}$$

In (16), (18), and (20), the functions $\tilde{\psi}(s)$ are the memory functions of the master equations governing the time evolution of the corresponding currents; that is,

$$\dot{j}(t) = -\int_0^t d\tau \, \psi(t - \tau) j(\tau) \tag{22}$$

An a priori calculation of these memory functions is quite difficult. Although the formal expressions for the ψ functions are analytically known, their explicit evaluation requires model calculations.

On the other hand, the ω's can be computed from (17), (19), and (21), and the explicit expressions for the different currents, $j(0)$, and generalized forces, as well as some numerical computations, can each be found in the literature.

The set (8)–(10) constitute the generalized hydrodynamic equations. To make contact with classical hydrodynamics, it suffices to take the long-time limit,

$$\lim_{s \to 0} \tilde{\psi}_k(s) = \tau(k) \tag{23}$$

where τ is the appropriate relaxation time for the transport process considered. Then one finds[511]

$$\lim_{k\to 0} [\omega_l^2(k) - \omega_0^2(k)]k^{-2}\tau_l = \nu_0 \tag{24}$$

$$\lim_{k\to 0} \omega_t^2(k)k^{-2}\tau_t = \nu_0' \tag{25}$$

$$\lim_{k\to 0} \omega_\lambda^2(k)k^{-2}\tau_\lambda = \lambda_0 \tag{26}$$

and

$$\lim_{k\to 0} \frac{k^2 k_B T}{S(k)} = k^2\left(\frac{\partial p}{\partial \rho}\right)_T^{-1} = U_T^2 k^2 = \gamma^{-1} U_s^2 k^2 \tag{27}$$

where the usual quantities are denoted by $U_s = [(\partial p/\partial \rho)_s]^{1/2}$, the adiabatic sound velocity; $\gamma = C_p/C_v$, the ratio of the specific heats; and $\beta = -(\partial \rho/\partial T)_p/\rho_0$, the coefficient of thermal expansion. The transport coefficients, ν_0, ν_0', λ_0, are the total kinematic viscosity, the shear kinematic viscosity and the thermometric conductivity, respectively:

$$\nu_0 = \frac{1}{\rho_0}(\tfrac{4}{3}\eta_s + \eta_B) \tag{28}$$

$$\nu_0' = \frac{\eta_s}{\rho_0} \tag{29}$$

$$\lambda_0 = \frac{\kappa}{\rho_0 C_v} = \gamma\lambda' \tag{30}$$

Different models may be constructed for the memory functions by assuming different mathematical forms for $\psi_k(t)$. These forms can be chosen and adjusted to fit the experimental data or may be computed from the sum rules. For the sake of simplicity, we restrict our attention here to the mathematically most convenient model, introducing the exponential ansatz

$$\tilde{\psi}_k^l(s) = [s + \tau_l^{-1}(k)]^{-1} \tag{31}$$

$$\tilde{\psi}_k^\lambda(s) = [s + \tau_\lambda^{-1}(k)]^{-1} \tag{32}$$

Here the relaxation times, $\tau_l(k)$ and $\tau_\lambda(k)$, should be considered as adjustable parameters to be determined, for example, by fitting the power spectrum to the molecular dynamics data.

However, when one is interested in the hydrodynamic regime, τ_l and τ_λ can be computed from the hydrodynamic limit equations, (24) and (26), combined with (17), (21), and (15) taken in the limit $k \to 0$. In the domain of light scattering, one may restrict oneself to the set of (8)–(10), wherefrom one obtains a dispersion equation of the fifth degree. Indeed it is expected that two additional modes will arise from the introduction of the generalized transport coefficients when cast in the form of the exponential model, (31) and (32). To the lowest order in $\omega_i\tau_i (i = l, \lambda)$, the five roots of the dispersion equation read[32]:

$$S_0 = -\frac{1}{\gamma}\,\omega_\lambda^{\,2}\tau_\lambda \qquad \text{(thermal diffusivity mode)} \qquad (33)$$

$$S_\lambda = -\tau_\lambda^{-1} \qquad \text{(thermal relaxation mode)} \qquad (34)$$

$$S_l = -\tau_l^{-1} \qquad \text{(viscous relaxation mode)} \qquad (35)$$

$$S_\pm = \pm i\gamma^{1/2}\omega_0 - \Omega \qquad \text{(Brillouin modes)} \qquad (36)$$

with

$$\Omega = \frac{1}{2}\left[(\omega_l^{\,2} - \omega_0^{\,2})\tau_l + \left(1 - \frac{1}{\gamma}\right)\omega_\lambda^{\,2}\tau_\lambda\right] \qquad (37)$$

Note that the mode S_l could still be decomposed into a contribution from shear relaxation and a contribution from structural relaxation (related to the existence of the generalized bulk viscosity coefficient).

In a simple liquid the equations yield two additional shear modes

$$S_{i=1,2} = -\nu' q^2$$

which will not be considered in more detail here, because they have not been observed in simple liquids to date. Such modes would probably not be detectable by light scattering in simple isotropic liquids since there is no direct coupling between the transverse component of the velocity and the density fluctuations. However, in fluids with anisotropic molecules, such coupling does occur and has been invoked to explain the low-frequency depolarized spectrum observed in such cases. More will be said of these spectra in Section III.

In general, coupling occurs between these different modes [(33)–(36)] resulting in deviations from the simple Lorentzian shape of the classical spectral components. Such deviations are indications of the existence of dynamical structure in the system, which structure will then manifest itself as hypersonic dispersion, as modifications to the shape and the intensity of the spectral lines, as well as deviations from the classical Landau-Placzek ratio. Although in general these changes are expected to be small, several liquids have been studied experimentally, whose spectra could not be fully explained by the classical theory. In the next sections, we consider two examples of situations where the present treatment has been used for the interpretation of (1) a new central component and (2) the possible existence of hypersound dispersion in simple liquids. Similar cases may occur also in fluid mixtures, as we shall see when considering the extension of the theory to multicomponent systems.

C. Internal Degrees of Freedom and Relaxation Processes

Energy transfer may occur between internal degrees of freedom and the translational degrees of freedom, in which case there will be coupling between a mode characteristic of structural relaxation and the collective modes describing the density fluctuations. This problem can be regarded as a particular case of the general prescription displayed in the previous section. There one restricts oneself to the limit $k \to 0$ and solely the bulk viscosity coefficient is generalized into the form of a transport function, with the following simple ansatz, assuming a single relaxation time process[131] [see (16) and (31)]:

$$\lim_{k\to 0} k^2 \tilde{\varphi}_k{}^B(s) = \lim_{k\to 0} [\omega_B{}^2(k) - \omega_0{}^2(k)] \tilde{\varphi}_k{}^B(s)$$

$$= \eta_B \frac{k^2}{\rho} + \frac{(U_\infty{}^2 - U_s{}^2)\tau_B k^2}{1 + s\tau_B} \tag{38}$$

with U_∞, the infinite frequency sound speed. Here the index B indicates that solely the contribution from the bulk viscosity is considered, and it has been found convenient to separate off the zero-frequency part of the function (i.e., the usual kinematic bulk viscosity coefficient, η_B/ρ_0). The dispersion equation can then be cast into the form

$$\left(s + \frac{\lambda k^2}{\gamma}\right)(s + i\omega_s + \Omega')(s - i\omega_s + \Omega')\left(s + \left(\frac{U_s}{v_s}\right)^2 \tau_B{}^{-1}\right) = 0 \tag{39a}$$

Here

$$\omega_s = v_s k = \frac{\tau_B^{-1}}{\sqrt{2}} [\Delta + \sqrt{\Delta^2 + 4(U_s k\tau_B)^2}]^{1/2} \tag{39b}$$

$$\Delta = [(U_\infty k\tau_B)^2 - 1]$$

$$\Omega' = \Omega + \frac{1}{2}\left[1 - \left(\frac{U_s}{v_s}\right)^2\right]\frac{\lambda k^2}{\gamma} + \frac{1}{2}\frac{(U_\infty{}^2 - U_s{}^2)k^2}{\omega_s{}^2 + \tau_B{}^{-2}}(\tau_B{}^{-1} - \lambda k^2) \tag{39c}$$

Now from the dispersion equation (39), the spectral distribution can be obtained and may be *approximately* represented as a sum of *four* Lorentzian curves[131] (with $\Gamma_p = \lambda_0/\gamma$):

$$\frac{I(\omega)}{I_0} = \left(1 - \frac{1}{\gamma}\right)\frac{2\Gamma_p k^2}{\omega^2 + (\Gamma_p k^2)^2} + \frac{2\left(\dfrac{U_s}{v_s}\right)^2 \tau_B^{-1}}{\omega^2 + \left(\dfrac{U_s}{v_s}\right)^4 \tau_B^{-2}}$$

$$\times \frac{[1 - (v_s/U_s)^2][(1 - 1/\gamma)(U_s k)^2 + (U_s/v_s)^4 \tau_B^{-2}] + (U_\infty^2 - U_s^2)k^2}{\omega_s^2 + (U_s/v_s)^4 \tau_B^{-2}}$$

$$+ \frac{[1 - (U_s/v_s)^2(1 - 1/\gamma)][\omega_s^2 + (U_s/v_s)^2 \tau_B^{-2}] - (U_\infty^2 - U_s^2)^2 k^2}{\omega_s^2 + (U_s/v_s)^4 \tau_B^{-2}}$$

$$\times \left\{\frac{\Omega'}{\Omega'^2 + (\omega + v_s k)^2} + \frac{\Omega'}{\Omega'^2 + (\omega - v_s k)^2}\right\} \quad (40)$$

which reduces to the classical hydrodynamic spectrum when there is no dispersion ($U_s = U_\infty = v_s$). From the above spectrum, one obtains the ratio of the integrated intensity of the unshifted central components, $I_R^{(1)} + I_R^{(2)}$, to the integrated intensity of the Brillouin components to read

$$\frac{[I_R^{(1)} + I_R^{(2)}]}{2I_B} = (\gamma - 1)\frac{1 + A}{B} \quad (41)$$

where

$$A = \frac{\gamma(U_\infty^2 - U_s^2)k^2 - [(v_s/U_s)^2 - 1]\left[\gamma\left(\dfrac{U_s}{v_s}\right)^4 \tau_B^{-2} + (kU_s)^2(\gamma - 1)\right]}{(\gamma - 1)\left[\omega_s^2 + \left(\dfrac{U_s}{v_s}\right)^4 \tau_B^{-2}\right]}$$

$$B = \frac{[\gamma - (U_s/v_s)^2(\gamma - 1)][\omega_s^2 + (U_s/v_s)^2 \tau_B^{-2}] - \gamma k^2(U_\infty^2 - U_s^2)}{\omega_s^2 + (U_s/v_s)^4 \tau_B^{-2}}$$

which, at low phonon frequencies, that is, for $\omega_s \ll \tau_B^{-1}$, reduces to the classical Landau-Placzek ratio: $(\gamma - 1)$. In the limit of high frequencies, that is, $\omega_s \gg \tau_B^{-1}$, one retrieves Rytov's result[66]

$$(\gamma - 1)\left\{1 + \frac{\gamma}{\gamma - 1}\frac{U_\infty^2 - U_s^2}{U_s^2}\right\} \quad (42)$$

The most striking feature of the spectrum given by (40) is, of course, the existence of a second central component due to structural relaxation with

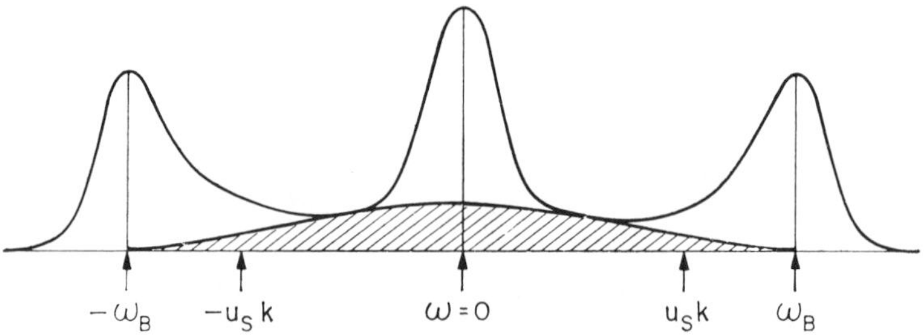

Fig. 4. Spectrum of density fluctuations for a singly relaxing fluid. The relaxation central component (shaded area) causes the Brillouin peaks (at $\pm\omega_B$) to be displaced from their zero-order positions (at $\pm U_s k$).

relaxation time τ_B. The spectral distribution is illustrated in Fig. 4, where the shaded area represents the new spectral component, which has been referred to as the "Mountain peak." Actually, in most experimental spectra, the existence of such a relaxation mode will manifest itself as a broad background between the Rayleigh component and the Brillouin peaks, which in turn will appear asymmetrical and will also be displaced from their classical position (at $\omega = U_s k$) to a new position, $\omega_s = v_s k$ according to (39a) and (39b). Notice, however, that ω_s/k measures the propagation velocity, whereas the maximum of the Brillouin component occurs at ω_B, for which $\partial I(\omega)/\partial\omega = 0$, and because of the asymmetrical terms [see (57)] in the spectral distribution, one has $\omega_B < \omega_s$. On the other hand, the effect of the broad central component (due to structure relaxation) results in a pushing of the Brillouin peaks away from the Rayleigh line, that is, $\omega_s > u_s k$. Thus considerable care must be taken to extract the sound velocity from the Brillouin spectrum in a relaxing fluid.

The present method can of course be extended to multiply relaxing systems. The case of both singly and doubly relaxing liquids has been investigated by Nichols and Carome.[134] In the latter case, a second relaxation time was assigned to the generalized longitudinal viscosity, with the assumption of an exponential ansatz.

D. Transport Function and High-Frequency Effects

In this section, we shall consider the case of simple monatomic liquids, uncomplicated by the relaxation processes associated with internal degrees of freedom. Even so the question of the possible frequency dependence of the transport coefficients in the "intermediate domain" characterized here as $k \to 0$, $\omega/\omega_c \sim 10^{-2}$ becomes of interest. ω_c is a characteristic frequency, of order of the collision frequency above which translational relaxation effects are important. The "intermediate domain" thus lies between the hydrodynamic domain ($k \to 0$, $\omega \to 0$) and the generalized hydrodynamic domain ($ka \sim 1$, $\omega\tau_c \sim 1$). Indeed it has been shown in transport theory that the linear response of a classical fluid becomes wholly nondissipative in the very high frequency limit; that is, when $\omega \to \infty$, the transport coefficients become purely imaginary.[110] We recall that in general, any transport coefficient at finite frequency, that is, any *transport function*, may be cast into a correlation expression as[505]

$$\sigma(\omega) = c \int_0^\infty dt\, e^{i\omega t} \langle J(0)J(t) \rangle \tag{43}$$

where c is a thermodynamic constant and $J(t)$ is the flux related to the transport property considered. As a result, the transport function is generally complex

$$\sigma(\omega) = \sigma_R(\omega) + i\sigma_I(\omega) \tag{44}$$

The zero frequency term, $\sigma_R(0)$, is merely the usual transport coefficient, and the first term in the ω expansion appears to be imaginary, constituting the first nondissipative contribution at high frequencies. Because for light scattering the accessible ω values place us in the intermediate domain, we may legitimately restrict ourselves to the lowest order in ω, in which case the transport functions take the form

$$\nu(\omega) = \lim_{\substack{k\to 0\\ \varepsilon\to 0}} \tilde{\Phi}_k(s) = \nu_0(1 - i\alpha_l\omega) \tag{45}$$

$$\lambda(\omega) = \lim_{\substack{k\to 0\\ \varepsilon\to 0}} \tilde{\varphi}_k(s) = \lambda_0(1 - i\alpha_\lambda\omega) \tag{46}$$

where ν_0 and λ_0 are the zero frequency transport coefficients as given by (28) and (30). α_l and α_λ are of the order of ω_c^{-1} and can be calculated explicitly [see (17) and (21)]; for example, one finds indeed that for the shear viscosity, $\alpha\omega \simeq 10^{-2}$. With these assumptions the dispersion equation must now be solved to second order in quantities like $\lambda k/U_s \equiv a$ and $\nu k/U_s \equiv b$, because, although these second-order terms are small, they are no longer negligible with respect to the $\alpha\omega$ terms in (45) and (46). The modes are now given by[32]

$$s_0' = -\Gamma_p' k^2 = -\Gamma_p k^2(1 - \delta_1) \tag{47}$$

$$s_\pm' = \pm i v_s k - \Gamma_s' k^2 = \pm i U_s k(1 - \delta_2) - \Gamma_s k^2(1 - \delta_3) \tag{48}$$

where Γ_p and Γ_s are given by $\Gamma_p = \lambda_0/\gamma$; $\Gamma_s = \nu_0 + (\gamma - 1)\lambda_0/\gamma$ [see (28) and (30)],

$$\delta_1 = \left(1 - \frac{1}{\gamma}\right)\left(\frac{b}{a} - \frac{1}{\gamma}\right)a^2 \tag{49}$$

$$\delta_2 = \delta_2^{(\omega)} + \delta_2^{(2)} \tag{50}$$

$$\delta_2^{(\omega)} = -\frac{1}{2}\left[\left(1 - \frac{1}{\gamma}\right)\mu_\lambda a + \mu_l b\right] \tag{51}$$

$$\delta_2^{(2)} = \tfrac{1}{2}[\tilde{\lambda}^2 - \delta_1] \tag{52}$$

$$\delta_3 = \delta_3^{(\omega)} + \delta_3^{(2)} \tag{53}$$

$$\delta_3^{(\omega)} = \frac{a\mu_\lambda}{\gamma} + \delta_2^{(\omega)}\left(1 + \frac{a}{\tilde{\lambda}\gamma}\right) \tag{54}$$

$$\delta_3^{(2)} = \frac{1}{\gamma}\left(\frac{1}{\gamma} - \frac{b}{2\tilde{\lambda}}\right)a^2 \tag{55}$$

with the reduced quantities

$$\tilde{\lambda} = \frac{\Gamma_s k}{U_s}; \qquad \mu_i = \alpha_i U_s k (i = \lambda, \ell) \tag{56}$$

The "high-frequency" effects arising from the nondissipative part of the transport functions have been separated off and appear in those terms with superscript (ω), whereas the other correction terms, labeled with superscript (2), represent pure second-order effects. Notice that the latter are the only corrections which affect the diffusion mode, whereas both the frequency and the lifetime of the thermal phonons are modified by the frequency effects.

With the correction terms discussed above, the spectrum of the scattered light reads[32]

$$\begin{aligned}\frac{I(\omega)}{I_0} = {} & \left(1 - \frac{1}{\gamma}\right)(1 + \theta_1)\frac{2\Gamma_p' k^2}{\omega^2 + (\Gamma_p' k^2)^2} \\ & + \frac{1}{\gamma}(1 + \theta_2)\left\{\frac{\Gamma_s' k^2}{(\omega + v_s k)^2 + (\Gamma_s' k^2)^2} + \frac{\Gamma_s' k^2}{(\omega - v_s k)^2 + (\Gamma_s' k^2)^2}\right\} \\ & + \frac{1}{\gamma}\theta_3\left\{\frac{\omega + v_s k}{(\omega + v_s k)^2 + (\Gamma_s' k^2)^2} - \frac{\omega - v_s k}{(\omega - v_s k)^2 + (\Gamma_s' k^2)^2}\right\}\end{aligned} \tag{57}$$

with

$$\theta_1 = 2\delta_2^{(\omega)} + \frac{a^2}{\gamma}\left(1 + 2\frac{b}{a} - \frac{3}{\gamma}\right) \tag{58}$$

$$\theta_2 = 2\delta_2^{(\omega)} + \mu_\lambda a - \left(1 - \frac{1}{\gamma}\right)\left(1 + \frac{2b}{a} - \frac{3}{\gamma}\right)a^2 \tag{59}$$

$$\theta_3 = \tilde{\lambda} + a\left(1 - \frac{1}{\gamma}\right) \tag{60}$$

It is easily recognized that neglecting the correction terms $\theta_i (i = 1, 2)$ and $\delta_j (j = 1, 2, 3)$ in (57) leads to the usual expression for the spectral distribution in the hydrodynamic limit. Furthermore, if one simply ignores the frequency dependence of the transport functions, that is, setting $\alpha_l = \alpha_\lambda = 0$, one obtains the spectrum involving the second-order terms.[30,132,134] These different effects are illustrated in Fig. 5, where they are considerably amplified for the sake of illustration. They induce the following spectral changes with respect to the first-order spectrum: a slight narrowing of the Rayleigh line, that is,

$$\Delta\omega_R' = \Delta\omega_R(1 - \delta_1) \tag{61}$$

a slight asymmetry and broadening of the Brillouin components, that is,

$$\Delta\omega_B' = \Delta\omega_B(1 - \delta_3)\,, \qquad \delta_3 < 0 \tag{62}$$

a slight increase of the Landau-Placzek ratio, that is,

$$\frac{I_R}{2I_B} = (\gamma - 1)\frac{1 + \theta_1}{1 + \theta_2}\,; \qquad \theta_2 < \theta_1 \tag{63}$$

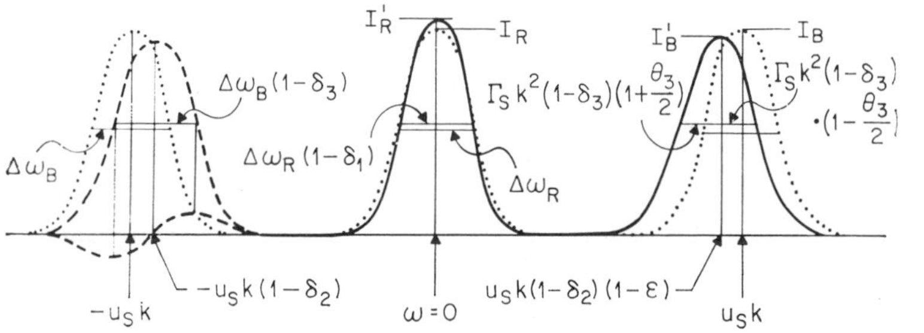

Fig. 5. Modifications of the Rayleigh-Brillouin spectrum for a simple nonrelaxing fluid due to finite frequency and to second-order effects. Dotted curves represent spectrum in their absence. Dashed curves show their effect on the Brillouin components.

and a modification of the Brillouin shift inducing hypersound dispersion, that is,

$$\omega_B' = \pm U_s k(1 - \delta_2)(1 - \varepsilon) \tag{64}$$

with

$$\varepsilon = \frac{3\theta_2^2 + \theta_3^2}{2\theta_3} \frac{1 - \delta_3}{1 - \delta_2} \tilde{\lambda} \tag{65}$$

Now (64) may be rewritten as

$$v_s \simeq U_s(1 - \delta_2^{(\omega)} - \varepsilon^{(2)}) \tag{66}$$

with

$$\delta^{(2)} + \varepsilon = \varepsilon^{(2)} \tag{67}$$

so as to separate the effects of the second-order terms, $\varepsilon^{(2)}$, from pure dispersion effects arising from the high-frequency corrections, $\delta_2^{(\omega)}$. Since the α's ($\simeq\tau_c$) are positive, $\delta_2^{(\omega)} < 0$, which as a consequence induces positive dispersion. Conversely, the effects due to second-order correction, with $\varepsilon^{(2)} > 0$, manifest themselves as a negative dispersion. The importance of these effects for the interpretation of Brillouin spectra observed in simple monatomic liquids like argon and neon are mentioned in Section III.

E. Light Scattering and Ultrasonics

Brillouin scattering may appear as a logical extension of ultrasonics to probe excitations in fluid systems by extending the domain of accessible frequencies to the GHz range. Such extension of course involves the problems discussed in this section, like dispersion effects, in particular for systems exhibiting relaxation processes, as is the case, for example, when internal degrees of freedom play an important role. Care should be taken, however, in the interpretation of light scattering results as compared to ultrasonic data, because of the boundary conditions which are not necessarily equivalent in both cases. Indeed when solving the dispersion equation, attention must be paid to the fact that either of two different boundary conditions may be appropriate. (1) When the frequency of the excitation is real and fixed by the experimental conditions, one faces the phenomenon of spatial absorption, commonly encountered in most ultrasonics experiments, whereas (2) temporal absorption occurs when the experimental conditions fix the wave vector, as in light scattering experiments.[102,131]

It was predicted in the previous section that the phase velocity measured in a light scattering experiment, for a system without structural or thermal relaxation, should exhibit negative dispersion. To compare this result with the prediction for the reciprocal situation of a spatially damped wave, we may, for the sake of clarity, consider the simplified case of a system with zero thermal conductivity; then the dispersion equation reduces to

$$s^2 + \nu k^2 s + (U_0 k)^2 = 0 \tag{68}$$

where U_0 denotes the zero frequency and infinite wavelength limit of the sound velocity. For k real, and $s = i\omega + \varepsilon$, the solution of (68) reads

$$s = \pm i v_s k - \tfrac{1}{2}\nu k^2 \tag{69}$$

with the hypersonic velocity

$$v_s = \pm U_0\left[1 - \frac{1}{4}\left(\frac{\nu k^2}{U_0 k}\right)^2\right]^{1/2} \tag{70}$$

(here we use the notation v_s for the hypersound velocity, and U_s for the ultrasonic velocity) exhibiting indeed negative dispersion. On the other hand, for $s = i\omega$ (real frequency) and $k = k' + ik''$, the dispersion equation yields the set

$$k'^2 - k''^2 = \left(\frac{\omega}{U_0}\right)^2 \bar{U}_0^{-1} \tag{71}$$

$$2k'k'' = -\left(\frac{\omega}{U_0}\right)^2 \frac{\nu}{u_0}\, \bar{U}_0^{-1} \tag{72}$$

with

$$\bar{U}_0 = 1 + \left(\frac{\nu\omega}{U_0^2}\right)^3 \tag{73}$$

from which one obtains the "ultrasonic" velocity:

$$U_s = \frac{\omega}{k'} = \pm U_0\left[\frac{2\bar{U}_0}{1 + \bar{U}_0^{1/2}}\right]^{1/2} \tag{74}$$

The latter, for low frequencies, that is, $\omega \ll (uk)^2/\nu k^2$, may be expanded to yield

$$U_s \simeq \pm U_0\left[1 + \frac{3}{4}\left(\frac{\nu\omega}{U_0^2}\right)^2\right]^{1/2} \tag{75}$$

from which it is clear that for complex k the dispersion is positive.

From (70) and (75) one observes that v_s is always smaller than U_s except for the limiting case $k \to 0$, $\omega \to 0$, for which one finds the expected equality $v_s(k \to 0) = U_s(\omega \to 0) = U_0$.

Considering next the case of a relaxing system, one may introduce a generalized bulk viscosity to account for thermal relaxation. Such an analysis for the comparison of ultrasonics with light scattering can be performed along the same lines as for the simple case treated above; we merely quote here the results of the analysis for the sound velocity in order to exhibit the dispersion effects.[102]

$$v_s = \pm \frac{U_0}{\sqrt{2}}\left[1 + \frac{\omega\eta_2''(\omega)}{\rho_0 U_0^2} + \frac{|c|^2}{U_0^2}\right]^{1/2} \tag{76}$$

$$U_s = \pm\sqrt{2}\,\frac{|c|^2}{U_0}\left[1 + \frac{\omega\eta_2''(\omega)}{\rho_0 U_0^2} + \frac{|c|^2}{U_0^2}\right]^{-1/2} \tag{77}$$

with

$$\frac{|c|^2}{U_0^2} = \left\{\left[1 + \frac{\omega\eta_2''(\omega)}{\rho_0 U_0^2}\right]^2 + \left[\frac{\omega\eta_2'(\omega)}{\rho_0 U_0^2}\right]^2\right\}^{1/2} \tag{78}$$

Here, $\eta_2'(\omega)$ and $\eta_2''(\omega)$ are the real and imaginary parts of the second viscosity, respectively. Using, for example, a single relaxation time model, for $\eta_2(\omega)$, with relaxation time τ_2, one has

$$\eta_2(\omega) = \frac{(U_\infty^2 - U_0^2)\tau_2}{1 + i\omega\tau_2}\,\rho_0 = \eta_2' + i\eta_2'' \tag{79}$$

It is then easily verified that in the limit $\omega \to \infty$, one obtains $v_s = U_s = U_\infty$ (whereas for $\omega \to 0$, of course, $v_s = U_s = U_0$). For the intermediate-frequency range, that is, when $\omega\tau \sim 1$, the use of a simple model permits the explicit computation as well as the numerical comparison of the dispersion effects. Probably the most interesting feature is that it is clear from (76) and (77) that relaxation effects induce positive dispersion for both U_s and v_s. Furthermore, here again one observes that one has always $U_s > v_s$, except in the limits discussed above, for which both the ultrasonic velocity and the hypersonic velocity tend to the same asymptotic values.

F. Solutions and Noncritical Mixtures

The spectra of multicomponent fluids are much more complex than for pure fluids. Advantage has been taken of this additional spectral information to measure mass diffusion coefficients,[356–359] chemical reaction rates,[346,347] departures from solution ideality,[310] etc., in solutions and binary liquid mixtures at thermal equilibrium. Although no experiments have yet been reported, recent theoretical discussions on fluids and mixtures *not* in equilibrium have suggested this as a fruitful area for future investigations. We present here in brief summary some of the principal theoretical predictions for both cases.

Mountain and Deutch[330] have calculated the spectrum of the coupled pressure, (p), entropy, (φ), and concentration (c) fluctuations in a binary fluid mixture at equilibrium. For their hydrodynamic analysis one must add to the basic conservation equations (8)–(10), the diffusion equation:

$$\frac{\partial c}{\partial t} = D\left[\nabla^2 c + \frac{k_T}{T}\nabla^2 T + \frac{k_p}{P}\nabla^2 p\right] \tag{80}$$

where c is the minority concentration, D the mass diffusion coefficient, k_T the thermal diffusion ratio $= c(1-c)(D'T/D)$ (D' = thermal diffusion coefficient), and $k_p = c(1-c)(D''/D)p$, in terms of the barodiffusion coefficient, D''. The resulting spectrum $S(\mathbf{k}, \omega)$ is[330]

$$\begin{aligned}
&\left(\frac{\partial\alpha}{\partial p}\right)^2_{\varphi,c}\frac{k_B T\rho}{\beta_s}\left\{\frac{\Gamma k^2}{(\Gamma k^2)^2 + (\omega + U_0 k)^2} + \frac{\Gamma k^2}{(\Gamma k^2)^2 + (\omega - U_0 k)^2}\right\}\\
&+ \left(\frac{\partial\alpha}{\partial\varphi}\right)^2_{c,p}\left(\frac{k_B T^2}{C_p}\right)\left(\frac{2}{s_2 - s_1}\right)\left\{\frac{(Dk^2 - s_1)s_1}{s_1^{\,2} + \omega^2} - \frac{(Dk^2 - s_2)s_2}{s_2^{\,2} + \omega^2}\right\}\\
&+ \left(\frac{\partial\alpha}{\partial c}\right)^2_{\varphi,p}\left(\frac{k_B T}{(\partial\mu/\partial c)_{s,T}}\right)\left(\frac{2}{s_2 - s_1}\right)\left(\frac{(s_2 - Dk^2)s_1}{s_1^{\,2} + \omega^2} + \frac{(s_1 - Dk^2)s_2}{s_2^{\,2} + \omega^2}\right)\\
&+ \left(\frac{\partial\alpha}{\partial\varphi}\right)_{p,c}\left(\frac{\partial\alpha}{\partial c}\right)_{p,\varphi}\left(\frac{2k_B T}{C_p}\right)\left(\frac{s_1}{s_1^{\,2} + \omega^2} - \frac{s_2}{s_2^{\,2} + \omega^2}\right)\left(\frac{2Dk^2 k_T}{s_1 - s_2}\right)
\end{aligned} \tag{81}$$

where Γk^2 is the Brillouin width:

$$\Gamma = \Gamma_0 + \frac{DU_0^{\,2}}{\rho(\partial\mu/\partial c)_{p,T}}\left\{\left(\frac{\partial\rho}{\partial c}\right)_{p,T} + \frac{k_T}{C_p}\left(\frac{\partial\rho}{\partial T}\right)_{p,c}\left(\frac{\partial\mu}{\partial c}\right)_{p,T}\right\}^2 \tag{82}$$

$\Gamma_0 = \nu_0 + (\gamma - 1)\lambda_0$ from (28) and (30). The roots s_1 and s_2 refer to the central components and are

$$s_1 = \tfrac{1}{2}\Lambda k^2 + \tfrac{1}{2}[\Lambda^2 k^4 - 4\lambda_0 D k^4]^{1/2} \tag{83}$$

$$s_2 = \tfrac{1}{2}\Lambda k^2 - \tfrac{1}{2}[\Lambda^2 k^4 - 4\lambda_0 D k^4]^{1/2} \tag{84}$$

$$\Lambda \equiv \lambda_0 + \mathscr{D} \tag{85}$$

$$\mathscr{D} = D\left[1 + \frac{k_T^{\,2}}{TC_p}\left(\frac{\partial\mu}{\partial c}\right)_{p,T}\right] \tag{86}$$

Here μ is the chemical potential.

The spectrum (81) contains in addition to the two propagating Brillouin modes, whose line width Γ is slightly modified, a complicated central feature due to the coupled entropy–concentration fluctuations. Only under certain simplifying conditions can this be decomposed into two Lorentzians corresponding to the thermal diffusivity and mass diffusion processes. In the very dilute limit, $c \ll 1$, we have $k_T \to 0$ and the two Lorentzians emerge with $s_1 \to \lambda_0 k^2/\gamma$, $s_2 \to -Dk^2$. For small but finite c these peaks are quasi-Lorentzian with somewhat modified line widths. In particular the thermal diffusivity (Rayleigh) peak is *broadened* by an amount

$k^2(k_T/TC_p)(\partial\mu/\partial c)_{p,T}$, whereas the diffusion (concentration) peak is *narrowed* by $k^2(\partial\mu/\partial c)_{p,T}D(k_T/TC_p\lambda_0)$. Although several experiments on fluid mixtures of low c have been reported,[28,310] the central spectra have been interpreted as two Lorentzians. Thus neither the Brillouin line width modifications nor the central line shape modifications discussed here have yet been measured for liquids. However, recent work of Gornall et al.[308] has revealed striking interaction effects in disparate gas mixtures. And Lekkerkerker and Boon[512] have reinterpreted the coupling effects observed in such mixtures.

Recently Boon[297] has extended these considerations to include the very interesting *nonequilibrium* case of a horizontal binary fluid layer subjected to an adverse vertical temperature gradient. When a fluid layer is heated from below, the system becomes topheavy and potentially unstable (provided that the fluid has a positive thermal expansion coefficient). Therefore the fluid has a tendency to redistribute itself. This natural tendency is inhibited, however, by the viscosity of the fluid and by its thermal conductivity which seeks to relieve the imposed gradient. As a consequence the temperature gradient must exceed a certain value before the instability can manifest itself. When the temperature gradient reaches this critical value (depending on the properties of the fluid and on the thickness of the layer) convective motion begins which permits the fluid to adjust. Convection arises in a peculiar way, however, in that a stationary instability occurs, which results in an arrangement of spatially ordered convection cells or rolls, first demonstrated by Bénard in 1900.[510]

In many ways the onset of this hydrodynamic instability, although a manifestly nonequilibrium phenomenon, bears close analogies to several well-studied second-order equilibrium phase transitions. Particularly intriguing is the behavior of fluctuations very near the instability point. Although in principle amenable to light scattering examination, the Bénard instability presents tremendous experimental challenges and no successful dynamic experiments have yet been reported. Nevertheless, Boon's theoretical predictions deserve summary here. He considers the dilute fluid mixture ($c \ll 1$) to form a layer of thickness, d, between two free boundaries and to be subjected to a temperature gradient $\beta = (T_h - T_c)/d$. β is conveniently expressed in dimensionless form as the Rayleigh number:

$$R = \frac{\alpha\beta g}{\lambda_0 \nu_0} d^4 \tag{87}$$

where α is the thermal expansion coefficient and g the gravitational acceleration. The instability occurs for a particular value of R, called R_c, which depends only on the fluid layer boundary conditions. Boon has examined the behavior of both the Rayleigh and concentration peaks as $R \to R_c$. He finds that the Rayleigh peak is most strongly affected when the spectrum (for a very small range of $k_\perp$'s near $k_c = \pi/2d$) approaches zero line width and infinite peak height as $R \to R_c$. In particular, the Rayleigh width is $(\lambda_0 k^2/\gamma)[(1 - R/R_c)/(1 + 2R/R_c)^{1/2}]$ for a pure fluid.[297,512] For a mixture near the instability point, the concentration mode is narrowed and the Rayleigh peak broadened slightly.

The application of light scattering to the study of hydrodynamic instabilities has thus far been rather limited. However, the recent experiments by Bertolotti et al.[272] on electric field induced instabilities in liquid crystals suggest how important a role light scattering can be expected to play in examining the dynamical aspects of nonequilibrium phenomena in fluids.

III. RECENT ADVANCES IN EXPERIMENTAL TECHNIQUE AND APPARATUS

Since the physically interesting phenomena amenable to examination by light scattering span so broad a range of characteristic lengths (10^{-5}–1 cm) and times ($\sim$1–10^{-15} sec) and exhibit such a wide variety of scattering efficiencies (10^{-1}–10^{-12} cm^{-1}-Sr^{-1}) the experimental problems encountered are manifold indeed. In its most general form the basic experimental problem consists in measuring the spectral content of the usually weak inelastically scattered light in the presence of a much stronger signal arising from elastic or parasitic scattering. One is often faced with the problem of simultaneously optimizing the spectral resolution and contrast of the detection system in a manner determined by the strength and time domain of the fluctuations to be studied. The spectral domain can be conveniently divided into the three areas, delineated above with regard to Fig. 1. In this section we present a brief description of experimental techniques and recent advances therein applicable to each frequency range.

Although some interesting progress has been made in the area of laser sources (such as tunable dye lasers, the He–Cd laser, and ultrashort pulse lasers) in recent years, the great majority of laser spectroscopy in fluids continue to utilize the He–Ne or argon ion laser. Single-frequency stabilized versions of both lasers are now commercially available, so no further discussion of sources is required here.

A. High-Frequency Shifts

Typical meter-long grating spectrometers are routinely capable of studying frequency shifts in the 0.2 to $\gtrsim$2000 cm^{-1} range (a maximum resolving power of $\sim$$10^5$). Double-grating instruments consisting of two such spectrometers in series are available which increase the contrast to

$\sim 10^{10}$ with virtually no sacrifice in resolving power. Such spectrometers have long been used for studies of rotational and vibrational Raman spectra in liquids, gases, and solids. However, some fluid phenomena of recent interest lie in this frequency range and their study has been carried out using grating spectrometers. Both anisotropy scattering from fluctuations in molecular orientation and intermolecular or "collision-induced" scattering fall into this category. The latter is closely related to the recently observed but very weak scattering of light by pairs of rotons in superfluid helium. About the only recent instrumental improvement in grating spectrometers has been the addition of a third stage—usually a lower-resolution premonochromator—to further enhance the contrast while leaving the resolution unaffected. Primarily of importance for solids where parasitic scattering is much more severe than in fluids, such improvements will not be discussed further here.

B. Intermediate-Frequency Shifts

The second spectral area denoted in Fig. 1 lies between 10^6 Hz and 10^{11} Hz (10^{-4}–$\sim 1\ \mathrm{cm}^{-1}$) and necessitates the use of optical interferometers. By far the most widely used is the Fabry-Perot (FP) interferometer, which in its simplest embodiment consists of a sharp tunable interference filter. Complete discussions of its operation are available in a variety of references.[9,509] There are two basic versions generally employed for fluid experiments, the flat FP and the spherical FP. The former consists of two flat plane-parallel low-loss mirrors of reflectivity R separated by an optical path length nd (n = refractive index of the medium separating the plates). The resolving characteristics of the FP are summarized[9] by the fraction of light I_p/I_0 traveling parallel to the interplate axis, passed at frequency ν:

$$\frac{I_p}{I_0} = \left[1 + \frac{4R}{(1-R)^2}\sin^2\left(2\pi\nu\frac{nd}{c}\right)\right]^{-1} \tag{88}$$

The transmission is periodic in $\Delta\nu = 2c/nd \equiv$ free spectral range $\equiv f$. When the $\sin^2$ term equals unity, the minimum I_p/I_0 obtains and is $\approx (1-R)^2/4R$, which determines the contrast. For $R = 0.98$ the contrast for a single FP is $\sim 10^4$. The resolving power depends on the free spectral range and on the "finesse," which is the ratio of f to $\delta\nu$. $\delta\nu$ is the full frequency width at half maximum of the transmission peaks in the function (1). Typical values of finesse are $\lesssim 100$. Thus for an f of 100 MHz the resolving power may approach 10^9. This value of f requires a plate separation d of order 150 cm. More typical for fluid studies are 1–10 GHz for spectral ranges corresponding to conveniently smaller values of d. Scanning of the pass frequency of the flat FP is most often achieved by varying the gas pressures (and hence n) between the interferometer plates. This is desirable because no physical motion of the plates is required which could easily misalign the optical cavity and degrade the instrument's spectral characteristics.

The incipient instability of the flat optical cavity can be obviated by using instead a spherical FP, two identical spherical mirrors separated by a distance r (r = radius of curvature). This confocal configuration is much more stable and permits reliable scanning of the FP by slight physical modulation of the mirror separation, as well as by the pressure variation previously mentioned. The light gathering properties of the spherical FP are superior to those of the flat FP for large r. The main disadvantage lies in the lack of adjustable f for the spherical FP. Finesse and resolving power considerations are quite similar to those for the flat FP.[9]

For many fluid applications either of these simple FP's may be quite adequate. However, some situations arise (such as diverging Rayleigh intensity near the critical point, or quasielastic scattering from particles in suspension) where a greater contrast than 10^4 is required. There are two ways in which this has been achieved: first, utilization of compound interferometers; and second, resonant reabsorption of the elastically scattered light.

1. *Compound Interferometers*

The most obvious direction for improvement by this method is the simple series arrangement of two or more FP's. This indeed has been discussed in some detail by Mack et al. nearly ten years ago.[25] Problems with simultaneity of scan, intercavity resonances, and overall instrumental line shape are of course considerably more complicated than with the single FP. Three different types of compound interferometers have been demonstrated and deserve mention here. In 1969 Langley and Ford[22] introduced in front of their FP a narrow prefilter consisting of an unequal-arm Michelson interferometer to prevent over 99% of the elastically scattered light from reaching the FP. Since the Michelson is not scanned, the system behaves very much like a simple FP with an enhanced contrast ($\sim 10^6$) and a more complicated instrumental function.

Cannell[90] has operated a tandem spherical FP interferometer for his Brillouin experiments near the critical point in Xe. Two FP's of differing f's ($f_1 = 1510$ MHz, $f_2 = 991$ MHz) were placed in series inside a common pressure can to ensure perfect tracking during the frequency scan. One FP incorporated a piezoelectrically tunable offset to allow initial simultaneous tuning of both FP's for peak transmission. The instrumental transmissions of each component as well as the compound instrument are shown in Fig. 6. The contrast is thus essentially squared ($\sim 3 \times 10^6$) and the

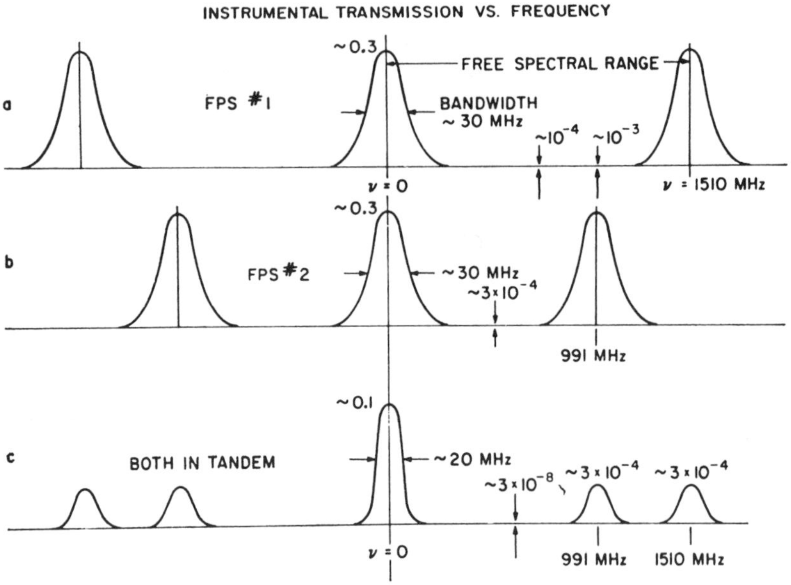

Fig. 6. Transmission patterns of single (*a* and *b*) Fabry-Perots individually and (*c*) in a tandem arrangement. After D. S. Cannell, Thesis, Massachusetts Institute of Technology, Cambridge, Mass., 1970, unpublished.

resolving power improved as well. The instrumental response function is considerably more complicated and construction more difficult than for a single FP, but the improvements permit experiments not otherwise possible. Cannell's application of his tandem FP to the study of the critical point in xenon is discussed in Section IV below.

A third and perhaps simpler method of enhancing contrast is that of the multiply passed FP, utilized recently by Sandercock.[26] By sending the optical signal through the same flat FP several times he has demonstrated

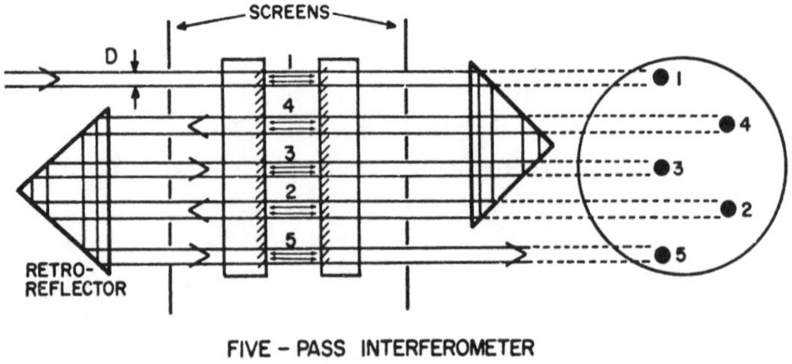

Fig. 7. Design of multiple-pass flat FP (Sandercock[26]).

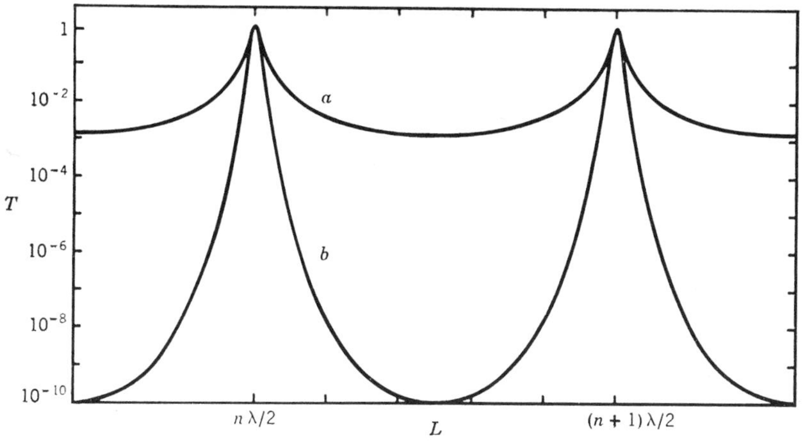

Fig. 8. Theoretical transmission curves for the above instrument used in (*a*) single-pass and (*b*) five-pass arrangements, assuming single-pass finesse of 40 (Sandercock[26]). Note log scale.

enormous contrast enhancements together with an increased finesse. Since only one pair of plates is involved, scanning, alignment and stability requirements are essentially those of a single FP. The requirement that each beam pass must use a different area of the plates provides some restrictions on light gathering power and overall instrumental transmission. The combination of various design requirements, including mirror loss, reflectivity, flatness, and overall transmission, has been shown to suggest a five-pass instrument as a practical optimum. Such an arrangement together with its theoretical transmission curve (compared to a single FP) are sketched in Figs. 7 and 8. These are for flatness of $\lambda/100$, 0.2% mirror absorption, and $0.86 = R$. The overall finesse is 40 and the contrast greater than 10^9. This kind of instrument should open several areas to experimental perusal.

2. *Resonant Reabsorption*

The *effective* contrast of a detection system can also be enhanced by reducing the intensity of the elastic component relative to that of the inelastic components prior to detection of the spectrum. In principle a very narrow absorption line, were it to coincide with the exciting laser frequency, would accomplish this. Unfortunately such coincidences are rare. However, nature has been kind in providing a strong and very narrow absorption line in the rotation-vibration spectrum of I_2 vapor which falls within the gain curve for the 5145 A line of the argon ion laser. Because it is so narrow (<300 MHz) and slightly shifted in frequency from the argon gain curve, the I_2 absorption must be used with a single-frequency argon laser tuned appropriately away from gain curve center. For long-term operation the laser frequency should be locked either directly or indirectly to the I_2 absorption frequency. Under such conditions attenuation of the elastic component by $\sim 10^8$ can be achieved,[20] with only about a tenfold increase in the absorption for the general inelastic spectrum more than 300 MHz outside of ν_{I_2}. Figure 9 shows the absorption of $\sim$5145 A laser output both on and off resonance by a 4.5 cm I_2 cell as functions of I_2 temperature and pressure. Some caution must be exercised in using the I_2 absorber for spectral shape and intensity analysis because the absorption spectrum of the I_2 molecule is quite complex and exhibits several weaker components in the near vicinity of the line discussed above, which can distort or obscure the *inelastic* spectrum. For use with grating spectrometer we have compared the observed quasicontinuous second-order Raman spectra in $SrTiO_3$ with and without the I_2 absorption cell (see Fig. 10). Similar calibration of the I_2 absorption in the lower-frequency range has been obtained by doing the same comparison on the Rayleigh wing spectrum of CS_2. In both cases the size, shape, and position of the "holes" eaten out of the spectrum reveal the effects of the resonant I_2 absorber. For studying weak inelastic spectra, fluorescence from the I_2 cell, induced by the absorption of the elastic component, can produce spurious lines in the spectral region of interest. So the improvement in *effective* contrast achieved with an I_2 cell is accompanied by a considerable complication in the overall instrumental response function.

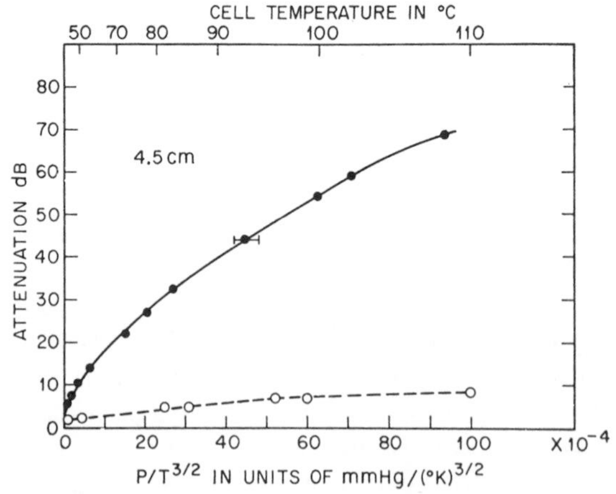

Fig. 9. Temperature and pressure dependence of resonant I_2 absorption for suitably tuned single-mode Ar laser at 5145 A, on (solid) and off (dashed) resonance. After Devlin et al.[20]

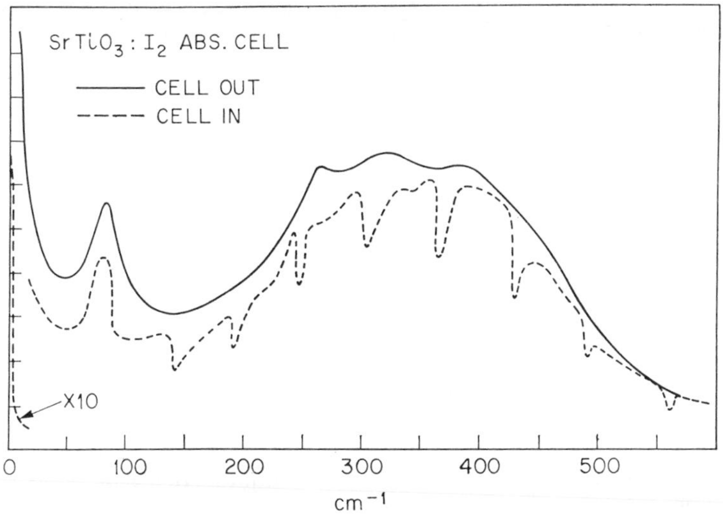

Fig. 10. Effect of subsidiary I_2 absorptions on inelastically scattered light shifted $0 \rightarrow 500\ cm^{-1}$ from 5145 A laser line. Illustrated by room-temperature second-order Raman spectrum of $SrTiO_3$.

Finally it should be mentioned that other absorbers exist for other laser lines (CH_4 for 3.39 μ HeNe; SF_6 for 10.6 μ CO_2, etc.). Naturally the combination of the I_2 absorber and one of the compound interferometers discussed earlier would be an obvious extension.

C. Low-Frequency Shifts

For the highest-resolution experiments of interest in fluids, optical spectrometers or interferometers are inadequate, and direct electronic processing of the optical signal must be employed. Basically these involve measuring the intensity correlation function (or its Fourier transform) of the scattered light via a nonlinear detector (most often a photomultiplier tube). Frequencies ranging from <1 Hz to >1 MHz can thus be measured, corresponding to a resolving power between 10^9 and 10^{15} for visible laser sources. Whereas the optical spectrometers and interferometers discussed above were widely employed before the advent of laser sources, the electronic spectrometers considered here require for successful operation source brightnesses obtainable only with lasers. Thus the literature on such spectrometers is relatively sparse and many of the concepts are unfamiliar to students of more traditional spectroscopy.

Recently several very useful papers have appeared which explicate the theory and practice of electronic spectroscopy and from which we shall incorporate several salient features into this section. However, for the researcher wishing to obtain a working knowledge of the subject, a consultation of the original articles is essential.

Electronic spectroscopy is done in one of two ways: (a) mixing of the scattered light either with itself or with some monochromatic local oscillator optical signal on a photosurface and frequency analysis of the resulting photocurrent to obtain the spectral content of the scattered field, and/or (b) digital processing of the photocurrent pulses to obtain directly the intensity correlation function of the scattered light.

The original applications of electronic spectroscopy to light scattering employed both the self (homodyne) and the local oscillator (heterodyne) mixing techniques. An excellent thorough discussion of the theory and applications of both techniques has been given recently by Cummins and Swinney.[3] We shall mention below some of their main conclusions.

The most striking advances in recent years, however, have been in the area of direct correlation spectroscopic techniques. An elegant analysis of intensity correlation spectrometers and their performance in comparison with light beating spectrometers has been given by Degiorgio and Lastovka.[19] Additional considerations on the theory of correlators by Jakeman and Pike[21] have also appeared. Although much of this analysis is for experts, we shall list some of these basic conclusions here as well.

Let us first consider the basic expression for the scattered intensity in a form amenable to electronic spectroscopy.[3] The utility of electronic spectroscopy rests upon the quantitative relationship between the spectrum or statistical properties of the scattered optical field and the spectrum or count distribution of the photoelectric current. This relationship is simple for an optical field which is a Gaussian random process and not so generally otherwise. Fortunately even when the incident light is coherent, the light scattered from thermal fluctuations is Gaussian, so we can proceed on that assumption.

We note that the instantaneous intensity at the photocathode $I(t)$ is just $E^*(t)E(t)$. Assuming the field to be spatially coherent over the illuminated photocathode area the photocurrent will be $i(t) = \sigma E^*(t)E(t)$ or

$$\langle i(t)\rangle = \sigma\langle i\rangle = \sigma\langle E^*(t)E(t)\rangle \tag{89}$$

It is also important to define the correlation function between the photocurrent at t and at $t + \Delta$

$$\begin{aligned}\langle i^{(2)}(t, t + \Delta t)\rangle &= \sigma^2\langle E^*(t)E(t)E^*(t + \Delta t)E(t + \Delta t)\rangle \\ &= \sigma^2\langle I\rangle^2 g^{(2)}(\Delta t)\end{aligned} \tag{90}$$

$$g^{(2)}(\Delta t) \equiv \frac{\langle E^*(t)E(t)E^*(t + \Delta t)E(t + \Delta t)\rangle}{\langle E^*E\rangle^2} \tag{91}$$

The scattered spectrum is just the Fourier transform of

$$\langle E^*(t)E(t + \Delta\tau)\rangle = \langle I\rangle g^{(1)}(\Delta\tau) \tag{92}$$

Now for Gaussian fields the correlation functions $g^{(1)}$ and $g^{(2)}$ are simply related

$$g^{(2)}(\tau) = 1 + |g^{(1)}(\tau)|^2 \tag{93}$$

The photocurrent correlation including the self-term can then be expressed[3]:

$$\begin{aligned} C_i(\tau) &= e\langle i\rangle\delta(\tau) + \langle i^2\rangle g^{(2)}(\tau) \\ &= e\langle i\rangle\delta(\tau) + \langle i^2\rangle(1 + |g^{(1)}(\tau)|^2) \end{aligned} \tag{94}$$

The photocurrent *spectrum* $P_i(\omega)$ is just the Fourier transform of $C_i(\tau)$, and the homodyne and heterodyne situations may be distinguished as follows. For homodyning the total field consists of the scattered field alone (no local oscillator). Its correlation function is given by $g^{(1)}(\Delta\tau)$ in (92), which is simply derivable from the photocurrent correlation function (94). Traditionally it has been the power spectrum of the photocurrent, that is, the Fourier transform of $C_i(\tau)$, which has been measured. This measurement involves feeding the photocurrent into a spectrum analyzer which then sweeps a narrow frequency window across the frequency range, viewing sequentially the various portions of $P_i(\omega)$. In the last few years, however, the direct measurement of $C_i(\tau)$ itself through the use of digital correlators has become increasingly popular. Although there are many subtleties in comparing the two approaches, it is generally true that, all other things being equal, the rate of data accumulation is faster in the correlation approach by a factor of order $(\Delta\omega\tau_c)^{-1}$ where τ_c is the correlation time and $\Delta\omega$ the frequency window width for the spectrum analyzer. Some other aspects of direct correlation will be discussed below.

For the heterodyne method the field appearing in (92) consists of the sum of the scattered field $\mathbf{E}_s(t)$ and a coherent local oscillator field $\mathbf{E}_0 e^{i\omega_0 t}$ usually derived from the incident laser beam itself. The current correlation $C_i(\tau)$ is therefore a more complicated expression than for the homodyne case ($\mathbf{E}_0 = 0$). In the limit where $E_0 \gg E_s$ we have[3]

$$\begin{aligned} C_i(\tau) &= e^2\sigma I_0\,\delta(\tau) + e^2\sigma^2 I_0^2 + e^2\sigma^2 I_0\langle I_s\rangle \\ &\quad\times \{e^{i\omega_0\tau}\langle E_s{}^*(t)E_s(t + \tau)\rangle + e^{-i\omega_0\tau}\langle E_s(t)E_s{}^*(t + \tau)\rangle\} \\ &= ei_0\,\delta(\tau) + i_0^2 + i_0\langle i_s\rangle\{e^{i\omega_0\tau}g^{(1)}(\tau) + e^{-i\omega_0\tau}g^{*(1)}(\tau)\} \end{aligned} \tag{95}$$

Notice in this case that the relation between $C_i(\tau)$ and $g^{(1)}(\tau)$ does not proceed through $g^{(2)}(\tau)$ and thus holds even for non-Gaussian statistics. The $P_i(\omega)$ resulting from (95) is an exact replica of the optical spectrum centered at a frequency $(\omega_1 - \omega_0)$, which is zero if the incident laser is used as the local oscillator. These latter attractive features of heterodyning are somewhat offset in practice by the requirement that the fields $\mathbf{E}_0$ and $\mathbf{E}_s$ must be parallel and coincident at the photosurface to prevent drastic decrease in the mixing efficiency. This presents severe alignment difficulties not present in the homodyne scheme.

As with homodyning the photocurrent may be either Fourier analyzed for $P_i(\omega)$ or processed directly to obtain the correlation function $C_i(\tau)$. Again the rate of data accumulation is much faster with the correlator than with the spectrum analyzer scheme. Recently multichannel spectrum analyzers have become available which permit simultaneous observation of all frequency intervals, rather than requiring sequential sampling of a swept, "single-channel" analyzer. Such devices recover the factor $(\Delta\omega\tau_c)^{-1}$ in the data accumulation rate mentioned above. The operation of spectrum analyzers is discussed more fully by Cummins and Swinney[3] and by Benedek.[217]

Let us now turn briefly to a description of digital correlators so as to appreciate the recent advance represented by the so-called "clipped" correlator.[19] The photocurrent in a photomultiplier is generated by the arrival of each photon at the photocathode causing the emission of a photoelectron, which is then accelerated through the dynode chain where it produces additional (roughly simultaneous) electrons, all of which, perhaps a million in all, arrive in a bunch at the tube anode, giving rise to a "pulse" of photocurrent. If the photon arrival rate (light intensity) is not too great each photon will give rise to a photoelectron pulse arriving at the anode at a precisely delayed time after the photon struck the photocathode. Ignoring the statistical complications of (a) the initial photoemission process and (b) the photoelectron amplification process, one would expect the arrival rate and distribution of photocurrent pulses at the anode to replicate accurately those of the photon arrivals at the cathode. The photocurrent, conveniently viewed as an analogue signal for light beating purposes, is in reality a digital signal, quite amenable to digital correlation processing. Degiorgio and Lastovka[19] have described the operation of an ideal digital correlator as follows: divide the total measurement time T into M_0 equal intervals of duration Δt. Define η_j as the number of counts (photocurrent pulses assumed identical in size and shape) recorded between $j\,\Delta t$ and $j\,\Delta t + \delta t$ (with $\delta t < \Delta t$). The photocurrent autocorrelation function (at $l\,\Delta t$) is then

$$R_l = \frac{1}{M_0}\sum_{j=1}^{M_0}\eta_j\eta_{j+l} \tag{96}$$

For a Gaussian field this has the form

$$R_l = \langle n\rangle^2(1 + e^{-l\Delta t/\tau_c}) \tag{97}$$

where τ_c is the correlation time. Naturally to achieve good representation of R_l in a reasonable T, one arranges for $\Delta t \ll \tau_c \lessdot T/5$. The time and circuiting required for formation of the full product pairs in (96) would make a many-channel ($\sim$100) full correlator quite complicated, expensive, and slower than the simple clipped correlator described below.

Present-day electronic components can perform simple logical operations as quickly as $\sim 10^{-9}$ sec and permit in principle comparably small values of Δt and measurements of τ_c as small as 3×10^{-8} sec. This corresponds to frequencies as high as 10 MHz and clearly overlaps the low-frequency range of Fabry-Perot spectrometers. Measurement of longer τ_c, or lower frequencies, down to <1 Hz are easy with correlators. However, a machine that will accurately perform the full correlation operation indicated in (96) requires many more logical operations and is thus slower than the simpler, nearly as accurate, device we now describe.

It has recently been shown that essentially the same information and accuracy are available from a much simpler device called a clipped correlator.[19] The basic simplifying operation is to introduce a positive integer clipping level, k, for the number of counts collected in interval δt. Thus we can define a clipped count $n^{(k)}$ as

$$\begin{aligned} n^{(k)} &= 1 \qquad n > k \\ &= 0 \qquad n \leq k \end{aligned}$$

The clipped correlator then operates in the following truncated manner[19]: "at $t = 0$ the correlator begins sampling photocurrent with gates of duration $\delta t \leq \Delta t$. The first gate in which the collected number of counts exceeds k triggers an N-channel scaler which records sequentially the number of counts during each of the N gates immediately following the trigger gate. After N gates the trigger is again started on the next $n > k$ count at the first channel. The process is then repeated a total of M_0 times with the number of counts in each channel added to the total in that channel from previous triggers. After M_0 triggers the lth channel contains the sum of all counts registered in all the M_0 gates occurring $l\,\Delta t$ after the trigger." The clipped correlation function thus formed is

$$R_l^k = \frac{1}{M_0} \sum_{m=1}^{M_0} n^{(k)}(t_m) n(t_m + l\,\Delta t) \tag{98}$$

where n is the full count and n^k the clipped count. The operations involved in forming the sum, then, multiplication by only 1 or 0 and addition of the results to the contents of a register, are much less complicated than in the full correlator. Jakeman and Pike[21] have computed the clipped correlation function for a Gaussian exponential field and found

$$R_k(t) = \langle n \rangle \left(\frac{\langle n \rangle}{1 + \langle n \rangle} \right)^{k+1} \left[1 + \left(\frac{1 + k}{1 + \langle n \rangle} \right) e^{-t/\tau_c} \right] \tag{99}$$

The similarity in form to that of (96) is striking and suggests how τ_c may be extracted from either form with equal ease.

A detailed analysis and comparison of the statistical errors inherent with both the ideal and clipped digital correlators, as well as the homodyne spectrum analyzer, have been given by Degiorgio and Lastovka.[19] One of their comparisons is summarized in Fig. 11, where the fractional rms statistical errors involved in determining τ_c are plotted for the ideal and the clipped correlators and for single-channel and 100-channel spectrum analyzers as a function of the counting rate (expressed in number of counts per correlation time $= \eta\tau_c > 1$). The clipped correlator is better than the spectrum analyzer and only slightly worse than the full correlator.

Finally, one crucial concept which must be mentioned is that of the "coherence area." All the above discussion has presumed that the optical field is spatially coherent over the surface area of the photodetector. As long as this is true the signal to shot noise ratio increases linearly with detector area. For a signal derived from an incoherent scattering process, the scattered field source is spatially incoherent. The result is that the scattered field is only coherent over a limited area $A_c = 2\lambda^2/\Omega$, where λ is the optical wavelength and Ω is the solid angle subtended by the source

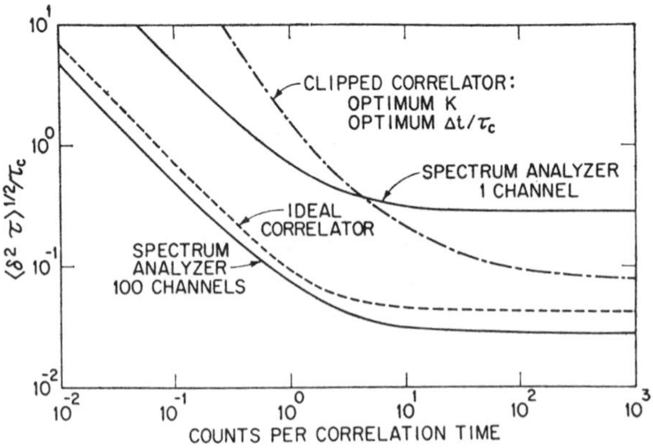

Fig. 11. RMS fractional errors vs. counting rate per correlation time for clipped and ideal correlators and for single- and 100-channel spectrum analyzers. After Degiorgio and Lastovka.[19]

at the detector. For a detector area $A = \alpha A_c$, the α elements act independently so far as mixing is concerned, with the result that increase in A beyond A_c causes shot noise to increase as rapidly as signal with no change in the ratio. Under certain conditions, in fact, the error in the $C_i(\tau)$ measurement actually *increases* with larger α whereas the error in $P_i(\omega)$ is independent of α.[19]

IV. EXPERIMENTAL RESULTS

So active is the research on light scattering in fluids that several new advances have been made which have not been discussed in even quite recent review articles. In this section we shall emphasize these, referring to earlier work only where necessary to provide a suitable context in which to view the more recent results. As we have seen in the earlier theoretical sections, the richness of spectral information obtainable from a fluid system increases as the system becomes more complex. In Fig. 12 we have tried to represent this situation schematically for fluid systems ranging in complexity from a simple monatomic gas to a mixture of polyatomic liquids. In the latter case the spectra are so complex, especially when one considers the possibility of interaction effects among the various degrees of freedom, that theoretical treatments are still on the level of phenomenology. Nevertheless it is hoped that by comparing in sufficient detail the spectra of successively more complex fluid systems we may be able to "boot strap" our way to a more fundamental understanding of the liquid state.

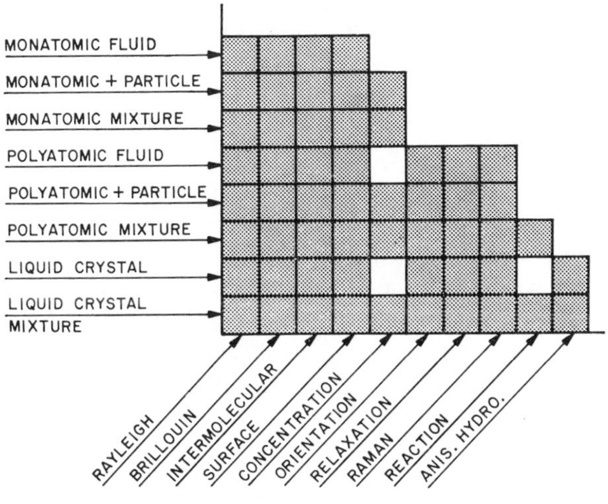

Fig. 12. Schematic illustration of contributions to spectra (shaded blocks) from various processes (horizontal axis) for fluids of increasing complexity (vertical axis).

As can be seen from Fig. 12 all fluid systems support density fluctuations, which in general give rise to Brillouin scattering from the thermal sound waves (adiabatic density fluctuations) and to Rayleigh scattering from isobaric entropy fluctuations. The Brillouin spectrum measures the velocity and attenuation of high-frequency sound, whereas the Rayleigh spectrum, centered at zero frequency shift, has a width proportional to the thermal diffusivity, $\kappa/\rho C_p$. In superfluid helium the isobaric entropy fluctuations are propagating rather than purely dissipative, and their spectra provide a measure of the velocity and attenuation of second sound.

A. Liquid Gas Critical Point

Much of the recent work on the Rayleigh-Brillouin spectra of pure fluids has been devoted to the vicinity of the liquid–gas critical point (C.P.). It has long been realized that near C.P. the Rayleigh peak will narrow in width and increase in intensity (critical opalescence), due mainly to the divergence in the specific heat and density correlation length. The precise manner of this behavior has been of considerable interest in testing the so-called "scaling laws" proposed for second-order phase transitions. According to the scaling hypothesis sufficiently near C.P. there is only one fundamental length whose behavior depends only on general symmetry properties of the phase transition and not on the details of interparticle interactions. This length, called the "correlation length," ξ, and its behavior in terms of $|X - X_c|$ (where X is some appropriate thermodynamic variable, like temperature or density) provide a convenient means of unifying the behavior of such outwardly diverse systems as pure fluids, magnets, fluid mixtures, ferroelectrics, superconductors, etc., when they undergo phase transformation. It also serves to connect the near-critical behavior of fluctuations having different wave vectors, k, through the scaling requirement that the correlation functions depend only on dimensionless product $k\xi$.

Another theoretical expectation regarding critical behavior is that very near C.P. (the "asymptotic region") the divergences of the various correlation functions are describable by simple power laws in $|X - X_c|$. The value of the power appropriate to a given quantity is the so-called "critical exponent." Based on general assumptions regarding the homogeneity of the system's free energy function, simple relations may be derived between these various critical exponents. Thus by measuring the rates of divergences of various thermodynamic and transport coefficients near C.P. one may obtain these exponents and thus test the predictions of the "scaling law" hypothesis.

Very recently a rather long-standing controversy regarding the apparently different critical behavior of SF_6 from that of simpler fluids like Xe and

CO_2 has been resolved. In the hydrodynamic regime and for measurements along the critical isochore, the Rayleigh line width measures the ratio $\kappa/\rho C_p$ and thus a combination of critical exponents $(\gamma - \psi) \equiv \varphi$ since $C_p \propto \varepsilon^{-\gamma}$ and $\kappa \propto \varepsilon^{-\psi}$ (where $\varepsilon = |T - T_c|/T_c$). Additional information is provided by the frequency integral of the scattered intensity $I(k) = [AK_T \sin^2 \alpha/(1 + k^2\xi^2)]$. The angular dependence of $I(k)$ measures ξ, and the angular integrated intensity I measures K_T, the isothermal compressibility. These are described by $\xi = \xi_0\varepsilon^{-\nu}$ and $K_T = K_0\varepsilon^{-\gamma}$, respectively.

Several years ago Saxman and Benedek[217] obtained in SF_6 a Rayleigh line width which vanished as $\Gamma_R \propto \varepsilon^{1.26}$; that is, $\varphi = 1.26$. Later measurements for the simple fluids CO_2 and Xe revealed apparent φ's of 0.73[81,82] and 0.751,[58] respectively. This striking disagreement appeared to contradict the expected universal behavior of critical exponents. Thus many other experiments were done on these three fluids to isolate the source of the discrepancy. Direct measurements of γ showed values for all three fluids of $\sim 1.22 \pm 0.02$; for ν the results were all consistent with $\sim 0.63 \pm 0.07$. The original value for φ for SF_6 became increasingly suspect. Recently three independent remeasurements of the Rayleigh line width in SF_6 were made[40,51,63] which yielded values for φ quite close to those for CO_2 and Xe. It now appears that the original SF_6 data were in error, although, because they represent *too narrow* a Rayleigh width for a given value of ε, the exact experimental source of that error remains unknown.

Paradoxically the existence of the error itself has probably had a positive effect on our understanding of critical phenomena, since the anomalous φ for SF_6 forced more careful consideration of critical phenomena and the subtleties of data analysis than might have otherwise occurred. One result of this effort has been the realization that the Rayleigh width $Dk^2(D = \kappa/\rho C_p)$ receives a non-negligible contribution from the nonsingular or background thermal conductivity κ_B. Thus $k^2\kappa_B/\rho C_p$ must be subtracted from the measured line width in order to obtain $k^2\kappa_S/\rho C_p$, the true singular part. When done for the above experiments, values of 0.61 ± 0.04,[51,63] 0.62,[81,82] and 0.64 ± 0.04[58] for φ are obtained for SF_6, CO_2, and Xe, respectively. Not only are these exponents all essentially the same, but furthermore they are in remarkably close agreement with a recent theoretical prediction by Kawasaki[508] that the singular part should be given by $k^2kT/6\pi\eta\xi$. That is, the singular contribution to the Rayleigh width is just that corresponding to the mass diffusion of a fluid sphere with radius $r = \xi =$ the correlation length. Thus φ should be equal to $-\nu$ according to Kawasaki, and the above corrected φ values are quite consistent with the previously quoted $-\nu = 0.63 \pm 0.07$.

B. Critical Mixing

The behavior of Brillouin spectra near C.P. has also been studied in detail for some pure fluids. However, before discussing these results we shall briefly consider the critical opalescence studies made in two-component fluid mixtures. As seen in Section II the spectrum of a solution or a fluid mixture exhibits an additional central frequency component due to fluctuations in the relative concentrations of the two species. Away from critical points where correlation lengths are small, this component is Lorentzian in shape with a width $\Gamma_c = Dk^2$, where D is (a) the solute mass diffusion coefficient for a dilute solution or (b) $\alpha(\partial\mu/\partial c)$ for a concentrated mixture, where α is a transport coefficient, c is the concentration, and μ is the chemical potential. For binary mixtures near the critical mixing point, concentration fluctuations are correlated over a sufficient range that $k\xi$ may no longer by $\ll 1$. Then the line width is $\Gamma_c = D_c k^2(1 + k^2\xi^2)$. Chu has done extensive studies of line width as well as of angular and total intensity for several binary mixtures and has found that the critical exponents agree well with their counterparts in the liquid–gas critical transition.[2,305,306,354] Swift[507] has predicted that $k^2 D$ should behave for binary mixtures with the same exponent that the Rayleigh line width does in a pure fluid near C.P.; that is, $D \propto \varepsilon^{\varphi=-\nu}$. The line widths measured for a variety of binary mixtures[2] yield exponents of $\sim 0.66 \pm 0.05$ on the average, in good agreement with Swift's modest proposal. Of course the critical behavior of ξ may also be measured by angular dependence of the scattered intensity. The results thus obtained are in good agreement with the line width measurements.

C. Brillouin Scattering

1. *Simple Fluids*

A few Brillouin experiments in simple monatomic liquids away from C.P. have been performed with the aim of possibly observing some of the departures from Navier-Stokes theory predicted by the generalized hydrodynamic analysis of Section II. The studies of Fleury and Boon[107] on argon and neon initially suggested the existence of a small negative dispersion in the hypersonic velocity for sound frequencies in the 1–3 GHz range. Although close in size to the experimental error, the observed velocity difference from low-frequency (<1 MHz) values subsequently measured in neon[506] supports the need for more accurate Brillouin and ultrasonic measurements in simple liquids. A simple corresponding-states argument (see Section II.3) suggests that if the $\sim 1\%$ negative dispersion in neon were due to quantum effects, rather than to finite frequency effects, then normal liquid helium should exhibit an even larger dispersion.[159] Very recently Pike et al.[78] have observed the Brillouin spectrum from normal liquid helium, and find no dispersion greater than $\sim 0.5\%$ relative to velocities obtained at much lower frequencies. The origin, if not the existence, of small velocity dispersion in argon and neon then remains to be understood.

2. *Complex Fluids*

In fluids where there exist "internal" energy-storing degrees of freedom (such as vibrational, rotational, or configurational), relaxation processes can introduce quite large frequency dependence to both sound velocity and attenuation. These effects (discussed theoretically in Section II.3) were first studied in detail using Brillouin scattering by Chiao and Fleury[95,108] and have since been extended by several others.[37,38,105,137,158] Because relaxation effects are most noticeable for frequencies $\omega \approx \tau^{-1}$ and because for many relaxation processes in fluids τ^{-1} is of order 0.1–10 GHz, Brillouin scattering is an ideal way to investigate such phenomena. With increasing precision and sophistication in spectral line shape measurement and analysis has come corresponding increase in the accuracy of determining the relaxation parameters, τ and $(U_\infty^2 - U_0^2)/U_0^2$.

Perhaps the most interesting aspect to Rayleigh-Brillouin spectra arising from relaxation process is an additional central peak of width approximately τ^{-1}. Implicit in the phenomenological theory of Rytov, this feature was discussed explicitly in Mountain's hydrodynamic theory of 1966[11,130] and was first observed experimentally in CCl_4 at about the same time.[115] Although since observed in several relaxing fluids,[121] this relaxation mode has not proved as helpful in providing a direct measure of relaxational parameters as was initially hoped. This is primarily due to its rather broad shape, difficult to extract precisely in the presence of the stronger Rayleigh and Brillouin components, at least with the instrumental resolution applied thus far.

It has been realized recently by a number of people that the relaxational peak and the form derived by Mountain are special-case manifestations of a rather general phenomenon not restricted to relaxing liquids, but present in magnets, ferroelectrics, superconductors, etc., as well. The existence of an additional central peak in the scattering response function $S(k, \omega)$ of a system can generally be ascribed to a frequency-dependent damping of the mode of interest. The shape and strength of this central peak are determined by the precise shape and strength of this frequency dependence. One can write the susceptibility $\chi(\omega) = [\omega_0^2 - \omega^2 + i\Gamma(\omega)\omega]^{-1}$. Then

$$S(\omega) = -\frac{n+1}{\pi}\,\mathrm{Im}\,\chi = \frac{(n+1)\Gamma'\omega}{\pi([\omega_0^2 - \omega^2 - \omega\Gamma'']^2 + \omega^2\Gamma'^2)}$$

where $\Gamma = \Gamma' + i\Gamma''$. For relaxation behavior $\Gamma(\omega) = \Gamma_0/(1 + i\omega\tau)$. The resulting $S(\omega)$ takes on different forms depending on the values of $\omega\tau$ and $\Gamma_0\tau^{-1}$. The limit most appropriate to Brillouin scattering in relaxing fluids has $\tau^{-1} < U_\infty q$ and $(U_\infty^2 - U_0^2)/U_\infty^2 < 1$. In this case the Brillouin peaks contribute a fraction $\sim[1 - (U_0^2/U_\infty^2)]$ of their strength to the relaxation peak and it will be approximately Lorentzian in shape, of width τ^{-1}. [$(n + 1)$ is the thermal population factor.]

The relaxation peak would appear much more striking in the limits $(U_\infty^2 - U_0^2)/U_0^2 > 1$; $\tau^{-1} < U_\infty q$. Then the central peak will carry the major fraction of the strength, $[1 - (U_0^2/U_\infty^2)]$, and will have a width $\omega_0^2/\Gamma_0 = U_0^2q^2/\Gamma_0$.

Passage from one region to another is usually not possible for Brillouin scattering in a relaxing liquid, because τ^{-1} and Γ_0 are not usually sufficiently strong functions of temperature, pressure, etc. An exception to this occurs in the vicinity of phase transitions. In particular, at the C.P. in a simple monatomic fluid, the fluid correlation volume elements themselves play the role of the internal degrees of freedom, giving rise theoretically to two distinct relaxation times τ_R and τ_B. $\tau_R^{-1} = (\kappa/\rho C_p)\xi^{-2}$ corresponds to the inverse time required for thermal diffusion to occur over a distance of one correlation length. $\tau_B = \xi/U$ corresponds to the time required for a sound wave to cross the correlation length. Both these relaxations vary greatly as $T \to T_c$. Cannell[90] has probed the region $\omega\tau_B \approx 1$ in his Brillouin studies near xenon's C.P. and has observed the central relaxational peak quite near C.P. The sound velocity dispersion accompanying the τ_B relaxation has been well documented in this and earlier ultrasonic studies. Other Brillouin studies near C.P. have been carried out by Gammon et al.[111] and by Ford et al.[109] in CO_2, and by Mohr et al.[127] in SF_6.

All these experiments obtained similar results and had to overcome similar difficulties, mainly arising from the diverging Rayleigh intensity as the C.P. is approached. In all experiments effects of relaxation phenomena on the hypersonic critical behavior were evident, and thus complicated considerably comparisons with low-frequency sound experiments. The correlation-length-related relaxation processes in Xe already mentioned are present in CO_2 and SF_6 as well. However, these polyatomic fluids exhibit in addition vibrational and rotational relaxations. The result is that although "exponents" describing both the Brillouin splitting and line width dependences on $(T - T_c)$ can be extracted from these data, their values are generally frequency dependent, are not the same for the different fluids, and thus are not directly relevant to tests of theoretical scaling predictions. Xenon presents by far the simplest case, and Cannell[90] has used the full spectral shapes to determine not only the hypersound velocity and attenuation, but several other parameters as well, including C_p/C_v, τ_B,

$\kappa/\rho C_v$, and U_0. His discussion clearly demonstrates the much greater difficulty in understanding Brillouin spectra near C.P. than the Rayleigh spectrum. There is room for much work on this problem.

Even more complicated to understand microscopically is the Brillouin spectrum near the solution critical point of a binary liquid mixture. Arefev[86] has observed the Brillouin spectrum of triethylamine (44.6% wt) and water near its critical mixing temperature, 17.9°C. A strongly temperature-dependent velocity dispersion was observed, but no definitive interpretation could be made. Similar results were obtained for the nitrobenzene–*n*-hexane mixture studied by Chen and Polonsky.[360] In contrast to the behavior in the pure fluid C.P. experiments, their results show the Brillouin line width continuing to diverge for $T \rightarrow T_c$ rather than leveling off for $\Delta T < \sim 0.7°C$, as observed in pure fluids.[109,111,127] Again the spectra receive contributions from relaxation processes associated with the correlation range of the concentration fluctuations, as well as from the noncritical molecular relaxations due to the polyatomic nature of the liquid constituents. Again, much more detailed experimental work on Brillouin spectra of critical mixtures is required to put understanding on a quantitative basis. It seems appropriate to point out here that the Rayleigh-Brillouin spectrum of a critical mixture should exhibit an additional central "relaxation" component. No such component has yet been reported, and a search for it might be of interest.

D. Depolarized Scattering

Thus far the processes discussed have arisen from density fluctuations (of either a single or a double liquid species) which can produce only diagonal elements to the polarizability tensor. Therefore the polarization of the scattered light is in the *same* direction as that of the incident light. It has long been known that the spectrum of light scattered from fluid systems contains appreciable depolarized components. These arise from off-diagonal elements of the polarizability tensor and require dynamic physical structures which are optically anisotropic. Perhaps the most familiar such example is a fluid of nonspherical molecules, like CS_2. The depolarized spectra of such fluids consist mainly of a Lorentzian central component whose width lies typically in the 1–10 cm^{-1} range. The simplest interpretation of this feature is that it measures the correlation function describing the angular reorientation of individual anisotropic molecules in the viscous environment provided by the surrounding fluid. The characteristic reorientation time was estimated by Debye in 1929 to be $\tau = \eta_s V/kT$, where η_s is the fluid's shear viscosity and V the effective molecular volume. This single-particle Brownian rotational diffusion picture dominated spectroscopists' interpretation of depolarized spectra until a very few years ago, when two more sophisticated concepts were recognized. For identification purposes we shall call these (1) "dynamic shear waves," and (2) "intermolecular light scattering," respectively.

1. *Dynamic Shear Waves*

Whatever ultimate physical interpretation is ascribed to it, the former process definitely illustrates the inadequacy of the Debye model cited above. In 1967 high-resolution depolarized spectra of nitrobenzene and quinolene revealed an anomalous doublet structure[224,256] (see Fig. 13). Further experiments by Stoicheff and his co-workers[259,269] revealed similar structure in a wide variety of anisotropic molecular liquids. Since these fluids are "viscoelastic" media, one is tempted to interpret the observed structure as scattering from high-frequency shear waves in the liquid. It

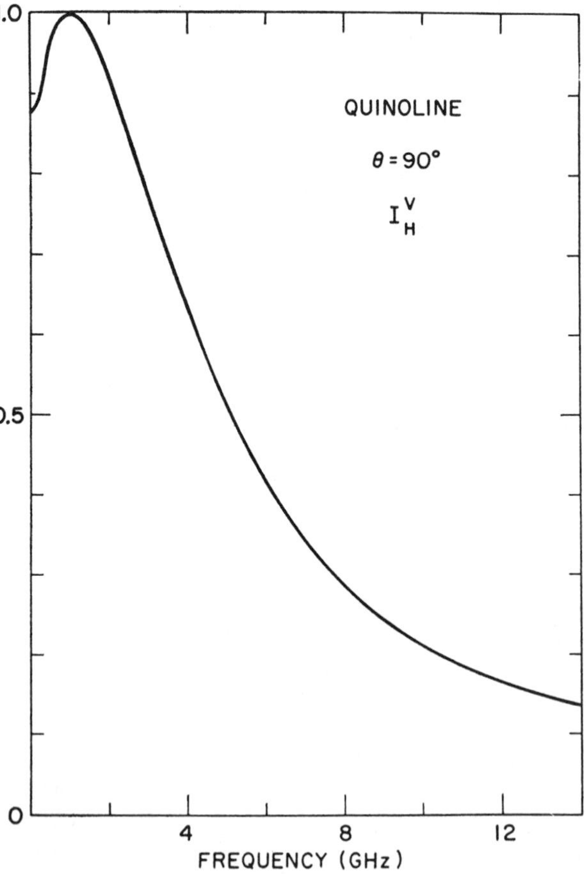

(a)

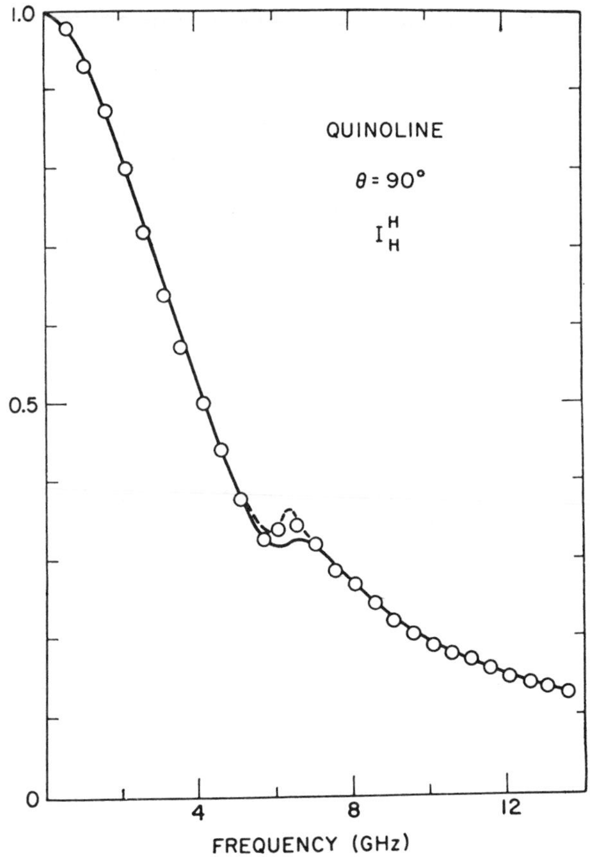

(*b*)

Fig. 13. Depolarized spectra for quinoline observed[259] and calculated [44] for 90° scattering in two polarization configurations. (*a*) Theory and experiment coincide perfectly with solid line. (*b*) Theories (solid and dashed curves) disagree slightly with experiment (circles).

is well known that although liquids exhibit no restoring force to the imposition of a static shear, at sufficiently high frequencies they do exhibit a nonzero (dynamic) shear modulus. Stoicheff's group has cited additional evidence for this interpretation in the k-dependence of the doublet peak splitting, which can be interpreted as $\omega_s = kV_s$, with a definite shear wave velocity V_s. This interpretation suggests that the observed doublet is associated with a collective orientation fluctuation, for which the Debye model provides the corresponding single-particle behavior.

Volterra[261] has produced a theory which quantifies these ideas and which is capable of reproducing the observed line shapes and polarization selection rules in terms of several adjustable parameters. Others have chosen to view the spectrum observed not as a doublet, displaced from zero frequency, but rather as a *central dip* arising from an *interference* between two central components of different widths. A recent paper by Chung and Yip[44] not only presents this viewpoint clearly and attractively, but contains a critical discussion of similar work by others.[218,219,220] Their theory is also phenomenological, but has the attractive feature of unifying both the Brillouin and depolarized spectra into a single description. We will not present here any of their rather complicated spectral shape calculations. Their results rest on two basic assumptions: (1) the anisotropic fluctuations in the polarizability are describable by an effective stress tensor which contains both elastic and molecular reorientational components; and (2) the dynamics of the polarizability fluctuations are described by viscoelastic equations which simulate effects of shear, orientational, and thermal relaxation processes. The resulting equations of motion for the stress tensor ($\sigma_{\alpha\beta}$), the density fluctuations [$\rho(\mathbf{r}, t)$], and the velocity field [$v_\alpha(\mathbf{r}, t)$] are all linearly coupled. These couplings among the various degrees of freedom modify the spectral shapes significantly. Most significant for this discussion is the interaction between the shear stress and the molecular reorientations. In their model, the shear viscosity relaxes with a characteristic time, τ_s, and gives rise to a relaxational central peak of width $\sim\tau_s^{-1}$, whereas the simple molecular reorientation time τ_0 would cause another Debye-like central peak of width τ_0^{-1}. The coupling of these processes permits an *interference* in their contributions to the overall spectrum, which under certain conditions can be destructive, causing a central dip. This result is based on the physically reasonable assumption that $\tau_s \ll \tau_0$. For ordinary organic liquids $\tau_s \sim 10^{-12}$ sec and $\tau_0 \sim 10^{-10}$–10^{-11} sec. Figure 13 shows the results calculated by Chung and Yip for quinoline in two geometries. The *a* geometry does not couple to density fluctuations and the agreement with experiment is essentially perfect. The *b* geometry does couple with density fluctuations and correspondingly exhibits an interference near ω_B rather than near zero frequency. In this case agreement with experiment (open circles) is less perfect. Although there is good agreement with both polarization and scattering angle dependences of the observed spectra, this and other theories contain at least three or four adjustable parameters and must therefore be considered incomplete.

Before discussing the second type of contribution to depolarized fluid spectra, we must consider the very rapidly growing field of study represented by liquid crystals. These fluids are typically composed of very anisotropic organic molecules which under certain conditions exhibit varying degrees of long-range orientational and translational order. These fluids may exist in several phases exhibiting increasing degrees of order: isotropic, nematic, smectic, and cholesteric. Most attention has been paid in light scattering experiments to the isotropic and the nematic phases. The former describes the normal liquid state exhibited in the anisotropic fluids discussed immediately above. The nematic phase occurs at lower temperatures and describes an arrangement where all the molecules are *orientationally* ordered (aligned), whereas their centers are translationally disordered with respect to each other. Below T_c, because of this molecular alignment the nematic liquid crystal bears many close analogies to the ferromagnetic solid. As T_c is approached from above, fluctuations in the orientational alignment Q become correlated over longer distances. Q is thus conveniently identified with the order parameter for the isotropic–nematic phase transition. ($Q = \frac{3}{2}\langle\cos^2\theta - \frac{1}{3}\rangle$, where θ = angle between molecules' axis and local optic axis.) The spectrum of depolarized light scattered from a liquid crystal measures quite directly the fluctuations in this order parameter. Although strictly speaking this is a first-order transition, the isotropic–nematic phase transition exhibits striking pre-transitional (second-order) phenomena in both its static (correlation length) and dynamic (characteristic frequencies) properties. These have been observed by measuring the integrated intensity and the frequency spectrum of the scattered light, respectively.

Among the most popular materials for optical study are *p*-methoxy-benzylidene, *p*-*n*-butylaniline (MBBA), and *p*-azoxyanisole (PAA). Using a high-resolution Fabry-Perot, Litster and Stinson[279] have measured the critical slowing down of the orientation fluctuations in MBBA above the nematic–isotropic transition. According to DeGennes' molecular field theory, the Q fluctuations in the isotropic phase decay exponentially with a characteristic time $\tau_1 = \nu/A$, where ν is proportional to the shear viscosity and $A = a(T - T_c^*)$. Stinson and Litster[279] have observed the quasielastic Lorentzian shaped scattering whose line width $\Gamma = \tau_1^{-1}$ is shown in Fig. 14. Assuming $\nu = \nu_0 \exp(-2800°/T)$ the temperature dependence of A is indeed rather linear in $T - T_c^*$. Γ was also independent of scattering angle. Another prediction of the molecular field theory borne out in these experiments pertains to the integrated intensity I_k.

$$I_k = \text{Const.}\langle Q^2(k)\rangle = \text{Const.}\,\frac{k_B T}{Va(T - T_c^*)}\,\frac{1}{1 + k^2\xi^2}$$

Figure 15 shows the reciprocal intensity to be linear in $|T - T_c^*|$ for $k \simeq 10^5\ \text{cm}^{-1}$ (i.e., $k^2\xi^2 \ll 1$).

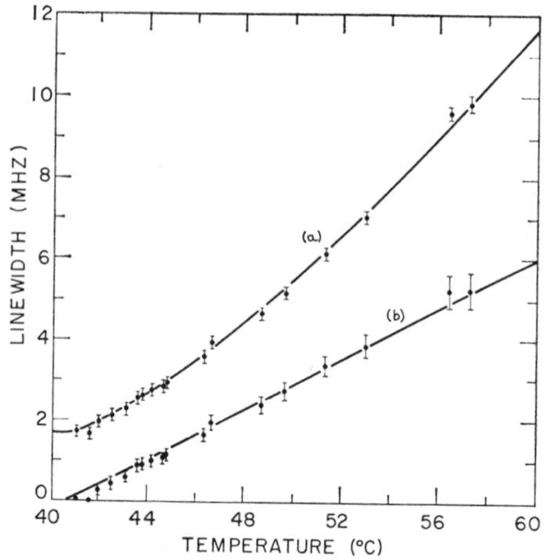

Fig. 14. Half width of central Lorentzian observed in isotropic phase of MBBA. (*a*) Raw data including 1.65 MHz instrumental line width. (*b*) Width corrected for instrumental effects and for assumed T dependence of shear viscosity mentioned in text. After Stinson and Litster.[279]

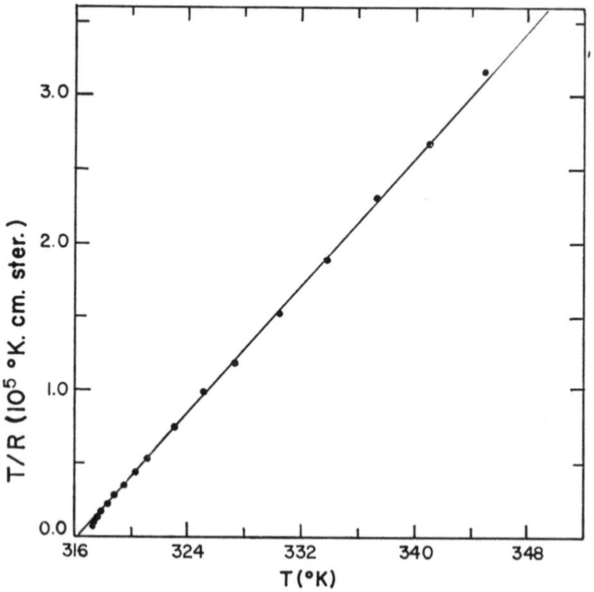

Fig. 15. Reciprocal intensity of the scattered light in isotropic MBBA vs. T. T_c is approximately 316°K. Solid line is a fit to mean field prediction.[280]

In the ordered phase a nematic liquid is conveniently described by a unit vector in the direction of molecular alignment $\hat{n}_0(r, t)$, often called the "director." Dynamical fluctuations of $\hat{n}_0$ are quite analogous to spin waves in a ferromagnet and have been studied in both PAA[277] and MBBA.[284] The PAA studies verified at constant temperature (125°C) that fluctuations $\delta\mathbf{n}_k$ can be decomposed into two uncoupled overdamped normal modes whose line widths are proportional to $a_i k^2$, where a_i depends on various geometry dependent combinations of the elastic and viscous coefficients of the nematic. Line widths (between 0.1 and 200 KHz) varied as k^2 for scattering angles between 1° and 40°.[277] The MBBA experiments concentrated on the temperature dependence of both intensity and line widths for these modes observed at fixed k. On approaching T_c from below, definite line narrowing within ~5° of T_c and a slight increase in scattered intensity was observed.[284] These results have been compared to a number of phenomenological theories for the dynamics of liquid crystals, but completely satisfactory agreement does not yet exist.

Some additional discrepancies with existing theory were found in the frequency dependent depolarization ratio for isotropic MBBA by Stinson, Litster, and Clark,[280] This is shown for two temperatures in Fig. 16 together with theoretical predictions (solid curves). The phenomenological

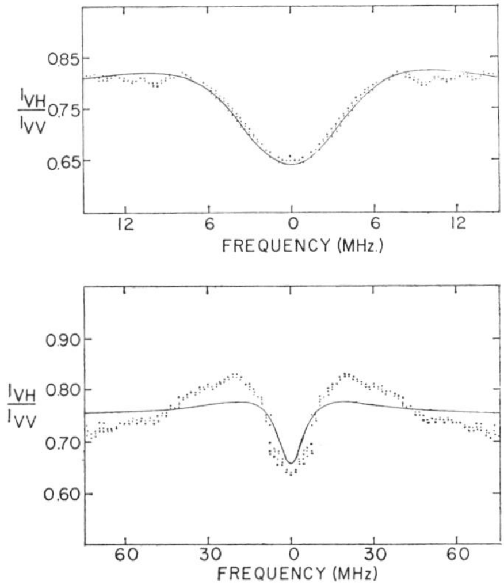

Fig. 16. Ratios of depolarized to polarized spectra in isotropic MBBA for $\theta = 6.55°$. Upper curve and lower curves taken 10.54 and 24.14°C, respectively, above isotropic–nematic transition. Solid curves are best fit to phenomenological theory discussed by Stinson et al.[280]

theory of DeGennes ascribes the central dip to coupling between orientational fluctuations and hydrodynamic shear waves.[280] In many ways the observations and explanations are similar to those discussed earlier for "shear waves" in normal anisotropic liquids, but a unified theoretical description covering both cases has yet to appear.

2. *Intermolecular Light Scattering*

A second important contribution to the spectra of all fluids was recently recognized for the first time. Since it arises from polarizability variations caused by *interactions among* fluid particles, we shall refer to it as "intermolecular light scattering." For dilute gases, however, the names "collision induced" and "translational Raman" scattering have often been applied. Traditionally, simple monatomic fluids were expected to exhibit no Raman or depolarized Rayleigh wing effects since they possess no vibrational or rotational degrees of freedom to modulate the polarizability. Levine and Birnbaum[176] first pointed out that a time-dependent polarizability results from the transient interaction between pairs of even the simplest atoms (i.e., binary collisions). The first observations of the spectra thus produced in dilute rare gases argon and krypton were made soon thereafter,[178,179] which verified that the spectrum was (1) largely depolarized; (2) quasielastic of the form $I(\nu) = I_0 e^{-\nu/\Delta}$; with (3) $\Delta \simeq V_{TH}/r_0$ and Δ independent of density. (For V_{TH} = thermal velocity and r_0 = interaction distance, $\Delta \simeq 5\text{–}10\ \text{cm}^{-1}$ for typical rare gases.) Also (4) the integrated intensity increased nearly as ρ^2, implying the two-particle source of the polarizability fluctuation.

Since then more complete studies[168,178,187,200] have been made of dilute rare gases ($\leqslant 100$ amagat) which demonstrate the possibility of quantitative calculation of their intermolecular scattering from (1) a knowledge of the intermolecular potential (binary collision dynamics), and (2) reasonable assumptions regarding the pair polarizability dependence on intermolecular separation. This work points out the wealth of information on the short-time dynamics (since $1\ \text{cm}^{-1} \rightarrow 5 \times 10^{-12}$ sec) contained in the intermolecular spectrum, and further, has shown how to extract it explicitly for the case of a dilute gas, where binary collisions exhaust the short-time dynamics.

The observation of intermolecular scattering in dense fluids and liquids has stimulated considerable effort to explicate similar connections between the spectra and the short-time dynamics for these more difficult cases. The original experiments on liquid argon showed several interesting contrasts with the vapor at the same pressure and temperature.[167,180] These experiments showed (i) the liquid spectrum is also nearly exponential in shape, (ii) Δ is nearly four times greater in the liquid than in the gas at the same

T, P, and (iii) the integrated intensity in the liquid is much smaller than expected from a density squared dependence. Thus, not surprisingly, the binary collision dynamics are demonstrably inadequate for liquids. McTague et al.[180] presented an interpretation of the liquid argon spectra based on a generalization of the familiar solid state concept of second-order Raman scattering.

Essentially a polarizability fluctuation of wave vector $\mathbf{q}$ can be produced by a suitably paired set of fluctuations in some dynamic variable δU; that is, $\delta\alpha_q \sim \delta U_{-\mathbf{k}}\delta U_{\mathbf{k}+\mathbf{q}}$. δU may describe a fluctuation in vibration, rotation, magnetization, density, etc. Second-order Raman scattering in solids usually arises from pairs of vibrational fluctuations (phonons), predominantly those with $k \simeq \pi/a$, since the phonon density of states is largest near Brillouin zone boundaries. Halley[199] and Stephen[201] have applied similar ideas to the well-defined phonon–roton dispersion curve in superfluid helium. Greytak and Yan[197] observed this second-order Raman scattering from superfluid helium, thus verifying the basic applicability of Stephen's theory (see Figs. 17 and 18). Stephen calculates the polarizability $\delta\alpha_q$ produced by a density fluctuation $\delta\rho_{-\mathbf{k}}$ in the presence of the optical field and a second fluctuation $\delta\rho_{\mathbf{k}+\mathbf{q}}$. Recall from Section II that generally the scattered spectrum measures $\langle\delta\alpha_q(t)\,\delta\alpha_q{}^*(t+\tau)\rangle$, so that for this process it measures a fourth-order density correlation function. To the extent that interactions between the fluctuations are negligible this may be factored into $S(-\mathbf{k}, \Omega - \omega)S(\mathbf{k}+\mathbf{q}, \omega)$, where $S(q, \omega) = \langle\delta\rho_{-\mathbf{q}}(t=0)\,\delta\rho_{\mathbf{q}}(\omega)\rangle$ is the familiar dynamic structure factor. For most solids and for superfluid helium, this situation obtains at least approximately. The intermolecular

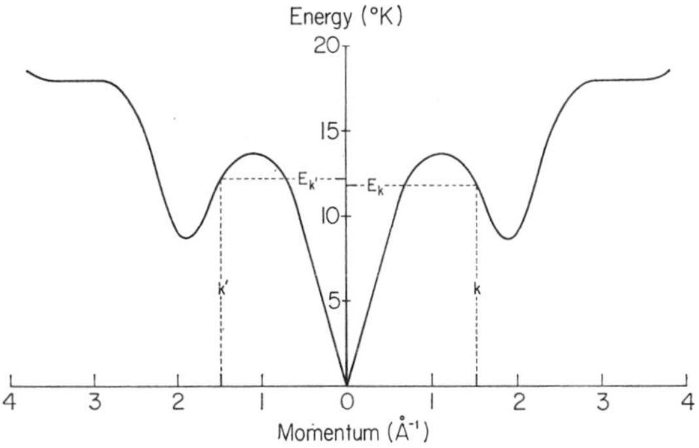

Fig. 17. Phonon–roton dispersion curves in superfluid helium. Intermolecular scattering receives its dominant contributions from roton pairs whose individual momenta are near 1 A^{-1} and 2 A^{-1} where extrema occur in the dispersion curve. These should give peaks near 26°K and 18°K, respectively, in the roton pair spectrum. After Greytak and Yan.[197]

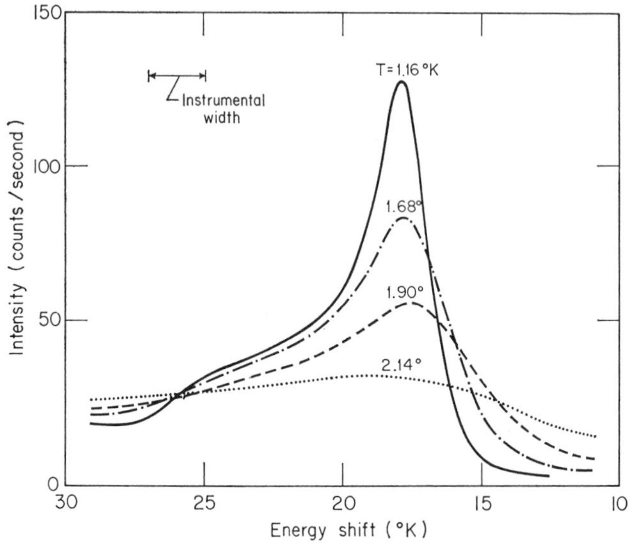

Fig. 18. Observed spectra in superfluid helium at various temperatures. The peak due to the roton minimum a 2 × 9°K is clear. The higher energy peak is suppressed by interaction effects. After Greytak and Yan.[197]

spectrum can then be reasonably described by the factored correlation functions. (Note, however, that even small interaction effects have important implications for the roton spectra.[198]) For a classical liquid like argon, McTague et al.[180] argued that the spectrum still measures the fourth-order density correlation function and have used the measured $S(\mathbf{k}, \omega)$ for argon in calculating the intermolecular spectral shape. Comparison with experiment shows that the factorization introduces about a 25% error into the resulting calculated shape, but confirms the basic correctness of the second-order scattering concept. Others who have studied intermolecular scattering in liquids[169,163,186] have utilized the binary collision formalism to interpret their results.

Various approaches have likewise been taken to explore the effects of increasing density on intermolecular spectra. McTague and Birnbaum[179] have described their spectra (0–200 amagat) in the rare gases in terms of density-dependent contributions at different frequencies, identifying ρ^n components with n-particle collisions, etc. Fleury et al.[166] have measured the intermolecular spectra of argon and neon as functions of temperature and density over a continuous range extending from the dilute gas limit to densities as high as 50% in excess of the normal liquid density. They were thus able to follow the evolution of the spectra throughout these ranges. Typical results for argon are shown in Fig. 19, where it can be seen that

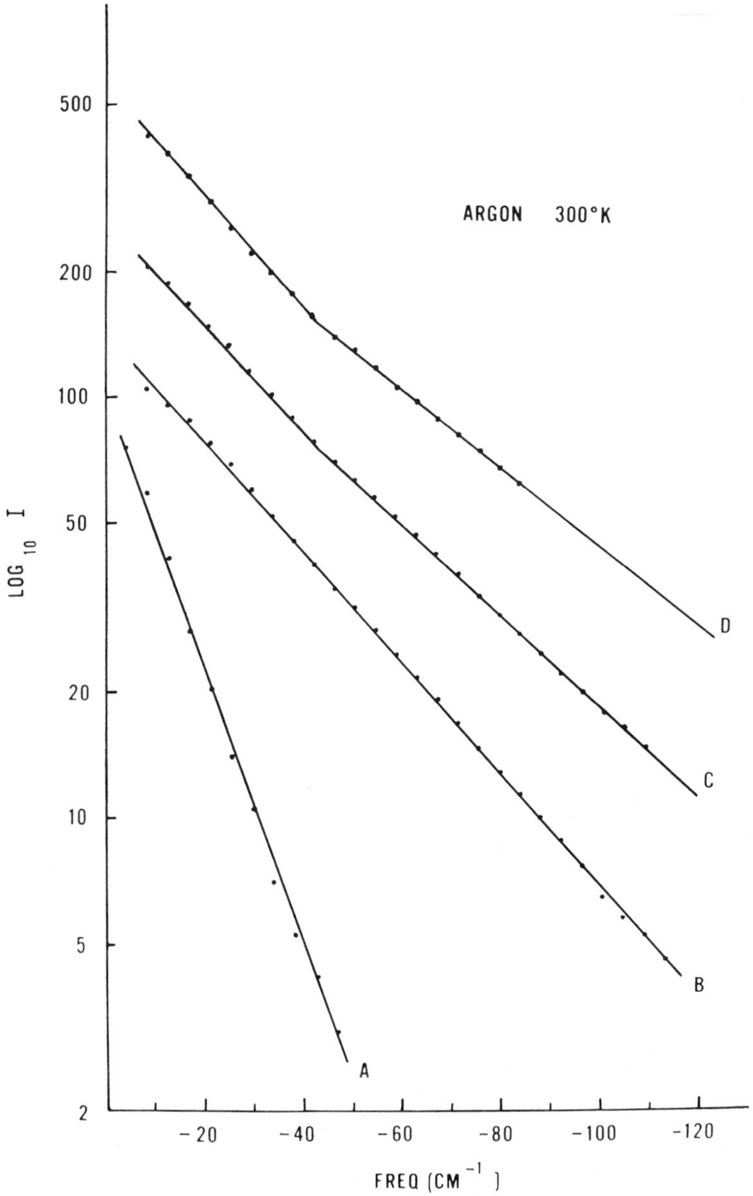

Fig. 19. Depolarized Stokes spectra for argon at 300°K for different densities. For display the curves are arbitrarily displaced along the vertical scale. The slopes of the straight lines drawn through the data points give the values of Δ plotted in Fig. 20. Curve A, 200; curve B, 750; curve C, 825; and curve D, 905 amagats. Note log scale for intensity. After Fleury et al.[166]

the spectral exponent Δ and shape are quite density dependent. Figure 20 summarizes the behavior of Δ with both ρ and T. For moderate reduced temperatures the behavior of Δ over a large range of T and ρ is adequately described by the remarkably simple form $\Delta(\rho, T) = \Delta_0(1 + (\rho/\rho_0)^2)$, where $\Delta_0 = V_{TH}/r_0$, $r_0 = 2.8$ A, and $\rho_0 = 500$ amagat for argon. These data exhibit the smooth but dramatic quantitative changes in the short-time dynamics of a fluid as it is compressed from a dilute gas to a superdense liquid. Were neutron scattering measurements of $S(k, \omega)$ available over a similar range of T and ρ, an empirical separation of the two- and

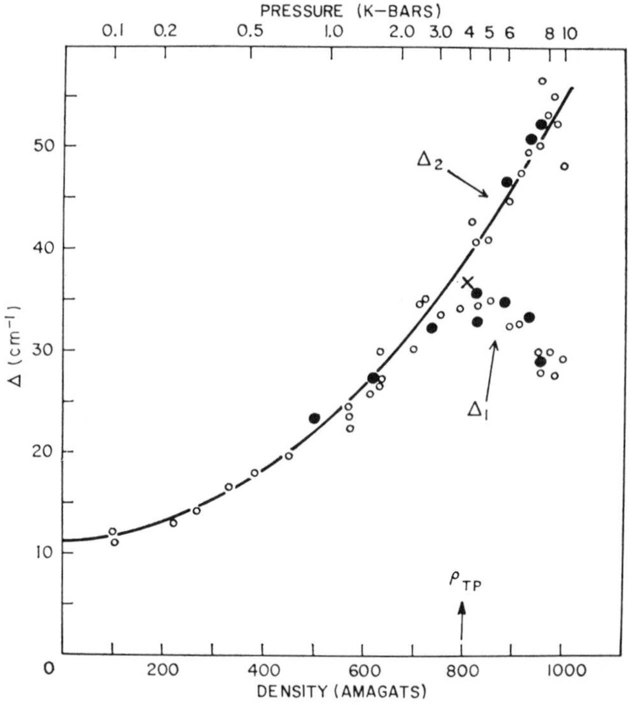

Fig. 20. Density and temperature dependences of the spectral exponents in argon. Δ_1 and Δ_2 represent the slopes in the low- (0–50 cm^{-1}) and high- (> 50 cm^{-1}) frequency regions of the spectra, respectively. Open circles are data at 300°K; closed circles are data at 180°K scaled by $(kT)^{1/2}$; the cross is the liquid at 90°K scaled by $(kT)^{1/2}$. The upper scale gives the pressure at 300°K corresponding to the density ρ indicated along the bottom scale. The solid line represents the simple empirical function discussed in the text. After Fleury et al.[166]

four-point density correlation functions would be possible. In their absence we must rely on developing theories of the dynamics of dense fluids to provide the quantitative connection between intermolecular spectra and the microscopic dynamic fluid structure. Despite the embryonic nature of our present understanding, it seems likely that further cooperative theoretical and experimental work on intermolecular light scattering will contribute significantly to a successful theory of the liquid state.

Finally, the intermolecular spectra of both He^3 and He^4 have been observed by Slusher and Surko[184] under pressure up to 150 atm. In the normal fluid ranges they are qualitatively similar to those observed in the classical fluids. The same workers have verified the second-order Raman effect in the solid phases of He^3 and He^4. As with the other rare gas solids, the spectral shapes change surprisingly little upon solidification.

E. Surface Scattering

Over half a century elapsed between the realization that light should be scattered by a surface and the measurement of the spectral content of that scattered light.[207,208] Thermally excited capillary waves or "ripplons" are describable through a rather complicated dispersion relation[203,214] whose essential physical content we now discuss. Surface tension σ provides the restoring force, and shear viscosity η_s the major contribution to damping for these waves. For wave vectors $k > k_c \equiv 2\pi\sigma\rho/3.65\eta_s^2$, the ripplons are overdamped. There is a maximum frequency for surface wave propagation $\omega_m = 0.375\sigma^2\rho/\eta_s^3$ regardless of wave vector. These waves, of course, are also sustained at the interface between two fluids and exhibit quite interesting behavior near the liquid–vapor critical point or near the critical mixing point. In particular, several experiments on liquid–gas interface which measure the critical slowing down of the surface waves near C.P. have now been reported.[213–215] This slowing down is essentially a result of the vanishing of the surface tension σ as $T \to T_c$. Typical k values explored in these experiments are $\ll k_c$, so that the ripplons are not overdamped. Although generally rather complicated, the dispersion relation for waves on the interface between two fluids of densities ρ_1 and ρ_2 and viscosities η_1 and η_2 can be approximately written (in the incompressible limit[214])

$$y = -\frac{2\rho_1\rho_2}{(\rho_1+\rho_2)^2} S(1+2S)^{1/2}[1+(1+2S)^{1/2}] - \frac{(\rho_1-\rho_2)^2}{(\rho_1+\rho_2)^2}[(1+S)^2-(1+2S)^{1/2}] \tag{100}$$

where

$$y = \frac{[\sigma k^2 + g(\rho_1 - \rho_2)][\rho_1 + \rho_2]}{4k^3(\eta_1 + \eta_2)^2} \tag{101}$$

and

$$S = (-\Gamma + i\omega)\frac{\rho_1 + \rho_2}{2k^2(\eta_1 + \eta_2)} \tag{102}$$

ω and Γ thus give the frequency and damping of the ripplons. Typically $\omega \sim 5$ KHz and $\Gamma \sim 1$ KHz for $k \simeq 2000$ cm^{-1}, σ of $\sim$0.1 dyne/cm, and normal liquid ρ's and η's. Thus light beating spectroscopy has usually been employed to study them.

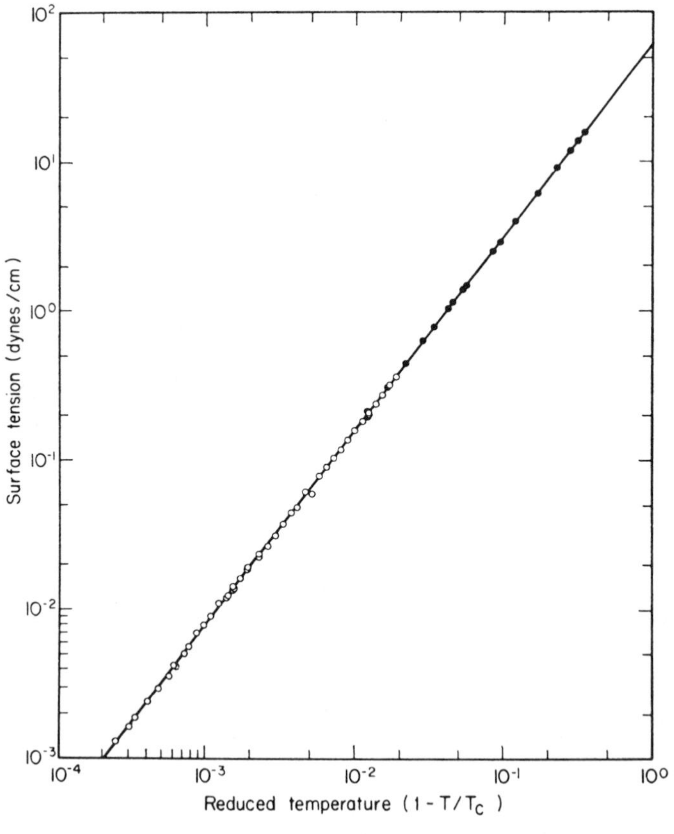

Fig. 21. Surface tension of xenon as a function of reduced temperature $1 - T/T_c$. Open circles inferred from surface wave scattering spectra. Solid circles from conventional capillary rise experiments. After Zollweg et al.[213]

By analyzing the k and T dependence of the ripplon spectra, one may deduce both $(\rho_1 - \rho_2)$, the C.P. order parameter, and σ, the vanishing surface tension. In fact, light scattering has extended by orders of magnitude the range of σ's that can be measured. Figure 21 shows for xenon both light scattering and conventional capillary rise data for σ over several decades in ε.[213] σ obeys a power law in ε over a remarkable range with ($\sigma = \sigma_0\varepsilon^{\mu}$) a value of $\mu = 1.302 \pm 0.006$. Similar experiments on CO_2 and SF_6 have produced μ's of 1.25 ± 0.01 and 1.30 ± 0.07, respectively.[214,215] Ripplon spectra also contain information on the correlation length, ξ, and on the closely related interfacial thickness L, both of which diverge at C.P.[213]

Langevin and Bouchiat[282] have studied surface wave scattering in the nematic liquid crystals MBBA and PAA and have deduced the temperature dependences of the *three* independent shear viscosity parameters as well as the surface tension σ near the isotropic–nematic transition.

Clearly the field of two-dimensional fluid behavior as viewed through surface and thin film[212,205,273] waves is just opening up and we can expect light scattering to play an important role in its promising future.

F. Solutions and Macromolecules

The simplest dynamic information obtainable in a light scattering experiment is the uncorrelated motion of single polarizable particles. In very dilute gases the particle motions obey a Maxwell-Boltzmann velocity distribution which produces an appropriately Doppler broadened (Gaussian) frequency profile in the scattered light. If these polarizable particles are embedded in a nonpolarizable fluid with which they make frequent collisions, they will still scatter light but with a different frequency profile (Lorentzian) which describes their diffusive or Brownian motion. The line width $\Gamma = Dk^2$ then measures the translational mass diffusion coefficient of the particles in the surrounding medium.[307] Macromolecules of biological interest exhibit D's of order 10^{-6} cm^2/sec, producing Γ's in the KHz range for visible light scattering ($k \lesssim 10^5$ cm^{-1}). This fact and the rapid development of electronic spectrometers has made the light scattering technique a standard one for rapid determination of diffusion coefficients, particularly in dilute solutions (~0.01 %). However, since most interesting molecules are anisotropic, and often long enough to contain segments comparable to optical wavelengths, the above simple picture of translational diffusion is almost always inadequate. In addition even the translational diffusion coefficient may be different for motion parallel ($D_{\parallel}$) and perpendicular ($D_{\perp}$) to the long molecular axis. Further, the molecules may interact with each other or with a charge compensating cloud of counterions arising from any charge which may reside on the dissolved macromolecules. All these effects greatly complicate the light scattering spectra in real systems. If the solute molecules also react or polymerize, or change conformation, further spectral complexity results from the accompanying polarizability modulations and/or changes in constituent mass, volume, or shape. Although, in principle, each of these effects produces a calculable change in the observed spectrum, not enough is known at present about the values of the necessary parameters to infer uniquely their separate contributions from a given set of experimental spectra. Nevertheless the considerable experimental and theoretical effort now being expended on such complications is cause for optimism, and even in the absence of perfectly quantitative spectral interpretation, considerable valuable information has already been obtained from light scattering experiments on solutions of macromolecules.

In view of the recent review article on this subject by Pecora,[14] our discussion here will be very brief. Most experiments[316,361] have deduced the translation D_T and rotational D_R diffusion coefficients from the spectral line shapes. If these diffusion processes are uncoupled, they can be inferred from the polarized (I_{zz}) and depolarized (I_{zy}) line shapes[14]:

$$I_{zz}(\omega, k) = \frac{A}{9\pi}(\alpha_{\parallel} + \alpha_{\perp})^2 \frac{k^2 D_T}{\omega^2 + (k^2 D_T)^2} + \tfrac{4}{3} I_{zy}(\omega, k)$$

$$I_{zy}(\omega, k) = \frac{A}{15\pi}(\alpha_{\parallel} - \alpha_{\perp})^2 \frac{k^2 D_T + L^{-2} D_R}{\omega^2 + (k^2 D_T + L^{-2} D_R)^2}$$

where L = molecule rod length; $\alpha_{\parallel}$, $\alpha_{\perp}$ molecular polarizabilities; A = Const. Several workers[316,361] have used these formulas to describe their data on solutions of tobacco mosaic virus (TMV). Schaeffer et al. have derived more complicated expressions which additionally account for anisotropic translational diffusion to apply to their TMV results.[363] They also mention possible complications arising from counterion clouds interacting with the macromolecules, but conclude these complications do not apply to their experiments. It would be of considerable interest, however, to observe these effects experimentally. Stephen has shown that counterion association will result in an enhanced apparent $D_T' = D_T(1 + \alpha)$ under certain circumstances.[343] (α depends on the Debye screening length of the macromolecular charge relative to the scattering wave vector, k.) This effect provides in principle a means to measure residual charge on macromolecules in solution.

Diffusion measurements on DNA,[362] R-17 virus, lysozyme, myosin, and *E. coli* bacteria have also been reported.[356,359]

Recently some intriguing studies of the nonequilibrium motion of bacteria by light scattering have been done by Nossal and Chen.[359] Such techniques provide exciting possibilities for microscopic studies of chimotaxis, bacterial association, and related phenomena. Indeed these experiments are very similar in technique and spirit to the Doppler shifted fluid flow measurements carried out by hydrodynamicists and plasma physicists.[313]

A final remark about molecular solutions should be made. It has been pointed out by several authors that light will be inelastically scattered by the concentration fluctuations associated with formation and destruction of components participating in a chemical reaction.[286–289,324] In the simplest case of optically isotropic constituents reacting at a rate $K = k_{12} + k_{21}$ much greater than the diffusion rate, $k^2\bar{D}_T$, the polarized spectrum has the simple form[14]:

$$I_{zz} = C\,|\bar{\alpha}|^2 \frac{k^2\bar{D}_T}{\omega^2 + (k^2\bar{D}_T)^2} + \frac{C_1C_2}{C}\,|\alpha_1 - \alpha_2|^2 \frac{K}{\omega^2 + K^2}$$

where α_i and C_i are polarizability and equilibrium concentrations of species i; and $\bar{\alpha} = (1/C)[C_1\alpha_1 + C_2\alpha_2]$; $C = C_1 + C_2$.

The dearth of experimental observations of chemical reaction scattering stems from the scarcity of simple systems[346,347] having α_1 much different from α_2 and C_1, C_2 both of order C. This scarcity somewhat attenuates the initial attractiveness of light scattering for studying fast chemical reactions. (Rates $K \gtrsim 10^{+8}$ sec^{-1}, for example, would be easily studied with Fabry Perot techniques.)

As with the film and surface wave studies mentioned above, the field of light scattering from chemical reactions and macromolecular solutions has a promising future.

Literature Search and Bibliography

This section consists of a relatively complete list of references, including titles on the general subject of inelastic light scattering in fluids. The list is a result of a literature search, conducted with the assistance of Bernard Stevens of Bell Laboratories, concentrating on the period beginning with the introduction of lasers to spectroscopy (1964) to approximately the beginning of 1972. Because no truly centralized and complete collection of all physics, chemistry, and biology literature exists, even for this recent period, the bibliography cannot be absolutely complete. However, it is the most extensive one yet compiled on the subject and will hopefully prove useful to both students and researchers.

The bibliography is arranged as follows: papers are divided into twelve categories. In addition to the reference information, the title of each paper is cited in full. In the event that a paper may fall into more than one category, it is listed under the one we judge the most applicable. To avoid having to duplicate those papers actually cited in the text, all papers whose primary content places them in a given category are numbered consecutively beginning with the first in category *1* until the last in category *12*. These consecutive numbers are also used to designate textual citations when appropriate. Although this results in the textual references not being numbered in the order of their citation, we feel it is the most concise way of having the bibliography serve two purposes without introducing unnecessary confusion. Within the categories papers are listed alphabetically by first author. A few papers appear in more than one category but are numbered only when appearing in their primary category. Their other listings are unnumbered, but the primary category to which they belong is then indicated by the number in brackets at the end of each such entry.

As mentioned in the introduction, the subjects of ordinary Raman scattering, scattering in solids and plasmas, and stimulated scattering were defined to be outside the scope of this review. With the exceptions of the latter topic, these exclusions apply as well to the bibliography. Nevertheless, we include stimulated scattering in the bibliography primarily because such a large number of references were revealed in the search that it was deemed useful to present them in a concise collection.

The categories defined are as follows, (*1*) review articles or books on various aspects of light scattering in fluids; (*2*) instrumentation and experimental technique; (*3*) Rayleigh scattering from entropy fluctuations; (*4*) Brillouin scattering; (*5*) intermolecular or collision induced scattering; (*6*) scattering from surfaces and thin films; (*7*) molecular anisotropy and orientation effects; (*8*) liquid crystals; (*9*) mixtures and particles in solution; (*10*) critical phenomena; (*11*) stimulated scattering; and (*12*) nonlinear scattering and miscellaneous references.

1. *General Reviews*

1. G. B. Benedek, "Thermal fluctuations and the scattering of light," in *Brandeis Lectures in Theoretical Physics, 1966*, Vol. 2, M. Chretien et al., Eds., Gordon and Breach, 1968.
2. B. Chu, "Laser Light Scattering," *Ann. Rev. Phys. Chem.*, **21**, 145 (1970).
3. H. Z. Cummins and H. L. Swinney, "Light Beating Spectroscopy," in *Progress in Optics*, E. Wolf, Ed., Vol. 8, 1970, pp. 133–200 (535/P96. 143190).
4. H. Z. Cummins, "Experimental Methods in Light Scattering Spectroscopy," in *Light Scattering in Solids*. M. Balkanski, Ed., Flammarion. Paris. 1971, p. 3.
5. D. E. Evans, "Laser Light Scattering in Laboratory Plasmas Review," *Rept. Progr. Phys.*, **32**, 207 (1969).
6. I. L. Fabelinskii, *Molecular Scattering of Light*, Plenum Press, New York, 1968. 622 pp. (535.4 F11: E, 125088).

7. I. L. Fabelinskii, "Problems of Molecular Scattering of Light in Liquids," *Adv. Phys. Sci.*, **63** (2), 474–551 (1957).
8. R. Figgins, "Inelastic Light Scattering in Liquids Brillouin Scattering," *Contemp. Phys.*, **12** (3), 283–297 (1971).
9. P. A. Fleury, "Light Scattering as a Probe of Phonons and other Excitations," in *Physical Acoustics*, Vol. 6. W. P. Mason, Ed., Academic Press, New York, 1970, pp. 1–64 (534 M41, 141506).
10. D. McIntyre and J. V. Sengers, "Light Scattering in Simple Liquids," in *Physics of Simple Liquids*. H. N. V. Temperley et al., Eds., North-Holland, Amsterdam, 1968 p. 449 (532. 1/T28. 136401).
11. R. D. Mountain, "Spectral Distribution of Scattered Light in a Simple Fluid," *Rev. Mod. Phys.*, **38** (1), 205 (1966).
12. R. D. Mountain, "Liquids: Dynamics of Liquid Structure," *CRC Crit. Rev. Solid State Sci.*, **1,** 5–46 (1970).
13. R. Pecora, "Fine Structure of Rayleigh Scattered Light," in *Interdisciplinary Conf. on Electromag Scattering. 2nd, 1965, Proc.*, R. L. Rowell and R. S. Stein, Eds., Gordon and Breach, New York, 1967, pp. 503–517 (538.3/159. 125401).
14. R. Pecora, "Laser Light Scattering from Macromolecules," *Ann. Rev. Biophys. Bioeng.*, **1972,** 1.
15. C. E. Quate, C. D. W. Wilkinson, and D. K. Winslow, *Proc. IEEE*, **53,** 1604–1623 (1965).
16. R. A. Smith, "Review Lecture. Lasers and Light Scattering," *Proc. Roy. Soc. (London) Ser. A*, **323,** 305–320 (50 Refs) (June 22, 1971).

2. *Instrumentation and Experimental Techniques*

— G. B. Benedek, "Optical Mixing Spectroscopy and Applications to Problems in Physics, Chemistry, Biology and Engineering," in *Polarisation, Matiere et Rayonnement. Presses Universitaire de France*. 1969, pp. 49–84 (539.1/P76. 138226) [7].
— L. Boyer and L. Cecchi, "Optical Heterodyning for Large Angle Brillouin Scattering," *J. Phys.*, **30,** (5–6), 477–481 (1969) (in French) [*4*].
17. S. J. Candau, "Inelastic Scattering of Light in Liquids," *Ann. Phys.*, **4** (1), 21–54 (in French).
— H. Z. Cummins and H. L. Swinney, "Light Beating Spectroscopy," in *Progress in Optics*. E. Wolf, Ed., Vol. 8, 1970, pp. 133–200 (535/P96. 143190 [*1*].
— H. Z. Cummins, "Experimental Methods in Light Scattering Spectroscopy," in *Light Scattering in Solids*. M. Balkanski, Ed., Flammarion, Paris. 1971, p. 3 [*1*].
18. H. G. Danielmeyer, "Aperture Corrections for Sound Absorption Measurements with Light Scattering," *J. Acoust. Soc. Am.*, **47,** 151–154 (1970).
19. V. Degiorgio and J. B. Lastovka, "Intensity Correlation Spectroscopy," *Phys. Rev. A*, **4,** 2033 (1971).
20. G. E. Devlin, J. L. Davis, L. Chase, and S. Geschwind, "Absorption of Unshifted Scattered Light by a Molecular I_2 Filter in Brillouin and Raman Scattering," *Appl. Phys. Lett.*, **19,** 138 (1971).
21. E. Jakeman and E. R. Pike, "Spectrum of Clipped Photon Counting Fluctuations of Gaussian Light," *J. Phys. A*, **2,** 411 (1969).
22. K. H. Langley and N. C. Ford, "Attenuation of the Rayleigh Component in Brillouin Spectroscopy Using Interferometric Filtering," *J. Opt. Soc. Am.* **59** (3), 281–284 (1969).

23. J. B. Lastovka and G. B. Benedek, "Light-Beating Techniques for the Study of the Rayleigh-Brillouin Spectrum," in *Physics of Quantum Electronics*, McGraw-Hill, New York, 1966, pp. 231–240.
24. H. W. Leidecker and J. T. La Macchia, "Instrumental Effects on Brillouin Line Shapes," *J. Acoust. Soc. Am.*, **43,** 143–151 (1968).
25. J. E. Mack, D. P. McNutt, F. L. Roseler, and R. Chabbal, "Pepsios: Purely Interferometric High Resolution Spectrometer," *Appl. Opt.*, **2,** 873 (1963).
26. J. R. Sandercock, "Design and Use of a Stabilized Multipass Interferometer of High Contrast Ratio," *Light Scattering in Solids*. M. Balkanski, Ed., Flammarion, Paris. 1971, p. 9.
27. E. R. Pike, "The Accuracy of Diffusion-Constant Measurements by Digital Correlation of Clipped Photon-Counting Fluctuations," in *Colloque Int. CNRS Diffusion De La Lumiere Par Les Fluides*. 1971. *J. Physique Suppl.*, **33**C1, 177 (1972).

3. *Rayleigh Scattering from Entropy Fluctuations*

— G. B. Benedek, "Thermal Fluctuations and the Scattering of Light," in *Brandeis Lectures in Theor Phys 1966*. Vol. 2, M. Chretien et al., Eds. Gordon and Breach, 1968 [*1*].
— G. B. Benedek, "Optical Mixing Spectroscopy and Applications to Problems in Physics, Chemistry, Biology and Engineering," in *Polarisation. Matiere et Rayonnement. Presses Universitaire de France*. 1969, pp. 49–84 (539.1/P76. 138226). [*7*].
28. P. Berge, P. Calmettes, M. Dubois, and C. Loj, "Experimental Observation of the Complete Rayleigh Central Component of the Light Scattered by a Two-Component Fluid," *Phys. Rev. Lett.*, **24** (3), 89–90 (1970).
29. P. Berge and M. Dubois, "Spectral Study of the Central Rayleigh Component of Benzene," *Compt. Rend. B* **269** (17), 842–845 (1969) (in French).
30. A. B. Bhatia and E. Tong, "Brillouin Scattering in Multiply Relaxing Liquids," *J. Acoust. Soc. Am.*, **49,** 1437–1441 (May Pt. 2, 1971).
31. H. Blend, "Free-field Technique for Measuring Ultrasonic Dispersion and Absorption in Gases," *J. Acoust. Soc. Am.*, **47** (3), 757–761 (1970).
32. J. P. Boon and P. Deguent, "Transport Functions and Light Scattering in Simple Dense Fluids," *Phys. Rev. A*, **2,** 2542 (1970).
33. I. M. Arefev, B. D. Kopylovskii, D. I. Mash, and I. L. Fabelinskii, "Determination of the Diffusion Coefficient by Heterodyning Light Scattered by Liquid Solutions," *JETP LETT.*, **5** (12), 355–358 (1967); **10** (11), 340–342 (1969).
34. I. M. Arefev and H. V. Shilin, "Gravitational Effect in Interference Spectra of the Fine Structure of the Rayleigh Line," *JETP Lett.*, **10** (3), 87–90 (1969).
35. E. W. Aslaksen, "Correlation Length of Density Fluctuations in Liquids," *Phys. Rev.*, **182,** 316–318 (1969).
36. A. K. Atakhodzhaev and E. K. H. Tukhvatuilin, "Influence of Hydrogen Bonds on the Linewidth of Rayleigh and Raman Scattering," *Izv. Vuz. Fig*. **11,** 7–11 (1970) (in Russian).
37. F. Barocchi, "Evidence of Multiple Vibrational-Translational Relaxation in Carbon Tetrachloride," *J. Chem. Phys.*, **51** (1), 10–14 (1969).
38. F. Barocchi, M. Mancini, and R. Vallauri, "Volume Relaxation in Carbon Disulfide-Carbon Tetrachloride Mixtures," *J. Chem. Phys.*, **49** (4), 1935–1937, (1968).
39. P. Belland, C. De Michelis, M. Mattioli et al., "Spectral Analysis of the Backscattered Light from a Laser-Produced Plasma," *Appl. Phys. Lett.* **18,** 542–544 (June 15, 1971).

40. P. Braun, D. Hammer, W. Tscharnuter, and P. Weinzierl, "Rayleigh Scattering by Sulfur Hexafluoride in the Critical Region," *Phys. Lett. A*, **32** (6), 390–391 (1970).
41. A. Broz, M. Harrigan, R. Kasten, and A. Monkewicz, "Light Scattered from Thermal Fluctuation in Gases," *J. Acoust. Soc. Am.*, **49** (3, Pt. 3), 550–553 (1971).
42. J. A. Bucaro, "Rayleigh Scattering: Molecular Motions and Interactions in Liquids," Thesis, Catholic University, 1971 (Univ. Microfilms 71-14077).
43. A. V. Chalyi and A. D. Alekhin, "Investigation of Light Scattering with Allowance for Gravitational Effect by Scaling Method and Determination of the Critical Indices on Basis of Light Scattering Data," *Zh. Eksperim i Teor. fiz.*, **59** (2), 337–345 (1970) (in Russian).
44. C. H. Chung and S. Yip, "Relaxation Equations for Depolarized Rayleigh and Brillouin Scattering in Liquids," *Phys. Rev. A*, **4** (3), 928–939 (1971).
45. H. Z. Cummins and R. W. Gammon, "Rayleigh and Brillouin: Scattering in Benzene: Depolarization Factors," *Appl. Phys. Lett.*, **6**, 171–173 (1965).
46. H. Z. Cummins and R. W. Gammon, "Rayleigh and Brillouin Scattering in Liquids: The Landau- Placzek Ratio," *J. Chem. Phys.*, **44** (7), 2785–2796 (1966).
47. H. Z. Cummins, N. Knable, and Y. Yeh, "Observation of Diffusion Broadening of Rayleigh Scattered Light," *Phys. Rev. Lett.*, **12**, 150–153 (1964).
48. H. Z. Cummins and H. L. Swinney, "Critical Opalescence: The Rayleigh Linewidth," *J. Chem. Phys.*, **45** (12), 4438–4444 (1966).
49. D. R. Dietz, C. W. Cho, and T. A. Wiggins, "Density Dependence in Thermal Rayleigh Scattering in Gases," *Phys. Rev.*, **182** (1), 259–261 (1969).
50. R. H. Enns and R. R. Haering, "Theory of Rayleigh Scattering in Isotropic Media," *Phys. Lett.*, **21** (5), 534–535 (1966).
51. G. T. Feke, G. A. Hawkins, J. B. Lastovka, and G. B. Benedek, "Spectrum and Intensity of the Light Scattered from Sulfur Hexafluoride Along the Critical Isochore," *Phys. Rev. Lett.*, **27**, 1780–1783 (Dec. 27, 1971).
52. N. C. Ford and G. B. Benedek, "Observation of the Spectrum of Light Scattered from a Pure Fluid Near its Critical Point," *Phys. Rev. Lett.*, **15**, 649–653 (1965).
53. N. C. Ford and G. B. Benedek, "Spectrum of Light Inelastically Scattered by a Fluid Near its Critical Point," in *Conf. Phenom. Neighborhood Critical Points, 1965 Proc., Natl. Bur. Std. (U.S.) Misc. Publ.*, **273**, 150–156 (1966).
54. B. N. Ganguly and A. Griffin, "Scattering of Light from Entropy Fluctuations in Helium-3 and Helium-4 Mixtures," *Can. J Phys.* **46** (17), 1895–1904 (1968).
55. M. Giglio and G. B. Benedek, "Angular Distribution of the Intensity of Light Scattered from Xenon Near its Critical Point," *Phys. Rev. Lett.*, **23**, 1145–1149 (1969).
— W. S. Gornall, C. S. Wang, C. C. Yang, and N. Bloembergen, "Coupling Between Rayleigh and Brillouin Scattering in a Disparate Mass Gas Mixture," *Phys. Rev. Lett.*, **26** (18), 1094–1097 (1971) [9]
56. C. C. Gravatt, "Applications of Light Scattering (Including Rayleigh and Brillouin)," *Appl. Spectry.*, **25**, 509–516 (1971).
57. T. J. Greytak and G. B. Benedek, "Spectrum of Light Scattered from Thermal Fluctuations in Gases," *Phys. Rev. Lett.*, **17**, 179–182 (1966).
— E. H. Hara, A. D. May, and H. F. P. Knaap, "Rayleigh-Brillouin Scattering in Compressed Hydrogen, Deuterium and Hydrogen Deuterium (mixture)," *Can. J. Phys.*, **49** (4), 420–431 (1971) [9].
58. D. L. Henry, H. L. Swinney, and H. Z. Cummins, "Rayleigh Linewidth in Xenon Near the Critical Point," *Phys. Rev. Lett.*, **25** (17), 1170–1173 (1970).

59. D. A. Jackson and D. M. Paul, "Measurement of Hypersonic Velocities and Turbulence by Spectral Analysis of Doppler Shifted Laser Light," *Phys. Lett. A*, **32,** 77–78 (1970).

60. R. A. J. Keijser, M. Jansen, V. G. Cooper, and H. F. P. Knaap, "Depolarized Rayleigh Scattering in Carbon Dioxide, Carbon Oxysulfide and Carbon Disulfide," *Physica*, **51** (4), 593–600 (1971).

61. J. B. Lastovka and G. B. Benedek, "Spectrum of Light Scattered Quasielectrically from a Normal Liquid," *Phys. Rev. Lett.*, **17,** 1039–1042 (1966).

62. A. Laubereau, W. Englisch, and W. Kaiser, "Hypersonic Absorption of Liquids Determined from Spontaneous and Stimulated Brillouin Scattering," *IEEE J. Quantum Electron.*, **5** (8), 410–415 (1969).

— R. C. C. Leite, R. S. Moore, S. P. S. Porto, and J. E. Ripper, "Angular Dependence of the Rayleigh Scattering from Low Turbidity Molecular Liquids," *Phys. Rev. Lett.*, **14,** 7–9 (1965) [*7*].

— R. C. C. Leite, R. S. Moore, and S. P. S. Porto, "Use of a Gas Laser in Studies of the Depolarization of the Rayleigh Scattering from Simple Liquids," *J. Chem. Phys.*, **40**, 3741–3842 (1964) [*7*].

63. T. K. Lim, H. L. Swinney, K. E. Laugley, and T. A. Kachnowski, "Rayleigh Linewidth in Sulfur Hexafluoride Near the Critical Point," *Phys. Rev. Lett.*, **27,** 1776–1780 (Dec. 27, 1971).

64. A. Litan, "Fluctuation Theory of Light Scattering from Liquids," *J. Chem. Phys.*, **48,** 1052–1058 (1968).

65. A. Litan, "Fluctuation Theory of Light Scattering from Pure Water," *J. Chem. Phys.*, **48,** 1059–1063 (1968).

66. A. D. May, E. G. Rawson, and H. L. Welsh, "Rayleigh Scattering from Low-Density Gases," *Int. Conf. Phys. Quantum Electron, 1966 Proc.* McGraw-Hill, New York, 1967. pp. 260–264 (537.1 I61).

67. A. Olivei, "Light Scattering by Liquids in Narrow Capillaries," *Optik*, **32** (3), 218–236 (1970).

68. C. J. Palin, W. F. Vinen, E. R. Pike, and J. M. Vaughn, "Rayleigh and Brillouin Scattering from Superfluid Helium-3 Helium-4 Mixtures," *J. Phys. C Solid State*, **4,** L225–228 (1971).

69. G. P. Roshchina and G. L. Gudimenko, "Investigation of the Fine Structure of the Rayleigh Line in *N*-Paraffins by Means of Gas Laser," *Ukr. Fiz. Zh.*, **15** (6), 888–892 (1970) (In Russian).

70. S. M. Rytov, "Relaxation Theory of Rayleigh Scattering," *Zh. Eksperim. i Teor. Fiz.*, **58** (6), 2154–2170 (1970) (in Russian).

71. T. Sato, "Brillouin Scattering in Plasmas as Applied to Ionospheric Irregularities," *J. Geophys. Res.*, **73** (9) 2941–2949 (1968).

72. A. Sugawara and S. Yip, "Kinetic Model Analysis of Light Scattering by Molecular Gases," *Phys. Fluids*, **10** (9), 1911–1921 (1967).

73. M. Tanaka, "Molecular Theory of Scattering of Light From Liquids," *Progr. Theoret. Phys.*, **40** (5), 975–989 (1968); *J. Phys. Soc. Japan* **26** (Suppl.), 298–300 (1969).

74. M. L. Ter-Mikaelyan and A. O. Melikyan, "Rayleigh and Combinational Scattering in the Field of an Intense Wave," *Zh. Eksperim. i Teor. Fiz.*, **58** (1), 281–290 (1970) (in Russian).

— W. F. Vinen, "Light Scattering by Superfluid Helium Under Pressure," *J. Phys. C Solid State*, **4,** L287 (1971) [*4*].

75. Y. Yeh, "Observation of the Long Range Correlation Effect in the Rayleigh Linewidth Near the Critical Point of Xenon," *Phys. Rev. Lett.*, **19,** 1043–1046 (1967).
76. S. Yip, "Rayleigh Scattering in Dilute Gases," *J. Acoust. Soc. Am.*, **49** (3, Pt. 3), 941–949 (1971).
77. R. M. Yulmetev, "Theory of Rayleigh Scattering of Light in Liquids," *Opt. Spectry.*, **19** (6), 532–535 (1965).
78. E. R. Pike, "Rayleigh and Brillouin Scattering from Liquid Helium," in *Colloque Int. CNRS, Diffusion de la Lumiere par les Fluides,* 1971. **33**C1, 25 (1972). *J. Physique Suppl.*,
— R. C. Desai, "Rayleigh-Brillouin Scattering from Fluid Mixtures," in *Colloque Int. CNRS. Diffusion de la Lumiere par les Fluides.* 1971. **33**C1, 27 (1972) [*9*]. *J. Physique Suppl.*,
— C. D. Boley and S. Yip, "Spectral Distributions of Light Scattered in Dilute Gases and Gas Mixtures," in *Colloque Int. CNRS, Diffusion de la Lumiere par les Fluids.* 1971. **33**C1, 43 (1972) [*9*]. *J. Physique Suppl.*,
79. A. D. May and E. H. Hara, "Rayleigh-Brillouin Scattering in Compressed Hydrogen, Deuterium and HD," in *Colloque Int. CNRS, Diffusion de la Lumiere par les Fluides,* 1971. **33**C1, 50 (1972). *J. Physique Suppl.*,
— W. L. Gornall and C. S. Wang, "Light Scattering from Thermal Fluctuations in Disparate-Mass Gas Mixtures," in *Colloque Int. CNRS. Diffusion de la Lumiere par les Fluides.* 1971. **33**C1, 51 (1972) [*9*]. *J. Physique Suppl.*,
80. A. M. Cazaeat-Longequeue and P. Lallemand, in *Colloque Int. CNRS, Diffusion de la Lumiere par les Fluides.* 1971. **33**C1, 57 (1972). *J. Physique Suppl.*,
81. H. L. Swinney and H. Z. Cummins, "Rayleigh Linewidth in CO_2 near the Critical Point," *Phys. Rev.*, **171,** 152 (1968).
82. H. L. Swinney and H. Z. Cummins, "The Rayleigh Linewidth in Carbon Dioxide and Xenon Near the Critical Point," in *Colloque Int. CNRS, Diffusion de la Lumiere par les Fluides.* 1971. To be published, *J. Physique Suppl.*, 1972.
83. D. S. Cannell and J. Lunacek, "Long Range Correlation Length and Isothermal Compressibility of Carbon Dioxide Near its Critical Point," in *Colloque Int. CNRS, Diffusion de la Lumiere par les Fluides.* 1971. **33**C1, 91 (1972). *J. Physique Suppl.*,
— J. Zollwegg, G. Hawkins, I. W. Smith, M. Giglio, and G. B. Benedek, in *Colloque Int. CNRS, Diffusion de la Lumiere par les Fluides.* 1971. **33**C1, 135 (1972) [*10*]. *J. Physique Suppl.*,

4. *Brillouin Scattering*

— G. B. Benedek, "Thermal Fluctuations and the Scattering of Light," in *Brandeis Lectures in Theor Phys 1966*, Vol. 2, M. Chretien et al., Eds. Gordon and Breach, 1968 [*1*].
84. G. B. Benedek, J. B. Lastovka, K. Fritsch, and T. J. Greytak, "Brillouin Scattering in Liquids and Solids Using Low Power Lasers," *J. Opt. Soc. Am.*, **54,** 1284–1285 (1964).
85. M. Bertoiotti, D. De Pasquale, and D. Sette, "Effects of Second Sound Excitations on the Photostatistics of a Laser Beam," *Nuovo Cimento, B*, **52** (2), 560–564 (1967).
86. I. M. Arefev, "Velocity of Hypersound and Dispersion of Speed of Sound Near the Critical Stratification Point of a Binary Solution of Triethylamine in Water," *JETP Leit.*, **7,** 285–287 (1968).
— E. W. Aslaksen, "Correlation Length of Density Fluctuations in Liquids," *Phys. Rev.*, **132,** 316–318 (1969) [*3*].

87. L. Boyer and L. Cecchi, "Optical Heterodyning for Large Angle Brillouin Scattering," *J. Phys.*, **30** (5–6), 477–481 (1969) (in French).
88. L. Brillouin, "Diffusion of Light and X-rays by a Transparent Homogeneous Body," *Ann. Phys.* (*Paris*), **17,** 88–122 (1922).
— A. Broz, M. Harrigan, R. Kasten, and A. Monkewicz, "Light Scattered from Thermal Fluctuation in Gases," *J. Acoust. Soc. Am.*, **49** (3, Pt. 3), 550–553 (1971) [*3*].
89. M. Caloin and S. J. Candau, "Thermal Relaxation in Benzene," *Compt. Rend. B*, **269** (10), 423–426 (1969) (in French).
90. D. S. Cannell and G. B. Benedek, "Brillouin Spectrum of Xenon Near its Critical Point," *Phys. Rev. Lett.*, **25** (17), 1157–1161 (1970).
91. E. F. Carome, W. H. Nichols, C. R. Kunsitis-Swyt, and S. P. Singal, "Ultrasonic and Light Scattering Studies of Carbon Tetrachloride," *J. Chem. Phys.*, **49** (3), 1013–1017 (1968).
92. A. M. Cazabat-Longequeue, "Study of the Rotation-Translation Relaxation in Gases by the Brillouin Effect," *Compt. Rend. Ser B, Sci. Phys.*, **272,** 607–610 (March 8, 1971) (in French).
93. V. Chandrasekharan, "Exact Equation for Brillouin Shifts," *J. Phys.*, **26** (11), 655–658 (1965) (in French).
94. A. N. Chernets, A. I. Timchenko, and V. V. Silin, "Mandelshtam-Brillouin Scattering of Coherent Light by Coherent Hypersound," *Ukr. Fiz. Zh.*, **16,** 678–680 (April, 1971) (in Russian).
95. R. Y. Chiao and P. A. Fleury, "Brillouin Scattering and the Dispersion of Hypersonic Waves," *Physics of Quantum Electronics*. McGraw-Hill, New York, 1966, pp. 241–252.
96. R. Y. Chiao and B. P. Stoicheff, "Brillouin Scattering in Liquids Excited by the He–Ne Maser," *J. Opt. Soc. Am.*, **54,** 1286–1287 (1964).
— C. H. Chung and S. Yip, "Relaxation Equations for Depolarized Rayleigh and Brillouin Scattering in Liquids," *Phys. Rev. A*, **4** (3), 928–939 (1971) [*3*].
97. M. G. Cohen and E. I. Gordon, "Acoustic Scattering of Light in a Fabry-Perot Resonator," *Bell. Syst. Tech. J.*, **45,** 945 (1966); *Bell Syst. Monogr.* **5234.**
98. M. G. Cohen and E. I. Gordon, "Acoustic Beam Probing Using Optical Techniques," *Bell. Syst. Tech. J.*, **44,** 693–721 (1965).
99. M. G. Cohen, M. A. Woolf, and P. M. Platzman, "Brillouin Scattering in Liquid Helium (2)," *Phys. Rev. Lett.*, **17,** 294–297 (1966).
— H. Z. Cummins and R. W. Gammon, "Rayleigh and Brillouin Scattering in Benzene: Depolarization Factors," *Appl. Phys. Lett.*, **6,** 171–173 (1965) [*3*].
— H. Z. Cummins and R. W. Gammon, "Rayleigh and Brillouin Scattering in Liquids: The Landau-Placzek Ratio," *J. Chem. Phys.*, **44** (7), 2785–2796 (1966) [*3*].
100. H. Z. Cummins and H. L. Swinney, "Dispersion of the Velocity of Sound in Xenon in the Critical Region," *Phys. Rev. Lett.*, **25** (17), 1165–1169 (1970).
— H. G. Danielmeyer, "Aperture Corrections for Sound Absorption Measurements With Light Scattering ," *J. Acoust. Soc. Am.*, **47,** 151–154 (1970) [*2*].
101. A. Daniels, *Mandelshtam-Brillouin Scattering of Carbon Dioxide Laser Light From the Atmosphere as a Remote Sensing Probe*, Michigan University High Altitude Engineering Laboratory, 1971 (103P N71-34439).
102. P. Deguent and J. P. Boon, "Generalized Regression of Fluctuations Theory as Applied to the Second Viscosity," *J. Chem. Phys.*, **54,** 4443 (1971).
103. R. Demicheli and L. Giulotto, "Effect of Lack of Acoustic Wave Coherence on Brillouin Scattering," *Nuovo Cimento B*, **46** (2), 277–281 (1966).

104. G. Durand and Dvgln Rao, "Brillouin Scattering of Light in a Liquid Crystal," *Phys. Lett. A*, **27** (7), 455–456 (1968).
105. D. P. Eastman, A. Hollinger, J. Kenemuth, and D. H. Rank, "Temperature Coefficient of Hypersonic Sound and Relaxation Parameters for Some Liquids," *J. Chem. Phys.*, **50** (4), 1567–1581 (1969).
106. D. P. Eastman, T. A. Wiggins, and D. H. Rank, "Spontaneous Brillouin Scattering in Gases," *Appl. Opt.*, **5** (5), 879–880 (1966).
107. P. A. Fleury and J. P. Boon, "Brillouin Scattering in Simple Liquids: Argon and Neon," *Phys. Rev.*, **186**, 244–254 (1969).
108. P. A. Fleury and R. Y. Chiao, "Dispersion of Hypersonic Waves in Liquids," *J. Acoust. Soc. Am.*, **39**, 751–752 (1966).
109. N. C. Ford, K. H. Langley, and V. G. Puglielli, "Brillouin Linewidths in Carbon Dioxide Near the Critical Point," *Phys. Rev. Lett.*, **21**, 9–12 (1968).
110. H. L. Frisch, "High Frequency Linear Response of Classical Fluids," *Physics*, **2**, 209–215 (1966).
— K. Fritsch, C. J. Montrose, J. L. Hunter, and J. F. Dill, "Relaxation Phenomena in Electrolytic Solutions," *J. Chem. Phys.*, **52** (5), 2242–2252 (1970) [*9*].
111. R. W. Gammon, H. L. Swinney, and H. Z. Cummins, "Brillouin Scattering in Carbon Dioxide in the Critical Region," *Phys. Rev. Lett.*, **19**, 1467–1469 (1967).
112. B. N. Ganguly, "Line Shape of Brillouin Light Spectrum in Helium-3 Helium-2 System," *Phys. Lett. A*, **29** (5), 234–235 (1969).
113. S. Gewurtz, W. S. Gornall, and B. P. Stoicheff, "Brillouin Spectra of Ethyl Ether and Carbon Disulfide," *J. Acoust. Soc. Am.*, **49**, 994–1000 (March, Pt. 3, 1971).
114. E. I. Gordon and M. G. Cohen, "High-Resolution Brillouin Scattering," *Phys. Rev.*, **153** (1), 201–207 (1967).
115. W. S. Gornall, G. I. A. Stegeman, B. P. Stoicheff, R. H. Stolen, and V. Volterra, "Identification of a New Spectral Component in Brillouin Scattering of Liquids," *Phys. Rev. Lett.*, **17**, 297–299 (1966).
— W. S. Gornall, C. S. Wang, C. C. Yang, and N. Bloembergen, "Coupling Between Rayleigh and Brillouin Scattering in a Disparate Mass Gas Mixture," *Phys. Rev. Lett.*, **26** (18), 1094–1097 (1931) [*9*].
— Gravatt, C. C., "Applications of Light Scattering (Including Rayleigh and Brillouin)," *Appl. Spectry.*, **25**, 509–516 (1971) [*3*].
— T. J. Greytak and G. B. Benedek, "Spectrum of Light Scattered from Thermal Fluctuations in Gases," *Phys. Rev. Lett.*, **17**, 179–182 (1966) [*3*].
116. H. Grimm and K. Dransfeld, "Scattering of Laser Light in Superfluid Helium," *Z. Naturforsch. A*, **22**, 1629–1630 (1967).
117. G. L. Gudimenko and G. P. Roschchina, "Temperature Investigation of Fine Structure of the Rayleigh Line and Hyperacoustic Properties of Benzole. Tetrachloromethane and Cyclohexane," *Ukr. Fiz. Zh.*, **15** (12), 2054–2059 (1970) (in Russian).
118. G. R. Hanes, R. Turner, and H. E. Piercy, "Hypersonic Absorption and Velocity from Measurement of Light Scattering Dichloromethane," *J. Acoust. Soc. Am.*, **38** (6), 1057–1058 (1965).
— E. H. Hara, A. D. May, and H. E. P. Knaap, "Rayleigh-Brillouin Scattering in Compressed Hydrogen, Deuterium and Hydrogen Deuterium (Mixture)," *Can. J. Phys.*, **49** (14), 420–431 (1971) [*9*].
119. D. A. Jackson, R. Jones, and E. R. Pike, "Measurement of Brillouin Line Widths in Benzene Derivatives Using a Super-Mode Laser and Frequency Stabilized, Digital Fabry. Perot Interferometer," *Phys. Lett., A*, **28** (4), 272–273 (1968).

— D. A. Jackson and D. M. Paul, "Measurement of Hypersonic Velocities and Turbulence by Spectral Analysis of Doppler Shifted Laser Light," *Phys. Lett. A*, **32**, 77–78 (1970) [*3*].

120. E. Kato and Y. Saji, "Vibrational Relaxation in Benzene Near the Melting Point," *Japan J. Appl. Phys.*, **10**, 1472–1473 (1971).

121. H. E. P. Knaap, W. S. Gornall, and P. P. Stoicheff, "Evidence of a Fourth Component in the Brillouin Spectrum of Liquid Glycerine," *Phys. Rev.*, **166** (1), 139–141 (1968).

122. P. Lallemand, "Study of Sound Velocity in Some Dense Gases by Brillouin Scattering," *Compt. Rend. B*, **270** (6), 389–392 (1970) (in French).

123. P. Lallemand and A. M. Longequeue, "Measurement of Sound Velocity in Gases Under Pressure by the Study of the Brillouin Scattering Spectrum," *Compt. Rend. B*, **269** (21), 1101–1104 (1969) (in French).

— J. B. Lastovka and G. B. Benedek, "Light-Beating Techniques for the Study of the Rayleigh-Brillouin Spectrum," in *Physics of Quantum Electronics*. McGraw-Hill, New York, 1966, pp. 231–240 [*2*].

124. A. M. Longequeue and P. Lallemand, "Study of the Ultrasound Dispersion in Gases Under Pressure by Brillouin Scattering," *Compt. Rend. B*, **269** (22), 1173–1176 (1969) (in French).

— J. A. Mann, J. F. Baret, F. J. Dechow, and R. S. Hansen, "Interfacial Forces and the Brillouin Spectrum of the Interfacial Light Scattering Power," *J. Colloid Interface Sci.*, **37**, 14–32 (1971) [*6*].

125. D. I. Mash, V. S. Starunov, and I. L. Fabelinskii, "Attenuation of Hypersound in Liquids by an Optical Method," *Soviet. Phys. JETP*, **20**, 523–524 (1965).

126. G. A. Miller and C. S. Lee, "Brillouin Spectra of Dilute Solutions and the Landau-Placzek Formula," *J. Phys. Chem.*, **72** (13), 4644–4650 (1968).

127. R. Mohr, K. H. Langley, and N. C. Ford, "Brillouin Scattering from Sulfur Hexafluoride in the Vicinity of the Critical Point," *J. Acoust. Soc. Am.*, **49** (3, Pt. 3), 1030–1032 (1971).

128. C. J. Montrose and K. Fritsch, "Hypersonic Velocity and Absorption in Aqueous Electrolytic Solutions, *J. Acoust. Soc. Am.*, **47** (3), 786–790 (1970).

129. C. J. Montrose, V. A. Solovev, and T. A. Litovitz, "Brillouin Scattering and Relaxation in Liquids," *J. Acoust. Soc. Am.*, **43** (1), 117–130 (1968).

130. R. D. Mountain, "Interpretation of Brillouin Spectra," *J. Chem. Phys.*, **44** (2), 832–833 (1966).

131. R. D. Mountain, "Thermal Relaxation and Brillouin Scattering in Liquids," *J. Res. Nat. Bur. Std. A*, **70**, 207–220 (1966).

132. R. D. Mountain and T. A. Litovitz, "Negative Dispersion and Brillouin Scattering," *J. Acoust. Soc. Am.*, **42**, 576–577 (1967).

133. M. Nelkin and S. Yip, "Brillouin Scattering by Gases as a Test of the Boltzmann Equation," *Phys. Fluids*, **9** (2), 380–381 (1966).

134. W. H. Nichols and E. F. Carome, "Light Scattering from a Multiply Relaxing Fluid, *J. Chem. Phys.*, **49**, 1000–1012 (1968).

135. W. H. Nichols, C. R. Kunsitis-Swyt, and S. P. Singal, "Ultrasonic and Light Scattering Studies of Benzene," *J. Chem. Phys.*, **51** (12), 5659–5662 (1969).

136. O. Nomoto, "Diffraction of Light by Ultrasound: Extension of the Brillouin Theory," *Japan J. Appl. Phys.*, **10**, 611–622 (May 1971).

137. C. L. O'Connor and J. P. Schlupf, "Brillouin Scattering and Thermal Relaxation in Benzene," *J. Acoust. Soc. Am.* **40** (3), 663–666 (1966).

— C. J. Palin, W. F. Vinen, E. R. Pike, and J. M. Vaughan, "Rayleigh and Brillouin Scattering from Superfluid Helium-3 Helium-4 Mixtures," *J. Phys. C Solid State*, **4,** L225–228 (1971) [*3*].

138. M. S. Pesin, "Intensity of the Fine Structure Components in the Scattering of Light by a Transverse Elastic Wave," *Izv. Vuz. Fiz.*, **1967** (1), 26–29 (in Russian).

139. J. E. Piercy and G. R. Hanes, "Discussion of the Physical Bases of Brillouin Scattering," *J. Acoust. Soc. Am.*, **49** (3, Pt. 3), 1001–1012 (1971).

140. E. R. Pike, J. M. Vaughn, and W. F. Vinen, "Brillouin Scattering from First and Second Sound in a Superfluid Helium-3 Helium-4 Mixture," *Phys. Lett. A*, **30** (7), 373–374 (1969).

141. E. R. Pike, J. M. Vaughn, and W. F. Vinen, "Brillouin Scattering from Superfluid Helium-4," *J. Phys. C*, **3** (2), 40–43 (1970).

142. D. A. Pinnow, S. J. Candau, J. T. La Macchia, and T. A. Litovitz, *J. Acoust. Soc. Am.*, **43,** 131 (1968).

143. D. H. Rank, "Brillouin Effect in Liquids in the Prelaser Era," *J. Acoust. Soc. Am.*, **49** (3, Pt. 3), 937–940 (1971).

144. D. H. Rank, "Brillouin Scattering: Stimulated and Spontaneous," *Int. Conf. Raman Spectry. 1970, 2nd Proc. (Abstr.)*.

145. D. H. Rank, E. M. Kiess, and U. Fink, "Brillouin Spectra of Viscous Liquids," *J. Opt. Soc. Am.*, **56,** 163–166 (1966).

146. E. G. Rawson, E. H. Hara, A. D. May, and H. L. Welsh, "Normal Brillouin Scattering from Compressed Gases," *J. Opt. Soc. Am.*, **56,** 1403–1405 (1966).

— G. P. Roshchina and G. L. Gudimenko, "Investigation of the Fine Structure of the Rayleigh Line in *N*-Paraffins by Means of Gas Laser," *Ukr. Fiz. Zh.*, **15** (6), 888–892 (1970) (in Russian) [*3*].

— S. M. Rytov, "Relaxation Theory of Rayleigh Scattering," *Zh. Eksperim i Teor. Fiz.*, **58** (6), 2154–2170 (1970) (in Russian) [*3*].

147. I. M. Sabirov, V. S. Starunov, and I. L. Fabelinskii, "Determination of Velocity and Absorption of Hypersound in Viscous Liquids by Spectra of Scattered Light," *Soviet Phys. JETP*, **33** (1), 82–89 (1971).

148. G. M. Searby and G. W. Series, "Brillouin Scattering at Small Angles in Carbon Tetrachloride and Carbon (6) Helium (6)," *Opt. Commun.*, **1** (4), 191–194 (1969).

149. S. L. Shapiro, M. McClintock, D. A. Jennings, and R. L. Barger, "Brillouin Scattering in Liquids at 4880 A," *IEEE J. Quantum Electron.*, **2** (5), 89–93 (1966).

150. M. S. Soren and A. L. Schawlow, "Dispersion in Carbon (6) Hydrogen (6) and Carbon (6) Deuterium (6)," *Phys. Lett. A*, **33** (5), 268–269 (1970).

151. R. L. St. Peters, T. J. Greytak, and G. B. Benedek, "Brillouin Scattering Measurements of the Velocity and Attenuation of High Frequency Sound Waves in Superfluid Helium," *Opt. Commun.*, **1** (9), 412–416 (1970).

152. G. I. A. Stegeman, W. S. Cornall, V. Volterra et al., "Brillouin Scattering and Dispersion and Attenuation of Hypersonic Thermal Waves in Liquid Carbon Tetrachloride," *J. Acoust. Soc. Am.*, **49,** 979–993 (March Pt. 3. 1971).

— A. Sugawara and S. Yip, "Kinetic Model Analysis of Light Scattering by Molecular Gases," *Phys. Fluids*, **10** (9), 1911–1921 (1967) [*3*].

— M. Tanaka, "Molecular Theory of Scattering of Light from Liquids," *Progr. Theor. Phys. (Kyoto)*, **40** (5), 975–989 (1968), *J. Phys. Soc. Japan* **26** (Suppl.), 298–300 (1969) [*3*].

— M. L. Ter-Mikablyan and A. O. Meiikyan, "Rayleigh and Combinational Scattering in the Field of an Intense Wave," *Zh. Eksperim i Teor. Fiz.*, **58** (1), 281–290 (1970) (in Russian) [*2*].

153. N. Uchida, "Elasto-Optic Coefficient of Liquids Determined by Ultrasonic Light Diffraction Method," *Japan J. Appl. Phys.*, **7,** 1259–1267 (1968).
154. C. S. Venkateswaran, "Interferometric Studies of Light Scattering in Viscous Liquids and Glasses. Hypersonic Velocities in Liquids," *Proc. Ind. Acad. Sci.*, **15,** 362–365 (1942).
155. W. F. Vinen, "Light Scattering by Superfluid Helium Under Pressure," *J. Phys. C Solid State*, **4**, L287 (1971).
156. M. A. Woolf, P. M. Platzman, and M. G. Cohen, "Brillouin Scattering in Liquid Helium II," *Phys. Rev. Lett.*, **17**, 294–297 (1966).
— S. Yip, "Rayleigh Scattering in Dilute Gases," *J. Acoust. Soc. Am.*, **49** (3, Pt. 3), 941–948 (1971) [*3*].
157. W. Low, "Brillouin Scattering in Ordinary Fluids," in *Colloque Int. CNRS, Diffusion de la Lumiere par les Fluids*, 1971. *J. Physique Suppl.*, **33**C1, 1 (1972).
158. S. U. Candau and M. Caloin, "Study of Vibrational Relaxation in Some Liquids from Analysis of Brillouin Spectra," in *Colloque Int. CNRS, Diffusion de la Lumiere par les Fluids*, 1971. *J. Physique Suppl.*, **33**C1, 7 (1972).
159. J. P. Boon and P. A. Fleury, "First Sound and Quantum Effects in Liquid Neon," in *Colloque Int. CNRS, Diffusion de la Lumiere par les Fluides*, 1971. *J. Physique Suppl.*, **33**C1, 19 (1972).
— E. R. Pike, "Rayleigh and Brillouin Scattering from Liquid Helium," in *Colloque Int. CNRS, Diffusion de la Lumiere par les Fluids*, 1971. *J. Physique Suppl.*, **33**C1, 25 (1972) [*3*].
— R. C. Desai, "Rayleigh-Brillouin Scattering from Fluid Mixtures," in *Colloque Int. CNRS, Diffusion de la Lumiere par les Fluids*, 1971. *J. Physique Suppl.*, **33**C1, 27 (1972) [*9*].
— C. D. Boley and S. Yip, "Spectral Distributions of Light Scattered in Dilute Gases and Gas Mixtures," in *Colloque Int. CNRS, Diffusion de la Lumiere par les Fluides*, 1971. *J. Physique Suppl.*, **33**C1, 43 (1972) [*9*].
— A. D. May and E. H. Hara, "Rayleigh-Brillouin Scattering in Compressed Hydrogen. Deuterium and HD," in *Colloque Int. CNRS, Diffusion de la Lumiere par les Fluides*, 1971. *J. Physique Suppl.*, **33**C1, 50 (1972) [*3*].
— W. L. Gornall and C. S. Wang, "Light Scattering from Thermal Fluctuations in Disparate-Mass Gas Mixtures," in *Colloque Int. CNRS, Diffusion de la Lumiere par les Fluides*, 1971. *J. Physique Suppl.*, **33**C1, 51 (1972) [*9*].
— A. M. Cazabat-Longequeue and P. Lallemand, "Light Scattering by Polyatomic Gases," in *Colloque Int. CNRS, Diffusion de la Lumiere par les Fluides*, 1971. *J. Physique Suppl.*, **33**C1, 57 (1972) [*3*].
160. R. Mohr and K. H. Langley, "Brillouin Scattering from Sulfur Hexafluoride in the Vicinity of the Critical Point," in *Colloque Int. CNRS, Diffusion de la Lumiere par les Fluides*, 1971. *J. Physique Suppl.* **33**, Cl, 97, (1972).
— S. H. Chen and N. Polonsky, "Anomalous Damping and Dispersion of Hypersound in a Binary Liquid Mixture Near the Solution Critical Point," *Phys. Rev. Lett.*, **20**, 809 (1968) [*9*].

5. *Intermolecular or Collision Induced Scattering*

161. R. Bersohn, Y. H. Pao, and H. L. Frisch, "Double Quantum Light Scattering by Molecules," *J. Chem. Phys.*, **45,** 3184 (1966).
— J. A. Bucaro, "Rayleigh Scattering: Molecular Motions and Interactions in Liquids," Thesis, Catholic University, 1971 (Univ. Microfilms 71-14077) [*3*].

162. J. A. Bucaro and T. A. Litovitz, "Rayleigh Scattering: Collisional Motions in Liquids," *J. Chem. Phys.*, **54,** 3846–3853 (May 1, 1971).
163. J. V. Champion, G. H. Meeten, and C. D. Whittle, "Electrooptic Kerr Effect in *N*-Alkane Liquids," *Trans. Faraday Soc.*, **66,** (575, Pt. 11), 2671–2680 (1970).
164. V. G. Cooper, A. D. May, E. H. Hara, and H. F. P. Knaap, "Depolarized Rayleigh Scattering in Gases as a New Probe of Intermolecular Forces," *IEEE J. Quantum Electron.*, **4** (11), 720–722 (1968).
165. H. C. Craddock, D. A. Jackson, and J. G. Powles, "Spectrum of Depolarized Light Scattered by Liquid Benzene Derivatives," *Mol. Phys.*, **14** (4), 373–380 (1968).
166. P. A. Fleury, W. B. Daniels, and J. M. Worlock, "Density and Temperature Dependence of Intermolecular Scattering in Simple Fluids," *Phys. Rev. Lett.*, **27,** 1493–1496 (1971).
167. P. A. Fleury and J. P. McTague, "Effects of Molecular Interactions on Light Scattering by Simple Fluids," *Opt. Commun.*, **1** (4), 164–166 (1969).
168. J. I. Gersten, R. E. Slusher, and C. M. Surko, "Relation of Collision Induced Light Scattering in Rare Gases to Atomic Collision Parameters," *Phys. Rev. Lett.*, **25,** 1739–1742 (1970).
169. W. S. Gornall, H. E. Howard-Lock, and B. P. Stoicheff, "Induced Anisotropy and Light Scattering in Liquids," *Phys. Rev. A*, **1** (5), 1288–1289 (1970).
170. C. G. Gray and H. I. Ralph, "Virial Expansion for the Depolarization Ratio of Rayleigh Scattering from Monatomic Gases," *Phys. Lett. A*, **33** (3), 165–166 (1970).
171. F. Iwamoto, "Raman Scattering in Liquid Helium," *Progr. Theoret. Phys.*, **44,** 1121–1134 (1970).
172. D. A. Jackson and B. Simic-Glavaski, "Study of Depolarized Rayleigh Scattering in Liquid Benzene Derivatives," *Mol. Phys.*, **18** (3), 393–400 (1970).
— R. A. J. Keijser, M. Jansen, V. G. Cooper, and H. F. P. Knaap, "Depolarized Rayleigh Scattering in Carbon Dioxide, Carbon Oxysulfide and Carbon Disulfide," *Physica*, **51** (4), 593–600 (1971) [*3*].
— S. Kielich, "Molecular Light Scattering in Dense Mixtures," *Acta Phys. Polon.*, **33,** 63–79 (1968) [*9*].
173. S. Kielich, "Relation Between Nonlinear Refractive Index and Molecular Light Scattering in Liquids," *Chem. Phys. Lett.*, **2,** 112–115 (1968).
174. S. Kielich, "Many Molecular Correlation Induced Anisotropic Light Scattering and Birefringence in Simple Fluids," *Opt. Commun.*, **4,** 135 (1971).
— T. Keyes and D. Kivelson, "Light Scattering and the Coupling of Molecular Reorientation and Hydrodynamic Modes," *J. Chem. Phys.*, **54** (4), 1786–1798 (1971) [*7*].
— T. Keyes, D. Kivelson, and J. P. McTague, "Theory of K Independent Depolarized Rayleigh Wing Scattering in Liquids Composed of Anisotropic Molecules," *J. Chem. Phys.*, **55,** 4096–4100 (1971) [*7*].
— S. Kielich, "Depolarization of Light Scattering by Atomic and Molecular Solutions With Strongly Anisotropic Translational Orientational Fluctuations," *Chem. Phys. Lett.*, **10,** 516–521 (1971) [*9*].
175. H. F. P. Knaap, V. G. Cooper, and R. A. J. Keijser, "Depolarized Rayleigh Scattering in Gases," *Symp. Mol. Structure Spectry. 1970, 25th Proc.* (*Abstr.*), p. 96.
— M. D. Levenson and A. L. Schawlow, "Depolarized Light Scattering from Liquid Bromine," *Opt. Commun.*, **2** (4), 192–195 (1970) [*7*].
176. H. B. Levine and G. Birnbaum, "Collision-Induced Light Scattering," *Phys. Rev. Lett.*, **20** (9), 439–441 (1968).

— H. C. Lucas and D. A. Jackson, "The Intensity and Spectra of Depolarized Light Scattered from Benzene Nitrobenzene Mixtures," *Mol. Phys.*, **20,** 801–810 (May 1971) [9].

— H. C. Lucas, D. A. Jackson, J. G. Powles, and B. Simic-Glavaski, "Temperature Variation of Polarized and Depolarized Scattered Light Spectra from Liquid Benzene Derivatives," *Mol. Phys.*, **18** (4) 505–521 (1970) [7].

177. F. B. Martin and J. R. Lalanne, "Agreement Between Depolarized Rayleigh Scattering and Optical Kerr Effect Induced by Q-Switched Laser Waves in some Liquids," *Phys. Rev. A*, **4** (3), 1275–1278 (1971).

178. J. P. McTague and G. Birnbaum, "Collision Induced Light Scattering in Gaseous Argon and Krypton," *Phys. Rev. Lett.*, **21,** 661–664 (1968).

179. J. P. McTague and G. Birnbaum, "Collision-Induced Light Scattering in Gases. Part-1: Rare Gases: Argon Krypton and Xenon," *Phys. Rev. A*, **3** (4), 1376–1383 (1971).

180. J. P. McTague, P. A. Fleury, and D. B. Du Pre, "Intermolecular Light Scattering in Liquids," *Phys. Rev.*, **188,** 303–309 (1969).

181. S. Nakajima, "Elementary Quantum Theory of Light in Liquid Helium," *Progr. Theoret. Phys.*, **45** (2), 353–364 (1971).

— P. T. Nikolaenko and A. I. Frorvin, "Oscillation of Liquid Molecules and the Wing of the Rayleigh Line," *Izv. Vuz. Fiz.*, **1967** (8), 107–111 (in Russian) [7].

182. A. V. Sechkarev and P. T. Nikolaenko, "Investigation of Intermolecular Dynamics in the Condensed States of Matter by the Method of Vibrational Spectroscopy. Part-1: Intensity Distribution and Intermolecular Light Scattering Spectra in the Neighborhood of the Rayleigh Line," *Izv. Vuz. Fiz.*, **1969** (4), 104–110 (in Russian).

183. B. Simic-Glavaski, D. A. Jackson, and J. G. Powles, "Cabannes-Daure Effect in Liquids," *Phys. Lett.*, **A34,** 255–256(1971).

184. R. E. Slusher and C. M. Surko, "Raman Scattering from Condensed Phases of He^3 and He^4," *Phys. Rev. Lett.*, **27,** 1699 (1971).

— B. Simic-Glavaski, D. A. Jackson, and J. G. Powles, "Low Frequency Raman Lines in the Depolarized Rayleigh Wing of Liquid Toluene and Benzene," *Phys. Lett. A*, **32** (5), 329–330 (1970) [7].

185. M. Thibeau, G. C. Tabisz, B. Oksengorn, and B. Vodar, "Pressure Induced Polarization of the Rayleigh Light Scattering from a Gas of Optically Isotropic Molecules," *J. Quantum Spectry. Radiat. Transf.*, **10** (8), 839–856 (1970).

186. V. Volterra, J. A. Bucaro, and T. A. Litovitz, "Molecular Motion and Light Scattering in Liquids," *Ber. Bunsenges. Phys. Chem.*, **75** (3/4), 309–315 (1971).

187. V. Volterra, J. A. Bucaro, and T. A. Litovitz, "Two Mechanisms for Depolarized Light Scattering from Gaseous Argon," *Phys. Rev. Lett.*, **26** (2), 55–57 (1971).

188. C. H. Wang, and R. B. Wright, "Is an Exponential Spectral Shape a Criterion for Collision Induced Light Scattering in Hydrogen Bonded Molecular Fluids," *Chem. Phys. Lett.*, **11**, 277–280 (1971).

189. A. Zawadowski, J. Ruvalds, and J. Solana, "Bound Roton Pairs in Superfluid Helium," *Phys. Rev. A*, **5,** 399–421 (1972).

— B. Simic-Glavaski and D. A. Jackson, "Rayleigh Depolarized Light Scattered from Isotropic and Anisotropic Molecular Liquids," in *Colloque Int. CNRS, Diffusion de la Lumiere par les Fluides*, 1971. *J. Physique Suppl.*, **33**C1, 183 (1972) [7].

— S. Kielich, J. R. Lalanne, and F. B. Martin, "Relevance of Depolarized Rayleigh Scattering Optical Kerr Effect to the Study of Radial and Orientational Correlations and Their Thermal Variations in Liquids," in *Colloque Int. CNRS, Diffusion de la Lumiere par les Fluides*, 1971. *J. Physique Suppl.*, **33**C1, 191 (1972) [7].

190. J. P. McTague, W. Ellenson, and L. H. Hall, "Spectral Virial Expansion for Collision Induced Light Scattering in Gases," in *Colloque Int. CNRS, Diffusion de la Lumiere pas les Fluides*, 1971. *J. Physique Suppl.*, **33**C1, 241 (1972).
191. M. Thiebeau and B. Oksengorn, "Experimental and Theoretical Study of Compression Induced Polarization of Rayleigh Scattered Light in Optically Isotropic Gases," in *Colloque Int. CNRS, Diffusion de la Lumiere par les Fluides*, 1971. *J. Physique Suppl.*, **33**C1, 247 (1972).
192. P. Lallemand, "Depolarized Light Scattering due to Collisions," in *Colloque Int. CNRS, Diffusion de la Lumiere par les Fluides*, 1971. *J. Physique Suppl.*, **33**C1, 257 (1972).
193. P. A. Fleury, "Dynamics of Simple Fluids by Intermolecular Light Scattering," in *Colloque Int. CNRS, Diffusion de la Lumiere par les Fluides*, 1971. *J. Physique Suppl.*, **33**C1, 264 (1972).
194. R. D. Mountain, "Depolarized Scattering and Long Range Correlations for a Simple Fluid Near the Critical Point," in *Colloque Int. CNRS, Diffusion de la Lumiere par les Fluides*, 1971. *J. Physique Suppl.*, **33**C1, 265 (1972).
195. T. J. Greytak and R. L. Woerner, "A Bound State of Two Rotons," in *Colloque Int. CNRS, Diffusion de la Lumiere par les Fluides*, 1971. *J. Physique Suppl.*, **33**C1, 269 (1972).
196. L. Blum and H. L. Frisch, "Polarization State of Light Scattered from Dense Fluids," *J. Chem. Phys.* **55,** 1188 (1971).
197. T. J. Greytak and J. Yan, "Light Scattering from Rotons in Liquid Helium," *Phys. Lett. Rev.* **22,** 987 (1969).
198. T. J. Greytak, R. Woerner, J. Yan, and R. Benjamin, "Experimental Evidence for a Two Roton Bound State in Liquid Helium, *Phys. Rev. Lett.* **25,** 1547 (1970).
199. J. W. Halley, "Theory of Optical Processes in Liquid Helium," in *Light Scattering in Solids*, G. B. Wright, Ed., Springer-Verlag, 1969, p. 175.
200. P. Lallemand, "Spectral Distribution of Double Light Scattering by Gases," *Phys. Rev. Lett.*, **25,** 1079 (1970).
201. M. J. Stephen, "Raman Scattering in Liquid Helium," *Phys. Rev.*, **187,** 279 (1969).

6. *Scattering from Surfaces and Thin Films*

202. M. A. Bouchiat and J. Meunier, "Light Scattering from Surface Waves on Carbon Dioxide near the Critical Point," *Phys. Rev. Lett.*, **23,** 752–755 (1969).
— J. Adams, W. Haar, and J. Wysocki, "Light Scattering Properties of Cholesteric Liquid Crystal Films," *Mol. Cryst. Liq. Cryst.*, **8,** 9 (1969) [*8*].
203. M. A. Bouchiat and J. Meunier, "Spectrum Characteristics of Light Scattered by Capillary Waves Present on a Liquid Surface," *Compt. Rend., B*, **266** (5), 301–304 (1968) (in French).
204. M. A. Bouchiat, J. Meunier, and J. Brossel, "Thermal Excitation of Capillary Waves at the Surface of a Liquid by the Inelastic Scattering of Light," *Compt. Rend, B*, **266** (4), 255–258 (1968) (in French).
205. P. G. Degennes, "Peristaltic Modes of a Soap Film," *Compt. Rend., B*, **268**, 1207–1209 (1969) (in French).
206. J. S. Huang and W. W. Webb, "Viscous Damping of Thermal Excitations on the Interface of Critical Fluid Mixtures," *Phys. Rev. Lett.*, **23** (4), 160–163 (1969).
207. R. H. Katyl and K. U. Ingard, "Line Broadening of Light Scattered from a Liquid Surface," *Phys. Rev. Lett.*, **19** (2), 64 66 (1967).
208. R. H. Katyl and K. U. Ingard, "Scattering of Light by Thermal Ripplons," *Phys. Rev. Lett.*, **20** (6), 248–249 (1968).

209. J. A. Mann, J. F. Baret, F. J. Dechow, and R. S. Hansen, "Interfacial Forces and the Brillouin Spectrum of the Interfacial Light Scattering Power," *J. Colloid Interface Sci.*, **31,** 14–32 (1971).
210. J. Meunier, D. Cruchon, and M. A. Bouchiat, "Experimental Study of Thermally Excited Capillary Waves at the Surface of Pure Liquids for Hydrodynamics States Near Critical Damping," *Compt. Rend. B*, **268** (5), 422–425 (1969) (in French).
211. M. Papoular, "Critical Surface Opalescence," *Compt. Rend. B*, **266,** 592–595 (1968) (in French).
212. V. Vrij and J. T. H. Overdeck, "Rupture of Thin Liquid Films due to Spontaneous Fluctuations in Thickness," *J. Am. Chem. Soc.*, **90** (12), 3074–3078 (1968).
213. J. Zollweg, G. A. Hawkins, and G. B. Benedek, "Surface Tension and Viscosity of Xenon Near its Critical Point. (Optical Inelastic Scattering Method)," *Phys. Rev. Lett.*, **27,** 1132–1135 (1971).
— D. Langevin and M. A. Bouchiat, "Light Scattering from the Free Surface of a Nematic Liquid Crystal," in *Colloque Int. CNRS, Diffusion de la Lumiere par les Fluides*, 1971. *J. Physique Suppl.*, **33**C1, 77 (1972) [*8*].
— J. Zollweg, G. Hawkins, I. W. Smith, M. Giglio, and G. B. Benedek, in *Colloque Int. CNRS, Diffusion de la Lumiere par les Fluides,* 1971. *J. Physique Suppl.*, **33**C1, 135 (1972) [*10*].
214. M. A. Bouchiat and J. Meunier, "Light Scattering from Surface Waves on Carbon Dioxide Near the Critical Point," in *Colloque Int. CNRS, Diffusion de la Lumiere par les Fluides*, 1971. *J. Physique Suppl.*, **33**C1, 141 (1972).
215. E. S. Wu and W. W. Webb, "Light Near the Critical Point Scattered by the Liquid-Vapor Interface of Sulfur-Hexafluoride," in *Colloque Int. CNRS, Diffusion de la Lumiere par les Fluides*, 1971. *J. Physique Suppl.*, **3**C1, 149 (1972).

7. *Molecular Anisotropy and Orientation Effects*

216. A. Ben-Reuven and N. D. Gershon, "Light Scattering by the Tight Binding of Molecular Reorientation to Collective Motions in Liquids," *Ber. Bunsenges. Phys. Chem.*, **75** (3/4): 340–342 (1971).
217. G. B. Bendek, "Optical Mixing Spectroscopy and Applications to Problems in Physics, Chemistry, Biology and Engineering" in *Polarisation, Matiere et Rayonnement*, Presses Universitaire de France, 1969, pp. 49–84 (539, 1/P76, 138226).
218. N. K. Ailawadi, B. J. Berne, and D. Forster, "Light Scattering from Shear Waves: The Role of Angular Momentum Fluctuations in Light Scattering," *Phys. Rev.*, **3** (4), 1472–1483 (1971).
219. H. C. Andersen and R. Pecora, "Kinetic Equations for Orientational and Shear Relaxation and Depolarized Light Scattering in Liquids," *J. Chem. Phys.*, **54,** 2584–2596 (March 15, 1971).
220. A. Ben-Reuven and N. D. Gershon, "Light Scattering by Coupling of Orientational Motion to Sound Waves in Liquids," *J. Chem. Phys.*, **54** (3), 1049–1053 (1971).
221. A. Ben-Reuven and N. D. Gershon, "Light Scattering by Orientational Fluctuations in Liquids," *J. Chem. Phys.*, **51** (3), 693–902 (1969).
— M. A. Bouchiat and J. Meunier, "Light Scattering from Surface Waves on Carbon Dioxide Near the Critical Point," *Phys. Rev. Lett.*, **23,** 752–755 (1969) [*6*].
222. J. P. Chaerat, J. Rouch, C. Voucamps, "New Measurements on the Anisotropic Rayleigh Scattering of Polar and Nonpolar Liquids and Their Solutions in Cyclohexane," *Compt. Rend. B*, **270** (24), 1556–1558 (1970) (in French).

223. D. J. Coumou, "Light Scattering in Pure Liquids and the Polarizibility Tensor of the Molecules," *Trans. Faraday Soc.*, **65** (10), 2654–2662 (1969).
— H. C. Craddock, D. A. Jackson, and J. G. Powles, "Spectrum of Depolarized Light Scattered by Liquid Benzene Derivatives," *Mol. Phys.* **14** (4), 373–380 (1968) [*5*].
224. I. L. Fabelinskii, I. M. Sabirov, and V. S. Starunov, "Fine Structure of Rayleigh Line Wing and Propagation of Transversal Hypersound in Liquids," *Phys. Lett. A*, **79** (7), 414–415 (1969).
225. S. F. Faizulloyev, A. K. Atakhodzhaev, and Z. K. H. Rakhimov, "Rate of Reorientation of Molecules of the Series of Disubstituted Benzol Derivatives," *Ukr. Fiz. Zh.*, **12** (1), 171–173 (1967) (in Russian).
226. J. A. Finnigan and D. J. Jacobs, "Light Scattering from Benzene, Toluene, Carbon Disulfide and Carbon Tetrachloride," *Chem. Phys. Lett.*, **6** (3), 141–143 (1970).
227. N. D. Gershon, E. Zamir, and A. Ben-Reuven, "Rayleigh-Wing Scattering from Liquids of Anisotropic Molecules," *Ber. Bunsenges. Phys. Chem.*, **75** (3/4), 316–319 (1971).
228. E. F. Gross, V. P. Romanov, V. A. Solovev, and E. O. Chernysheva, "Structure of the Depolarized Optical Scattering Spectrum in Viscoelastic Media," *Fiz. Tverd. Tela*, **11** (12), 3686–3688 (1969) (in Russian).
229. S. Hess, "Depolarized Rayleigh Light Scattering and Tensor Polarization Resonance," *Phys. Lett. A*, **29** (8), 108–109 (1969).
230. S. Hess, "Light Scattering by Polyatomic Fluids," *Z. Naturforsch. A*, **24** (11), 1675–1687 (1969).
231. S. Hess, "Spectrum of the Depolarized Rayleigh Light Scattered by Gases of Linear Molecules," *Z. Naturforsch. A*, **25** (3), 350–362 (1970).
— D. A. Jackson and B. Simic-Glavaski, "Study of Depolarized Rayleigh Scattering in Liquid Benzene Derivatives," *Mol. Phys.*, **18** (3), 393–400 (1970) [*5*].
— R. A. J. Keijser, M. Jansen, V. G. Cooper, and H. F. P. Knaap, "Depolarized Rayleigh Scattering in Carbon Dioxide, Carbon Oxysulfide and Carbon Disulfide," *Physica*, **51** (4), 593–600 (1971) [*3*].
232. T. Keyes and D. Kivelson, "Light Scattering and the Coupling of Molecular Reorientation and Hydrodynamic Modes," *J. Chem. Phys.*, **54** (4), 1786–1798 (1971).
233. T. Keyes, D. Kivelson, and J. P. McTague, "Theory of K Independent Depolarized Rayleigh Wing Scattering in Liquids Composed of Anisotropic' Molecules," *J. Chem. Phys.*, **55**, 4096–4100 (1971).
— S. Kielich, "Depolarization of Light Scattering by Atomic and Molecular Solutions with Strongly Anisotropic Translational Orientational Fluctuations," *Chem. Phys. Lett.*, **10**, 516–521 (1971) [*9*].
— H. F. P. Knaap, V. G. Cooper, R. A. J. Keijser, "Depolarized Rayleigh Scattering in Gases," in *Symp. Mol. Structure Spectry, 1970, 25th Proc.* (*Abstr.*), p. 96 [*5*].
234. J. R. Lalanne, "Contribution of Depolarized Rayleigh Scattering and Optical Anisotropy Measurement Techniques to the Determination of the Intermolecular Angular Correlation Parameter of Liquids," *J. Phys.*, **30** (8–9), 643–653 (1969) (in French).
235. J. R. Lalanne and P. Botherol, "Near Infrared Laser Rayleigh Scattering Measurements. Optical Anisotropy Dispersion Studies and the Determination of the Anisotropy of Colored Compounds," *Mol. Phys.*, **19** (2), 227–231 (1970).
236. P. Lallemand, "Comparison of Depolarized Raman and Rayleigh Lines Scattered By Chloroform," *Compt. Rend. B*, **72** (7), 429–432 (1971) (in French).

237. R. C. C. Leite, R. S. Moore, S. P. S. Porto, and J. E. Rippler, "Angular Dependence of the Rayleigh Scattering from Low Turbidity Molecular Liquids," *Phys. Rev. Lett.*, **14,** 7–9 (1965).
238. R. C. C. Leite, R. S. Moore, and S. P. S. Porto, "Use of a Gas Laser in Studies of the Depolarization of the Rayleigh Scattering from Simple Liquids," *J. Chem. Phys.*, **40,** 3741–3742 (1964).
239. M. D. Levenson and A. L. Schawlow, "Depolarized Light Scattering from Liquid Bromine," *Opt. Commun.*, **2** (4), 192–195 (1970).
— H. C. Lucas and D. A. Jackson, "The Intensity and Spectra of Depolarized Light Scattered from Benzene Nitrobenzene Mixtures," *Mol. Phys.*, **20,** 801–810 (May 1971) [*9*].
240. H. C. Lucas, D. A. Jackson, J. G. Powles, and B. Simić-Glavaski, "Temperature Variation of Polarized and Depolarized Scattered Light Spectra from Liquid Benzene Derivatives," *Mol. Phys.*, **18** (4), 505–621 (1970).
— A. G. Marshall and R. Pecora, "Depolarized Light Scattering from a Helix Coil System," *J. Chem. Phys.*, **55,** 1245–1248 (1971) [*9*].
— F. B. Martin and J. R. Lalanne, "Agreement Between Depolarized Rayleigh Scattering and Optical Kerr Effect Induced by Q-Switched Laser Waves in some Liquids," *Phys. Rev. A*, **4** (3), 1275–1278 (1971) [*5*].
241. P. T. Nikolaenko and A. I. Prorbin, "Oscillation of Liquid Molecules and the Wing of the Rayleigh Line," *Izv. Vuz. Fiz.*, **1967** (8), 107–111 (in Russian).
242. R. Pecora, "Spectrum of Light Scattered from Optically Anisotropic Macromolecules," *J. Chem. Phys.*, **49,** 1036–1043 (1968).
— R. Pecora, "Spectral Distribution of Light Scattered from Flexible Coil Molecules," *J. Chem. Phys.*, **49,** 1032–1035 (1968) [*9*].
243. R. Pecora and W. A. Steele, "Scattering from Fluids of Non-Spherical Molecules. Part-2. Light," *J. Chem. Phys.*, **42,** 1872–1879 (1965).
— M. S. Pesin, "Intensity of the Fine Structure Components in the Scattering of Light by a Transverse Elastic Wave," *Izv. Vuz. Fiz.* **1967** (1): 26–29 (in Russian) [*4*].
244. D. A. Pinnow, S. J. Candau, and T. A. Litovitz, "Rayleigh Scattering: Orientational Relaxation in Liquids," *J. Chem. Phys.*, **49,** 347–357 (1968).
245. A. I. Prorvin, "Anisotropic Component of the Rayleigh Scattering and the Frenkel Model of Molecular Oscillations in Liquids," *Opt. Spectry.*, **27** (1), 97–106 (1969).
246. A. I. Prorvin, "Relation Between the Oscillations of Liquid Molecules and the Distribution of the Continuous Spectrum of the Rayleigh Line," *Opt. Spectry.*, **24** (5), 380–382 (1968).
247. A. I. Prorvin, A. P. Bogachev, and L. M. Romanova, "Comparison Between Integral Background Intensities and the Wing of the Rayleigh Line," *Izv. Vuz. Fiz.* **1969** (7), 117–119 (in Russian).
248. D. H. Rank, A. Hollinger, and D. P. Eastman, "Depolarization of the Components of Rayleigh Scattering in Liquids," *J. Opt. Soc. Am.*, **56** (8), 1057–1058 (1966).
249. R. L. Rowell and G. M. Aval, "Rayleigh-Raman Depolarization of Laser Light Scattered by Gases," *J. Chem. Phys.*, **54** (5), 1960–1964 (1971).
250. N. B. Rozhdestvenskaya and L. A. Zubkov, "Study of the Spectral Structure of the Anisotropic Portion of Light Scattered by A-Chloronaphthalene," *Opt. Spectry.*, **28** (3), 599–600 (1970).
— S. M. Rytov, "Relaxation Theory of Rayleigh Scattering," *Zh. Eksperim. i Teor. Fiz.*, **58** (6), 2154–2170 (1970) (in Russian) [*3*].
251. S. M. Rytov, "Shear Doublet in Rayleigh Scattering of Light in Liquids with two Relaxation Times," *Zh. Eksperim. i Teor. Fiz.* **59** (6), 2130–2139 (1970) (in Russian).

252. E. V. Sagitova, F. K. M. Tukhvatullin, and A. K. Atakhodzhaev, "Study of Intermolecular Interactions by the Line Widths of Rayleigh and Raman Lines," *Opt. Spectry.*, **26** (1), 41–44 (1969).
253. S. L. Shapiro and H. P. Broida, "Light Scattering from Fluctuations in Orientations of Carbon Disulfide in Liquids," *Phys. Rev.*, **154,** 129–138 (1967).
— B. Simic-Glavaski, D. A. Jackson, and J. G. Powles, "Cabannes-Daure Effect in Liquids," *Phys. Lett. A*, **34** (5), 255–256 (1971) [5].
254. B. Simic-Glavaski, D. A. Jackson, and J. G. Powles, "Low Frequency Raman Lines in the Depolarized Rayleigh Wing of Liquid Toluene and Benzene," *Phys. Lett. A*, **32** (5), 329–330 (1970).
255. V. S. Starunov, "Interpretation of the Spectral Composition of Light Scattered by Fluctuations of Anisotropy in Liquids," *Soviet Phys. Doklady*, **8**, 1206–1208 (1964).
256. V. S. Starunov, E. V. Tiganov, and I. L. Fabelinskii, "Fine Structure in the Spectrum of the Thermal Rayleigh Line Wing in Liquids," *JETP Lett.* **5**, 260–262 (1967).
257. V. S. Starunov, E. V. Tiganov, and I. L. Fabelinskii, "Spectrum of Light Scattered by Density and Anisotropy Fluctuations in Liquid Nitrobenzene," *JETP Lett.*, **4** (7), 176–179 (1966).
258. W. A. Steele, "Orientational Correlations in Liquids and the Depolarization of Scattered Light," in *Statistical Mechanics, Foundations and Application, 1956*, Benjamin, New York, 1967, pp. 385–400.
259. G. I. A. Stegeman and B. P. Stoicheff, "Spectrum of Light Scattering from Thermal Shear Waves in Liquids," *Phys. Rev. Lett.*, **21,** 202–206 (1968).
260. A. Szoke, E. Courtens, and A. Pen-Reuven, "Orientational Order in Dipolar Liquids," *Chem. Phys. Lett.*, **1** (3), 87–90 (1967).
261. V. Volterra, "Theory of Light Scattering from Shear Waves in Liquids," *Phys. Rev.*, **180,** 156–166 (1969).
— V. Volterra, J. A. Bucaro, and T. A. Litovitz, "Molecular Motion and Light Scattering in Liquids," *Ber. Bunsenges. Phys. Chem.*, **75** (3/4), 309–315 (1971) [5].
— C. H. Wang and R. B. Wright, "Is an Exponential Spectral Shape a Criterion for Collision Induced Light Scattering in Hydrogen Bonded Molecular Fluids," *Chem. Phys. Lett.*, **11,** 277–280 (1971) [5].
262. G. L. Zaitsev, "Wing of the Rayleigh Line in Viscous Liquids," *Opt. Spectry.*, **23** (2), 175–176 (1967).
263. G. I. Zaitsev and V. S. Starunov, "Study of the Rayleigh-Line Wing in Liquids at Different Temperatures," *Opt. Spectry.*, **22** (3), 221–223 (1967).
264. E. Zamir, N. D. Gershon, and A. Ben-Reuven, "Rayleigh Wing Scattering by Aromatic Liquids," *J. Chem. Phys.*, **55**, 3397–3403 (1971).
265. L. A. Zubkov, N. B. Rozhdestvenskaya, and A. S. Khromov, "Structure of Depolarized Component of Scattered Light in A-Chloronaphthalene," *JETP Lett.*, **11** (10), 373–374 (1970).
266. T. A. Litovitz and F. J. Bartoli, "Depolarized Rayleigh and Raman Scattering: Mechanism of Orientational Motion in Liquids," in *Colloque Int. CNRS, Diffusion de la Lumiere par les Fluides*, 1971. *J. Physique Suppl.*, **33**C1, 181 (1972).
267. B. Simic-Glavaski and D. A. Jackson, "Rayleigh Depolarized Light Scattered from Isotropic and Anisotropic Molecular Liquids," in *Colloque Int. CNRS, Diffusion de la Lumiere par les Fluides*, 1971. *J. Physique Suppl.*, **33**Cl, 183 (1972).
268. S. Kielich, J. R. Lalanne, and F. B. Martin, "Relevance of Depolarized Rayleigh Scattering Optical Kerr Effect to the Study of Radial and Orientational Corre-

lations and Their Thermal Variations in Liquids," in *Colloque Int. CNRS, Diffusion de la Lumiere par les Fluides*, 1971. *J. Physique Suppl.*, **33**C1, 191 (1972).

269. G. I. A. Stegeman, B. P. Stoicheff, and G. Enright, "Spectrum of Light Scattering from Thermal Shear Waves in Liquids," in *Colloque Int. CNRS, Diffusion de la Lumiere par les Fluides*, 1971. *J. Physique Suppl.*, **33**C1, 207 (1972).
270. N. Ailawadi and B. J. Berne, "Theories of Depolarized Light Scattering," in *Colloque Int. CNRS, Diffusion de la Lumiere par les Fluides*, 1971. *J. Physique Suppl.* **33**C1, 221 (1972).
271. T. Keyes and D. Kivelson, "Low Frequency Depolarized VH-Scattering from Liquids Composed of Anisotropic Molecules," in *Colloque Int. CNRS, Diffusion de la Lumiere par les Fluides*, 1971. J. Physique Suppl., **33**C1, 231 (1972).

8. *Liquid Crystals*

272. M. Bertolotti, B. Daino, P. DiPorto et al., "Light Scattering by Electrodynamic Fluctuations in Nematic Liquid Crystals," *J. Phys. A Gen. Phys.*, **4,** L97–101 (1971).
273. J. Adams, W. Haar, and J. Wysocki, "Light Scattering Properties of Cholesteric Liquid Crystal Films," *Mol. Cryst. Liq. Cryst.*, **8,** 9 (1969).
274. M. A. Anisimov, I. M. Arefev, A. V. Voronel et al., "Propagation of Sound Near the Binary Mixture Stratification Point. (Mandelshtam Brillouin Scattering)," *Zh. Eksperim. i Teor. Fiz.*, **61,** 1526–1536 (1971).
275. C. Deutsch and P. N. Keating, "Scattering of Coherent Light from Nematic Liquid Crystals in the Dynamic Scattering Mode," *J. Appl. Phys.*, **40,** 4049–4054 (1969).
— S. Durand and Dvgln Rao, "Brillouin Scattering of Light in a Liquid Crystal," *Phys. Lett. A*, **27** (7), 455–456 (1968) [4].
276. Orsay Liquid Crystal Group, "Dynamics of Fluctuations in Nematic Liquid Crystals," *J. Chem. Phys.*, **51** (2), 816–822 (1969).
277. Orsay Liquid Crystal Group, "Quasielastic Rayleigh Scattering in Nematic Liquid Crystals," *Phys. Rev. Lett.* **22** (25), 1361–1363 (1969).
278. M. Bertolotti, B. Daino, F. DiPorto, F. Scudier, and D. Sette, "Light Scattering by Fluctuations Induced by an Applied Electric Field in a Nematic Liquid Crystal (APAPA)," in *Colloque Int. CNRS, Diffusion de la Lumiere par les Fluides*, 1971. *J. Physique Suppl.*, **33**C1, 63 (1972).
279. T. W. Stinson and J. D. Litster, "Pretransitional Phenomena in the Isotropic Phase of a Nematic Liquid Crystal," *Phys. Rev. Lett.*, **25,** 503 (1970).
280. T. W. Stinson, J. D. Litster, and N. A. Clark, "Static and Dynamic Behavior near the Order Disorder Transition of Nematic Liquid Crystals," in *Colloque Int. CNRS, Diffusion de la Lumiere Par les Fluides*, 1971. *J. Physique Suppl.*, **33**C1, 69 (1972).
281. Orsay Liquid Crystal Group, "Quasi Elastic Rayleigh Scattering in a Smectic Liquid Crystal," in *Colloque Int. CNRS, Diffusion de la Lumiere par les Fluides*, 1971. *J. Physique Suppl.*, **33**C1, 76 (1972).
282. D. Langevin and M. A. Bouchiat, "Light Scattering from the Free Surface of a Nematic Liquid Crystal," in *Colloque Int. CNRS, Diffusion de la Lumiere par les Fluides*, 1971. *J. Physique Suppl.*, **33**C1, 77 (1972).
283. M. Bertolotti, B. Davis, F. Scudieri, and D. Sette, "Spatial Distribution of Light Scattered by P Azoxyamsole in Applied Electric Field," *Mol. Cryst. Liq. Cryst.*, **15,** 133 (1971).

284. I. Haller and J. D. Litster, "Temperature Distribution of Normal Modes in a Nematic Liquid Crystal," *Phys. Rev. Lett.*, **25**, 1550 (1970).
285. M. J. Stephen, "Hydrodynamics of Liquid Crystals," *Phys. Rev. A*, **2**, 1558 (1970).
— P. Berge, P. Calmettes, M. Dupois, and C. Loj, "Experimental Observation of the Complete Rayleigh Central Component of the Light Scattered by a Two-Component Fluid," *Phys. Rev. Lett.*, **24** (3), 89–90 (1970) [*3*].

9. *Mixtures and Particles in Solution, Including Flow Measurements*

286. B. J. Berne and H. L. Frisch, "Light Scattering as a Probe of Fast Reaction Kinetics," *J. Chem. Phys.*, **47**, 3675–3676 (1967).
287. L. Blum, "Light Scattering from Reactive Fluids. Part 3. Intensity Calculation," *J. Chem. Phys.*, **51** (11), 5024–5028 (1969).
288. L. Blum and Z. W. Salsburg, "Light Scattering from a Chemically Reactive Fluid. Part 1. Spectral Distribution," *J. Chem. Phys.*, **48**, 2292–2309 (1968).
289. L. Blum and Z. W. Salsburg, "Light Scattering from Chemically Reactive Fluids. Part 2. Case with Diffusion," *J. Chem. Phys.*, **50** (4), 1654–1660 (1969).
290. I. I. Adamenko and I. O. Chernyauska, "Investigation of Brownian Molecular Motion in Liquids with Similar Molecular Structure," *Ukr. Fiz. Zh.*, **11** (3), 336–340 (1966) (in Ukrainian).
291. S. S. Alpert, "Time Dependent Concentration Fluctuations near the Critical Temperature," *Critical Phenomena Conf. Proc., 1965*, M. S. Green and J. V. Seugers, Eds., *Natl. Bur. Std. Misc. Publ.*, **273**, 157–162 (1966) (530.8/G79, 123312).
292. S. S. Alpert, Y. Yeh, and E. Lipworth, "Observation of Time-Dependent Concentration Fluctuations in a Binary Mixture Near the Critical Temperature Using A He–He Laser," *Phys. Rev. Lett.*, **14**, 486–488 (1965).
— M. A. Anisimov, I. M. Arefev, A. V. Voronel et al., "Propagation of Sound Near the Binary Mixture Stratification Point. (Mandelshtam Brillouin Scattering)," *Zh. Eksperim. i Teor. Fiz.*, **61**, 1526–1536 (1971) [*8*].
— I. M. Arefev, "One New Method of Determining the Diffusion Coefficient in Gas Mixtures," *JETP Lett.*, **10** (11), 340–342 (1969) [*3*].
— I. M. Arefev, "Velocity of Hypersound and Dispersion of Speed of Sound Near the Critical Stratification Point of a Binary Solution of Triethylamine in Water," *JETP Lett.*, **7**, 285–287 (1968) [*4*].
— I. M. Arefev and V. N. Biryukov, "Stimulated Mandelstham-Brillouin Scattering in the Region of the Critical Lamination Point of Solutions," *JETP Lett.*, **12** (7), 240–242 (1970) [*11*].
— I. M. Arefev and V. V. Morozov, "Stimulated Concentration Scattering of Light," *JETP Lett.*, **9** (8), 269–271 (1969) [*10*].
293. C. S. Bak and W. I. Goldburg, "Light Scattering in an Impure Binary Liquid Mixture Near the Critical Point," *Phys. Rev. Lett.*, **23** (21), 1218–1220 (1969).
294. C. S. Bak, W. I. Goldburg, and P. N. Pusey, "Light Scattering Study of the Critical Behavior of a Three-Component Liquid Mixture," *Phys. Lett. Rev.*, **25** (20), 1420–1422 (1970).
295. Y. Balta and C. C. Gravatt, "Elastic Scattering of Light by Binary Mixtures above and Below the Critical Solution Temperature. Part 1: Methanol Cyclohexane," *J. Chem. Phys.*, **48**, 3839 (1968).
296. V. G. Baranov, "Small Angle Light Scattered by Ordered Polymer Solutions," *Discuss. Faraday Soc.*, **49**, 137–143 (1970).

— F. Barocchi, M. Mancini, and R. Vallauri, "Volume Relaxation in Carbon Disulfide-Carbon Tetrachloride Mixtures," *J. Chem. Phys.*, **49** (4), 1935–1937 (1968) [*3*].
297. J. P. Boon, "Spectral Analysis of a Fluid under Thermal Constraint," to be published in *Phys. Chem. Liq.* (1972).
298. G. W. Brady, D. McIntyre, M. Myer, and A. Wims, "Critical Scattering of the Perfluoroheptane-1 Octane System," *J. Chem. Phys.*, **44**, 2197–2198 (1966).
299. B. J. Berne and R. Pecora, "Light Scattering as a Probe of Fast Reaction Kinetics: The Depolarized Spectrum of Rayleigh Scattered Light from a Chemically Reacting Medium," *J. Chem. Phys.*, **50** (2), 783–791 (1969).
300. C. Caroli and C. Parodi, "Frequency Spectrum of the Depolarized Light Scattered by a Rigid Molecule in Solution," *J. Phys. B*, **2** (11), 1229–1234 (1969).
— J. P. Chabrat, J. Rouch, and C. Voucamps, "New Measurements on the Anisotropic Rayleigh Scattering of Polar and Nonpolar Liquids and Their Solutions in Cyclohexane," *Compt. Rend. B*, **270** (24), 1556–1558 (1970) (in French) [7].
301. R. F. Chang, P. H. Keyes, J. V. Sengers, and C. O. Adey, "Dynamics of Concentration Fluctuations Near the Critical Point of a Binary Fluid. (3-Methylpentane–Nitroethane)," *Phys. Rev. Lett.*, **27**, 1706–1709 (1971).
302. S. H. Chen and N. Polonsky, "Elastic and Inelastic Scattering of Light from a Binary Mixture Near the Critical Point," *J. Phys. Soc. Japan.*, **26** (Suppl.), 179–182 (1969).
303. S. H. Chen and N. Polonsky, "Intensity Correlation Measurement of Light Scattered From a Two-Component Fluid Near the Optical Mixing Point," *Opt. Commun.*, **1** (2), 64–66 (1969).
304. B. Chu, "Observation of Time-Dependent Concentration Fluctuations in Critical Mixtures," *Phys. Rev. Lett.*, **18**, 200–202 (1967).
305. B. Chu and F. J. Schoenes, "Diffusion Coefficient of the Isobutyric Acid–Water System in the Critical Region," *Phys. Rev. Lett.*, **21**, 6–9 (1968).
306. B. Chu and E. J. Schoenes, "Light Scattering and Pseudospinodal Curves: The Isobutyric Acid–Water System in the Critical Region," *Phys. Rev.*, **185** (1), 219–226 (1969).
307. N. A. Clark, J. H. Lunacek, and G. B. Benedek, "Study of Brownian Motion Using Light Scattering," *Am. J. Phys.*, **38** (5), 575–585 (1970).
308. P. Debye, "Spectral Width of the Critical Opalescence Due to Concentration Fluctuations," *Phys. Rev. Lett.*, **14**, 783–784 (1965).
309. P. G. Degennes, "Critical Opalescence of Macromolecular Solutions," *Phys. Lett. A*, **26**, 313–314 (1968).
310. M. Dubois and P. Berge, "Experimental Study of Rayleigh Scattering Related to Concentration Fluctuations in Binary Solutions: Evidence of a Departure from Ideality," *Phys. Rev. Lett.*, **26** (3), 121–124 (1971).
311. S. B. Dubin, N. A. Clark, and G. B. Benedek, "Measurement of Rotational Diffusion Coefficient of Lysozyme by Depolarized Light Scattering. Configuration of Lysozyme in Solution," *J. Chem. Phys.*, **54**, 5158–5164 (1971).
312. M. Dubois, P. Berge, and C. Loj, "Measurement of the Diffusion Coefficient of Small Molecules by means of Quasi-Elastic Scattering of Light," *Chem. Phys. Lett.*, **6** (3), 227–230 (1970).
313. R. V. Edwards, J. C. Angus, M. J. French, and J. W. Dunning, "Spectral Analysis of the Signal from a Doppler Flowmeter. Time Independent Systems," *J. Appl. Phys.*, **42**, 837–850 (1971).

314. N. C. Ford, F. E. Karasz, and J. E. M. Owen, "Rayleigh Scattering from Polystyrene Solutions," *Discuss Faraday Soc.*, **49,** 228–237 (1970).
315. K. Fritsch, C. J. Montrose, J. L. Hunter, and J. F. Diel, "Relaxation Phenomena in Electrolytic Solutions," *J. Chem. Phys.*, **52** (6), 2242–2252 (1970).
316. S. Fujime, "Quasielastic Light Scattering from Solutions of Macromolecules. Part 1. Doppler Broadening of Light Scattered from Solutions of Tobacco Mosaic Virus Particles," *J. Phys. Soc. Japan*, **29,** 416–430 (1970).
— B. N. Ganguly, "Line Shape of Brillouin Light Spectrum in Helium-3 Helium-2 System," *Phys. Lett. A*, **29** (5), 234-235 (1969) [*4*].
— B. N. Ganguly and A. Griffin, "Scattering of Light from Entropy Fluctuations in Helium-3 and Helium-4 Mixtures," *Can. J. Phys.*, **46** (17), 1895–1904 (1968) [*3*].
317. W. S. Gornall, C. S. Wang, C. C. Yang, and N. Bloembergen, "Coupling Between Rayleigh and Brillouin Scattering in a Disparate Mass Gas Mixture," *Phys. Rev. Lett.*, **26** (18), 1094–1097 (1971).
318. E. H. Hara, A. D. May, and H. F. P. Knaap, "Rayleigh-Brillouin Scattering in Compressed Hydrogen, Deuterium and Hydrogen Deuterium (Mixture)," *Can. J. Phys.* **49** (4), 428–431 (1971).
319. W. Heller, R. Tabibian, M. Nakagaki, and L. Papazian, "Flow Light Scattering. Part 1. Principles of the Effect and Apparatus for its Measurement," *J. Chem. Phys.*, **52,** 4294–4305 (1970).
320. M. Ieda, M. Kosaki, and J. Mezutani, "Effect of Electric Field on Light Scattering of Binary Mixture of Dielectric Liquid," *Japan J. Appl. Phys.*, **9** (9), 1182 (1970).
321. H. C. Kelley, "Velocity Measurement of Small Particles by Photon Counting," *Appl. Phys. Lett.*, **17,** 453–455 (1970).
322. S. Kielich, "Molecular Light Scattering in Dense Mixtures," *Acta Phys. Polon.*, **33,** 63–79 (1968).
323. S. Kielich, "Depolarization of Light Scattering by Atomic and Molecular Solutions with Strongly Anisotropic Translational Orientational Fluctuations," *Chem. Phys. Lett.*, **10,** 516–521 (1971).
324. D. L. Knirk and Z. W. Salsburg, "Light Scattering from Chemically Reactive Fluids. Part 4. Intensity Calculations for the Eulerian Fluid with one Reaction," *J. Chem. Phys.*, **54,** 1251–1270 (1971).
325. S. M. Lo and K. Kawasaki, "Vortex Correction Contribution to the Decay Rate of Concentration Fluctuations in Binary Liquid Critical Mixtures," *Phys. Rev. A*, **5,** 421–424 (1972).
326. H. C. Lucas and D. A. Jackson, "The Intensity and Spectra of Depolarized Light Scattered from Benzene Nitrobenzene Mixtures," *Mol. Phys.* **20,** 801–810 (May 1971).
327. H. Maeda and N. Salto, "Spectral Distribution of Light Scattered from Rod Like Macromolecules in Solution " *J. Phys. Soc. Japan*, **27,** 984–991 (1969).
328. A. G. Marshall and R. Pecora, "Depolarized Light Scattering from a Helix Coil System," *J. Chem. Phys.*, **55,** 1245–1248 (1971).
329. B. N. Miller, "Electromagnetic Theory of Light Scattering From an Inhomogeneous Fluid," *J. Comput. Phys.*, **7** (3), 576–591 (1971).
— G. A. Miller and C. S. Lee, "Brillouin Spectra of Dilute Solutions and the Landau-Placzek Formula," *J. Phys. Chem.*, **72** 4644–4650 (1968) [*4*].
— C. J. Montrose and K. Fritsch, "Hypersonic Velocity and Absorption in Aqueous Electrolytic Solutions," *J. Acoust. Soc. Am.*, **47** (3), 786–790 (1970) [*4*].
330. R. D. Mountain and J. M. Deutch, "Light Scattering from Binary Solutions," *J. Chem. Phys.*, **50,** (3), 1103–1108 (1969).

— C. J. Palin, W. F. Vinen, E. R. Pike, and J. M. Vaughan, "Rayleigh and Brillouin Scattering from Superfluid Helium-3 Helium-4 Mixtures," *J. Phys. C Solid State*, **4,** L225–228 (1971) [*3*].

331. R. Pecora, "Doppler Shifts in Light Scattering from Pure Liquids and Polymer Solutions," *J. Chem. Phys.*, **40,** 1604–1614 (1964).

332. R. Pecora, "Spectral Distribution of Light Scattered by Monodisperse Rigid Rods," *J. Chem. Phys.*, **48,** 4126–4128 (1968).

333. R. Pecora, "Spectral Distribution of Light Scattered from Flexible Coil Molecules," *J. Chem. Phys.*, **49,** 1032–1033 (1968).

334. R. Pecora, "Laser Light Scattering and Macromolecular Brownian Motion," *Nature Phys. Sci.*, **231,** 73–75 (May 24 1971).

335. R. Pecora, "Light Scattering Spectra and Dynamic Properties of Macromolecular Solutions," *Discuss. Faraday Soc.*, **49,** 222–227 (1970).

336. E. R. Pike, D. A. Jackson, P. J. Bourke, and D. I. Page, "Measurement of Turbulent Velocities from the Doppler Shift in Scattered Laser Light," *J. Sci. Instrum. J. Phys. E*, **1,** 727–730 (1968).

— E. R. Pike, J. M. Vaughn, and W. F. Vinen, "Brillouin Scattering from First and Second Sound in a Superfluid Helium-3 Helium-4 Mixture," *Phys. Lett. A*, **30** (7), 373–374 (1969) [*4*].

337. P. N. Pusey and W. I. Goldburg, "Light Scattering Measurement of Concentration Fluctuations in Phenol–Water Near Its Critical Point," *Phys. Rev. A*, **3** (2), 766–776 (1971); *Phys. Rev. Lett.*, **23** (2), 67–70 (1969).

338. V. P. Romanov, V. A. Solovev, and L. S. Filatova, "Correlation Theory of Light Scattering in Solutions," *Opt. Spectry.*, **28,** 447–449 (1970).

339. G. P. Roshchina, E. V. Koncvalov, and S. P. Makarenko, "Spectroscopic Investigation of Rotational Mobility of Molecules of Some Fluoro-Organic Compounds," *Ukr. Fiz. Zh.*, **15** (5), 769–774 (1970) (in Russian).

340. N. Saito and S. Ito, "Rayleigh Linewidth of Light Scattered from Flexible Polymers in Solution," *J. Phys. Soc. Japan*, **25,** 1447–1451 (1969).

341. T. Sato and S. Sakamoto, "Measurement of Dielectric Liquid Motions Under Electrostatic Stress by a Laser Doppler Method," *J. Phys. D Appl. Phys.*, **4,** L53–55 (1971).

342. D. W. Schaefer, G. B. Benedek, P. Schofield, and E. Bradford, "Spectrum of Light Quasielastically Scattered from Tobacco Mosaic Virus," *J. Chem. Phys.*, **55,** 3684 (1971).

343. M. J. Stephen, "Spectrum of Light Scattered from Charged Macromolecules in Solution," *J. Chem. Phys.*, **55,** 3878 (1971).

344. S. P. Stolyov and S. Sokerov, "Transient Electric Light Scattering. Part 1. Method for Determination of Rotational Diffusion Constant," *J. Colloid Interface Sci.*, **24,** 235–240 (1967).

345. M. F. Vuks, "Fluctuations of the Dielectric Constant and Light Scattering in Solutions," *Opt. Spectry.*, **28** (1), 71–74 (1970).

346. Y. Yeh, "Experimental Study of Reaction Kinetics by Light Scattering. Part 2: Helix-Coil Transition of the Copolymer Deoxyadenylate–Deoxythymidylate (DAT)," *J. Chem. Phys.*, **52** (12), 6218–6224 (1970).

347. Y. Yeh and R. N. Keeler, "Experimental Study of Reaction Kinetics by Light Scattering. Part 1. Polarized Rayleigh Component," *J. Chem. Phys.*, **51,** 1120–1127 (1969).

348. Y. Yeh and R. N. Keeler, "Flow Light Scattering. Part 1. Theoretical Reaction Kinetics Probed by Dynamic Light Scattering," *J. Comput-Phys.*, **7,** 566 (1971).

349. R. C. Desai, "Rayleigh-Brillouin Scattering from Fluid Mixtures," in *Colloque Int. CNRS, Diffusion de la Lumiere par les Fluides*, 1971. *J. Physique Suppl.*, **33**C1, 27 (1972).
350. M. Dubois and P. Berge, "Quasi-Elastic Rayleigh Scattering in Binary Mixtures. Non Ideal Solutions. Measurement of Correlation Length," in *Colloque Int. CNRS. Diffusion de la Lumiere par les Fluides*, 1971. *J. Physique Suppl.*, **33**C1, 37 (1972).
351. C. D. Boley and S. Yip, "Spectral Distributions of Light Scattered in Dilute Gases and Gas Mixtures," in *Colloque Int. CNRS, Diffusion de la Lumiere par les Fluides*, 1971. *J. Physique Suppl.*, **33**C1, 43 (1972).
352. W. L. Gornall and C. S. Wang, "Light Scattering from Thermal Fluctuations in Disparate-Mass Gas Mixtures," in *Colloque Int. CNRS, Diffusion de la Lumiere par les Fluides*, 1971. *J. Physique Suppl.*, **33**C1, 51 (1972).
353. W. I. Goldburg and P. N. Pusey, "Light Scattering Study of the Critical Behavior of a Three Component Liquid Mixture," in *Colloque Int. CNRS, Diffusion de la Lumiere par les Fluides*, 1971. *J. Physique Suppl.*, **33**C1, 105 (1972).
354. B. Chu, D. Thiel, W. Tscharnuter, and D. V. Fenby, "Critical Opalescence of Perfluoromethylcyclohexane in Carbon Tetrachloride," in *Colloque Int. CNRS, Diffusion de la lumiere par les Fluides*, 1971. *J. Physique Suppl.*, **33**C1, 111 (1972).
355. P. Calmettes, I. Lagues, and C. Laj, "Measurement of Turbidity and Scattered Intensity in a Binary Critical Mixture," in *Colloque Int. CNRS, Diffusion de la Lumiere par les Fluides*, 1971. *J. Physique Suppl.*, **33**C1, 125 (1972).
356. F. D. Carlson and T. Herbert, "A Study of the Self-Association of Myosin by Intensity Fluctuation Spectroscopy," in *Colloque Int. CNRS, Diffusion de la Lumiere par les Fluides*, 1971. *J. Physique Suppl.*, **33**C1, 157 (1972).
357. P. N. Pusey, D. W. Schaefer, D. E. Koppel, R. D. Camerini-Otero, and R. M. Franklin, "Study of the Diffusion Properties of R17 Virus by Time-Dependent Light Scattering," in *Colloque Int. CNRS, Diffusion de la Lumiere par les Fluides*, 1971. *J. Physique Suppl.*, **33**C1, 163 (1972).
358. Y. Yeh, T. M. Schuster, and D. A. Yphantis, "Quasi-Elastic Light Scattering of Studies of the Kinetics of Lysozyme Dimerization," in *Colloque Int. CNRS, Diffusion de la Lumiere par les Fluides*, 1971. *J. Physique Suppl.*, **33**C1, 169 (1972).
359. R. Nossal and S. H. Chen, "Light Scattering from Motile Bacteria," in *Colloque Int. CNRS, Diffusion de la Lumiere par les Fluides*, 1971. *J. Physique Suppl.*, **33**C1, 171 (1972).
360. S. H. Chen and N. Polonsky, "Anomalous Damping and Dispersion of Hypersound in a Binary Liquid Mixture Near the Solution Critical Point," *Phys. Rev. Lett.*, **20**, 909 (1968).
361. H. Z. Cummins, P. D. Carlson, J. J. Herbert, and G. Woods, "Light Scattering from Tobacco Mosaic Virus," *Biophys. J.*, **9**, 518 (1969).
362. S. B. Dubin, J. H. Lunacek, and G. B. Benedek, "Light Scattering from DNA and Tobacco Mosaic Virus," *Proc. Natl. Acad. Sci., U.S.*, **57**, 1164 (1967).
363. D. W. Schaefer, G. B. Benedek, P. Schofield and E. Bradford, "Spectrum of Light Quasi Elastically Scattered from Tobacco Mosaic Virus," *J. Chem. Phys.*, **55**, 3884 (1971).

10. *Critical Phenomena*

364. A. D. Alekhin, N. P. Kruiskii, and Y. U. B. Mynchenko, "Temperature Dependence of the Isothermal Compressibility Near the Liquid Vapor Critical Point," *Ukr. Fiz. Zh.*, **15** (3), 509–510 (1970) (in Russian).

— S. S. Alpert, "Time Dependent Concentration Fluctuations Near the Critical Temperature," in *Critical Phenomena Conf. Proc., 1965*, M. S. Green and J. V. Seugers, *Natl. Bur. Std. Misc. Publ.* **273,** pp. 157–162, (1966) (530.8/G79, 123312) [*9*].

— M. A. Anisimov, I. M. Arefev, A. V. Voronel et al., "Propagation of Sound Near the Binary Mixture Stratification Point. (Mandelshtam Brillouin Scattering)," *Zh. Eksperim. i Teor. Fiz.*, **61,** 1526–1536 (1971) [*8*].

— I. M. Arefev, "Velocity of Hypersound and Dispersion of Speed of Sound Near the Critical Stratification Point of a Binary Solution of Triethylamine in Water," *JETP Lett.*, **7,** 285–287 (1968) [*4*].

— I. M. Arefev and V. N. Biryukov, "Stimulated Mandelshtam-Brillouin Scattering in the Region of the Critical Lamination Point of Solutions," *JETP Lett.*, **12** (7), 240–242 (1970) [*11*].

365. I. M. Arefev and V. V. Morozov, "Stimulated Concentration Scattering of Light," *JETP Lett.*, **9** (8), 269–271 (1969).

— C. S. Bak and W. I. Goldburg, "Light Scattering in an Impure Binary Liquid Mixture Near the Critical Point," *Phys. Rev. Lett.*, **23** (21), 1218–1220 (1969) [*9*].

— C. S. Bak, W. I. Goldburg, and P. N. Pusey, "Light Scattering Study of the Critical Behavior of a Three-Component Liquid Mixture," *Phys. Rev. Lett.*, **25** (20), 1420–1422 (1970) [*9*].

— Y. Balta and C. C. Gravatt, "Elastic Scattering of Light by Binary Mixtures above and Below the Critical Solution Temperature. Part 1: Methanol Cyclohexane," *J. Chem. Phys.*, **48,** 3835 (1968) [*9*].

— G. W. Brady, D. McIntyre, M. Myer, and A. Wims, "Critical Scattering of the Perfluoroheptane-1 Octane System," *J. Chem. Phys.*, **44,** 2197–2198 (1966) [*9*].

— P. Braun, D. Hammer, W. Tscharnuter, and P. Weinzierl, "Rayleigh Scattering by Sulfur Hexafluoride in the Critical Region," *Phys. Lett. A*, **32** (6), 390–391 (1970) [*3*].

— D. S. Cannell and G. B. Benedek, "Brillouin Spectrum of Xenon Near its Critical Point," *Phys. Rev. Lett.*, **25** (17), 1157–1161 (1970) [*4*].

— A. V. Chalyi and A. D. Alexhin, "Investigation of Light Scattering with Allowance for Gravitational Effect by Scaling Method and Determination of the Critical Indices on Basis of Light Scattering Data," *Zh. Ekspermin. i Teor. Fiz.*, **59** (2), 337–345 (1970) (in Russian) [*3*].

— R. F. Chang, P. H. Keyes, J. V. Sengers, and C. O. Adey, "Dynamics of Concentration Fluctuations Near the Critical Point of a Binary Fluid. (3-Methylpentane-Nitroethane)," *Phys. Rev. Lett.*, **27,** 1706–1709 (1971) [*9*].

— S. H. Chen and N. Polonsky, "Elastic and Inelastic Scattering of Light from a Binary Mixture Near the Critical Point," *J. Phys. Soc. Japan*, **26** (Suppl.), 179–182 (1969) [*9*].

— S. H. Chen and N. Polonsky "Intensity Correlation Measurement of Light Scattered from a Two-Component Fluid Near the Critical Mixing Point," *Opt. Commun.*, **1** (2), 64–65 (1969) [*9*].

— B. Chu, "Observation of Time-Dependent Concentration Fluctuations in Critical Mixtures," *Phys. Rev. Lett.*, **18,** 200–202 (1967) [*9*].

— B. Chu and F. J. Schoenes, "Diffusion Coefficient of the Isobutyric Acid-Water System in the Critical Region," *Phys. Rev. Lett.*, **21,** 6–9 (1968) [*9*].

— B. Chu and F. J. Schoenes, "Light Scattering and Pseudospinodal Curves: The Isobutyric Acid-Water System in the Critical Region," *Phys. Rev.*, **195** (1), 219–226 (1969) [*9*].

— H. Z. Cummins and H. L. Swinney, "Critical Opalescence: The Rayleigh Linewidth," *J. Chem. Phys.*, **45** (12), 4438–4444 (1966) [*3*].

— H. Z. Cummins and H. L. Swinney, "Dispersion of the Velocity of Sound in Xenon in the Critical Region," *Phys. Rev. Lett.*, **25** (17), 1165–1169 (1970) [*4*].

— P. Debye, "Spectral Width of the Critical Opalescence Due to Concentration Fluctuations," *Phys. Rev. Lett.*, **14**, 783–784 (1965) [*9*].

— P. G. Degennes, "Critical Opalescence of Macromolecular Solutions," *Phys. Lett. A*, **26**, 13–14 (1968) [*9*].

— G. T. Feke, G. A. Hawkins, J. B. Lastovka, and G. B. Benedek, "Spectrum and Intensity of the Light Scattered from Sulfur Hexafluoride Along the Critical Isochore," *Phys. Lett. Rev.*, **27**, 1780–1783 (Dec. 27, 1971) [*3*].

366. M. Fixman, "Ultrasonic Attenuation in the Critical Region," *J. Chem. Phys.*, **33**, 1363–1370 (1960).

— N. C. Ford and G. B. Benedek, "Observation of the Spectrum of Light Scattered From a Pure Fluid Near Its Critical Point," *Phys. Rev. Lett.*, **15**, 649–653 (1965) [*3*].

— N. C. Ford and G. B. Benedek, "Spectrum of Light Inelastically Scattered by a Fluid Near its Critical Point," in *Conf. Phenom. Neighborhood Critical Points, 1965, Proc., Natl. Bur. Std. Misc. Publ.*, **273**, 150–156 (1966) [*3*].

— N. C. Ford, K. H. Langley, and V. G. Puglielli, "Brillouin Linewidths in Carbon Dioxide Near the Critical Point," *Phys. Rev. Lett.*, **21**, 9–12 (1968) [*4*].

367. H. L. Frisch and G. W. Brady, "Small Angle Critical Scattering," *J. Chem. Phys.*, **37**, 1514–1521 (1962); *Bell Syst. Monogr.*, **4366.**

— R. W. Gammon, H. L. Swinney, and H. Z. Cummins, "Brillouin Scattering in Carbon Dioxide in the Critical Region," *Phys. Rev. Lett.*, **19**, 1467–1469 (1967) [*4*].

— M. Giglio and G. B. Benedek, "Angular Distribution of the Intensity of Light Scattered from Xenon Near Its Critical Point," *Phys. Rev. Lett.*, **23**, 1145–1149 (1969) [*3*].

— D. L. Henry, H. L. Swinney, and H. Z. Cummins, "Rayleigh Linewidth in Xenon Near the Critical Point," *Phys. Rev. Lett.*, **25** (17), 1170–1173 (1970) [*3*].

— J. S. Huang and W. W. Webb, "Viscous Damping of Thermal Excitations on the Interface of Critical Fluid Mixtures," *Phys. Rev. Lett.*, **23** (4), 160–163 (1969) [*6*].

368. Y. U. K. Kolpakov and V. P. Skripov, "Measurement of the Degree of Depolarization of Scattered Light in the Vicinity of the Carbon Dioxide Critical Point Using a Helium-Neon Laser," *Opt. Spectry.*, **29**, 405–407 (Oct. 1970).

369. P. Lambropoulos, S. Kern and R. K. Mueller, "Theory of Stimulated Brillouin Scattering with Stokes Feedback," *IEEE J. Quantum Electron*, **2** (9), 649–658 (1966); *Int. Quantum Electron Conf. 1966, 46th Proc.*

— T. K. Lim, H. L. Swinney, K. H. Langley, and T. A. Kachnowski, "Rayleigh Linewidth in Sulfur Hexafluoride Near the Critical Point," *Phys. Rev. Lett.*, **27**, 1776–1780 (Dec. 27, 1971) [*3*].

— S. M. Lo and K. Kawasaki, "Vortex Correction Contribution to the Decay Rate of Concentration Fluctuations in Binary Liquid Critical Mixtures," *Phys. Rev. A*, **5**, 421–424 (1972) [*9*].

— R. Mohr, K. H. Langley, and N. C. Ford, "Brillouin Scattering from Sulfur Hexafluoride in the Vicinity of the Critical Point," *J. Acoust. Soc. Am.*, **49** (3, PT. 3), 1030–1032 (1971) [*4*].

— M. Papoular, "Critical Surface Opalescence," *Compt. Rend. B*, **266**, 592–595 (1968) (in French) [*6*].

370. D. Pohl and W. Kaiser, "Time-Resolved Investigations of Stimulated Brillouin Scattering in Transparent and Absorbing Media; Determination of Phonon Lifetimes," *Phys. Rev. B*, **1** (1), 31–43 (1970).

371. D. Pohl, M. Maier, and W. Kaiser, "Phonon Lifetimes Measured in Amplifiers for Brillouin Radiation," *Phys. Rev. Lett.*, **20** (8), 366–369 (1968).
— P. N. Pusey and W. I. Goldburg, "Light Scattering Measurement of Concentration Fluctuations in Phenol–Water Near its Critical Point," *Phys. Rev. A*, **3** (2), 766–776 (1971): *Phys. Rev. Lett.*, **23** (2), 67–70 (1969) [*9*].
372. J. P. Webb, "Critical Opalescent Light Scattering in Helium-3," Thesis, Stanford University, 1968, 164 pp. (Univ. Microfilms Order No. 68-8294).
— Y. Yeh, "Observation of the Long Range Correlation Effect in the Rayleigh Linewidth Near the Critical Point of Xenon," *Phys. Rev. Lett.*, **18,** 1043–1046 (1967) [*3*].
— J. Zollweg, G. A. Hawkins, and G. B. Benedek, "Surface Tension and Viscosity of Xenon Near its Critical Point. (Optical Inelastic Scattering Method)," *Phys. Rev. Lett.*, **27,** 1182–1185 (1971) [*6*].
— H. L. Swinney, D. L. Henry, and H. Z. Cummins, "The Rayleigh Linewidth in Carbon Dioxide and Xenon near the Critical Point," in *Colloque Int. CNRS, Diffusion de la Lumiere par les Fluides*, 1971. *J. Physique Suppl.*, **33**C1, 81 (1972) [*3*],
— D. S. Cannell and J. H. Lunacek, "Long Range Correlation Length and Isothermal Compressibility of Carbon Dioxide Near its Critical Point," in *Colloque Int. CNRS, Diffusion de la Lumiere par les Fluides*, 1971. *J. Physique Suppl.*, **33**C1, 91 (1972) [*3*].
— R. Mohr and K. H. Langley, "Brillouin Scattering from Sulfur Hexafluoride in the Vicinity of the Critical Point," in *Colloque Int, CNRS, Diffusion de la Lumiere pvr les Fluides*, 1971. *J. Physique Suppl.*, **33**C1, 97 (1972) [*4*].
— W. I. Goldburg, and P. N. Pusey, "Light Scattering Study of the Critical Behavior of a Three Component Liquid Mixture," in *Colloque Int. CNRS, Diffusion de la Lumiere par les Fluides*, 1971. *J. Physique Suppl.*, **33**C1, 105 (1972) [*9*].
— B. Chu, D. Thiel, W. Tscharnuter, and D. V. Fenby, "Critical Opalescence of Perfluoromethylcyclohexane in Carbon Tetrachloride," in *Colloque Int. CNRS, Diffusion de la Lumiere par les Fluides*, 1971. *J. Physique Suppl.*, **33**Cl, 111 (1972) [*9*].
— P. Calmettes, I. Lagues, and C. Laj, "Measurement of Turbidity and Scattered Intensity in a Binary Critical Mixture," in *Colloque Int. CNRS, Diffusion de la Lumiere par les Fluides*, 1971. *J. Physique Suppl.*, **33**C1, 125 (1972) [*9*].
373. J. Zollweg, G. Hawkins, I. W. Smith, M. Giglio, and G. B. Benedek, "On the Spectrum and Intensity of Light Scattered from the Bulk and the Interface of Xenon near its Critical Point," in *Colloque Int. CNRS, Diffusion de la Lumiere par les Fluides*, 1971. *J. Physique Suppl.*, **33**C1, 135 (1972).
— M. A. Bouchiat and J. Meunier, "Light Scattering from Surface Waves on Carbon Dioxide near the Critical Point," in *Colloque Int. CNRS, Diffusion de la Lumiere par les Fluides*, 1971. *J. Physique Suppl.*, **33**C1, 141 (1972) [*9*].
— E. S. Wu and W. W. Webb, "Light Near the Critical Point Scattered by the Liquid Vapor Interface of Sulfur-Hexafluoride," in *Colloque Int. CNRS, Diffusion de la Lumiere par les Fluides*, 1971. *J. Physique Suppl.*, **33**C1, 149 (1972) [*6*].
— R. D. Mountain, "Depolarized Scattering and Long Range Correlations for a Simple Fluid Near the Critical Point," in *Colloque Int. CNRS, Diffusion de la Lumiere par les Fluides*, 1971. *J. Physique Suppl.*, **33**C1, 265 (1972) [*5*].
374. I. W. Smith, M. Giglio, and G. B. Benedek, "Correlation Range and Compressibility of Xenon Near its Critical Point," *Phys. Rev. Lett.*, **27,** 1556 (1971).
375. M. Giglio and G. B. Benedek, "Angular Distribution of the Intensity of Scattered Light from Xenon Near the Critical Point," *Phys. Rev. Lett.*, **23,** 1145 (1969).

376. J. H. Lunacek and D. S. Cannell, "Long Range Correlation and Isothermal Compression of Carbon Dioxide near the Critical Point," *Phys. Rev. Lett.*, **27**, 841 (1971).

— S. H. Chen and N. Polonsky, "Anomalous Damping and Dispersion of Hypersound in a Binary Liquid Mixture near the Solution Critical Point," *Phys. Rev. Lett*, **20**, 909 (1968) [*9*].

377. V. G. Puglielli and N. C. Ford, "Turbidity Measurements in Sulfur Hexafluoride Near the Critical Point," *Phys. Rev. Lett.*, **25**, 143 (1970).

11. *Stimulated Scattering*

378. V. I. Bespalov and A. M. Kubarev, "Stimulated Rayleigh Scattering of Light in Liquid Solutions," *Soviet Phys. JETP*, **6** (2), 31–33 (1967).

379. V. I. Bespalov, A. M. Kubarev, and G. A. Pasmanik, "Stimulated Rayleigh Light Scattering," *Izv. Vuz. Radiofiz*, **13** (10), 1433–1466 (1970) (in Russian).

380. V. I. Bespalov, A. M. Kubarev, and G. A. Pasmanik, "Spectral Investigations of the Stimulated Mandelshtam Brillouin Scattering and Stimulated Temperature Light Scattering in Liquids," *Izv. Vuz. Radiofiz*, **14**, 1514–1517 (1971) (in Russian).

381. V. I. Bespalov and G. A. Pasmanik, "Stimulated Mandelshtam-Brillouin and Stimulated Entropy Back Scattering of Light Pulses," *Soviet Phys. JETP*, **31**, 168–174 (1970).

382. V. S. Starunov and I. L. Fabelinskii, "Stimulated Mandelshtam-Brillouin Scattering and Stimulated Entropy (Temperature) Scattering of Light," *Soviet Phys. Usp.*, **12** (4), 463–489 (1970).

383. N. Bloembergen and P. Lallemand, "Complex Intensity-Dependent Index of Refraction, Frequency Broadening of Stimulated Raman Lines and Stimulated Rayleigh Scattering," *Phys. Rev. Lett.*, **16** (3), 81–84 (1966).

384. N. Bloembergen, W. H. Lowdermilk, M. Matsuoka, and C. S. Wang, "Theory of Stimulated Concentration Scattering," *Phys. Rev. A*, **3** (1), 404–412 (1971).

385. I. I. Abrikosova and O. M. Bochkova, "Breakdown of Liquid and Gaseous Helium by a Laser Beam and Observation of Stimulated Mandelshtam Brillouin Scattering in Liquid Helium," *JETP Lett.*, **9** (5), 1679 (1969).

386. I. L. Abrikosova and N. G. Skrypnik, "Breakdown by a Laser Beam and Stimulated Mandelstam Brillouin Scattering in Liquid Helium-4," *Zh. Eksperim. i Teor. Fiz.*, **59** (1), 59–63 (1970) (in Russian).

387. A. J. Alcock and C. De Micheils, "Nanosecond Pulse Generation by Means of Stimulated Brillouin Scattering," *Appl. Phys. Lett.*, **11** (6), 185–186 (1967).

388. I. M. Arefev and V. N. Biryukov, "Stimulated Mandelshtam-Brillouin Scattering in the Region of the Critical Lamination Point of Solutions," *JETP Lett.*, **12** (7), 240–242 (1970).

389. F. Barocchi, "Stimulated Brillouin Scattering: Phonon-Frequency Dependence of Steady State Gain in Dispersive Liquids," *J. Appl. Phys.*, **40** (7), 2867–2873 (1969).

390. F. Barocchi and M. Zoppi, "Angular Distribution of Stimulated Brillouin Scattering Gain," *Opt. Commun.*, **3**, 335–339 (July 1971).

391. I. P. Batra and R. H. Enns, "Stimulated Thermal Scattering in Isotropic Media. Parts 1 & 2," *Can. J. Phys.*, **47** (12), 1283–1290 (1969); **47** (16), 1745–1750 (1969).

392. I. P. Batra and R. H. Enns, "Stimulated Thermal Scattering of Light," *Phys. Stat. Solidi*, **49**, 11–63 (1971).

393. I. P. Batra and R. H. Enns, "Stimulated Thermal Rayleigh Scattering in Liquids," *Phys. Rev.* **185** (1), 396–399 (1969).

394. I. P. Batra and R. H. Enns, "Value of the Critical Absorption Coefficient in Stimulated Thermal Rayleigh Scattering," *J. Chem. Phys.*, **51** (4), 1668–1669 (1969).
395. A. I. Bozhkov and F. V. Bunkin, "Stimulated Light Scattering at the Surface of a Highly Viscous Liquid," *Zh. Eksperim. i Teor. Fiz.*, **56** (6), 1976–1978 (1969) (in Russian).
396. A. I. Bozhkov, F. V. Bunkin, and M. V. Feborov, "Stimulated Light Scattering From the Surface of a Liquid of Arbitrary Viscosity," *Opt. Spectroskopiya*, **28** (1), 58–62 (1970).
397. R. G. Brewer, "Frequency Shifts in Self-Focussed Light," *Phys. Rev. Lett*, **19** (1), 8–9 (1967).
398. R. G. Brewer, "Growth of Optical Plane Waves in Stimulated Brillouin Scattering," *Phys. Rev. A*, **140** (3), 800–803 (1965).
399. R. G. Brewer, "Stimulated Brillouin Shifts by Optical Beats," *Appl. Phys. Lett.*, **9** (1), 51–53 (1966).
400. R. G. Brewer and B. Rieckhoff, "Stimulated Brillouin Scattering in Liquids," *Phys Rev. Lett.*, **13**, 334–336 (1964).
401. R. G. Brewer and D. C. Shapero, "Multiple Stimulated Brillouin Scattering," in *Physics of Quantum Electronics*, McGraw-Hill, New York, 1966, pp. 216–222.
402. E. Burlefinger and H. Puell, "Intensive Stimulated Brillouin Scattering in a Parallel Laser Beam," *Phys. Lett.*, **15**, 313–314 (1965).
403. R. Y. Chiao, "Polarization Dependence of Stimulated Rayleigh-Wing Scattering and the Optical Frequency Kerr Effect," *Phys. Rev.*, **185**, (2), 430–445 (1969).
404. C. W. Cho, N. D. Foltz, D. H. Rank, and T. A. Wiggins, "Stimulated Rayleigh Scattering," *Phys. Rev. Lett.*, **18** (4), 107–109 (1967).
405. C. W. Cho, N. D. Foltz, D. H. Rank, and T. A. Wiggins, "Stimulated Thermal Rayleigh Scattering in Liquids," *Phys. Rev.*, **175** (1), 271–274 (1968).
406. A. A. Chopan, "Concerning Induced Mandelshtam-Brillouin Scattering," *JETP Lett.*, **3** (2), 45–47 (1966).
407. G. W. Cohen-Solal, "Study of Ultrasonic Generation by Stimulated Brillouin Effect," *Comet. Rend. B*, **267** (26), 1431–1434 (1968) (in French).
408. M. J. Colles, "Efficient Stimulated Raman Scattering from Picosecond Pulses," *Opt. Commun.*, **1** (4), 169–172 (1969).
409. K. Daree and W. Kaiser, "Competition Between Stimulated Brillouin and Rayleigh Scattering in Absorbing Media," *Phys. Rev. Lett.*, **26**, 816–819 (April 5, 1971).
410. M. Denariez and G. Bret, "Investigation of Rayleigh Wings and Brillouin Stimulated Scattering in Liquids," *Phys. Rev.*, **171** (1), 160–171 (1968).
411. M. Denariez and G. Bret, "Measurements of Response in the Stimulated Rayleigh Effect," *Compt. Rend. B*, **265** (2), 144–147 (1967) (in French).
412. S. Du Martin, B. Oksemgorm, and B. Vodar, "Depolarization of Light Emitted by Stimulated Brillouin Effect in Compressed Gaseous Nitrogen," *Compt. Rend. B*, **268** (6), 471–474 (1969) (in French).
413. S. Dumartin, B. Oksengorn, and B. Vodar, "Stimulated Brillouin Effect in Compressed Gases and Inverted Brillouin Effect (in Absorption) in Acetone," *Compt. Rend, B*, **262** (26), 1680–1683 (1966) (in French).
414. Y. U. E. Dyakov, "Estimate of Line Width of Stimulated Mandelshtam-Brillouin and Raman Scattering of Light in Saturation," *JETP Lett.*, **10** (11), 347–350 (1969).
415. Y. U. E. Dyakov, "Influence of Nonmonochromatic Pumping on the Form of the Spectrum of Stimulated Mandelshtam-Brillouin Scattering," *JETP Lett.* **9** (8), 296–298 (1968).

416. J. L. Emmett and A. L. Schawlow, "Transverse Stimulated Emission in Liquids (Brillouin)," *Phys. Rev.*, **170** (2), 358–362 (1968).
417. R. H. Enns, "Approach to Steady State for Stimulated Thermal Rayleigh Scattering," *Can. J. Phys.*, **48** (6), 710–715 (1970).
418. R. H. Enns and S. S. Rangnekar, "Resolution of the Critical Absorption Coefficient Anomaly in Stimulated Thermal Rayleigh Scattering," *Phys. Lett., A*, **34** (4), 249–250 (1971).
419. I. L. Fabelinskii, D. I. Mash, V. V. Morozov and V. S. Starunov, "Stimulated Scattering of Light in Hydrogen Gas at Low Pressures," *Phys. Lett. A*, **27** (5), 253–254 (1968).
420. N. D. Foltz, C. W. Cho, D. H. Rank and T. A. Wiggins, "Stimulated Rayleigh Scattering in Liquids," *Phys. Rev.*, **165** (2), 396–400 (1968).
421. E. Garmire and C. H. Townes, "Stimulated Brillouin Scattering in Liquids," *Appl. Phys. Lett.*, **5**, 84–86 (1964).
422. C. R. Giuliano, "Time Resolved Interferometry in Stimulated Brillouin Scattering," *Appl. Phys. Lett.*, **7** (10), 279–281 (1965).
423. N. Goldblatt, "Stimulated Brillouin Scattering," *Appl. Opt.*, **8** (8), 1559–1566 (1969).
424. N. Goldblatt and M. Mercher, "Stimulated Brillouin Scattering Origins of Anti-Stokes Components," *Phys. Rev. Lett.*, **20** (7), 310–314 (1968).
425. N. Goldblatt and T. A. Litovitz, "Stimulated Brillouin Scattering: Measurement of Hypersonic Velocity in Liquids," *J. Acoust. Soc. Am.*, **41** (5), 1301–1307 (1967).
426. A. L. Golger, "Competition Between Stimulated Mandelshtam Brillouin Scattering and Raman Scattering in Liquids," *Vestn. Mosk. Univ. Fiz. Astron*, **1970**, (6), 693–698 (1970) (in Russian).
427. A. Z. Grasyuk, V. I. Popovichev, and V. V. Ragulskii, "Increase of Radiation Brightness in a Brillouin Laser," *JETP Lett.*, **12**, 193–195 (1970).
428. M. A. Gray, "Theory of Stimulated Thermal Rayleigh Scattering in Fluids," Thesis, Pennsylvania State University, 1969, 86 pp. (Univ. Microfilms 69-9760); *Phys. Abstr.*, **73**, 39484.
429. M. A. Gray and R. M. Herman, "Nonlinear Thermal Rayleigh Scattering in Gases," *Phys. Rev.*, **181** (1), 374–378 (1969).
— H. Grimm and K. Dransfeld, "Scattering of Laser Light in Superfluid Helium," *Z. Naturforsch. A*, **22**, 1629–1630 (1967) [*4*].
430. K. Groe, "Theory of Stimulated Brillouin Scattering in Liquids," *Z. Phys.*, **201** (1), 59–68 (1967) (in German).
431. B. S. Guberman and V. V. Morozov, "Induced Mandelshtam-Brillouin Scattering in Carbon Dioxide Near the Critical Point," *Opt. Spectry.*, **22** (4), 368 (1967).
432. E. E. Hagenlocker, R. W. Minck, and W. G. Rado, "Effects of Phonon Lifetime on Stimulated Optical Scattering in Gases," *Phys. Rev.*, **154** (2), 226–233 (1967).
433. W. Heinicke, G. Winterling, and K. Dransfeld, "Low-Temperature Applications of the Stimulated Brillouin Scattering," *J. Acous. Soc. Am.*, **49** (3), 954 (1971).
434. R. M. Herman, "Saturation Effects in Stimulated Rayleigh-Wing Scattering," *Phys. Rev.*, **164** (1), 200–206 (1967).
435. R. M. Herman and M. A. Gray, "Theoretical Prediction of the Stimulated Thermal Rayleigh Scattering in Liquids," *Phys. Rev. Lett.*, **19** (15), 824–828 (1967).
436. K. Inoue, "Study on Stimulated Brillouin Scattering by Varying Both Phonon Lifetime and Pulse Duration of Laser," *Japan. J. Appl. Phys.*, **9** (11), 1347–1355 (1970).

437. K. Inoue and T. Yajima, "Temperature Dependence of Stimulated Brillouin Scattering in Quartz," *Japan. J. Appl. Phys.*, **6** (11), 1346–1347 (1967).
438. T. Ito and H. Takuma, "Interference Between Stimulated Brillouin and Raman Scattering," in *Physics of Quantum Electronics*. McGraw-Hill, New York, 1966, pp. 200–206.
439. T. Ito and H. Takuma, "Threshold Condition of the Small and Large Angle Stimulated Brillouin Scattering," *J. Phys. Soc. Japan*, **24** (4), 965 (1968).
440. T. S. Jaseja, P. C. Pande, B. Kumar, and V. Parkash, "Stimulated Brillouin Scattering in Liquids of High Viscosities and Cyclohexane," *IEEE J. Quantum Electron.*, **7**, 537 (1971).
441. S. Jorma, "Atmospheric Depolarization and Stimulated Brillouin Scattering," *Appl. Optics*, **10**, 2661–2664 (1971).
442. W. Kaiser, "Quantitative Investigations of Stimulated Brillouin Scattering," *J. Acoust. Soc. Am.*, **49**, 959–963 (March Pt. 3, 1971).
443. R. N. Keeler, G. H. Bloom, and A. C. Mitchell, "Stimulated Brillouin Scattering in Shock Compressed Fluids," *Phys. Rev. Lett.*, **17** (16), 852–854 (1966).
444. V. V. Korcbkin, D. I. Mash, V. V. Morozov, I. D. Fabelinskii, and M. Y. A. Shchelev, "Development of Stimulated Mandelshtam-Brillouin Scattering with Time in Nitrogen Gas at 150 Atmospheres," *JETP Lett.*, **5** (10), 307–309 (1967).
445. Y. U. I. Kyzylasov and V. S. Stanunov, "Amplification of the Anti-Stokes and Stokes Components of Stimulated Mandelshtam-Brillouin Scattering as a Result of Four-Phonon Interaction," *Soviet Phys. JETP*, **7** (5), 123–125 (1968).
446. Y. U. I. Kyzylasov and V. S. Starunov, "Observation of Ultrashort Radiation Pulses in Stimulated Scattering of Light in the Rayleigh Line Wing," *JETP Lett.*, **9** (12), 401–403 (1969).
447. Y. U. I. Kyzylasov, V. S. Starunov, and I. L. Fabelinskii, "Stimulated Scattering of Light of the Rayleigh Line Wing in an External Resonator," *JETP Lett.*, **9** (7), 227–229 (1969).
448. Y. U. I. Kyzylasov, V. S. Starunov, and I. L. Fabelinskii, "Stimulated Mandelshtam-Brillouin Scattering and Fracture of Glasses by Giant Ruby Laser Pulses," *Fiz. Tverd. Tela*, **12** (1), 233–239 (1970) (in Russian).
449. N. N. Lavrinovich, "Stationary Theory of Induced Mandelshtam-Brillouin Scattering in a Medium with Weak Linear Acoustic Damping," *Zh. Eksperim. i Teor. Fiz.* **60** (1), 69–72 (1971) (in Russian).
450. M. E. Mack, "Stimulated Thermal Light Scattering in the Picosecond Regime," *Phys. Rev. Lett.*, **22** (1), 13–15 (1969).
451. W. K. Madigosky, A. Monkewicz, and T. A. Litovitz, "Stimulated Brillouin Scattering: Measurement of Hypersonic Velocities in Gases," *J. Acoust. Soc. Am.*, **41** (5), 1308–1311 (1967).
452. M. Maier, "Quasisteady State in the Stimulated Brillouin Scattering of Liquids," *Phys. Rev.*, **166** (1), 113–119 (1968).
453. M. Maier, O. Rahn, and G. Wendl, "Self-Focussing of Laser Light and Stimulated Scattering Processes in Transparent and Absorbing Liquids: Part 1: Self-Focussing," *Z. Naturforsch. A*, **25** (12), 1868–1879 (1970) (in German).
454. M. Maier and G. Renner, "Transient Threshold Power of Stimulated Brillouin Raman Scattering," *Phys. Lett. A*, **34**, 299–300 (April 5, 1971).
455. M. Maier, W. Rother, and W. Kaiser, "Time-Resolved Measurements of Stimulated Brillouin Scattering," *Appl. Phys. Lett.*, **10** (3), 80–82 (1967).
456. D. I. Mash, V. V. Morozov, V. S. Starunov, E. V. Tiganov, and I. L. Fabelinskii, "Induced Mandelshtam-Brillouin Scattering in Solid Amorphous Bodies and Liquids," *JETP Lett.*, **2** (5), 157–160 (1965).

457. D. I. Mash, V. V. Morozov, V. S. Starunov, and I. L. Fabelinskii, "Induced Mandelshtam-Brillouin Scattering in Gases," *JETP Lett.*, **2** (12), 349–351 (1965).
458. D. I. Mash, V. V. Morozov, V. S. Starunov, and I. L. Fabelinskii, "Stimulated Scattering of Light of the Rayleigh-Line Wing," *JETP Lett.*, **2** (1), 25–27 (1965).
459. D. I. Mash, V. S. Starunov, E. V. Tiganov, and I. L. Fabelinskii, "Intensity and Width of the Components of the Fine Structure Line of Scattered Light in Liquids and the Damping of Hypersound," *Zh. Eksperim. i Teor. Fiz.*, **49** (6), 1764–1773 (1965) (in Russian).
460. R. W. Minck, E. E. Hagenlocker, and W. G. Rado, "Simultaneous Occurrence of and Competition Between Stimulated Optical Scattering Processes in Gases," *J. Appl. Phys.*, **38** (5), 2254–2260 (1967).
461. A. S. Pine, "Stimulated Brillouin Scattering in Liquids," *Phys. Rev.*, **149**, 113–117 (1966).
462. D. Pohl, L. Reinhold, and W. Kaiser, "Experimental Observation of Stimulated Thermal Brillouin Scattering," *Phys. Rev. Lett.*, **20** (21), 1141–1143 (1968).
463. A. L. Pohyakova, "Elastic Nonlinearity in Stimulated Mandelshtam-Brillouin Scattering," *JETP Lett.*, **7** (2), 57–60 (1968).
464. D. H. Rank, "Some Stimulated Effects in Nonlinear Optics," *J. Opt. Soc. Am.*, **60** (4), 433–444 (1970).
465. D. H. Rank, C. W. Cho, N. D. Foltz, and T. A. Wiggins, "Stimulated Thermal Rayleigh Scattering," *Phys. Rev. Lett.*, **19** (15), 828–830 (1967).
466. D. H. Rank, T. A. Wiggins, D. R. Dietz, N. D. Faltz, and C. W. Cho, "Stimulated Scattering in Liquids and Gases," Pennsylvania State University, 1969, 78 pp. (AD63 4880); *Sci. Abstr. Sect. A* **73,** 39676 (1970).
467. D. H. Rank, T. A. Wiggins, R. V. Wick, D. P. Eastman, and A. H. Guenther, "Stimulated Brillouin Effect in High-Pressure Gases," *J. Opt. Soc. Am.*, **56** (2), 174–176 (1966).
468. Dvgln Rao, "Stimulated Brillouin Scattering in a Liquid Crystal," *Phys. Lett. A*, **32** (7), 533–534 (1970).
469. W. Rother and W. Kaiser, "Time and Frequency Dependence of Stimulated Thermal Rayleigh Scattering," *Phys. Rev. Lett.*, **22** (18), 915–918 (1969).
470. W. Rother, H. Meyer, and W. Kaiser, "Angular Dependence of Stimulated Thermal Rayleigh Scattering," *Phys. Lett. A*, **31** (5), 245–246 (1970).
471. W. Rother, H. Meyer, and W. Kaiser, "Amplification of Light in Absorbing Media: Stimulated Thermal Rayleigh Scattering," *Z. Naturforsch. A*, **25,** 1136–1143 (July 1970).
472. C. A. Sacchi, "Stimulated Scattering in the Far Wing of the Rayleigh Line and Low Frequency Raman Lines in Liquids," *Opt. Commun.* **4,** 83–87 (1971).
473. T. T. Saito, L. M. Peterson, D. H. Rank, T. A. Wiggins, "Measurement of Hypersound Speed in Gases by Stimulated Brillouin Scattering," *J. Opt. Soc. Am.*, **60** (6), 749–755 (1970), *Ann. Meet. Opt. Soc. Am., 1969*, p. 28.
474. Y. R. Shen and Y. J. Shaham, "Self Focussing and Stimulated Raman and Brillouin Scattering in Liquids," *Phys. Rev.* **168** (2), 224–231 (1967).
475. V. S. Starunov, "Problems Pertaining to the Theory of Stimulated Molecular Scattering of Light," *Zh. Eksierim. i Teor. Fiz.*, **57** (3), 1012–1023 (1969) (in Russian).
476. V. S. Starunov, "Theory of Stimulated Rayleigh-Wing Scattering," *Soviet Phys. Doklady*, **13** (3), 217–219 (1968).
477. T. S. Stepanova, L. D. Khazov, and I. K. Nikitin, "Generation of Stimulated Mandelshtam-Brillouin Scattering in Water," *Opt. Spectry.*, **29,** 514–517 (Nov. 1970).

478. H. Takuma and D. A. Jennings, "Stimulated Brillouin Scattering in the off Axis Resonator. Optical Heterodyne Detection of the Forward Stimulated Brillouin Scattering (Carbon Disulfide)," *Appl. Phys. Lett.*, **5**, 239–242 (1964).
479. C. L. Tang, "Saturation and Spectral Characteristics of the Stokes Emission in the Stimulated Brillouin Process," *J. Appl. Phys.*, **37** (8), 2945–255 (1966).
480. E. A. Tikhonov and M. T. Shpak, "Resonance Stimulated Rayleigh Scattering of Light by Organic Dye Solutions," *Ukr. Fiz. Zh.*, **14** (8), 1378–1385 (1969) (in Russian).
481. N. L. Tsintsadze, "Stimulated Mandelshtam-Brillouin Scattering in Helium-2," *Soviet Phys. JETP*, **28**, 950 (1969).
482. J. Walder, "Stimulated Brillouin Scattering," Thesis, Cornell University, 1968, 121 pp. (Univ. Microfilms Order N . 68-11647).
483. J. Walder and C. L. Tang, "Photoelastic Amplification of Light and the Generation of Hypersound by the Stimulated Brillouin Process," *Phys. Rev. Lett.*, **19** (17), 623–626 (1967).
484. J. Walder and C. L. Tang, "Stimulated Brillouin Scattering in Non Focussing Liquids," *Phys. Rev.*, **155**, 318–320 (1967).
485. C. S. Wang, "Effect of Strong Driving Field on the Frequency Shift of Stimulated Brillouin Scattering in Gases," *Phys. Rev. Lett.*, **24** (25), 1394–1395 (1970).
486. C. S. Wang, "Vanishing of the Forward First Stokes Line in Stimulated Brillouin Scattering," *Phys. Lett.*, *A*, **27**, 633 (1968).
487. R. V. Wick and A. H. Guenther, "Intracavity Brillouin Scattering from Positive Q-Spoiling Cells," *Appl. Opt.*, **7** (1), 73–76 (1968).
488. T. A. Wiggins, C. W. Cho, D. R. Dietz, and N. D. Foltz, "Stimulated Thermal Rayleigh Scattering in Gases," *Phys. Rev. Lett.*, **20** (16), 831–834 (1968).
489. T. A. Wiggins, R. V. Wick, and D. H. Rank, "Stimulated Effects in Nitrogen(2) and Carbon Hydrogen(4) Gases," *Appl. Opt.* **5** (6), 1069–1072 (1966).
490. G. Winterling, G. Walda, and W. Heinicke, "Stimulated Brillouin Scattering in Liquid Helium," *Phys. Lett. A*, **26** (7), 301–302 (1968).
491. G. I. Zaitsev, Y. U. I. Kyzylasov, V. S. Starunov, and I. L. Fabelinskii, "Experimental Investigation of Stimulated Light Scattering in the Wing of the Rayleigh Line," *Soviet Phys. JETP*, **6** (2), 35–38 (1967).
492. G. I. Zaitsev, Y. U. I. Kyzylasov, V. S. Starunov, and I. L. Fabelinsfiii, "Observations of Four-Photon Interactions in the Spectrum of Stimulated Scattering of the Light of the Rayleigh Line Wing," *JETP Lett.*, **6** (6), 180–182 (1967).
493. G. I. Zaitsev, Y. U. I. Kyzylasov, V. S. Starunov, and I. L. Fabelinskii, "Stimulated Temperature Scattering of Light in Liquids," *JETP Lett.*, **6** (8), 255–257 (1967).
494. G. Z. Zverev and A. D. Martynov, "Investigation of Stimulated Mandelshtam-Brillouin Scattering Thresholds for Different Media at Wavelengths 0.35, 0.69 and 1.06 Microns," *JETP Lett.*, **6** (11), 351–354 (1967).

12. *Nonlinear Scattering and Miscellaneous References*

495. Y. U. K. Danileiko et al., "Nonlinear Scattering of Light in Inhomogeneous Media," *Zh. Eksperim. i Teor. Fiz.*, **60**, 1245–1250 (April 1971) (in Russian, English Abstr.).
496. S. Kielich, "Changes in Rayleigh Scattering of Light Caused by Laser Optical Saturation," *Acta Phys. Polon. A*, **37** (5), 719–731 (1970).
497. S. Kielich, "Multiharmonic Molecular Light Scattering in Liquids," *Chem. Phys. Lett.*, **1** (10), 441–442 (1967).

498. S. Kielich, "Nonlinear Changes in Rayleigh Light Scattering Due to Electric and Magnetic Fields," *Opt. Commun.* **1** (7), 345–348 (1970).
499. S. Kielich and M. Kozierowski, "Depolarization Ratio of Third Harmonic Elastic Scattering of Laser Light in Gases and Liquids," *Acta Phys. Polon. A*, **38** (2), 271–273 (1970).
500. S. Kielich, J. R. Lalanne, and P. E. Martin, "Double-Photon Elastic Light Scattering by Liquids Having Centrosymmetric Molecules," *Phys. Rev. Lett.*, **26,** 1295–1298 (May 24, 1971).
501. C. R. Lalanne, "Anomaly Exhibited by the Depolarized Rayleigh Scattering of Some Liquids Subjected to an Intense Luminous Wave of a Pulsed Laser," *Compt. Rend., B*, **265** (21), 1181–1184 (1967) (in French).
502. P. D. Maker, "Nonlinear Light Scattering in Methane," in *Physics of Quantum Electronics*. McGraw-Hill, New York, 1966, pp. 60–66.
503. D. L. Weinberg, "Temperature Dependence, Orientation Correlation and Molecular Fields in Second-Harmonic Light Scattering from Liquids and Gases," *J. Chem. Phys.*, **47** (4), 1307–1313 (1967).
504. I. M. Wolinski and A. Sadownik-Wodzinska, "Effect of Molecular Expansion on Light Scattering. 1," *Optik*, **33,** (3), 282–292 (1971) (in German, English Abstr.).
505. B. J. Berne, J. P. Boon, and S. A. Rice, "Transport Phenomena in Simple Liquids," *J. Chem. Phys.*, **45,** 1086 (1966).
506. E. V. Larson, D. G. Naugle, and T. W. Adair, "Ultrasonic Velocity and Attenuation in Liquid Neon," *J. Chem. Phys.*, **54,** 2429 (1971).
507. J. Swift, "Transport Coefficients Near the Consolute Temperature of a Binary Liquid Mixture," *Phys. Rev.*, **173,** 257 (1968).
508. K. Kawasaki, "Sound Attenuation and Dispersion Near the Liquid-Gas Critical Point," *Phys. Rev. A*, **1,** 1750 (1970).
509. M. Born and E. Wolf, *Principles of Optics*, Pergamon Press, Oxford, 1964.
510. S. Chandrasekhar, *Hydrodynamic and Hydromagnetic Stability*, Oxford University Press, 1961.
511. N. K. Ailawadi, A. Rahman, and R. Zwanzig, "Generalized Hydrodynamics and Analysis of Current Correlation Functions," *Phys. Rev. A*, **4,** 1616 (1971).
512. H. N. Lekkerkerker and J. P. Boon, "Light Scattering in Gas Mixtures," *Phys. Lett. A*, **39A,** 9 (1972).
513. J. P. Boon and P. Deguent, "Light Scattering as a Probe of Convective Instability," *Phys. Lett. A*, **39A,** 315 (1972).

THERMODYNAMICS OF DISCRETE MECHANICAL SYSTEMS WITH MEMORY

BERNARD D. COLEMAN

Center for Special Studies, Mellon Institute of Science, Carnegie-Mellon University, Pittsburgh, Pennsylvania

DEDICATED TO MY FRIEND AND TEACHER, WALTER NOLL

CONTENTS

I. Introduction . . . 95
II. Preliminary Definitions . . . 102
III. Elementary Consequences of the Second Law . . . 106
IV. On the Theory of Fading Memory . . . 108
V. Smoothness of the Constitutive Functionals . . . 119
VI. Thermodynamic Restrictions on the Constitutive Functionals . . . 123
VII. Consequences of the Relaxation Property . . . 128
VIII. Lagrangians and Hamiltonians . . . 133
IX. Inflation of a Cylindrical Tube . . . 139
X. Inflation of a Spherical Shell . . . 145
XI. Appendix on Generalizations . . . 148
Acknowledgments . . . 152
References . . . 153

I. INTRODUCTION

The thermodynamics of mechanical systems with memory is a new and rapidly developing field. Research in the subject has emphasized applications to continuous media for which a process is a function mapping the time axis into a set of fields over a manifold.* This article deals with simpler applications. The tensor analysis required by any general approach to continuum thermodynamics is avoided here by limiting the discussion to systems whose processes can be described by giving the time course of a finite list of numerical variables. The systems considered do have long-range memory, however, and their dynamical equations are such that the future evolution of a system in a given environment is determined only

* See, for example, Coleman,[1–3] Coleman and Gurtin,[4–8] Coleman and Mizel,[9] Gurtin,[10] Day,[11,12] Owen,[13] Coleman and Owen,[14] and Coleman and Dill.[15,16]

after specification of the complete history of certain thermal and configurational variables.

The basic variables of the present theory are as follows: the *configurational coordinates* $q_1, \ldots, q_n$ which, like the "generalized coordinates" of classical analytical dynamics, are employed to describe the configuration or "mechanical state" of a system, the *applied forces* $F_1, \ldots, F_n$ which characterize the "mechanical interaction" of the system with its environment, the *rate h of addition of heat*, which characterizes the "nonmechanical" or "thermal" interaction of the system with its environment, and finally, three variables, the *total energy* E, the *temperature* θ, and the *entropy* η, which further describe the "state" of the system. Because of the existing diversity of approaches to the science of thermodynamics, I should emphasize, at the outset, that the real variables, $q_1, \ldots, q_n, F_1, \ldots, F_n$, h, E, θ, and η, are primitive concepts here. Once it is stated how these variables occur in the general laws of thermodynamics and in the class of constitutive assumptions to be considered, then the properties of configurational coordinates, applied forces, heat, energy, temperature, and entropy will be settled, and the theory can be developed by rigorous argument without recourse to operational definitions which, although of use in deciding the applicability of a theory to specific physical situations, have, I believe, no place in its mathematical development.

A *process* of a system is here a specification of the $2n + 4$ primitive variables $q_1, \ldots, q_n, F_1, \ldots, F_n, h, E, \theta$, and η as functions of time; the integer n may be called the system's *degree of mechanical freedom*. The basic laws of thermodynamics are, of course, two. The first, the *law of balance of energy*, states that the only processes to be considered are those for which

$$\dot{E} = \sum_{i=1}^{n} F_i \dot{q}_i + h \tag{1.1}$$

that is, at each time the rate of change of the total energy E equals the rate of working $\sum_i F_i \dot{q}_i$ of the applied forces plus the rate h of addition of heat.* Individual systems are described by stating constitutive assumptions. In the broadest sense of the term, a *constitutive assumption* is a preassigned rule, or list of rules, for selecting a particular set $\mathscr{A}$ of the processes that obey (1.1). The processes in $\mathscr{A}$ are said to be *admissible* in the system under consideration. The second law, sometimes called the *dissipation principle*, asserts that constitutive assumptions must be such that every process in $\mathscr{A}$ obeys the inequality

$$\dot{\eta} \geqslant \frac{h}{\theta} \tag{1.2}$$

* The superposed dots indicate one-sided time-derivatives. [See (2.1).]

at each instant of time.* That is, in an admissible process the rate of increase of entropy must never be less than the rate of addition of heat divided by the temperature.

I consider here only systems which have inertia. That is, I assume that there is assigned a function $\mathscr{K}(\cdot, \cdot)$ which for each specification of the coordinate vector $\mathbf{q} = (q_1, \ldots, q_n)$ gives a quadratic form $\mathscr{K}(\mathbf{q}, \cdot)$ on the velocity vectors $\dot{\mathbf{q}} = (\dot{q}_1, \ldots, \dot{q}_n)$; one-half the value of this quadratic form, that is, the number

$$T \overset{\text{def}}{=} \tfrac{1}{2}\mathscr{K}(\mathbf{q}; \dot{\mathbf{q}}) = \frac{1}{2} \sum_{i,j=1}^{n} K_{ij}(q_1, \ldots, q_n)\dot{q}_i\dot{q}_j \tag{1.3}$$

is called the *kinetic energy*. In a given process, T depends on time. The function $\mathscr{K}(\cdot, \cdot)$ is called the *inertia function* for the system under consideration. Of course, the quadratic form $\mathscr{K}(\mathbf{q}, \cdot)$ determines, in the usual way, a symmetric linear transformation $\mathbf{K}(\mathbf{q})$ of the space of n-tuples into itself. This linear transformation is called the *inertia tensor*, and the ith component of the vector $\mathbf{p} = \mathbf{K}(\mathbf{q})\dot{\mathbf{q}}$, that is, the number

$$p_i = \sum_{j=1}^{n} K_{ij}(q_1, \ldots, q_n)\dot{q}_j \tag{1.4}$$

is called the *momentum associated with the* ith *coordinate*. Once the inertia function is specified one may define, in terms of the basic primitive variables q_i, F_i, θ, η, and E, several new or "derived" variables. The momenta p_i and kinetic energy T are examples of such derived variables; others are the *internal energy* ϵ and the *Helmholtz free-energy* ψ, defined by

$$\epsilon = E - T \quad \text{and} \quad \psi = \epsilon - \theta\eta \tag{1.5}$$

The *inertial force associated with the* ith *coordinate* is defined by the relation†

$$Q_i = \frac{d}{dt}\frac{\partial T}{\partial \dot{q}_i} - \frac{\partial T}{\partial q_i} \tag{1.6}$$

in which T is the function of $q_1, \ldots, q_n$ and $\dot{q}_1, \ldots, \dot{q}_n$ shown in (1.3). The quantity

$$\Xi_i = F_i - Q_i \tag{1.7}$$

is called the *internal force associated with the* ith *coordinate*.

The basic constitutive assumption of the present theory is the following: each system is characterized by its degree n of mechanical freedom, its

* In this interpretation of the second law as a restriction on *constitutive assumptions*, I follow the point of view developed in an article[17] written with Noll in 1963.

† The motivation for this definition, which may be already clear from classical analytical dynamics, is discussed in Section II. (See Theorem 2.1.)

inertia function $\mathscr{K}(\cdot,\cdot)$, and $n+2$ functionals $\mathfrak{p}$, $\mathfrak{z}_i$, $\mathfrak{h}$, $i = 1, \ldots, n$, which give the present values of the Helmholtz free energy, the internal forces, and the entropy, when the history of the coordinates and temperature is known:

$$\left.\begin{aligned}
\psi(t) &= \mathop{\mathfrak{p}}_{s=0}^{\infty}(q_1(t-s), \ldots, q_n(t-s), \theta(t-s)) \\
\Xi_i(t) &= \mathop{\mathfrak{z}_i}_{s=0}^{\infty}(q_1(t-s), \ldots, q_n(t-s), \theta(t-s)) \\
\eta(t) &= \mathop{\mathfrak{h}}_{s=0}^{\infty}(q_1(t-s), \ldots, q_n(t-s), \theta(t-s))
\end{aligned}\right\} \tag{1.8}$$

The admissible processes of the system are those for which not only the balance law (1.1) but also the constitutive equations (1.8) hold, with ψ and Ξ_i given by (1.5)–(1.7).

In subsequent sections, particularly in Sections II and III, the definitions and assumptions hastily summarized in this introduction are restated with precision and in a more efficient, albeit less familiar notation.

The second law places certain restrictions on the choice of the constitutive functionals $\mathfrak{h}$, $\mathfrak{z}_1, \ldots, \mathfrak{z}_n$, and $\mathfrak{h}$. In Sections V–VII, employing an hypothesis of smoothness for constitutive functionals called the "principle of fading memory,"[1,18–23] I derive and discuss the conditions on $\mathfrak{p}$, $\mathfrak{z}_1, \ldots, \mathfrak{z}_n$, and $\mathfrak{h}$ necessary and sufficient for (1.2) to hold for all admissible processes. In Section IV, I describe an axiomatic approach to the theory of fading memory worked out with V. J. Mizel,[9,23] and that self-contained section is the only section of the article that rests heavily on the terminology and methods of modern functional analysis. Not all of Section IV need be read by those interested mainly in applications; of particular importance to the subsequent discussion, however, are Theorems 4.7 and 4.8 and the definitions of the "instantaneous derivative" and "past-history derivative" given in (4.20) and (4.21). The treatment of Sections V–VII is mainly a translation into the framework of the theory of discrete systems of results[1,2,9] obtained earlier in continuum thermodynamics. The results given in Section VIII, however, are new and are published here for the first time; they show that certain familiar classical theorems of analytical dynamics,* due to Lagrange, Poisson, and Hamilton, can be generalized to hold for systems with memory. In these generalizations, Lagrangian and Hamiltonian functions are replaced by appropriate functionals which can be constructed when the free energy functional $\mathfrak{p}$ is known, and ordinary derivatives with respect to coordinates are replaced by "instantaneous derivatives." In Sections IX and X, it is observed that certain classes of

* See, for example, Whittaker.[38]

motions* of continuous media with memory can be treated within the framework of the thermodynamics of discrete systems. These motions are used to illustrate the principal theorems of the present study, Theorems 6.1, 8.1, 8.2, and 8.3. The Appendix is concerned with an extension of the theory to systems whose memory need not fade in time.†

The aim of the present subject is close to that of analytical dynamics: to prove theorems about the evolution in time of the coordinates when information about the applied forces and either the temperature or the supply of heat is specified in advance. Because the present theory permits "memory effects," the dynamical equations encountered here, that is, the equations describing the evolution of the coordinates in time, differ markedly from those encountered in classical analytical dynamics; they are functional-differential equations rather than ordinary differential equations. Indeed, by (1.6)–(1.8), when θ and $F_1, \ldots, F_n$ are specified, the dynamical equations have the form

$$\frac{d}{dt}\frac{\partial T}{\partial \dot{q}_i} - \frac{\partial T}{\partial q_i} = F_i(t) - \mathop{\mathfrak{Z}_i}_{s=0}^{\infty}(q_1(t-s), \ldots, q_n(t-s), \theta(t-s)), \qquad i = 1, \ldots, n \tag{1.9}$$

with T as shown in (1.3). If h is assigned instead of θ, to the n equations (1.9) one must adjoin the equation

$$\frac{d}{dt}\mathop{\mathfrak{e}}_{s=0}^{\infty}(q_1(t-s), \ldots, q_n(t-s), \theta(t-s)) = h(t) + \sum_{i=1}^{n} \dot{q}_i(t)\mathop{\mathfrak{Z}_i}_{s=0}^{\infty}(q_1(t-s), \ldots, q_n(t-s), \theta(t-s)) \tag{1.10}$$

where $\mathfrak{e}$, the functional defined by

$$\mathop{\mathfrak{e}}_{s=0}^{\infty}(q_1(t-s), \ldots, q_n(t-s), \theta(t-s)) = \mathop{\mathfrak{p}}_{s=0}^{\infty}(q_1(t-s), \ldots, q_n(t-s), \theta(t-s)) + \theta(t)\mathop{\mathfrak{h}}_{s=0}^{\infty}(q_1(t-s), \ldots, q_n(t-s), \theta(t-s)) \tag{1.11}$$

has the internal energy $\epsilon(t)$ for its value; in Section II we shall see that (1.10) is a direct consequence of the law of balance of energy (1.1).‡ It is easier to study the functional-differential equations (1.9) and (1.10) than

* Compare Coleman and Dill.[26]
† Compare Owen[13] and Coleman and Owen.[14]
‡ See Theorem 2.1.

it is to study the evolution equations found in general continuum mechanics, just as the theory of ordinary differential equations is easier than the theory of partial differential equations. Unfortunately, there is not space here to discuss the advances recently made in the abstract qualitative theory of functional-differential equations of the type (1.9). I must content myself with remarking that there is a theory of existence and stability for such equations, and the theory uses the smoothness assumptions for functionals employed here in Sections V–VIII and discussed at length in Section IV.* In the theory of stability it is shown that the functional $\mathscr{H}$, here (in Sections VIII–X) called the "Hamiltonian functional," can serve as a Lyapunov functional for (1.9).

I hope that after he has patiently perused several theorems and examples the reader will agree that, even when the degree of freedom, n, is finite, the thermodynamics of systems with memory is a potentially rich mathematical subject with interesting physical applications.

A Note on the Word "System"

In spite of the lack of precision of the preliminary outline just given, it may be already clear that the word "system," so often used without rigorous definition, can here be given a mathematical meaning by saying that *a system $\mathscr{S}$ is a set, $\mathscr{S} = \{\mathfrak{p}, \mathfrak{h}, \mathfrak{z}_1, \ldots, \mathfrak{z}_n, \mathscr{K}\}$, with $\mathfrak{p}, \mathfrak{h}, \mathfrak{z}_1, \ldots, \mathfrak{z}_n$, functionals mapping $(n + 1)$-tuple-valued functions on $[0, \infty)$ into the real numbers and with $\mathscr{K}(\cdot, \cdot)$ a real-valued function of pairs of n-tuples such that $\mathscr{K}(\mathbf{q}, \cdot)$ is a quadratic form for each fixed n-tuple $\mathbf{q}$*. In some applications of the theory it is convenient to identify a system $\mathscr{S}$ with the set $\mathscr{A}$ of its "admissible processes" or with the functional-differential equations (1.9) and (1.10) which must be obeyed by each admissible process.

An Example: The Dangling Spider†

To have an elementary example of a physical situation covered by the theoretical framework introduced here, let us suppose that a rigid ball (or a spider) of mass M is hanging from a ceiling by a massless but extensible filament of length q. The forces acting on the ball are the tension in the filament, which tends to pull the ball upward, and a body force F acting downward. In the special case in which the only long-range force acting on the ball is that of gravity, $F = gM$, with g the gravitational constant; we may, however, seek greater generality and allow F to vary with t or q. Since the filament is supposed massless, at each instant the tension in it is spatially homogeneous. Let us assume, further, that the filament is composed of a homogeneous nonlinear viscoelastic material and

* Coleman and Mizel.[24,25] See also Coleman and Dill.[26]

† This example is discussed in an article[25] written with Mizel; I shall return to the example in Section VIII.

has, at each instant, a uniform temperature θ equal to the temperature of the ball.* If we assume that the ball executes only a vertical motion, then we have here a system with one degree of freedom. A process of the system is a specification of six variables, q, F, h, E, θ, and η, as functions of time. I have already given an interpretation to q, F, and θ; h is the total rate at which the filament and the ball are absorbing heat from their surroundings; E and η are, respectively, the total energy and the entropy of the system. Equation (1.1) here reads

$$\dot{E} = h + F\dot{q} \tag{1.12}$$

that is, the rate of change of the total energy equals the rate of addition of heat plus the rate of working of the body force F applied to the ball. The kinetic energy of the system is

$$T = \tfrac{1}{2}M\dot{q}^2 \tag{1.13}$$

Hence in this special case the inertia tensor $\mathbf{K}$ is just the number M. For the internal energy ϵ and Helmholtz free energy ψ defined in (1.5), we have

$$\epsilon = E - \tfrac{1}{2}M\dot{q}^2\,, \qquad \psi = E - \tfrac{1}{2}M\dot{q}^2 - \theta\eta \tag{1.14}$$

It follows from (1.13) that the definition (1.6) of inertial force here reduces to

$$Q = \frac{d}{dt}M\dot{q} = M\ddot{q} \tag{1.15}$$

Now, (1.7) tells us that the internal force Ξ should be the difference between the applied force F and the inertial force Q; that is,

$$\Xi = F - Q$$

or, by (1.15),

$$M\ddot{q} = F - \Xi \tag{1.16}$$

This equation clearly agrees with Newton's second law† if and only if *we identify the internal force* Ξ *with the tension in the filament.*

The constitutive equations (1.8) here take the form

$$\psi(t) = \underset{s=0}{\overset{\infty}{\mathfrak{p}}}\,(q(t-s),\, \theta(t-s)) \tag{1.17}$$

$$\Xi(t) = \underset{s=0}{\overset{\infty}{\mathfrak{z}}}\,(q(t-s),\, \theta(t-s)) \tag{1.18}$$

$$\eta(t) = \underset{s=0}{\overset{\infty}{\mathfrak{h}}}\,(q(t-s),\, \theta(t-s)) \tag{1.19}$$

* This requires that the ball and the filament be good conductors of heat.

The mathematical theory outlined here applies also if the filament does not exchange heat with the ball, provided the filament itself is a good conductor. In such an application, θ is to be interpreted as the temperature of the filament alone; h is the rate of addition of heat to the filment; ϵ, η, and ψ are properties of the filament; and E is the sum of the internal energy of the filament and the kinetic energy of the ball.

† Applied to the ball.

Of particular interest is (1.18), which tells us that, at each instant, the present value of the tension in the filament is determined by the history of the filament's length and temperature. Substitution of (1.18) into (1.16) yields the following special case of (1.9):

$$M\ddot{q}(t) = F(t) - \mathop{\mathfrak{Z}}_{s=0}^{\infty}(q(t-s), \theta(t-s)) \tag{1.20}$$

If θ and F are specified as functions of t or q, (1.20) becomes a functional-differential equation for q; this equation is called the *dynamical equation* for the system. As an example, one might wish to consider a situation in which $F(t) = Mg$ = constant at all times after some special time, say, $t = 0$, and $\theta(t)$ = constant at all times. This would correspond to the assumption that the process under consideration is isothermal and that from time $t = 0$ onward the only applied force is that due to gravity. In such a case, one might ask the following question: If the history of q up to time $t = 0$ and the value of $\lim_{t\to 0+} \dot{q}(t)$ are given, does (1.20) yield $q(t)$ for positive t? It follows from recent work in the theory of functional-differential equations that, granted reasonable regularity assumptions for the constitutive functional $\mathfrak{Z}$ and the initial history of q, the answer is *yes.**

Since (1.14) and (1.16) yield

$$\dot{E} = \dot{\epsilon} + M\dot{q}\ddot{q} = \dot{\epsilon} + F\dot{q} - \Xi\dot{q} \tag{1.21}$$

(1.12) is equivalent to the equation

$$\dot{\epsilon} = h + \Xi\dot{q} \tag{1.22}$$

that is, the rate of change of the internal energy equals the rate of addition of heat plus the rate of working of the internal force. If we define a functional $\mathfrak{e}$ by the formula (1.11) with $q_1, \ldots, q_n$ replaced by q, then (1.22) can be written

$$\frac{d}{dt}\mathop{\mathfrak{e}}_{s=0}^{\infty}(q(t-s), \theta(t-s)) = h(t) + \dot{q}(t)\mathop{\mathfrak{Z}}_{s=0}^{\infty}(q(t-s), \theta(t-s)) \tag{1.23}$$

If h and F are preassigned instead of θ and F, (1.20) and (1.23) together yield two equations for the two functions $q(\cdot)$ and $\theta(\cdot)$. If it is assumed that the process under consideration is adiabatic, that is, that the system neither absorbs nor transmits heat to its environment, then $h(t) \equiv 0$, and (1.23) reduces to

$$\frac{d}{dt}\mathop{\mathfrak{e}}_{s=0}^{\infty}(q(t-s), \theta(t-s)) = \dot{q}(t)\mathop{\mathfrak{Z}}_{s=0}^{\infty}(q(t-s), \theta(t-s)) \tag{1.24}$$

II. PRELIMINARY DEFINITIONS

A *smooth process* $\mathscr{P}$ is an ordered set of $2n + 4$ real-valued functions, $q_1(\cdot), \ldots, q_n(\cdot)$, $F_1(\cdot), \ldots, F_n(\cdot)$, $h(\cdot)$, $E(\cdot)$, $\theta(\cdot)$, and $\eta(\cdot)$, defined on an interval of the form $(-\infty, a]$ and such that $F(\cdot)$ and $h(\cdot)$ are measurable,

* See Coleman and Mizel.[25]

$q_1(\cdot), \ldots, q_n(\cdot)$, $E(\cdot)$, $\theta(\cdot)$, and $\eta(\cdot)$ are continuous, and $\dot{q}_1(t), \ldots, \dot{q}_n(t)$, $\ddot{q}_1(t), \ldots, \ddot{q}_n(t)$, $\dot{E}(t)$, $\dot{\theta}(t)$, and $\dot{\eta}(t)$ exist for each t in $(-\infty, a]$, where the superposed dot denotes the left-hand derivative; for example,

$$\dot{q}_i(t) = \lim_{s\to 0+} \frac{q_i(t) - q_i(t-s)}{s}, \qquad \ddot{q}_i = \lim_{s\to 0+} \frac{\dot{q}_i(t) - \dot{q}_i(t-s)}{s} \tag{2.1}$$

The independent variable t is, of course, the *time*. As I mentioned in the introduction, the values of the functions comprising a process have the following names: $q_i = q_i(t)$ $(i = 1, \ldots, n)$ is called the *ith configurational coordinate*, F_i the *applied force associated with the* i*th coordinate*, h the *rate of addition of heat*, E the *total energy*, θ the *temperature*, and η the *entropy*.

Although there is a branch of physics in which negative temperatures occur,* for simplicity of exposition I here assume that θ is always positive:

$$\theta(t) > 0 \tag{2.2}$$

The n-tuples $\mathbf{q} = (q_1, \ldots, q_n)$, $\dot{\mathbf{q}} = (\dot{q}_1, \ldots, \dot{q}_n)$, and $\mathbf{F} = (F_1, \ldots, F_n)$ are called, respectively, the *coordinate vector*, the *velocity vector*, and the *vector of applied forces*. These n-tuples are regarded as members of the vector space R^n endowed with the usual inner product "$\cdot$" for n-tuples. The number

$$w(t) = \mathbf{F}(t) \cdot \dot{\mathbf{q}}(t) = \sum_{i=1}^{n} F_i(t)\dot{q}_i(t) \tag{2.3}$$

is called the *total rate of working of the applied forces, while* $F_i\dot{q}_i$ is the *rate of working of the applied force associated with the* i*th coordinate.*

A smooth process $\mathscr{P}$ is called a *smooth thermodynamic process* if and only if it obeys the *law of balance of energy*

$$\dot{E}(t) = \mathbf{F}(t) \cdot \dot{\mathbf{q}}(t) + h(t) \tag{2.4}$$

at each time t in the domain $(-\infty, a]$ of $\mathscr{P}$. Throughout the discussion, attention will be restricted to such thermodynamic processes, and hence it will be assumed that the rate of change of the total energy is always equal to the rate of addition of heat plus the total rate of working of the applied forces.

An *inertia function* is a real-valued function $\mathscr{K}$ of $\mathbf{q}$ and $\dot{\mathbf{q}}$ such that $\mathscr{K}(\mathbf{q}, \dot{\mathbf{q}})$ is continuously differentiable in $\mathbf{q}$ for each specification of $\dot{\mathbf{q}}$ and is a homogeneous quadratic function of $\dot{\mathbf{q}}$ for each choice of $\mathbf{q}$. Such a function $\mathscr{K}$ has the general form

$$\mathscr{K}(\mathbf{q}, \dot{\mathbf{q}}) = \dot{\mathbf{q}} \cdot \mathbf{K}(\mathbf{q})\dot{\mathbf{q}} \tag{2.5}$$

* See, for example, Purcell and Pound,[27] Ramsey,[28] and Coleman and Noll.[29]

where, for each $\mathbf{q}$, $\mathbf{K}(\mathbf{q})$ is a symmetric linear transformation of R^n into itself, and $\mathbf{K}(\cdot)$ is continuously differentiable on a subset of R^n. The value $\mathbf{K}(\mathbf{q})$ of the function $\mathbf{K}(\cdot)$ is called the *inertia tensor*.

Let us suppose that an inertia function has been assigned. Then in each smooth process $\mathscr{P}$ we may compute, for each t in $(-\infty, a]$, the number

$$T(t) = \tfrac{1}{2}\mathscr{K}(\mathbf{q}(t), \dot{\mathbf{q}}(t)) \tag{2.6}$$

which is called the *kinetic energy* at time t in $\mathscr{P}$. Of course (2.6) can be written

$$T = \tfrac{1}{2}\dot{\mathbf{q}} \cdot \mathbf{K}(\mathbf{q})\dot{\mathbf{q}} \tag{2.7}$$

where the dependence of $\mathbf{q}$, $\dot{\mathbf{q}}$, and T on t is understood, as it is in (1.3). As the inertia tensor $\mathbf{K}(\mathbf{q})$ is a symmetric tensor on R^n, and the numbers $K_{ij}(\mathbf{q})$ shown in (1.3) are the components of $\mathbf{K}$ relative to an orthonormal basis, one has $K_{ij}(\mathbf{q}) = K_{ji}(\mathbf{q})$ for all pairs (i, j). In the present direct notation, the symmetry of $\mathbf{K}(\mathbf{q})$ as a linear transformation means that

$$\mathbf{u} \cdot \mathbf{K}(\mathbf{q})\mathbf{v} = \mathbf{v} \cdot \mathbf{K}(\mathbf{q})\mathbf{u} \tag{2.8}$$

for all $\mathbf{u}$, $\mathbf{v}$ in R^n.

The number Q_i, defined by*

$$Q_i = \frac{d}{dt}\frac{\partial T}{\partial \dot{q}_i} - \frac{\partial T}{\partial q_i} \tag{2.9}$$

is the *inertial force associated with the ith coordinate*. The *vector of inertial force* is the n-tuple $\mathbf{Q} = (Q_1, \ldots, Q_n)$. If we introduce the notation

$$\nabla_{\mathbf{q}} T = \left(\frac{\partial T}{\partial q_1}, \ldots, \frac{\partial T}{\partial q_2}\right), \qquad \nabla_{\dot{\mathbf{q}}} T = \left(\frac{\partial T}{\partial \dot{q}_1}, \ldots, \frac{\partial T}{\partial \dot{q}_2}\right) \tag{2.10}$$

then (2.9) becomes

$$\mathbf{Q} = \left[\frac{d}{dt}\nabla_{\dot{\mathbf{q}}} T\right] - \nabla_{\mathbf{q}} T \tag{2.11}$$

Now, by (2.7),

$$\nabla_{\dot{\mathbf{q}}} T = \mathbf{K}(\mathbf{q})\dot{\mathbf{q}} \tag{2.12}$$

and hence the chain rule,

$$\dot{T} = \sum_{i=1}^{n}\left\{\frac{\partial T}{\partial q_i}\dot{q}_i + \frac{\partial T}{\partial \dot{q}_i}\ddot{q}_i\right\}$$

can be written

$$\dot{T} = (\nabla_{\mathbf{q}} T) \cdot \dot{\mathbf{q}} + (\nabla_{\dot{\mathbf{q}}} T) \cdot \ddot{\mathbf{q}} = \dot{\mathbf{q}} \cdot \nabla_{\mathbf{q}} T + \ddot{\mathbf{q}} \cdot \mathbf{K}(\mathbf{q})\dot{\mathbf{q}} \tag{2.13}$$

If we differentiate (2.7) with respect to t, we obtain, upon use of (2.8),

$$\dot{T} = \tfrac{1}{2}\dot{\mathbf{q}} \cdot \mathbf{A}\dot{\mathbf{q}} + \ddot{\mathbf{q}} \cdot \mathbf{K}\dot{\mathbf{q}} \tag{2.14}$$

* Here, d/dt, like the superposed dot, denotes the left-hand derivative with respect to t.

where $\mathbf{K} = \mathbf{K}(\mathbf{q})$ and

$$\mathbf{A} \overset{\text{def}}{=} \frac{d}{dt} \mathbf{K}(\mathbf{q}(t)) \tag{2.15}$$

Clearly, (2.13) and (2.14) imply that in each smooth process

$$\dot{\mathbf{q}} \cdot \nabla_{\mathbf{q}} T = \tfrac{1}{2}\dot{\mathbf{q}} \cdot \mathbf{A}\dot{\mathbf{q}} \tag{2.16}$$

On taking the inner product of (2.11) with $\dot{\mathbf{q}}$ and making use of (2.12), (2.15), and (2.16), one finds that

$$\begin{aligned} \dot{\mathbf{q}} \cdot \mathbf{Q} &= \dot{\mathbf{q}} \cdot \left[\frac{d}{dt}[\nabla_{\dot{\mathbf{q}}} T]\right] - \dot{\mathbf{q}} \cdot \nabla_{\mathbf{q}} T = \dot{\mathbf{q}} \cdot \left[\frac{d}{dt}(\mathbf{K}\dot{\mathbf{q}})\right] - \tfrac{1}{2}\dot{\mathbf{q}} \cdot \mathbf{A}\dot{\mathbf{q}} \\ &= \dot{\mathbf{q}} \cdot \left[\frac{d}{dt}\mathbf{K}\right]\dot{\mathbf{q}} + \dot{\mathbf{q}} \cdot \mathbf{K}\ddot{\mathbf{q}} - \tfrac{1}{2}\dot{\mathbf{q}} \cdot \mathbf{A}\dot{\mathbf{q}} = \tfrac{1}{2}\dot{\mathbf{q}} \cdot \mathbf{A}\dot{\mathbf{q}} + \dot{\mathbf{q}} \cdot \mathbf{K}\ddot{\mathbf{q}} \end{aligned} \tag{2.17}$$

Since the tensor $\mathbf{K}$ is symmetric, the right-hand sides of (2.17) and (2.14) are the same, which proves

Theorem 2.1. *In a smooth process, at each instant t,*

$$\dot{T}(t) = \mathbf{Q}(t) \cdot \dot{\mathbf{q}}(t) \tag{2.18}$$

That is, the rate of change of the kinetic energy equals the total rate of working of the inertial forces.

Let us call the n-tuple defined by

$$\boldsymbol{\Xi} = \mathbf{F} - \mathbf{Q} \tag{2.19}$$

the *vector of internal force*. The internal force associated with the ith coordinate is the ith component, Ξ_i, of $\boldsymbol{\Xi} = (\Xi_1, \ldots, \Xi_n)$, and the total rate of working of the internal forces is $\boldsymbol{\Xi} \cdot \dot{\mathbf{q}} = \sum_{i=1}^{n} \Xi_i \dot{q}_i$.

The difference ϵ between the total energy E and the kinetic energy T is called the *internal energy*:

$$\epsilon = E - T \tag{2.20}$$

Since $E = T + \epsilon$ and $\mathbf{F} = \mathbf{Q} + \boldsymbol{\Xi}$, the law of balance of energy (2.4) can be written

$$\dot{\epsilon} + \dot{T} = \mathbf{Q} \cdot \dot{\mathbf{q}} + \boldsymbol{\Xi} \cdot \dot{\mathbf{q}} + h$$

and hence Theorem 2.1 has the corollary,

Theorem 2.2. *The following equation is equivalent to (2.4):*

$$\dot{\epsilon}(t) = \boldsymbol{\Xi}(t) \cdot \dot{\mathbf{q}}(t) + h(t) \tag{2.21}$$

Thus in a smooth thermodynamic process the rate of change of the internal energy always equals the rate of addition of heat plus the total rate of working of the internal forces.

The Helmholtz free energy ψ is defined to be

$$\psi = \epsilon - \theta\eta \tag{2.22}$$

and, in a given process, ψ is a known function of t.

The function $\mathbf{q}(\cdot)$ giving the coordinates $\mathbf{q}(t) = (q_1(t), \ldots, q_n(t))$ at each instant t is called the *motion.* If $f(\cdot)$ is a function on $(-\infty, a]$ and if t is in $(-\infty, a]$, then the *history of f up to t* is the function on $[0, \infty)$ defined by

$$f^t(s) = f(t - s)\,, \qquad 0 \leqslant s < \infty$$

Thus the *history up to time t of the motion* $\mathbf{q}(\cdot)$ is the function $\mathbf{q}^t$ mapping $[0, \infty)$ into R^n and obeying $\mathbf{q}^t(s) = \mathbf{q}(t - s)$. The ith component of $\mathbf{q}^t$, that is, the real-valued function $q_i^{\,t}$ on $[0, \infty)$ defined by $q_i^{\,t}(s) = q_i(t - s)$, is called the *history up to t of the ith coordinate.*

To characterize a system $\mathscr{S}$, one must specify not only the number n of its coordinates and its inertia function $\mathscr{K}$, but also certain functionals denoted by $\mathfrak{p}$, $\mathfrak{Z}$, and $\mathfrak{h}$. These functionals, called *constitutive functionals*, give the present values of the Helmholtz free energy, the vector of internal force, and the entropy, when the history $\mathbf{q}^t$ of the motion and the history θ^t of the temperature are specified:

$$\left.\begin{aligned} \psi(t) &= \mathfrak{p}(\mathbf{q}^t, \theta^t) \\ \Xi(t) &= \mathfrak{Z}(\mathbf{q}^t, \theta^t) \\ \eta(t) &= \mathfrak{h}(\mathbf{q}^t, \theta^t) \end{aligned}\right\} \tag{2.23}$$

The equations (2.23), which can be written

$$\left.\begin{aligned} \psi(t) &= \mathfrak{p}(q_1^{\,t}, \ldots, q_n^{\,t}, \theta^t) \\ \Xi_i(t) &= \mathfrak{Z}_i(q_1^{\,t}, \ldots, q_n^{\,t}, \theta^t)\,, \qquad i = 1, \ldots, n \\ \eta(t) &= \mathfrak{h}(q_1^{\,t}, \ldots, q_n^{\,t}, \theta^t) \end{aligned}\right\} \tag{2.24}$$

are called the *constitutive equations* of the system.*

A smooth process $\mathscr{P}$ is called an *admissible process* of the system under consideration if it is compatible with the law of balance of energy (2.4) and the constitutive equations (2.23).

III. ELEMENTARY CONSEQUENCES OF THE SECOND LAW

If $\mathscr{S}$ is a system and $\mathscr{A}$ the set of all admissible processes of $\mathscr{S}$, then the elements of $\mathscr{A}$ are smooth thermodynamic processes obeying constitutive equations of the form (2.23). The second law of thermodynamics

* Equations (2.23) and (2.24) are compact versions of (1.8) and are written using the modern notation for functionals in which the argument s of $\mathbf{q}^t$ and θ^t is not shown.

requires that every process in $\mathscr{A}$ obey the inequality

$$\dot{\eta}(t) \geqslant \frac{h(t)}{\theta(t)} \tag{3.1}$$

at each time t. This requirement restricts the choice of the response functionals appearing in the constitutive equations. The limitations which the second law places on our constitutive assumptions will be derived in Section VI after a review of recent results from the theory of fading memory and a precise statement of the hypotheses of smoothness for constitutive functionals. The present short section has a less ambitious goal; here I merely gather together for future use some immediate consequences of the inequality (3.1).

The quantity γ, defined by

$$\gamma = \dot{\eta} - \frac{h}{\theta} \tag{3.2}$$

is called the *internal dissipation* or the *rate of production of entropy* in $\mathscr{S}$. Employing the law of balance of energy in the form (2.21), one can write

$$\dot{\epsilon} = \mathbf{\Xi} \cdot \dot{\mathbf{q}} + \theta\dot{\eta} - \theta\gamma \tag{3.3}$$

or, in terms of the Helmholtz free energy, $\psi = \epsilon - \theta\eta$,

$$\dot{\psi} = \mathbf{\Xi} \cdot \dot{\mathbf{q}} - \eta\dot{\theta} - \theta\gamma \tag{3.4}$$

Since (3.1) can be written

$$\gamma(t) \geqslant 0 \tag{3.5}$$

and since θ is assumed positive, the second law requires that each admissible process be such that

$$\dot{\epsilon} \leqslant \mathbf{\Xi} \cdot \dot{\mathbf{q}} + \theta\dot{\eta} \tag{3.6}$$

and

$$\dot{\psi} \leqslant \mathbf{\Xi} \cdot \dot{\mathbf{q}} - \eta\dot{\theta} \tag{3.7}$$

A process in which the rate h of addition of heat is zero at all times is called *adiabatic*. An *isothermal process* is one in which the temperature θ is constant at all times.

It is clear from (3.1) that *in every admissible adiabatic process the entropy η is an increasing function of time*. Let us now construct a quantity which is a monotone function of time in admissible *isothermal* processes.

In a given smooth process, the total rate of working of the applied forces, $\mathbf{F} \cdot \dot{\mathbf{q}}$, is a function of time alone. Let a constant c and the origin of the time axis be chosen for convenience, and put

$$\zeta(t) = -\int_0^t \mathbf{F}(\tau) \cdot \dot{\mathbf{q}}(\tau)\, d\tau + c\,, \qquad \text{i.e.,} \quad \dot{\zeta}(t) = -\mathbf{F}(t) \cdot \dot{\mathbf{q}}(t) \tag{3.8}$$

$l\zeta$ may be called the *mechanical potential of the applied forces*. The *canonica* *free energy** is defined to be the sum of the Helmholtz free energy, the kinetic energy, and the mechanical potential of the applied forces:

$$\Phi(t) = \psi(t) + T(t) + \zeta(t) \tag{3.9}$$

Since (2.18) and (2.19) yield

$$\dot{T} = \mathbf{Q} \cdot \dot{\mathbf{q}} = \mathbf{F} \cdot \dot{\mathbf{q}} - \mathbf{\Xi} \cdot \dot{\mathbf{q}} \tag{3.10}$$

it follows from (3.8) and (3.9) that

$$\dot{\Phi} = \dot{\psi} - \mathbf{\Xi} \cdot \dot{\mathbf{q}} \tag{3.11}$$

When $\dot{\theta} = 0$, (3.4) reduces to

$$\dot{\psi} = \mathbf{\Xi} \cdot \dot{\mathbf{q}} - \theta\gamma$$

and (3.11) becomes

$$\dot{\Phi} = -\theta\gamma \tag{3.12}$$

Thus by (3.5) and (2.2), when $\dot{\theta}$ is zero we have $\dot{\Phi} \leqslant 0$. This proves

Theorem 3.1. *The canonical free energy Φ never increases in an admissible isothermal process.*

Note. To prove Theorem 3.1, one does not really need the relation (3.10). For, since $E = \epsilon + T$ and $\psi = \epsilon - \theta\eta$, the definition (3.9) can be written

$$\Phi = E - \theta\eta + \zeta \tag{3.13}$$

and hence (3.8) and (2.4) yield

$$\dot{\Phi} = h - \theta\dot{\eta} - \dot{\theta}\eta$$

or, by (3.2),

$$\dot{\Phi} = -\theta\gamma - \dot{\theta}\eta$$

Clearly, this equation reduces to (3.12) when $\dot{\theta} = 0$.

IV. ON THE THEORY OF FADING MEMORY

Often in physics, as here, when one uses the word *process* one means a collection $\mathscr{P}$ of functions $\mathbf{f}$, $\mathbf{g}$ over an interval of the form $(-\infty, a]$ or $(-\infty, \infty)$. The argument t of these functions is the time; the values $\mathbf{f}(t)$, $\mathbf{g}(t)$, . . . , lie in normed vector spaces, such as the real line or the space of n-tuples. Given any $\mathbf{f}$ in $\mathscr{P}$, the function $\mathbf{f}^t$ on $[0, \infty)$ defined by

$$\mathbf{f}^t(s) = \mathbf{f}(t - s)\,, \qquad 0 \leqslant s < \infty \tag{4.1}$$

* Coleman and Dill[26] [equation (2.13)] and Coleman and Mizel[25] [equation (6.4)]; closely related quantities are considered by Ericksen[30] and Coleman.[23]

is called the history of $\mathbf{f}$ up to t. Different materials and physical systems are distinguished by different constitutive assumptions, which restrict the class of processes. The constitutive assumptions of interest here are called constitutive equations and have the form

$$\mathbf{g}(t) = \mathfrak{g}(\mathbf{f}^t) \tag{4.2}$$

with $\mathfrak{g}$ a preassigned functional. In words, (4.2) asserts that the constitutive functional $\mathfrak{g}$, which characterizes the substance under consideration, gives for each t the "present value" of $\mathbf{g}$, $\mathbf{g}(t)$, when the history of $\mathbf{f}$ up to t is specified. Of course, the constitutive assumptions (2.23) are of the type (4.2) with $\mathbf{f}$ the ordered pair $(\mathbf{q}, \theta)$. An example of (4.2) from continuum mechanics is the constitutive equation of a simple material* without thermal influences; in that equation $\mathbf{g}(t)$ is the stress at a material point at time t and hence a symmetric tensor, whereas $\mathbf{f}^t(s)$ is the deformation gradient at the point at time $t - s$ and therefore an invertible tensor. In another subject $\mathbf{g}(t)$ may be the entropy density at a point at time t and $\mathbf{f}^t(s)$ the temperature at the point at time $t - s$ and hence a positive number. The sets of invertible tensors and of positive numbers are cones.† It is often the case that the values of $\mathbf{f}^t$ are restricted to a cone.

It is frequently possible to prove theorems in a branch of physics without completely specifying the form of the functional $\mathfrak{g}$, but usually one must assume something about the smoothness of $\mathfrak{g}$. For this reason several topologies have been proposed as appropriate for sets of histories.‡

Let us here suppose that the histories $\mathbf{f}^t$ of interest form a cone in a Banach function space $\mathfrak{B}$. The basic requirements of physical theories place certain limitations on the choice of the function space $\mathfrak{B}$ and its norm $\|\cdot\|$. I list below three of these requirements.[22,23]

1. Given an arbitrary history $\mathbf{f}^t$ in the domain $\mathscr{D}$ of a constitutive functional $\mathfrak{g}$ and a positive number σ, we expect to find in $\mathscr{D}$ the history of $\mathbf{f}$ up to $t + \sigma$ for processes in which $\mathbf{f}$ has the history $\mathbf{f}^t$ up to time t and subsequently is constant throughout the interval $[t, t + \sigma]$. The history up to time $t + \sigma$ of $\mathbf{f}$ in such a process is called the "static continuation of $\mathbf{f}^t$ by the amount σ." The static continuation of a history should be well defined even if we identify the history with the set of functions at zero distance from it in $\mathfrak{B}$.

* Noll.[31]

† A subset C of a vector space is called a *cone* if $\mathbf{u} \in C$ and $b > 0$ imply $b\mathbf{u} \in C$.

‡ For examples, see Coleman and Noll,[18–20] Coleman,[1,2] Wang,[21] Coleman and Mizel,[22,23,9] and Perzyna.[32] For a survey of the subject through 1964, see Truesdell and Noll.[33]

2. If the history $\mathbf{f}^t$ of $\mathbf{f}$ up to a time t is in the domain of $\mathfrak{g}$, then we expect to find in $\mathscr{D}$ the histories $\mathbf{f}^{t-\sigma}$ of $\mathbf{f}$ up to earlier times $t - \sigma$, $\sigma \geqslant 0$. These previous histories are called "σ-sections of $\mathbf{f}^t$."

3. Since it should be possible to evaluate $\mathfrak{g}$ at "equilibrium states," we expect the domain of $\mathfrak{g}$ to contain constant histories of the form $\mathbf{f}^t(s) \equiv \mathbf{a}$, $0 \leqslant s < \infty$.

V. J. Mizel and I[22,23] have found some apparently useful implications of these elementary physical requirements, and in this section I summarize our main results. Since it is not intended that this be a treatise on functional analysis, I omit those proofs which are readily available in the literature and which would take the discussion far afield from thermodynamics.

Let μ be an *influence measure*; that is, a nontrivial,* sigma-finite, positive, regular Borel measure on $[0, \infty)$ and let $\mathscr{s}$ be the set of all μ-measurable functions ϕ mapping $[0, \infty)$ into $[0, \infty)$. Let ν be a function on $\mathscr{s}$ such that for all ϕ (or ϕ_i) in $\mathscr{s}$:

(i) $0 \leqslant \nu(\phi) \leqslant \infty$, and $\nu(\phi) = 0$ if and only if $\phi(s) = 0$ μ-a.e.†;

(ii) $\nu(\phi_1 + \phi_2) \leqslant \nu(\phi_1) + \nu(\phi_2)$ and $\nu(a\phi) = a\nu(\phi)$ for all numbers $a \geqslant 0$;

(iii) if $\phi_1(s) \leqslant \phi_2(s)$ μ-a.e., then $\nu(\phi_1) \leqslant \nu(\phi_2)$;

(iv) there is at least one function ϕ in $\mathscr{s}$ with $0 < \nu(\phi) < \infty$;

(v) if $\phi_0, \phi_1, \phi_2, \ldots$ are in $\mathscr{s}$ and if $\phi_n(s) \uparrow \phi_0(s)$ μ-a.e., then $\nu(\phi_n) \uparrow \nu(\psi)$.

Such a function ν is called a *nontrivial function norm, relative to* μ, *with the sequential Fatou property*.

Let E^N be a real Euclidean vector space of finite dimension $N \geqslant 1$ and denote the norm on E^N by $|\cdot|$.‡ Let $\overline{\mathscr{V}}^N$ be the set of μ-measurable functions $\boldsymbol{\phi}$ mapping $[0, \infty)$ into E^N, and let $\|\cdot\|$ be the function on $\overline{\mathscr{V}}^N$ defined by

$$\|\boldsymbol{\phi}\| = \nu(|\boldsymbol{\phi}|) \qquad \text{for each} \quad \boldsymbol{\phi} \in \overline{\mathscr{V}}^N \tag{4.3}$$

I write $\mathscr{V}^N$ for the set of all functions $\boldsymbol{\phi}$ in $\overline{\mathscr{V}}^N$ with $\|\boldsymbol{\phi}\| < \infty$. In applications, the functions $\boldsymbol{\phi}$ in $\mathscr{V}^N$ are called *histories*; their independent variable is usually denoted by s and is called the *elapsed time*. The value $\boldsymbol{\phi}(0)$ of a history $\boldsymbol{\phi}$ at $s = 0$ is the *present value* of $\boldsymbol{\phi}$ and the *past values* are those for which $0 < s < \infty$. The function space $\mathfrak{B}^N$ obtained by calling two functions $\boldsymbol{\phi}_1$ and $\boldsymbol{\phi}_2$ in $\mathscr{V}^N$ the same whenever $\|\boldsymbol{\phi}_1 - \boldsymbol{\phi}_2\| = 0$ is easily

* That is, not identically zero.

† That is, for all s in $[0, \infty)$ except for a set X with $\mu(X) = 0$.

‡ Theorems 4.1–4.5 below remain valid if E^N is replaced by an arbitrary separable Banach space of finite or infinite dimension.

shown to be a Banach space; it is sometimes called a *history space* or, at length, a *Banach function space formed from histories with values in* E^N.

Let C^N be a cone in E^N, and suppose that $N \geqslant 1$ is minimal in the sense that each vector in E^N is a linear combination of elements of C^N; that is, no space of dimension less than N contains C^N. Let $\mathscr{C}^N$ be the set of functions $\boldsymbol{\phi}$ mapping $[0, \infty)$ into C^N such that $\|\boldsymbol{\phi}\|$ is finite, and let $\mathfrak{C}^N$ be the set of equivalence classes obtained by calling the same those elements $\boldsymbol{\phi}_\alpha, \boldsymbol{\phi}_\beta$ of C^N for which $\|\boldsymbol{\phi}_\alpha - \boldsymbol{\phi}_\beta\| = 0$. Clearly, $\mathscr{C}^N$ is a cone in $\mathscr{V}^N$, and $\mathfrak{C}^N$ is a cone contained in the Banach function space $\mathfrak{B}^N$. One takes $\mathfrak{C}^N$ to be the domain $\mathscr{D}$ of definition of the constitutive functional $\mathfrak{g}$ in (4.2).

If $\boldsymbol{\phi}$ is a function on $[0, \infty)$ and σ a positive number, then the *static continuation* of $\boldsymbol{\phi}$ by the amount σ is the function $\boldsymbol{\phi}^{(\sigma)}$ on $[0, \infty)$ defined by

$$\boldsymbol{\phi}^{(\sigma)}(s) = \begin{cases} \boldsymbol{\phi}(0), & 0 \leqslant s \leqslant \sigma \\ \boldsymbol{\phi}(s - \sigma), & \sigma < s < \infty \end{cases} \tag{4.4}$$

and the *σ-section* of $\boldsymbol{\phi}$ is the function $\boldsymbol{\phi}_{(\sigma)}$ on $[0, \infty)$ given by

$$\boldsymbol{\phi}_{(\sigma)}(s) = \boldsymbol{\phi}(s + \sigma), \qquad 0 \leqslant s < \infty \tag{4.5}$$

If $\mathbf{f}$ is a function on $(-\infty, \infty)$ and if $\boldsymbol{\psi}$ is the history of $\mathbf{f}$ up to t, then $\boldsymbol{\psi}_{(\sigma)}$ equals the history of $\mathbf{f}$ up to $t - \sigma$, whereas $\boldsymbol{\psi}^{(\sigma)}$ gives the history of $\mathbf{f}$ up to $t + \sigma$ assuming that $\mathbf{f}$ is held constant from t to $t + \sigma$. The physical requirements 1 and 2 stated above are made precise by laying down the following two postulates.

Postulate 1. *If a given function* $\boldsymbol{\phi}$ *is in* $\mathscr{C}^N$, *then all the static continuations* $\boldsymbol{\phi}^{(\sigma)}$, $\sigma \geqslant 0$, *of* $\boldsymbol{\phi}$ *are also in* $\mathscr{C}^N$. *Furthermore, if* $\boldsymbol{\phi}$ *and* $\boldsymbol{\psi}$ *in* $\mathscr{C}^N$ *are such that* $\|\boldsymbol{\phi} - \boldsymbol{\psi}\| = 0$, *then* $\|\boldsymbol{\phi}^{(\sigma)} - \boldsymbol{\psi}^{(\sigma)}\| = 0$ *for all* $\sigma \geqslant 0$.

Postulate 2. *If* $\boldsymbol{\phi}$ *is in* $\mathscr{C}^N$, *then so also are all its σ-sections,* $\boldsymbol{\phi}_{(\sigma)}$, $\sigma \geqslant 0$.

Employing Postulate 1, one can prove the following theorem which shows that the present value $\boldsymbol{\phi}(0)$ of a history $\boldsymbol{\phi}$ has a special status, in the sense that the norm $\|\boldsymbol{\phi}\| = \nu(|\boldsymbol{\phi}|)$ places greater emphasis on $\boldsymbol{\phi}(0)$ than on any individual past value.

Theorem 4.1.* *The influence measure* μ *must have an atom at* $s = 0$ *and be absolutely continuous on* $(0, \infty)$ *with respect to Lebesgue measure.*

Postulates 1 and 2, together, yield

Theorem 4.2.† *Either* $\mu((0, \infty)) = 0$ *or Lebesgue measure is absolutely continuous on* $(0, \infty)$ *with respect to* μ.

* Ref. 23 (Theorem 2.1).
† Ref. 23 (Theorem 2.2).

Thus the μ-measure of the singleton $\{0\}$ is not zero, and if $\mu((0, \infty))$ is not zero, then an arbitrary subset X of $(0, \infty)$ has zero μ-measure if and only if it has zero Lebesgue measure. So as to have a nontrivial theory, *let us assume that* $\mu((0, \infty))$ *is not zero*. Since the measure μ is used in the theory only to render precise the expression "μ-a.e." which occurs in the axioms (i), (iii), and (v) for ν, Theorems 4.1 and 4.2 tell us that we can replace μ with the Borel measure on $[0, \infty)$ that assigns the value 1 to the singleton $\{0\}$ and equals Lebesgue measure when restricted to Borel subsets of $(0, \infty)$.

If $\boldsymbol{\phi}$ is a function in $\mathscr{V}^N$, the restriction of $\boldsymbol{\phi}$ to $(0, \infty)$ is called the *past history* of $\boldsymbol{\phi}$ and is denoted by ${}_r\boldsymbol{\phi}$. The symbol $\mathscr{V}_r^N$ is employed for the set of functions ${}_r\boldsymbol{\phi}$ obtained by restricting members of $\mathscr{V}^N$ to $(0, \infty)$. We may define $\|\cdot\|_r$ on $\mathscr{V}_r^N$ by

$$\|{}_r\boldsymbol{\phi}\|_r = \|\boldsymbol{\phi}\chi_{(0,\infty)}\| = \nu(|\boldsymbol{\phi}|\,\chi_{(0,\infty)}) \tag{4.6}$$

where $\chi_{(0,\infty)}$ is the characteristic function of $(0, \infty)$.* The *space of past histories* is the function space $\mathfrak{B}_r^N$ obtained by calling the same those elements ${}_r\boldsymbol{\phi}$, ${}_r\boldsymbol{\psi}$ of $\mathscr{V}_r^N$ for which $\|{}_r\boldsymbol{\phi} - {}_r\boldsymbol{\psi}\|_r = 0$. It is easily verified that $\|\cdot\|_r$ is a norm on $\mathfrak{B}_r^N$ and that $\mathfrak{B}_r^N$, like $\mathfrak{B}^N$, is a Banach space. Let $\mathscr{C}_r^N$ be the set of functions in $\mathscr{V}_r^N$ with values in C^N, and let $\mathfrak{C}_r^N$ be the corresponding cone in $\mathfrak{B}_r^N$.

A consequence of Theorems 4.1 and 4.2 is

Theorem 4.3.† *The Banach space* $\mathfrak{B}^N$ *is algebraically and topologically the direct sum of* E^N *and* $\mathfrak{B}_r^N$; *that is,*

$$\mathfrak{B}^N = E^N \oplus \mathfrak{B}_r^N \tag{4.7}$$

and the norm $\|\cdot\|$ *on* $\mathfrak{B}^N$ *is equivalent to the norm* $\|\cdot\|'$ *defined by*

$$\|\boldsymbol{\phi}\|' = |\boldsymbol{\phi}(0)| + \|{}_r\boldsymbol{\phi}\|_r \tag{4.8}$$

Here $|\cdot|$ is the original norm on E^N, $\|\cdot\|$ is the norm on $\mathfrak{B}^N$ defined in (4.3), and $\|\cdot\|_r$ is the norm on $\mathfrak{B}_r^N$ defined in (4.6). The equivalence of $\|\cdot\|'$ and $\|\cdot\|$ means that there exist two positive numbers c_1, c_2 such that

$$c_1\,\|\boldsymbol{\phi}\| \leqslant \|\boldsymbol{\phi}\|' \leqslant c_2\,\|\boldsymbol{\phi}\| \tag{4.9}$$

for all $\boldsymbol{\phi}$ in $\mathfrak{B}^N$. Among other things, Theorem 4.3 tells us that, even after the elements of $\mathscr{V}^N$ are grouped together to form the equivalence classes which comprise $\mathfrak{B}^N$, each history $\boldsymbol{\phi}$ has a well-defined present value $\boldsymbol{\phi}(0)$.

* That is, $\chi_{(0,\infty)}$ is defined on $[0, \infty)$ so that $\chi_{(0,\infty)}(s) = 1$ for $s \in (0, \infty)$, while $\chi_{(0,\infty)}(0) = 0$.

† Ref. 23 (Theorem 3.1).

Furthermore, in terms more suggestive than precise, the present value $\boldsymbol{\phi}(0)$ of an element $\boldsymbol{\phi}$ of $\mathfrak{B}^N$ has approximately the same importance to $\boldsymbol{\phi}$ as its entire past history ${}_r\boldsymbol{\phi}$.

If $\mathbf{a}$ is a vector, the constant function on $[0, \infty)$ with value $\mathbf{a}$ is denoted by $\mathbf{a}^\dagger$:

$$\mathbf{a}^\dagger(s) = \mathbf{a}\,, \qquad 0 \leqslant s < \infty \tag{4.10}$$

The following postulate meets the third of the physical requirements listed at the beginning of this section.

Postulate 3. *For each vector* $\mathbf{a}$ *in* C^N, *the function* $\mathbf{a}^\dagger$ *defined by (4.10) is in* $\mathscr{V}^N$.

It follows immediately from this assumption that given any functional $\mathfrak{g}$ on $\mathfrak{C}^N$ we can define a function $\mathfrak{g}^\circ$ on C^N by the formula

$$\overset{\circ}{\mathfrak{g}}(\mathbf{a}) = \mathfrak{g}(\mathbf{a}^\dagger) \qquad \text{for all vectors } \mathbf{a} \text{ in } C^N \tag{4.11}$$

$\mathfrak{g}^\circ$ is called the *equilibrium response function corresponding to* $\mathfrak{g}$. It is not difficult to show that $\mathfrak{g}^\circ$ is a continuous function on C^N whenever $\mathfrak{g}$ is continuous on $\mathfrak{C}^N$.

Since, by assumption, the elements of C^N span the space E^N, Postulate 3 obviously implies that for each $\mathbf{x}$ in E^N, the corresponding constant function $\mathbf{x}^\dagger$ on $[0, \infty)$ is in $\mathscr{V}^N$.

In the terminology of Coleman and Mizel,[22] the norm $\|\cdot\|$ on $\mathfrak{B}^N$ is said to have the *relaxation property*, if for each function $\boldsymbol{\phi}$ in $\mathscr{V}^N$,

$$\lim_{\sigma \to \infty} \|\boldsymbol{\phi}^{(\sigma)} - \boldsymbol{\phi}(0)^\dagger\| = 0 \tag{4.12}$$

where $\boldsymbol{\phi}(0)^\dagger$ is the constant function on $[0, \infty)$ with value $\boldsymbol{\phi}(0)$. It is easy to show that $\|\cdot\|$ has the relaxation property if (and of course only if) (4.12) holds for each $\boldsymbol{\phi}$ in $\mathscr{C}^N$. Hence the assumption of the relaxation property is equivalent to the assertion that every continuous functional $\mathfrak{g}$ on $\mathfrak{C}^N$ obeys the relation

$$\lim_{\sigma \to \infty} \mathfrak{g}(\boldsymbol{\phi}^{(\sigma)}) = \mathfrak{g}(\boldsymbol{\phi}(0)^\dagger) = \overset{\circ}{\mathfrak{g}}(\boldsymbol{\phi}(0)) \tag{4.13}$$

for each $\boldsymbol{\phi}$ in $\mathscr{C}^N$; that is, in the limit of large σ, the response of $\mathfrak{g}$ to the static continuation $\boldsymbol{\phi}^{(\sigma)}$ of an arbitrary initial history $\boldsymbol{\phi}$ depends on only the present value of $\boldsymbol{\phi}$ and is given by the equilibrium response function $\mathfrak{g}^\circ$ defined in (4.11). Since this is a desirable property of constitutive functionals, let us add the following postulate to the list.

Postulate 4. *The norm* $\|\cdot\|$ *on* $\mathfrak{B}^N$ *has the relaxation property.*

Postulates 1–4 yield the following theorem.

Theorem 4.4.* *Let* $\mathbf{f}$ *and* $\mathbf{g}$ *be functions mapping* $(-\infty, \infty)$ *into* E^N *such that* $\mathbf{f}^t$ *and* $\mathbf{g}^t$ *are in* $\mathscr{V}^N$ *for all* t. *If* $\lim_{t\to\infty} |\mathbf{f}(t) - \mathbf{g}(t)| = 0$, *then* $\lim_{t\to\infty} \|\mathbf{f}^t - \mathbf{g}^t\| = 0.$

A C^N-valued function $\boldsymbol{\phi}$ with $\|\phi\| < \infty$ is called a *tame history* if

1. $\boldsymbol{\phi}$ is differentiable in the classical sense at $s = 0$; that is,

$$\dot{\boldsymbol{\phi}}(0) \overset{\text{def}}{=} -\frac{d}{ds}\boldsymbol{\phi}(s)\bigg|_{s=0} = \lim_{s\to 0+} \frac{\boldsymbol{\phi}(0) - \boldsymbol{\phi}(s)}{s} \tag{4.14}$$

exists;

2. ${}_r\boldsymbol{\phi}$, the past history of $\boldsymbol{\phi}$, is an absolutely continuous function on $(0, \infty)$;

3. $\mathfrak{V}^N$ contains an element $\dot{\boldsymbol{\phi}}$, called the *time-derivative of* $\boldsymbol{\phi}$, which obeys the equation

$$\dot{\boldsymbol{\phi}}(s) = -\frac{d}{ds}\boldsymbol{\phi}(s), \qquad \mu\text{-a.e.} \tag{4.15}$$

For technical reasons, one assumes

Postulate 5. *Tame histories with time-derivatives of compact support are dense in* $\mathfrak{C}^N$. *That is, given any* $\boldsymbol{\psi}$ *in* $\mathfrak{C}^N$ *and any* $\epsilon > 0$, *there exists a tame history* $\boldsymbol{\phi}$ *in* $\mathfrak{C}^N$ *such that*

(a) $\dot{\boldsymbol{\phi}}(s) = \mathbf{0}$ *for all* s *outside a closed bounded set in* $[0, \infty)$;
(b) $\|\boldsymbol{\psi} - \boldsymbol{\phi}\| < \epsilon$.

It follows from Postulate 5 that $\mathfrak{V}^N$ is separable, that continuous functions of compact support are dense in $\mathfrak{V}^N$, and that $\mathfrak{V}^N$ has the following *dominated-convergence property*† familiar in the theory of $\mathscr{L}_p$-spaces: If $\boldsymbol{\psi}$ belongs to $\mathfrak{V}^N$ and if $\boldsymbol{\phi}_j$ is a sequence of elements of $\mathfrak{V}^N$ with $|\boldsymbol{\phi}_j(s)| \leqslant |\boldsymbol{\psi}(s)|$, μ-a.e., such that $\boldsymbol{\phi}_j(s) \to \mathbf{0}$, μ-a.e., then $\|\boldsymbol{\phi}_j\| \to 0$.

Using Postulates 1 and 5, one can prove‡

Theorem 4.5. *The dependence of the static continuation operation,* $\boldsymbol{\phi} \mapsto \boldsymbol{\phi}^{(\sigma)}$, *on* $\sigma \geqslant 0$ *is one of strong continuity in the sense that for each* $\boldsymbol{\phi}$ *in* $\mathfrak{V}^N$ *and each* $\delta \geqslant 0$,

$$\lim_{\sigma\to\delta} \|\boldsymbol{\phi}^{(\sigma)} - \boldsymbol{\phi}^{(\delta)}\| = 0$$

* Ref. 23 (Theorem 5.1).

† For theorems of this type see Luxemburg[34] (Theorem 46.2, p. 241) and Luxemburg and Zaanen[35] (Theorem 2.2, p. 157). See also the discussion of Coleman and Mizel[23] (Remarks 3.1 and 3.2).

‡ Coleman and Mizel[9] (Appendix 1). The proof given there employs Postulate 3, but that postulate is not needed.

Let $\mathfrak{g}$ in (4.2) be a continuous function mapping $\mathfrak{C}^N$ into some metric space. It follows from Theorem 4.3 that $\mathfrak{g}$ can be regarded equally well as a function of ordered pairs $(\boldsymbol{\phi}(0); {}_r\boldsymbol{\phi})$ with $\boldsymbol{\phi}(0)$ in C^N and ${}_r\boldsymbol{\phi}$ in $\mathfrak{C}_r^N$; that is, for each $\boldsymbol{\phi}$ in $\mathfrak{C}^N$

$$\mathfrak{g}(\boldsymbol{\phi}) = \mathfrak{g}(\boldsymbol{\phi}(0); {}_r\boldsymbol{\phi})$$

Hence (4.1) can be written

$$\mathbf{g}(t) = \mathfrak{g}(\mathbf{f}^t) = \mathfrak{g}(\mathbf{f}^t(0); {}_r\mathbf{f}^t) = \mathfrak{g}(\mathbf{f}(t); {}_r\mathbf{f}^t) \tag{4.16}$$

where $\mathbf{f}(t) = \mathbf{f}^t(0)$ is the present value of $\mathbf{f}^t$, and ${}_r\mathbf{f}^t$ is the element of $\mathfrak{C}_r^N$ corresponding to the restriction of $\mathbf{f}^t$ to $(0, \infty)$. By Theorem 4.3, the continuity of $\mathfrak{g}$ over $\mathfrak{C}^N$ implies that $\mathfrak{g}(\mathbf{f}(t); \mathbf{f}_r^t)$ is jointly continuous in its two variables, $\mathbf{f}(t)$ in C^N and ${}_r\mathbf{f}^t$ in $\mathfrak{C}_r^N$.

Now, (4.2) can be viewed as a functional transformation mapping functions $\mathbf{f}$ on $(-\infty, \infty)$ into functions $\mathbf{g}$ on $(-\infty, \infty)$; that is, if we are given $\mathbf{f}(\tau)$ for all τ in $(-\infty, \infty)$, we can use (4.2) to calculate $\mathbf{g}(t)$ for each t in $(-\infty, \infty)$, provided of course, that $\mathbf{f}^t$ is in the domain $\mathfrak{C}^N$ of $\mathfrak{g}$ for each t. Mizel and I[23,22] have shown that this functional transformation $\mathbf{f} \mapsto \mathbf{g}$ preserves regularity in the following sense.*

Theorem 4.6. *Let $\mathfrak{g}$ be a continuous function mapping $\mathfrak{C}^N$ into a metric space, and suppose that $\mathbf{f}$ is a function on $(-\infty, \infty)$ with $\mathbf{f}^t$ in $\mathfrak{C}^N$ for each t. If $\mathbf{f}$ is a regulated function, that is, a function for which the limits $\lim_{\tau \to t+} \mathbf{f}(\tau)$ and $\lim_{\tau \to t-} \mathbf{f}(\tau)$ exist for each t in $(-\infty, \infty)$, then $\mathbf{g}$, given by (4.2), is also a regulated function. Furthermore, $\mathbf{g}$ can suffer discontinuities at only those times t_i at which $\mathbf{f}$ is discontinuous; at all other times $\mathbf{g}$ is continuous.*

To obtain this result one first shows that the mapping $t \mapsto {}_r\mathbf{f}^t \in \mathfrak{C}_r^N$ is continuous, for all t, even for those at which $\mathbf{f}(t)$ experiences a discontinuity; the theorem then follows immediately from (4.16) and the joint continuity of $\mathfrak{g}(\cdot; \cdot)$ in its two variables.

A constitutive functional $\mathfrak{g}$, with its range in a Banach space, is said to be *continuously Fréchet-differentiable* on $\mathfrak{C}^N$ if $\mathfrak{g}$ has $\mathfrak{C}^N$ for its domain of definition, and, for each $\boldsymbol{\psi}$ in $\mathfrak{C}^N$ and every $\boldsymbol{\phi}$ with $\boldsymbol{\psi} + \boldsymbol{\phi}$ in $\mathfrak{C}^N$,

$$\mathfrak{g}(\boldsymbol{\psi} + \boldsymbol{\phi}) = \mathfrak{g}(\boldsymbol{\psi}) + d\mathfrak{g}(\boldsymbol{\psi} \mid \boldsymbol{\phi}) + o(\|\boldsymbol{\phi}\|) \tag{4.17}$$

where $d\mathfrak{g}(\cdot \mid \cdot)$ is defined and continuous on $\mathfrak{C}^N \times \mathfrak{B}^N$ and such that $d\mathfrak{g}(\boldsymbol{\psi} \mid \boldsymbol{\phi})$ is a linear function of $\boldsymbol{\phi}$ for each $\boldsymbol{\psi}$. The linear functional $d\mathfrak{g}(\boldsymbol{\psi} \mid \cdot)$ is called the *Fréchet derivative* of $\mathfrak{g}$ at $\boldsymbol{\psi}$. Of course, in (4.17),

$$\lim_{\|\boldsymbol{\phi}\| \to 0} \frac{o(\|\boldsymbol{\phi}\|)}{\|\boldsymbol{\phi}\|} = 0$$

* This theorem does not require Postulate 4.

An argument given by Coleman and Mizel* here yields the following chain rule.

Theorem 4.7. *If* $\mathfrak{g}$ *is a continuously Fréchet-differentiable, real-valued, functional on* $\mathfrak{C}^N$, *then, for each tame history* $\boldsymbol{\phi}$ *in* $\mathfrak{C}^N$, *the derivative*

$$\dot{\mathfrak{g}} \overset{\text{def}}{=} \lim_{\sigma\to 0+} \frac{\mathfrak{g}(\boldsymbol{\phi}) - \mathfrak{g}(\boldsymbol{\phi}_{(\sigma)})}{\sigma} \tag{4.18}$$

exists and is given by

$$\dot{g} = d\mathfrak{g}(\boldsymbol{\phi} \mid \dot{\boldsymbol{\phi}}) \tag{4.19}$$

where $\dot{\boldsymbol{\phi}}$ *is the time-derivative of* $\boldsymbol{\phi}$ *defined in (4.15).*

Suppose $\mathfrak{g}$ is continuously Fréchet-differentiable on $\mathfrak{C}^N$, and recall that $\mathfrak{g}(\boldsymbol{\phi})$ can be written in the form

$$\mathfrak{g}(\boldsymbol{\phi}) = \mathfrak{g}(\boldsymbol{\phi}(0); {}_r\boldsymbol{\phi}) \tag{4.20}$$

where $\boldsymbol{\phi}(0)$ is in C^N and ${}_r\boldsymbol{\phi}$, the restriction of $\boldsymbol{\phi}$ to $(0, \infty)$, is in $\mathfrak{C}_r^{\ N}$. The assumed differentiability of $\mathfrak{g}$ on $\mathfrak{C}^N$ implies the existence, for each $\boldsymbol{\phi}$, of the *instantaneous derivative*† $D\mathfrak{g}(\boldsymbol{\phi})$ and the *past-history derivative* $\delta\mathfrak{g}(\boldsymbol{\phi} \mid \cdot)$, determined by the equations

$$\mathfrak{g}(\boldsymbol{\phi}(0) + \mathbf{u}; {}_r\boldsymbol{\phi}) = \mathfrak{g}(\boldsymbol{\phi}(0); {}_r\boldsymbol{\phi}) + D\mathfrak{g}(\boldsymbol{\phi}) \cdot \mathbf{u} + o(|\mathbf{u}|) \tag{4.21}$$

and

$$\mathfrak{g}(\boldsymbol{\phi}(0); {}_r\boldsymbol{\phi} + \boldsymbol{\zeta}) = \mathfrak{g}(\boldsymbol{\phi}(0); {}_r\boldsymbol{\phi}) + \delta\mathfrak{g}(\boldsymbol{\phi} \mid \boldsymbol{\zeta}) + o(\|\boldsymbol{\zeta}\|_r) \tag{4.22}$$

(4.21) holds for all $\mathbf{u}$ in E^N with $\boldsymbol{\phi}(0) + \mathbf{u}$ in C^N, whereas (4.22) holds for all $\boldsymbol{\zeta}$ in $\mathfrak{B}_r^{\ N}$ with ${}_r\boldsymbol{\phi} + \boldsymbol{\zeta}$ in $\mathfrak{C}_r^{\ N}$. For each $\boldsymbol{\phi}$ in $\mathfrak{C}^N$ the value $D\mathfrak{g}(\boldsymbol{\phi})$ of $D\mathfrak{g}$ is a vector in E^N, and $\delta\mathfrak{g}(\boldsymbol{\phi} \mid \cdot)$ is a linear function on $\mathfrak{B}_r^{\ N}$.

The functionals $D\mathfrak{g}$ and $\delta\mathfrak{g}$ determine the functional $d\mathfrak{g}$ through the relation

$$d\mathfrak{g}(\boldsymbol{\phi} \mid \boldsymbol{\psi}) = D\mathfrak{g}(\boldsymbol{\phi}) \cdot \boldsymbol{\psi}(0) + \delta\mathfrak{g}(\boldsymbol{\phi} \mid {}_r\boldsymbol{\psi}) \tag{4.23}$$

Hence the assumed continuity of $d\mathfrak{g}(\cdot \mid \cdot)$ on $\mathfrak{C}^N \times \mathfrak{B}^N$ implies that $D\mathfrak{g}(\cdot)$ is continuous on $\mathfrak{C}^N$, while $\delta\mathfrak{g}(\cdot \mid \cdot)$ is continuous on $\mathfrak{C}^N \times \mathfrak{B}_r^{\ N}$. Employing (4.23) one can rewrite the chain rule (4.19) in the form

$$\dot{g} = D\mathfrak{g}(\boldsymbol{\phi}) \cdot \dot{\boldsymbol{\phi}}(0) + \delta\mathfrak{g}(\boldsymbol{\phi} \mid {}_r\dot{\boldsymbol{\phi}}) \tag{4.24}$$

with $\dot{\boldsymbol{\phi}}(0)$ the present value and ${}_r\dot{\boldsymbol{\phi}}$ the past history of the time-derivative $\dot{\boldsymbol{\phi}}$ defined in (4.15).

* Ref. 9 (Remark 1 and Appendix II); see also Coleman,[1] Mizel and Wang,[36] and Coleman and Owen.[14] The proof given in Ref. 9 does not use Postulate 4.

† Coleman.[1,2]

The linear operator D taking $\mathfrak{g}$ into $D\mathfrak{g}$ plays a central role in the thermodynamics of systems with memory.*

A C^N-valued function $\boldsymbol{\psi}$ on $[0, \infty)$ is called a *tame and smooth history* if it is a tame history which is continuous on $[0, \infty)$ and such that the limits

$$\left.\begin{aligned} \boldsymbol{\psi}'(s) &= \lim_{\tau\to s+} \frac{\boldsymbol{\psi}(s+\tau) - \boldsymbol{\psi}(s)}{\tau} \\ \boldsymbol{\psi}''(s) &= \lim_{\tau\to s+} \frac{\boldsymbol{\psi}'(s+\tau) - \boldsymbol{\psi}'(s)}{\tau} \end{aligned}\right\} \tag{4.25}$$

exist at each s in $[0, \infty)$.

I record the following observation for future application.

Remark 4.1. It is a consequence of Postulate 5 that histories of compact support that are both tame and smooth are dense in $\mathfrak{C}^N$. Furthermore, it follows from Postulates 1 and 2 and the definitions given that if a history $\boldsymbol{\psi}$ is tame and smooth, then so also are all its σ-sections $\boldsymbol{\psi}_{(\sigma)}$ and static continuations $\boldsymbol{\psi}^{(\sigma)}$, $\sigma \geqslant 0$.

In terms more suggestive than precise, the next theorem tells us that if the cone C^N is open in E^N, then given a tame and smooth history $\boldsymbol{\phi}$ in $\mathfrak{C}^N$, we can arbitrarily change the present value $\dot{\boldsymbol{\phi}}(0)$ of its time-derivative $\dot{\boldsymbol{\phi}}$, without appreciably affecting either $\boldsymbol{\phi}$ or the past history $\dot{\boldsymbol{\phi}}_r$ of $\dot{\boldsymbol{\phi}}$. That is, we can find another tame and smooth history $\boldsymbol{\psi}$ such that $\dot{\boldsymbol{\psi}}(0)$ has any value $\boldsymbol{\omega}$ we wish in E^N, while $\boldsymbol{\psi}$ remains arbitrarily close in $\mathfrak{C}^N$ to $\boldsymbol{\phi}$, and ${}_r\dot{\boldsymbol{\psi}}$ remains arbitrarily close in $\mathfrak{B}_r{}^N$ to ${}_r\dot{\boldsymbol{\phi}}$. The importance of this for the thermodynamics of materials with fading memory was emphasized in my 1964 essay†; the proof given there for function spaces of $\mathscr{L}_p$-type may be readily extended to the more general Banach function spaces under consideration here. In a later work, Mizel and I‡ proved that $\boldsymbol{\psi}$ will have the desired properties if it is set equal to a "linear continuation" $\boldsymbol{\phi}^{[\zeta]}$ of $\boldsymbol{\phi}$, with $\zeta > 0$ chosen small. The linear continuation that serves is defined as follows:

$$\boldsymbol{\phi}^{[\zeta]}(s) = \begin{cases} \boldsymbol{\phi}(0) + (\zeta - s)\boldsymbol{\omega}\,, & \text{for} \quad 0 \leqslant s < \zeta \\ \boldsymbol{\phi}(s - \zeta)\,, & \text{for} \quad \zeta \leqslant s < \infty \end{cases} \tag{4.26}$$

The function $\boldsymbol{\phi}^{[\zeta]}$ may be regarded as the history up to time $t + \zeta$ of the

* Several formulas for instantaneous derivatives are given and discussed, along with physical interpretations, in Ref. 2.

† Ref. 1 (Remark 5, p. 17).

‡ Ref. 9 (Remark 2, p. 262).

function $\mathbf{f}$ on $(-\infty, t + \zeta]$ such that $\mathbf{f}^t = \boldsymbol{\phi}$ while, for $t < \tau \leqslant t + \zeta$,

$$\mathbf{f}(\tau) = f(t) + (\tau - t)\boldsymbol{\omega}$$

Here I give the proof worked out with Mizel, for I believe it furnishes a typical example of an easy application of the concepts introduced in this section. The proof does not employ Postulate 4.

Theorem 4.8. *If the cone C^N is an open subset of E^N, then given any tame and smooth history $\boldsymbol{\phi}$ in $\mathfrak{C}^N$, any vector $\boldsymbol{\omega}$ in E^N, and any positive number δ, one can find a tame and smooth history $\boldsymbol{\psi}$ in $\mathfrak{C}^N$ such that*

$$\dot{\boldsymbol{\psi}}(0) = \boldsymbol{\omega} \tag{4.27}$$

$$\|\boldsymbol{\psi} - \boldsymbol{\phi}\| < \delta \tag{4.28}$$

$$\|{}_r\dot{\boldsymbol{\psi}} - {}_r\dot{\boldsymbol{\phi}}\|_r < \delta \tag{4.29}$$

Proof. Let $\boldsymbol{\psi}$ be the linear continuation of $\boldsymbol{\phi}, \boldsymbol{\phi}^{[\zeta]}$, defined in (4.26); that is, put

$$\boldsymbol{\psi} = \boldsymbol{\phi}^{(\zeta)} + \boldsymbol{\omega} l_\zeta \tag{4.30}$$

where $\boldsymbol{\phi}^{(\zeta)}$ is the static continuation of $\boldsymbol{\phi}$ by amount ζ, and

$$l_\zeta(s) = \begin{cases} \zeta - s, & 0 \leqslant s < \zeta \\ 0, & \zeta \leqslant s < \infty \end{cases} \tag{4.31}$$

Since C^N is assumed open and $\boldsymbol{\phi}$ has all its values in C^N, if $\zeta > 0$ is chosen small enough, $\boldsymbol{\psi}$ is a C^N-valued function. Postulate 1 tells us that $\boldsymbol{\phi}^{(\zeta)}$ is in $\mathfrak{B}^N$. Because $|\boldsymbol{\omega} l_\zeta(s)|$ does not exceed the constant $|\zeta\boldsymbol{\omega}|$, Postulate 3 and property (iii) of ν imply that $\boldsymbol{\omega} l_\zeta$ is also in $\mathfrak{B}_r{}^N$. Hence for small $\zeta > 0$, $\boldsymbol{\psi}$ is in $\mathfrak{C}^N$. It is clear from (4.26) that $\dot{\boldsymbol{\psi}}(0)$, defined as in (4.14), exists and obeys (4.27). Furthermore, the absolute continuity of $\boldsymbol{\phi}$ implies that $\boldsymbol{\psi}$ is absolutely continuous, and we have

$$\dot{\boldsymbol{\psi}} = \dot{\boldsymbol{\phi}}^{(\zeta)} + \boldsymbol{\omega} g_\zeta \tag{4.32}$$

where $\dot{\boldsymbol{\phi}}$ and $\dot{\boldsymbol{\psi}}$ are the time-derivatives (4.15) of $\boldsymbol{\phi}$ and $\boldsymbol{\psi}$, $\dot{\boldsymbol{\phi}}^{(\zeta)}$ is the static continuation of $\dot{\boldsymbol{\phi}}$ by amount ζ, and

$$g_\zeta(s) = \begin{cases} 1, & 0 \leqslant s < \zeta \\ 0, & \zeta \leqslant s < \infty \end{cases} \tag{4.33}$$

Since $\boldsymbol{\phi}$ is a tame history, $\dot{\boldsymbol{\phi}}$ is in $\mathfrak{B}^N$, and hence $\dot{\boldsymbol{\phi}}^{(\zeta)}$ is in $\mathfrak{B}^N$. Of course, $|\boldsymbol{\omega} g_\zeta|$ is dominated by the constant $|\boldsymbol{\omega}|$, which implies that $\boldsymbol{\omega} g_\zeta$ is in $\mathfrak{B}^N$. Thus $\dot{\boldsymbol{\psi}}$ is in $\mathfrak{B}^N$, and hence $\boldsymbol{\psi}$ is a tame history. Because $\boldsymbol{\phi}$ is tame and smooth, it follows from (4.30) and (4.31) that $\boldsymbol{\psi}$ is continuous, and $\boldsymbol{\psi}'$ and $\boldsymbol{\psi}''$, defined in (4.25), exist everywhere, and therefore $\boldsymbol{\psi}$ is also both tame and smooth. By (4.30) and the triangle inequality, we have

$$\left.\begin{aligned} \|\boldsymbol{\psi} - \boldsymbol{\phi}\| &= \|\boldsymbol{\phi}^{(\zeta)} - \boldsymbol{\phi} - \boldsymbol{\omega} l_\zeta\| \leqslant \|\boldsymbol{\phi}^{(\zeta)} - \boldsymbol{\phi}\| + \|\boldsymbol{\omega} l_\zeta\| \\ \|{}_r\dot{\boldsymbol{\psi}} - {}_r\dot{\boldsymbol{\phi}}\|_r &= \|\dot{\boldsymbol{\phi}}_r{}^{(\zeta)} - {}_r\dot{\boldsymbol{\phi}} - \boldsymbol{\omega}_r g_\zeta\|_r \leqslant \|{}_r\dot{\boldsymbol{\phi}}^{(\zeta)} - {}_r\dot{\boldsymbol{\phi}}\|_r + \|\boldsymbol{\omega}_r g_\zeta\|_r \end{aligned}\right\} \tag{4.34}$$

It has been noted that for all s in $[0, \infty)$ and for sufficiently small $\zeta > 0$,

$$|\boldsymbol{\omega} l_\zeta(s)| \leqslant |\boldsymbol{\omega}|\,, \qquad |\boldsymbol{\omega} g_\zeta(s)| \leqslant |\boldsymbol{\omega}|$$

furthermore, at each s in $[0, \infty)$

$$\lim_{\zeta \to 0} l_\zeta(s) = \lim_{\zeta \to 0} g_\zeta(s) = 0$$

Therefore, because $\mathfrak{B}^N$ has the dominated-convergence property and constant functions are in $\mathfrak{B}^N$,

$$\lim_{\zeta \to 0} \|\boldsymbol{\omega} l_\zeta\| = 0 \tag{4.35}$$

and

$$\lim_{\zeta \to 0} \|\boldsymbol{\omega} g_\zeta\| = 0 \tag{4.36}$$

By property (iii) of ν and the definitions (4.3) and (4.6), one has, for each function $\boldsymbol{\xi}$ in $\mathfrak{B}^N$,

$$\|\boldsymbol{\xi}\| \geqslant \|_r\boldsymbol{\xi}\|_r \geqslant 0 \tag{4.37}$$

Hence (4.36) yields

$$\lim_{\zeta \to 0} \|\boldsymbol{\omega}_r g_\zeta\|_r = 0 \tag{4.38}$$

It is an immediate consequence of Theorem 4.5 that

$$\lim_{\zeta \to 0} \|\boldsymbol{\phi}^{(\zeta)} - \boldsymbol{\phi}\| = \lim_{\zeta \to 0} \|\dot{\boldsymbol{\phi}}^{(\zeta)} - \dot{\boldsymbol{\phi}}\| = 0 \tag{4.39}$$

and, by (4.37),

$$\lim_{\zeta \to 0} \|_r\dot{\boldsymbol{\phi}}^{(\zeta)} - \dot{\boldsymbol{\phi}}\|_r = 0 \tag{4.40}$$

The relations (4.34), (4.35), (4.38), (4.39), and (4.40) clearly imply that for each $\delta > 0$ there exists a number $\zeta' > 0$, depending on $\boldsymbol{\phi}$, $\boldsymbol{\omega}$, and δ, such that (4.28) and (4.29) hold for all $0 < \zeta < \zeta'$. Q.E.D.

It is evident from the proof just given that Theorem 4.8 remains valid if the word *smooth* is deleted, that is, if we require only that $\boldsymbol{\phi}$ and $\boldsymbol{\psi}$ be tame histories.

V. SMOOTHNESS OF THE CONSTITUTIVE FUNCTIONALS

To apply the theory outlined in Section IV to the constitutive equations (2.23), one may identify the Euclidean space E^N with the space R^{n+1} of $(n + 1)$-tuples. Let $\boldsymbol{\Lambda}(t)$ be the ordered pair $(\mathbf{q}(t), \theta(t))$, that is, the $(n + 1)$-tuple

$$\boldsymbol{\Lambda}(t) \stackrel{\text{def}}{=} (q_1(t), \ldots, q_n(t), \theta(t)) \tag{5.1}$$

The equations (2.23) may then be written

$$\left.\begin{aligned} \psi(t) &= \mathfrak{p}(\boldsymbol{\Lambda}^t) \\ \boldsymbol{\Xi}(t) &= \mathfrak{Z}(\boldsymbol{\Lambda}^t) \\ \eta(t) &= \mathfrak{h}(\boldsymbol{\Lambda}^t) \end{aligned}\right\} \tag{5.2}$$

with $\mathbf{\Lambda}^t$ the history of $\mathbf{\Lambda}(\cdot)$ up to t,

$$\mathbf{\Lambda}^t = (\mathbf{q}^t, \theta^t) = (q_1{}^t, \ldots, q_n{}^t, \theta^t) \tag{5.3}$$

It can happen in physical problems that some of the coordinates q_i are subject to restrictions of the type $q_i \neq 0$ or $q_i > 0$. For example, if q_i is the radius of an expanding and contracting sphere containing a fixed amount of matter which must have finite density, then q_i is constrained to be strictly positive at all times. Furthermore, it is assumed here that θ is always positive. Thus the physically realizable values of $\mathbf{\Lambda} = (q_1, \ldots, q_n, \theta)$ are not expected to fill all of R^{n+1}, but are instead restricted to a cone, which may be denoted by C^{n+1}. Let us suppose that the cone C^{n+1} is an *open* subset of R^{n+1}.

I have now assembled apparatus sufficient for a precise statement of the postulate of smoothness for the constitutive functionals $\mathfrak{p}$, $\mathfrak{Z}$, and $\mathfrak{h}$.

The Principle of Fading Memory. *There exists a Banach function space* $\mathfrak{V}^{n+1}$, *with norm* $\|\cdot\|$, *such that*

1. $\mathfrak{V}^{n+1}$ *is formed from functions mapping* $[0, \infty)$ *into* R^{n+1}, *as explained in Section IV.*
2. $\mathfrak{V}^{n+1}$ *and* $\mathfrak{C}^{n+1}$, *the cone in* $\mathfrak{V}^{n+1}$ *corresponding to functions mapping* $[0, \infty)$ *in* C^{n+1}, *obey Postulates 1–5 of Section IV with* $N = n + 1$;
3. *The functionals* $\mathfrak{p}$, $\mathfrak{Z}$, *and* $\mathfrak{h}$ *in (5.2) are defined and continuous on* $\mathfrak{C}^{n+1}$;
4. *The functionals* $\mathfrak{p}$ *and* $\mathfrak{h}$ *are continuously Fréchet-differentiable on* $\mathfrak{C}^{n+1}$ *in the sense of (4.17).*

In the discussion of the class $\mathscr{A}$ of admissible processes in Sections I–III, no mention was made of the nearly obvious requirement that the history $\mathbf{\Lambda}^t = (\mathbf{q}^t, \theta^t)$ be in the domain of definition of the constitutive functionals. To avoid a possible ambiguity, I now reformulate the original definition of an admissible process: *A smooth process* $\mathscr{P}$, *as defined in the first sentence of Section II, is called* **admissible** *if* $\mathscr{P}$ *is such that for each t in domain* $(-\infty, a]$ *of* $\mathscr{P}$ *the law of balance of energy (2.4) holds,* $\mathbf{\Lambda}^t = (\mathbf{q}^t, \theta^t)$ *is in the domain* $\mathfrak{C}^{n+1}$ *of* $\mathfrak{p}$, $\mathfrak{Z}$, *and* $\mathfrak{h}$, *and* $\psi(t)$, $\Xi(t)$, *and* $\eta(t)$ *are given by the constitutive equations (5.2).*

Let us now recall the definition of a *tame and smooth history* given in Section IV, and note that according to Remark 4.1 such histories are dense in the domain of definition of our constitutive functionals.

The following easy theorem tells us that to every sufficiently regular choice of the coordinate vector $\mathbf{q}$ and the temperature θ as functions on $(-\infty, \tau]$, there corresponds a single admissible process with domain $(-\infty, \tau]$. Here "sufficiently regular" means that $\mathbf{\Lambda}^\tau = (\mathbf{q}^\tau, \theta^\tau)$ is a tame and smooth history. It is for this reason that tame and smooth histories are important to our subject.

Theorem 5.1. *Let a number* τ *and a function* $\mathbf{\Psi}$ *mapping* $[0, \infty)$ *into* C^{n+1}, *with* $\|\mathbf{\Psi}\| < \infty$, *be given. If* $\mathbf{\Psi}$ *is a tame and smooth history, then there exists a unique admissible process with domain* $(-\infty, \tau]$ *such that the history of* $\mathbf{\Lambda} = (\mathbf{q}, \theta)$ *up to* τ *equals* $\mathbf{\Psi}$.

Proof. To construct an admissible process with

$$\mathbf{\Lambda}^{\tau} = \mathbf{\Psi} \tag{5.4}$$

one must start by putting

$$\mathbf{\Lambda}(t) \stackrel{\text{def}}{=} \mathbf{\Psi}(\tau - t), \qquad -\infty < t \leqslant \tau \tag{5.5}$$

Indeed, (5.5) states that $\mathbf{\Lambda}(\tau - s) = \mathbf{\Psi}(s)$ for $0 \leqslant s < \infty$, which is the content of (5.4). Of course, the function $\mathbf{\Lambda}(\cdot) = (\mathbf{q}(\cdot), \theta(\cdot))$ defined by (5.5) has all its values in C^{n+1}, and, since $\mathbf{\Psi}$ is tame and smooth, the numbers $\dot{q}_i$ and $\ddot{q}_i$ shown in (2.1) exist; so also do $\dot{\theta}$ and $\ddot{\theta}$, and $\theta(\cdot)$ and $\mathbf{q}(\cdot)$ are continuous functions. It follows from Postulate 2 and Remark 4.1 that all σ-sections of $\mathbf{\Psi}$ have finite norm and are tame and smooth. Thus (5.4) implies that for each $t \leqslant \tau$, $\mathbf{\Lambda}^t$ is a tame and smooth history in $\mathfrak{C}^{n+1}$, the domain of $\mathfrak{p}$, $\mathfrak{Z}$, and $\mathfrak{h}$. The constitutive equations (5.2) may now be used to calculate $\psi(t)$, $\Xi(t)$, and $\eta(t)$ for each $t \leqslant \tau$. Since $\mathfrak{p}$ and $\mathfrak{h}$ are continuously Fréchet-differentiable, $\mathfrak{Z}$ is continuous, and each history $\mathbf{\Lambda}^t$ is tame and smooth, Theorem 4.6 tells us that $\psi(\cdot)$, $\Xi(\cdot)$, and $\eta(\cdot)$ are continuous functions, and it follows from Theorem 4.7 that $\dot{\psi}(t)$ and $\dot{\eta}(t)$ exist for every $t \leqslant \tau$. Furthermore, $T(t)$ is given by (2.6), and the assumed smoothness of $\mathscr{K}$ is such that $T(\cdot)$ is continuous and $\dot{T}(t)$ exists for each t. Hence

$$E(t) = \epsilon(t) + T(t) = \psi(t) + \theta(t)\eta(t) + T(t)$$

is determined, $E(\cdot)$ is continuous, and $\dot{E}(t)$ exists for every $t \leqslant \tau$. The applied force vector $\mathbf{F}$ is given for each t by (2.11) and (2.19). Thus once one writes (5.5), all the functions comprising a process, except $h(\cdot)$, are determined via the constitutive equations, and these functions have the regularity assumed for them at the beginning of Section II. Clearly, one may pick $h(\cdot)$ so that (2.4) holds, and the function $h(\cdot)$ so chosen is unique. Q.E.D.

It sometimes simplifies matters to put

$$\mathbf{\Sigma} \stackrel{\text{def}}{=} (\Xi, -\eta) = (\Xi_1, \ldots, \Xi_n, -\eta) \tag{5.6}$$

the $(n + 1)$-tuple $\mathbf{\Sigma}(t)$ may be called the *tension-entropy vector* at time t. In this notation the equations (5.2) become

$$\left.\begin{aligned} \psi(t) &= \mathfrak{p}(\mathbf{\Lambda}^t) \\ \mathbf{\Sigma}(t) &= \mathfrak{S}(\mathbf{\Lambda}^t) \end{aligned}\right\} \tag{5.7}$$

Of course, the R^{n+1}-valued functional $\mathfrak{S}$ appearing here has $\mathfrak{Z}_1, \ldots, \mathfrak{Z}_n$, and $-\mathfrak{h}$ for its components; that is,

$$\mathfrak{S} = (\mathfrak{Z}, -\mathfrak{h}) = (\mathfrak{z}_1, \ldots, \mathfrak{z}_n, -\mathfrak{h}) \tag{5.8}$$

Since

$$\mathbf{\Sigma} \cdot \dot{\mathbf{\Lambda}} = \sum_{i=1}^{n+1} \Sigma_i \cdot \dot{\Lambda}_i = \sum_{i=1}^{n} \Xi_i \dot{q}_i - \eta\dot{\theta} = \mathbf{\Xi} \cdot \dot{\mathbf{q}} - \eta\dot{\theta} \tag{5.9}$$

(3.4) can be written

$$\dot{\psi} = \mathbf{\Sigma} \cdot \dot{\mathbf{\Lambda}} - \theta\gamma \tag{5.10}$$

Thus by (3.5), the second law of thermodynamics asserts that $\mathfrak{p}$ and $\mathfrak{S}$ must be such that, in every admissible process,

$$\dot{\psi}(t) \leqslant \mathbf{\Sigma}(t) \cdot \dot{\mathbf{\Lambda}}(t) \tag{5.11}$$

at each time t. In view of Theorem 5.1, the second law therefore requires that $\mathfrak{p}$ and $\mathfrak{S}$ obey (5.11) whenever $\mathbf{\Lambda}^t$ is a tame and smooth history.

Theorem 4.7 tells us that whenever $\mathbf{\Lambda}^t$ is a tame history in $\mathfrak{C}^{n+1}$,

$$\dot{\psi}(t) = d\mathfrak{p}(\mathbf{\Lambda}^t \,|\, \dot{\mathbf{\Lambda}}^t) \tag{5.12}$$

where $\dot{\mathbf{\Lambda}}^t$ is an element of $\mathfrak{B}^{n+1}$ such that

$$\dot{\mathbf{\Lambda}}^t(s) = -\frac{d}{ds}\mathbf{\Lambda}^t(s)\,, \qquad \mu\text{-a.e.} \tag{5.13}$$

It follows from (4.23) that if we employ the operators D and δ, defined in (4.21) and (4.22), we can write (5.12) in the form

$$\dot{\psi}(t) = D\mathfrak{p}(\mathbf{\Lambda}^t) \cdot \dot{\mathbf{\Lambda}}(t) + \delta\mathfrak{p}(\mathbf{\Lambda}^t \,|\, {}_r\dot{\mathbf{\Lambda}}^t) \tag{5.14}$$

where

$$\dot{\mathbf{\Lambda}}(t) = -\frac{d}{ds}\mathbf{\Lambda}^t(s)\bigg|_{s=0} = \dot{\mathbf{\Lambda}}^t(0) \tag{5.15}$$

and ${}_r\dot{\mathbf{\Lambda}}^t$ is the restriction of $\dot{\mathbf{\Lambda}}^t$ to $(0, \infty)$. Thus (5.10) becomes

$$\theta\gamma = [\mathbf{\Xi}(t) - D\mathfrak{p}(\mathbf{\Lambda}^t)] \cdot \dot{\mathbf{\Lambda}}(t) - \delta\mathfrak{p}(\mathbf{\Lambda}^t \,|\, {}_r\dot{\mathbf{\Lambda}}^t) \tag{5.16}$$

with $\mathbf{\Xi}(t)$ given by $(5.7)_2$. Hence we can assert

Remark 5.1. The second law of thermodynamics requires that the constitutive functionals $\mathfrak{p}$ and $\mathfrak{S}$ in (5.7) be such that the inequality

$$0 \leqslant [\mathfrak{S}(\mathbf{\Lambda}^t) - D\mathfrak{p}(\mathbf{\Lambda}^t)] \cdot \dot{\mathbf{\Lambda}}(t) - \delta\mathfrak{p}(\mathbf{\Lambda}^t \,|\, {}_r\dot{\mathbf{\Lambda}}^t) \tag{5.17}$$

hold for each tame and smooth history $\mathbf{\Lambda}^t$ in their domain of definition $\mathfrak{C}^{n+1}$.

The next section deals with necessary and sufficient conditions on $\mathfrak{p}$ and $\mathfrak{S}$ for the validity of (5.17).

VI. THERMODYNAMIC RESTRICTIONS ON THE CONSTITUTIVE FUNCTIONALS

At the conclusion of the preceding section we saw that the second law of thermodynamics requires that the inequality

$$0 \leqslant [\mathfrak{S}(\boldsymbol{\Psi}) - D\mathfrak{p}(\boldsymbol{\Psi})] \cdot \dot{\boldsymbol{\Psi}}(0) - \delta\mathfrak{p}(\boldsymbol{\Psi} \mid {}_r\dot{\boldsymbol{\Psi}}) \tag{6.1}$$

hold for each tame and smooth history $\boldsymbol{\Psi}$ in the domain $\mathfrak{C}^{n+1}$ of the functionals $\mathfrak{p}$ and $\mathfrak{S}$. Here $\dot{\boldsymbol{\Psi}}$ is an element of $\mathfrak{B}^{n+1}$ such that

$$\dot{\boldsymbol{\Psi}}(s) = -\frac{d}{ds}\boldsymbol{\Psi}(s), \qquad \mu - \text{a.e.}$$

${}_r\dot{\boldsymbol{\Psi}}$ is the past history of $\dot{\boldsymbol{\Psi}}$; and $\dot{\boldsymbol{\Psi}}(0)$ is the present value of $\dot{\boldsymbol{\Psi}}$. By (5.8) and the assumed principle of fading memory, the functional $\mathfrak{S}$ is continuous on $\mathfrak{C}^{n+1}$, and since $\mathfrak{p}$ is assumed to be continuously Fréchet-differentiable* on $\mathfrak{C}^{n+1}$, $D\mathfrak{p}$ is continuous on $\mathfrak{C}^{n+1}$, and $\delta\mathfrak{p}(\cdot \mid \cdot)$ is continuous on $\mathfrak{C}^{n+1} \times \mathfrak{B}_r^{n+1}$.

Now, let a tame and smooth history $\boldsymbol{\Phi}$ in $\mathfrak{C}^{n+1}$ be assigned, and let $\boldsymbol{\Omega}$ be an arbitrary element of R^{n+1}. Theorem 4.8 tells us that given any $\delta > 0$ one can find a tame and smooth history $\boldsymbol{\Psi}$ in $\mathfrak{C}^{n+1}$ such that

$$\dot{\boldsymbol{\Psi}}(0) = \boldsymbol{\Omega} \tag{6.2}$$

and

$$\|\boldsymbol{\Psi} - \boldsymbol{\Phi}\| < \delta, \qquad \|{}_r\dot{\boldsymbol{\Psi}} - {}_r\dot{\boldsymbol{\Phi}}\|_r < \delta \tag{6.3}$$

For this tame and smooth history, (6.1) reads

$$0 \leqslant [\mathfrak{S}(\boldsymbol{\Psi}) - D\mathfrak{p}(\boldsymbol{\Psi})] \cdot \boldsymbol{\Omega} - \delta\mathfrak{p}(\boldsymbol{\Psi} \mid {}_r\dot{\boldsymbol{\Psi}}) \tag{6.4}$$

By (6.3) and the continuity of $\mathfrak{S}$, $D\mathfrak{p}$, and $\delta\mathfrak{p}$, the inequality (6.4) can be written

$$0 \leqslant [\mathfrak{S}(\boldsymbol{\Phi}) - D\mathfrak{p}(\boldsymbol{\Phi})] \cdot \boldsymbol{\Omega} - \delta\mathfrak{p}(\boldsymbol{\Phi} \mid {}_r\dot{\boldsymbol{\Phi}}) + o(1) \tag{6.5}$$

where $o(1)$ is a number which depends on $\boldsymbol{\Phi}$ and $\boldsymbol{\Omega}$ but which, once these have been selected, can be made to have as small a magnitude as desired merely by choosing $\delta > 0$ small. When $\boldsymbol{\Phi}$ is given, (6.5) can hold for every $\boldsymbol{\Omega}$ in R^{n+1} only if

$$\mathfrak{S}(\boldsymbol{\Phi}) - D\mathfrak{p}(\boldsymbol{\Phi}) = \mathbf{0} \tag{6.6}$$

and

$$0 \leqslant -\delta\mathfrak{p}(\boldsymbol{\Phi} \mid {}_r\dot{\boldsymbol{\Phi}}) \tag{6.7}$$

This argument is clearly independent of the choice of the tame and smooth history $\boldsymbol{\Phi}$ in $\mathfrak{C}^{n+1}$. Furthermore, since $\mathfrak{S}$ and $D\mathfrak{p}$ are continuous on $\mathfrak{C}^{n+1}$,

* See the paragraphs containing equations (4.17) and (4.23).

and, as has been noted, tame and smooth histories are dense in $\mathfrak{C}^{n+1}$, (6.6) holds throughout the domain of $\mathfrak{S}$ and $\mathfrak{p}$, that is, for *every history* $\mathbf{\Phi}$ in $\mathfrak{C}^{n+1}$. The inequality (6.7) is not meaningful for all histories $\mathbf{\Phi}$ in $\mathfrak{C}^{n+1}$; for some histories, ${}_r\dot{\mathbf{\Phi}}$ does not exist. However, ${}_r\dot{\mathbf{\Phi}}$ does exist for tame histories, and, because of the continuity of $\delta\mathfrak{p}$ on $\mathfrak{C}^{n+1} \times \mathfrak{B}_r^{n+1}$, (6.7) holds for *every tame history* $\mathbf{\Phi}$, including those which are not also smooth on $[0, \infty)$. This proves the main theorem of our subject:

Theorem 6.1. *It follows from the second law of thermodynamics and the assumed principle of fading memory, that*

1. $\mathfrak{p}$ *determines* $\mathfrak{S}$ *through the relation*

$$\mathfrak{S} = D\mathfrak{p} \tag{6.8}$$

2. *for each tame history* $\mathbf{\Phi}$ *in* $\mathfrak{C}^{n+1}$

$$\delta\mathfrak{p}(\mathbf{\Phi} \,|\, {}_r\dot{\mathbf{\Phi}}) \leqslant 0 \tag{6.9}$$

In my essay of 1964 I obtained a theorem* that can be regarded as the natural analogue of Theorem 6.1 in continuum thermodynamics. The proof given there employed a form of the theory of fading memory less general than the present.† The constitutive functionals were assumed to have smoothness properties (i.e., continuity and Fréchet-differentiability) similar to those used here, but for the history space (i.e., the analogue of $\mathfrak{B}^{n+1}$), I employed a Hilbert space $\mathfrak{H}$ formed from functions $\mathbf{\Phi}$ mapping $[0, \infty)$ into a Euclidean space E, and I assumed that the norm $\|\cdot\|$ on $\mathfrak{H}$ has the form

$$\|\mathbf{\Phi}\|^2 = |\mathbf{\Phi}(0)|^2 + \int_0^\infty |\mathbf{\Phi}(s)|^2 k(s)\, ds$$

with $|\cdot|$ the norm on E; k, called the "influence function," was a fixed, positive, monotone decreasing function, assumed summable on $(0, \infty)$.‡ In 1967 Mizel and I[9] observed that the theorem remains valid for the more general history spaces discussed in Section IV.§

In view of $(5.7)_2$ and (5.16), Theorem 6.1 has the following corollary.

Remark 6.1. In each admissible process, at each time t,

$$\mathbf{\Sigma}(t) = D\mathfrak{p}(\mathbf{\Lambda}^t) \tag{6.10}$$

* Theorem 1, on p. 19 of Ref. 1.

† The theory of fading memory used in Refs. 1 and 2 was drawn from earlier work[18–20] done with Noll.

‡ No further assumptions on k are needed for the main theorems of Refs. 1 and 2. Ref. 22 gives an axiomatic approach to such history spaces.

§ This and related matters from the thermodynamics of continua are discussed in more detail in the Appendix.

and, if $\mathbf{\Lambda}^t$ is a tame history,

$$\gamma(t) = -\frac{1}{\theta(t)}\,\delta\mathfrak{p}(\mathbf{\Lambda}^t \mid {}_r\dot{\mathbf{\Lambda}}^t) \tag{6.11}$$

Hence *the relations (6.8) and (6.9) give not only necessary, but also sufficient conditions on* $\mathfrak{p}$ *and* $\mathfrak{S}$ *for the rate of production of entropy,* $\gamma(t)$, *to be non-negative in an admissible process with* $\mathbf{\Lambda}^t$ *tame.*

It is clear from (2.23) and (4.20) that $\psi(t)$ may be regarded as a function of the present values $\mathbf{q}$, θ of $\mathbf{q}^t$ and θ^t and the corresponding past histories ${}_r\mathbf{q}^t$, ${}_r\theta^t$; that is,

$$\psi(t) = \mathfrak{p}(\mathbf{q}, \theta; {}_r\mathbf{q}^t, {}_r\theta^t) \tag{6.12}$$

where $\mathbf{q} = \mathbf{q}^t(0) = \mathbf{q}(t)$, $\theta = \theta^t(0) = \theta(t)$, and ${}_r\mathbf{q}^t$, ${}_r\theta^t$ are the restrictions of $\mathbf{q}^t$ and θ^t to $(0, \infty)$. The principle of fading memory implies the existence of functionals $D_{\mathbf{q}}\mathfrak{p}$ and $D_\theta\mathfrak{p}$ obeying

$$\left.\frac{\partial}{\partial\alpha}\,\mathfrak{p}(\mathbf{q} + \alpha\mathbf{u}, \theta; {}_r\mathbf{q}^t, {}_r\theta^t)\right|_{\alpha=0} = D_{\mathbf{q}}\mathfrak{p}(\mathbf{q}^t, \theta^t)\cdot\mathbf{u} \tag{6.13}$$

$$\left.\frac{\partial}{\partial\alpha}\,\mathfrak{p}(\mathbf{q}, \theta + \alpha; {}_r\mathbf{q}^t, {}_r\theta^t)\right|_{\alpha=0} = D_\theta\mathfrak{p}(\mathbf{q}^t, \theta^t) \tag{6.14}$$

for each pair $(\mathbf{q}^t, \theta^t)$ in $\mathfrak{C}^{n+1}$. The functional $D_{\mathbf{q}}\mathfrak{p}$ has its values in R^n, and (6.13) holds for each vector $\mathbf{u}$ in R^n; $D_\theta\mathfrak{p}$ is a real-valued functional. Similarly, by writing (6.12) in the extended form

$$\psi(t) = \mathfrak{p}(q_1, \ldots, q_n, \theta; {}_rq_1{}^t, \ldots, {}_rq_n{}^t, {}_r\theta^t) \tag{6.15}$$

with $q_i = q_i{}^t(0)$, one can define "instantaneous partial derivatives," $D_{q_i}\mathfrak{p}$, as follows:

$$D_{q_i}\mathfrak{p}(\mathbf{q}^t, \theta^t) = \left.\frac{\partial}{\partial\alpha}\,\mathfrak{p}(q_1, \ldots, q_i + \alpha, \ldots, q_n, \theta; {}_rq_1{}^t, \ldots, {}_rq_n{}^t, {}_r\theta^t)\right|_{\alpha=0} \tag{6.16}$$

The R^{n+1}-valued functional $D\mathfrak{p}$, obeying

$$D\mathfrak{p}(\mathbf{\Lambda}^t)\cdot\mathbf{\Omega} = \left.\frac{\partial}{\partial\alpha}\,\mathfrak{p}(\mathbf{\Lambda} + \alpha\mathbf{\Omega}; {}_r\mathbf{\Lambda}^t)\right|_{\alpha=0} \tag{6.17}$$

for all $\mathbf{\Omega}$ in R^{n+1}, has $D_{q_1}\mathfrak{p}, \ldots, D_{q_n}\mathfrak{p}$, $D_\theta\mathfrak{p}$ for its components; that is,

$$D\mathfrak{p} = (D_{\mathbf{q}}\mathfrak{p}, D_\theta\mathfrak{p}) \tag{6.18}$$

and

$$D_{\mathbf{q}}\mathfrak{p} = (D_{q_1}\mathfrak{p}, \ldots, D_{q_n}\mathfrak{p}) \tag{6.19}$$

In view of (5.8), (6.18), and (6.19), one can assert

Remark 6.2. The relation (6.8) between $\mathfrak{S}$ and $\mathfrak{p}$ is equivalent to the $n + 1$ equations

$$\mathfrak{h} = -D_\theta \mathfrak{p} \tag{6.20}$$

$$\mathfrak{Z}_i = D_{q_i}\mathfrak{p}\,, \qquad i = 1, \ldots, n \tag{6.21}$$

In other words, in each admissible process $\mathfrak{p}$ determines the entropy η and the internal forces Ξ_i through the equations

$$\eta(t) = -D_\theta \mathfrak{p}(\mathbf{q}^t, \theta^t) \tag{6.22}$$

$$\Xi_i(t) = D_{q_i}\mathfrak{p}(\mathbf{q}^t, \theta^t)\,, \qquad i = 1, \ldots, n \tag{6.23}$$

Thus it is a consequence of the assumed principle of fading memory and the second law of thermodynamics that a system $\mathscr{S}$ is completely characterized once one has specified its inertia function $\mathscr{K}$ and a single scalar-valued functional $\mathfrak{p}$.

Of course, by (6.19), the n equations (6.21) can be written as the single vector equation

$$\boldsymbol{\mathfrak{Z}} = D_{\mathbf{q}}\mathfrak{p} \tag{6.24}$$

which we may call the *general internal force relation.*

The Banach function space $\mathfrak{V}^{n+1}$ can be regarded as the direct sum, $\mathfrak{V}^1 \oplus \mathfrak{V}^1 \oplus \cdots \oplus \mathfrak{V}^1$, of $n + 1$ identical function spaces $\mathfrak{V}^1$, each of which is formed from functions mapping $[0, \infty)$ into $(-\infty, \infty)$ (i.e., R^1) and is endowed with a norm $\|\cdot\|$ defined by (4.3) with $N = 1$. The space $\mathfrak{V}^{n+1}$ can also be regarded as a direct sum $\mathfrak{V}^n \oplus \mathfrak{V}^1$ of one of the spaces $\mathfrak{V}^1$ and a space $\mathfrak{V}^n$ which is formed from functions mapping $[0, \infty)$ into R^n and which has a norm $\|\cdot\|$ given by (4.3) with $N = n$. The past history spaces $\mathfrak{V}_r{}^1$ and $\mathfrak{V}_r{}^n$ corresponding to $\mathfrak{V}^1$ and $\mathfrak{V}^n$ are comprised of functions mapping $(0, \infty)$ into R^1 and R^n, respectively, and have norms defined by (4.6). The physically realizable values of the coordinate vector $\mathbf{q}$ form an open cone C^n in R^n. We may denote by $\mathfrak{C}^n$ and $\mathfrak{C}_r{}^n$ the cones in $\mathfrak{V}^n$ and $\mathfrak{V}_r{}^n$ corresponding to functions with values in C^n. Let $\mathfrak{C}^1_{r,+}$ be the cone in $\mathfrak{V}_r{}^1$ comprised of positive functions on $(0, \infty)$ and let $\mathfrak{C}^1_{r,i}$ be the cone in $\mathfrak{V}_r{}^1$ corresponding to functions whose range is contained in the set of physically realizable values of the ith coordinate q_i. The principle of fading memory implies the existence of linear functionals $\delta_{\mathbf{q}}\mathfrak{p}(\mathbf{q}^t, \theta^t \mid \cdot)$, $\delta_\theta\mathfrak{p}(\mathbf{q}^t, \theta^t \mid \cdot)$, $\delta_{q_i}\mathfrak{p}(\mathbf{q}^t, \theta \mid \cdot)$ which, in the notation of (6.12) and (6.15), obey the relations

$$\left.\begin{aligned}
\mathfrak{p}(\mathbf{q}, \theta; {}_r\mathbf{q}^t + \boldsymbol{\zeta}, {}_r\theta^t) &= \mathfrak{p}(\mathbf{q}, \theta; {}_r\mathbf{q}^t, {}_r\theta^t) + \delta_{\mathbf{q}}\mathfrak{p}(\mathbf{q}^t, \theta^t \mid \boldsymbol{\zeta}) + o(\|\boldsymbol{\zeta}\|_r) \\
\mathfrak{p}(\mathbf{q}, \theta; {}_r\mathbf{q}^t, {}_r\theta^t + \zeta) &= \mathfrak{p}(\mathbf{q}, \theta; {}_r\mathbf{q}^t, {}_r\theta^t) + \delta_{\theta}\mathfrak{p}(\mathbf{q}^t, \theta^t \mid \zeta) + o(\|\zeta\|_r) \\
\mathfrak{p}(q_1, \ldots, q_n, \theta; {}_rq_1{}^t, \ldots, {}_rq_i{}^t + \xi, \ldots, {}_r\theta^t) & \\
= \mathfrak{p}(\mathbf{q}, \theta; {}_r\mathbf{q}^t, {}_r\theta^t) &+ \delta_{q_i}\mathfrak{p}(\mathbf{q}^t, \theta^t \mid \xi) + o(\|\xi\|_r)
\end{aligned}\right\} \quad (6.25)$$

These equations hold for each $\boldsymbol{\zeta}$ in $\mathfrak{V}_r{}^n$ with ${}_r\mathbf{q}^t + \boldsymbol{\xi}$ in $\mathfrak{C}_r{}^n$, each ζ in $\mathfrak{V}_r{}^1$ with ${}_r\theta^t + \zeta$ in $\mathfrak{C}^1_{r,+}$ and each ξ in $\mathfrak{V}_r{}^1$ with ${}_rq_i{}^t + \xi$ in $\mathfrak{C}^1_{r,i}$.

Since $\dot{\mathbf{\Lambda}}_r{}^t = ({}_r\dot{\mathbf{q}}^t, {}_r\dot{\theta}^t) = ({}_r\dot{\mathbf{q}}_1{}^t, \ldots, {}_r\dot{\mathbf{q}}_n{}^t, {}_r\dot{\theta}^t)$, the following assertion is an immediate consequence of (6.11) and the linearity of the functionals $\delta\mathfrak{p}(\mathbf{\Lambda}^t \mid \cdot)$ and $\delta_{\mathbf{q}}\mathfrak{p}(\mathbf{q}^t, \theta^t \mid \cdot)$.

Remark 6.3. In an admissible process with $(\mathbf{q}^t, \theta^t)$ a tame history, the rate $\gamma(t)$ of production of entropy is given by the expressions

$$\left.\begin{aligned}
\theta\gamma(t) &= -\delta_{\mathbf{q}}\mathfrak{p}(\mathbf{q}^t, \theta^t \mid {}_r\dot{\mathbf{q}}^t) - \delta_{\theta}\mathfrak{p}(\mathbf{q}^t, \theta^t \mid {}_r\dot{\theta}^t) \\
&= -\sum_{i=1}^{n} \delta_{q_i}\mathfrak{p}(\mathbf{q}^t, \theta^t \mid {}_r\dot{q}_i^t) - \delta_{\theta}\mathfrak{p}(\mathbf{q}^t, \theta^t \mid {}_r\dot{\theta}^t)
\end{aligned}\right\} \quad (6.26)$$

If we add to item 4 of the Principle of Fading Memory the assumption that $\mathfrak{Z}$ (and hence each of its component functionals $\mathfrak{Z}_i$) is continuously Fréchet-differentiable on $\mathfrak{C}^{n+1}$, then the quantities

$$\left.\begin{aligned}
D_\theta\mathfrak{Z}_i(\mathbf{q}^t, \theta^t) &= \frac{\partial}{\partial\alpha}\, \mathfrak{Z}_i(\mathbf{q}, \theta + \alpha; q_r^{\,t}, \theta_r^{\,t}) \\
D_{q_j}\mathfrak{Z}_i(\mathbf{q}^t, \theta^t) &= \frac{\partial}{\partial\alpha}\, \mathfrak{Z}_i(q_1, \ldots, q_j + \alpha, \ldots, q_n, \theta; {}_rq_1{}^t, \ldots, {}_rq_n{}^t, {}_r\theta^t)
\end{aligned}\right\} \quad (6.27)$$

as well as the quantities $D_\theta\mathfrak{h}$ and $D_{q_i}\mathfrak{h}$, exist, and, by Remark 6.2, obey the relations

$$\left.\begin{aligned}
D_\theta\mathfrak{Z}_i &= D_\theta D_{q_i}\mathfrak{p}\,, & D_{q_j}\mathfrak{Z}_i &= D_{q_j}D_{q_i}\mathfrak{p} \\
D_\theta\mathfrak{h} &= -D_\theta D_\theta\mathfrak{p}\,, & D_{q_i}\mathfrak{h} &= -D_{q_i}D_\theta\mathfrak{p}
\end{aligned}\right\} \quad (6.28)$$

Indeed, by (6.8), Fréchet-differentiability of $\mathfrak{Z}$ and $\mathfrak{h}$ implies that the second-order instantaneous derivatives of $\mathfrak{p}$ exist, and, by the definition of instantaneous derivatives and the well-known symmetry of second-order gradients, there hold the relations

$$D_{q_j}D_{q_i}\mathfrak{p} = D_{q_i}D_{q_j}\mathfrak{p}\,, \qquad D_\theta D_{q_j}\mathfrak{p} = D_{q_j}D_\theta\mathfrak{p}$$

This proves

Theorem 6.2.* *If* $\mathfrak{Z}$ *is continuously Fréchet-differentiable on* $\mathfrak{C}^{n+1}$, *then the cross-relations*

$$D_{q_j}\mathfrak{Z}_i = D_{q_i}\mathfrak{Z}_j\,, \quad D_\theta\mathfrak{Z}_i = -D_{q_i}\mathfrak{h}\,, \qquad i,j = 1,\ldots,n \qquad (6.29)$$

hold throughout $\mathfrak{C}^{n+1}$.

In words: the instantaneous derivative of the ith internal force Ξ_i with respect to the jth coordinate q_j equals the instantaneous derivative of the jth internal force Ξ_j with respect to the ith coordinate q_i, and the instantaneous derivative of the ith internal force with respect to temperature equals the negative of the instantaneous derivative of the entropy with respect to the ith coordinate.

The results obtained in this and the preceding chapter do not require the assumption that $\mathfrak{V}^{n+1}$ has the relaxation property. That is, in proving Theorems 5.1, 6.1, and 6.2 and Remarks 6.1, 6.2, and 6.3, no use was made of Postulate 4 of Section IV. This postulate is needed, however, for the theorems proved in the next section. In Section VII we shall see that when (6.22) and (6.23) are specialized to equilibrium situations, they reduce to formulas familiar in classical thermostatics.† To obtain this result one first shows that it follows from Postulate 4 and the second law that equilibrium situations minimize the Helmholtz free energy in the following sense: Of all histories ending with given values of $\mathbf{q}$ and θ, that corresponding to constant values of $\mathbf{q}$ and θ for all times gives the smallest value to ψ.‡ One then shows that this extremum principle implies that the differential form $\delta_\tau\mathfrak{p}(\mathbf{\Lambda}^t \mid \boldsymbol{\zeta})$ vanishes for all $\boldsymbol{\zeta}$ in $\mathfrak{V}_\tau^{n+1}$ whenever the function $\mathbf{\Lambda}^t$ is a constant.§

VII. CONSEQUENCES OF THE RELAXATION PROPERTY

Let $\mathscr{P}_\tau$ be an arbitrary admissible process with domain $(-\infty, \tau]$. If $\mathscr{P}_{\tau+\sigma}$, with $\sigma > 0$, is an admissible process which has $(-\infty, \tau + \sigma]$ for its domain and which for each t in $(-\infty, \tau]$ gives the same values to $\mathbf{\Lambda}(t) = (\mathbf{q}(t), \theta(t))$ as $\mathscr{P}_\tau$, and for each t in $[\tau, \tau + \sigma]$ gives the constant value $\mathbf{\Lambda}(\tau)$ to $\mathbf{\Lambda}(t)$, then $\mathscr{P}_{\tau+\sigma}$ is said to be an *isothermal stationary extension* of $\mathscr{P}_\tau$ by the amount σ.

* Compare Coleman[2] (Theorem on p. 249), Coleman and Gurtin[5] (equations (3.11)–(3.13) on p. 274 and equations (4.12) and (4.13) on pp. 330 and 331), and Coleman and Dill[16] (Theorem 5 on p. 149). Discussions of the physical significance of "instantaneous derivatives" are given in Refs. 2, 6, and 8; see, particularly, pp. 240 and 241 of Ref. 2.

† Theorem 7.3 and Remark 7.4.

‡ Theorem 7.1.

§ Theorem 7.2.

The following remark is an immediate consequence of (4.1), (4.4), and the definition just given.

Remark 7.1. If $\mathscr{P}_{\tau+\sigma}$ and $\mathscr{P}_\tau$ are admissible processes with domains $(-\infty, \tau + \sigma]$ and $(-\infty, \tau]$, then $\mathscr{P}_{\tau+\sigma}$ is an isothermal, stationary extension of $\mathscr{P}_\tau$ if and only if $(\mathbf{q}^{\tau+\sigma}, \theta^{\tau+\sigma})$ for the process $\mathscr{P}_{\tau+\sigma}$ is a static continuation by amount σ of $(\mathbf{q}^\tau, \theta^\tau)$ for the process $\mathscr{P}_\tau$.

The proof given for Theorem 5.1 here yields

Remark 7.2. If $\mathscr{P}_\tau$ is an admissible process on $(-\infty, \tau]$, then for each $\sigma > 0$ there exists a unique admissible process $\mathscr{P}_{\tau+\sigma}$ which is an isothermal, stationary extension of $\mathscr{P}_\tau$ by the amount σ. During the "static part," $\tau < t \leqslant \tau + \sigma$, of the process $\mathscr{P}_{\tau+\sigma}$

$$\dot{\mathbf{\Lambda}}(t) = (\dot{\mathbf{q}}(t), \dot{\theta}(t)) = \mathbf{0} \tag{7.1}$$

and, by (3.4) and (3.5),

$$\dot{\psi}(t) = -\theta(t)\gamma(t) \leqslant 0 \tag{7.2}$$

Thus *the Helmholtz free energy cannot increase during the static part of an isothermal stationary extension of an admissible process.*

Let $\mathfrak{p}^\circ$ and $\mathfrak{S}^\circ$ denote the equilibrium response functions corresponding to the functionals $\mathfrak{p}$ and $\mathfrak{S}$. That is, for each $(n + 1)$-tuple $\mathbf{\Lambda}$ in $\mathfrak{C}^{n+1}$, let

$$\mathfrak{p}^\circ(\mathbf{\Lambda}) \overset{\text{def}}{=} \mathfrak{p}(\mathbf{\Lambda}^\dagger), \qquad \mathfrak{S}^\circ(\mathbf{\Lambda}) \overset{\text{def}}{=} \mathfrak{S}(\mathbf{\Lambda}^\dagger) \tag{7.3}$$

where $\mathbf{\Lambda}^\dagger$ is the constant function on $[0, \infty)$ with value $\mathbf{\Lambda}$. Because we have assumed that $\mathfrak{p}$ and $\mathfrak{S}$ are smooth functions on $\mathfrak{C}^{n+1}$, the relation (4.13), which is a direct consequence of Postulates 1 and 4 for $\mathfrak{B}^{n+1}$, here tells us that if $\mathbf{\Lambda}^{\tau+\sigma}$ is the static continuation by amount σ of a history $\mathbf{\Lambda}^\tau$ in $\mathfrak{C}^{n+1}$, then

$$\lim_{\sigma\to\infty} \mathfrak{p}(\mathbf{\Lambda}^{\tau+\sigma}) = \mathfrak{p}^\circ(\mathbf{\Lambda}^\tau(0)), \qquad \lim_{\sigma\to\infty} \mathfrak{S}(\mathbf{\Lambda}^{\tau+\sigma}) = \mathfrak{S}^\circ(\mathbf{\Lambda}^\tau(0)) \tag{7.4}$$

It follows from Theorem 5.1, Remark 7.1, and Remark 7.2 that if $\mathbf{\Lambda}^\tau = (\mathbf{q}^\tau, \theta^\tau)$ is a tame and smooth history in $\mathfrak{C}^{n+1}$ and if $\mathbf{\Lambda}^{\tau+\sigma}$ is a static continuation of $\mathbf{\Lambda}^\tau$, then

$$\mathfrak{p}(\mathbf{\Lambda}^{\tau+\sigma}) \leqslant \mathfrak{p}(\mathbf{\Lambda}^\tau) \tag{7.5}$$

From this observation and $(7.4)_1$ we may conclude that whenever $\mathbf{\Lambda}^\tau$ is a tame and smooth history

$$\mathfrak{p}^\circ(\mathbf{\Lambda}^\tau(0)) \leqslant \mathfrak{p}(\mathbf{\Lambda}^\tau) \tag{7.6}$$

However, since tame and smooth histories are dense in the domain $\mathfrak{C}^{n+1}$ of the continuous functional $\mathfrak{p}$, the relation (7.6), which may be written in the form

$$\text{if } \mathbf{\Lambda}^{\dagger}(s) \equiv \mathbf{\Phi}(0) \text{ for } s \text{ in } [0, \infty), \text{ then } \mathfrak{p}(\mathbf{\Lambda}^{\dagger}) \leqslant \mathfrak{p}(\mathbf{\Phi}) \tag{7.7}$$

is actually valid for *every* function $\mathbf{\Phi}$ in $\mathfrak{C}^{n+1}$. This proves the following theorem, which has several applications, both in the present subject and in the theory of stability of motion.

Theorem 7.1.* *Of all histories in $\mathfrak{C}^{n+1}$ ending with a given value of $\mathbf{\Lambda} = (\mathbf{q}, \theta)$, that corresponding to constant values of $\mathbf{\Lambda}$ for all times gives the smallest value to the Helmholtz free energy; that is,* (7.7) *holds for each history $\mathbf{\Phi}$ in the domain of $\mathfrak{p}$.*

There is another way of stating this theorem: For each fixed present value of $\mathbf{\Lambda}$ the functional $\mathfrak{p}$ has a minimum when the past history is the function ${}_r\mathbf{\Lambda}^{\dagger}$ defined by

$${}_r\mathbf{\Lambda}^{\dagger}(s) = \mathbf{\Lambda}\,, \qquad 0 < s < \infty \tag{7.8}$$

that is, for each $\mathbf{\Lambda}$ in C^{n+1},

$$\mathfrak{p}(\mathbf{\Lambda};\, {}_r\mathbf{\Lambda}^{\dagger}) \leqslant \mathfrak{p}(\mathbf{\Lambda}; \boldsymbol{\xi}) \tag{7.9}$$

for every $\boldsymbol{\xi}$ in $\mathfrak{C}_r^{n+1}$. One expects such an extremum property for $\mathfrak{p}$ to imply that the first variation of $\mathfrak{p}$ with respect to the past history, that is, $\delta\mathfrak{p}(\mathbf{\Phi} \mid \cdot)$, should vanish identically whenever $\mathbf{\Phi}$ reduces to a constant. In fact, we have

Theorem 7.2.† *For each constant function $\mathbf{\Lambda}^{\dagger}$ mapping $[0, \infty)$ into C^{n+1} and for all functions $\boldsymbol{\zeta}$ in $\mathfrak{B}_r^{n+1}$*

$$\delta\mathfrak{p}(\mathbf{\Lambda}^{\dagger} \mid \boldsymbol{\zeta}) = 0 \tag{7.10}$$

Proof. To establish the theorem by contradiction, let us suppose that for some constant function $\mathbf{\Lambda}^{\dagger}$ in $\mathfrak{C}^{n+1}$ and some $\boldsymbol{\zeta}_1$ in $\mathfrak{B}_r^{n+1}$,

$$\delta\mathfrak{p}(\mathbf{\Lambda}^{\dagger} \mid \boldsymbol{\zeta}_1) \neq 0 \tag{7.11}$$

In view of the observation made immediately after the statement of Postulate 5 in Section IV, $\mathfrak{B}_r^{n+1}$ contains a dense family of bounded functions, namely, the set of continuous functions in $\mathfrak{B}^{n+1}$ with compact support. The restrictions of these functions to $(0, \infty)$ obviously form another set of bounded functions, and, by Theorem 4.3, this latter set is dense in $\mathfrak{B}_r^{n+1}$. Therefore, we may conclude from (7.11) that for some bounded function $\boldsymbol{\zeta}_2$ in $\mathfrak{B}_r^{n+1}$

$$\delta\mathfrak{p}(\mathbf{\Lambda}^{\dagger} \mid \boldsymbol{\zeta}_2) = d \qquad \text{with} \qquad d \neq 0 \tag{7.12}$$

* Coleman[1] (Theorem 3, pp. 25 and 26).

† Coleman[1] (Corollary to Theorem 3, p. 26). The proof I give here is taken from a paper[9] written with Mizel (Theorem 3, p. 268).

Let ${}_r\mathbf{\Lambda}^\dagger$ be the restriction to $(0, \infty)$ of $\mathbf{\Lambda}^\dagger$. Since $\boldsymbol{\zeta}_2$ is bounded, and the set C^{n+1} is open in R^{n+1}, there is a number $\beta > 0$ such that ${}_r\mathbf{\Lambda}^\dagger + \alpha\boldsymbol{\zeta}_2$ is in $\mathfrak{C}_r^{n+1}$ for all α with $|\alpha| < \beta$. On putting ${}_r\mathbf{\Lambda}^\dagger + \alpha\boldsymbol{\zeta}_2$ for $\boldsymbol{\xi}$ in (7.9), we obtain

$$\mathfrak{p}(\mathbf{\Lambda}; {}_r\mathbf{\Lambda}^\dagger + \alpha\boldsymbol{\zeta}_2) \geqslant \mathfrak{p}(\mathbf{\Lambda}; {}_r\mathbf{\Lambda}^\dagger) \tag{7.13}$$

However, by (4.22),

$$\mathfrak{p}(\mathbf{\Lambda}; {}_r\mathbf{\Lambda}^\dagger + \alpha\boldsymbol{\zeta}_2) = \mathfrak{p}(\mathbf{\Lambda}; {}_r\mathbf{\Lambda}^\dagger) + \delta\mathfrak{p}(\mathbf{\Lambda}^\dagger \mid \alpha\boldsymbol{\zeta}_2) + o(\|\alpha\boldsymbol{\zeta}_2\|_r)$$

or, by (7.12) and the linearity of $\delta\mathfrak{p}(\mathbf{\Lambda}^\dagger \mid \cdot)$,

$$\mathfrak{p}(\mathbf{\Lambda}; {}_r\mathbf{\Lambda}^\dagger + \alpha\boldsymbol{\zeta}_2) = \mathfrak{p}(\mathbf{\Lambda}; {}_r\mathbf{\Lambda}^\dagger) + \alpha d + o(\alpha) \tag{7.14}$$

Since (7.13) and (7.14) hold for *all* α in $(-\beta, \beta)$, (7.13) and (7.14) are compatible only if $d = 0$, which contradicts (7.12). Thus (7.11) is impossible. Q.E.D.

Employing the definitions given in the paragraph containing (6.25), Theorem (7.2) may be paraphrased as follows:

Remark 7.3. For each pair $(\mathbf{q}, \theta)$ with $\mathbf{q}$ in C^n and $\theta > 0$, for each $\boldsymbol{\phi}$ in $\mathfrak{B}_r{}^n$, and for each ξ in $\mathfrak{B}_r{}^1$,

$$\delta_{\mathbf{q}}\mathfrak{p}(\mathbf{q}^\dagger, \theta^\dagger \mid \boldsymbol{\phi}) = 0\,, \qquad \delta_\theta\mathfrak{p}(\mathbf{q}^\dagger, \theta^\dagger \mid \xi) = 0$$

where $\mathbf{q}^\dagger$ and $\theta^\dagger$ are the constant functions on $[0, \infty)$ with values $\mathbf{q}$ and θ.

Let us return now to the equilibrium response functions $\mathfrak{p}^\circ$ and $\mathfrak{S}^\circ$ defined in (7.3). The gradient of $\mathfrak{p}^\circ$, $\nabla\mathfrak{p}^\circ$, is defined by the relation

$$\mathfrak{p}^\circ(\mathbf{\Lambda} + \mathbf{\Gamma}) = \mathfrak{p}^\circ(\mathbf{\Lambda}) + \nabla\mathfrak{p}^\circ(\mathbf{\Lambda}) \cdot \mathbf{\Gamma} + o(|\mathbf{\Gamma}|) \tag{7.15}$$

which holds for each $\mathbf{\Lambda}$ in C^{n+1} and all $\mathbf{\Gamma}$ in R^{n+1} such that $\mathbf{\Lambda} + \mathbf{\Gamma}$ is in C^{n+1}; the value $\nabla\mathfrak{p}^\circ(\mathbf{\Lambda})$ of $\nabla\mathfrak{p}^\circ$ is, of course, an element of R^{n+1}. The existence of $\nabla\mathfrak{p}^\circ$ is guaranteed by item 4 of the assumed principle of fading memory. Indeed, for each $\mathbf{\Gamma}$ in R^{n+1}

$$\begin{aligned}\nabla\mathfrak{p}^\circ(\mathbf{\Lambda}) \cdot \mathbf{\Gamma} &= \frac{\partial}{\partial\alpha}\,\mathfrak{p}^\circ(\mathbf{\Lambda} + \alpha\mathbf{\Gamma})\bigg|_{\alpha=0} \\ &= \frac{\partial}{\partial\alpha}\,\mathfrak{p}(\mathbf{\Lambda}^\dagger + \alpha\mathbf{\Gamma}^\dagger)\bigg|_{\alpha=0} \\ &= d\mathfrak{p}(\mathbf{\Lambda}^\dagger \mid \mathbf{\Gamma}^\dagger)\end{aligned} \tag{7.16}$$

where $\mathbf{\Lambda}^\dagger$ and $\mathbf{\Gamma}^\dagger$ are the constant functions on $[0, \infty)$ with values $\mathbf{\Lambda}$ and $\mathbf{\Gamma}$, respectively. Now, in view of (4.23), (7.16) can be written

$$\nabla\mathfrak{p}^\circ(\mathbf{\Lambda}) \cdot \mathbf{\Gamma} = D\mathfrak{p}(\mathbf{\Lambda}^\dagger) \cdot \mathbf{\Gamma} + \delta\mathfrak{p}(\mathbf{\Lambda}^\dagger \mid {}_r\mathbf{\Gamma}^\dagger)$$

with ${}_r\mathbf{\Gamma}^\dagger$ the restriction of $\mathbf{\Gamma}^\dagger$ to $(0, \infty)$, and it follows from (7.10) that this relation reduces to

$$\nabla\mathfrak{p}^\circ(\mathbf{\Lambda}) \cdot \mathbf{\Gamma} = D\mathfrak{p}(\mathbf{\Lambda}^\dagger) \cdot \mathbf{\Gamma}$$

Since this is true for every $\mathbf{\Gamma}$ in R^{n+1}, we have

$$\nabla \mathfrak{p}^\circ(\mathbf{\Lambda}) = D\mathfrak{p}(\mathbf{\Lambda}^\dagger) \tag{7.17}$$

for each $\mathbf{\Lambda}$ in C^{n+1}. By (7.3) and equation (6.8),

$$\mathfrak{S}^\circ(\mathbf{\Lambda}) = \mathfrak{S}(\mathbf{\Lambda}^\dagger) = D\mathfrak{p}(\mathbf{\Lambda}^\dagger) \tag{7.18}$$

and we have proven

Theorem 7.3.* *The equilibrium response functions defined in (7.3) are related by the formula*

$$\mathfrak{S}^\circ = \nabla \mathfrak{p}^\circ \tag{7.19}$$

The equilibrium response functions $\mathfrak{p}^\circ$, $\mathfrak{z}_i^\circ$, and $\mathfrak{h}^\circ$ corresponding to the response functionals $\mathfrak{p}$, $\mathfrak{z}_i$, and $\mathfrak{h}$ in (2.24) are defined by

$$\left.\begin{aligned} \mathfrak{p}^\circ(q_1, \ldots, q_n, \theta) &= \mathfrak{p}(q_1^\dagger, \ldots, q_n^\dagger, \theta^\dagger) \\ \mathfrak{z}_i^\circ(q_1, \ldots, q_n, \theta) &= \mathfrak{z}_i(q_1^\dagger, \ldots, q_n^\dagger, \theta^\dagger), \qquad i = 1, \ldots, n \\ \mathfrak{h}^\circ(q_1, \ldots, q_n, \theta) &= \mathfrak{h}(q_1^\dagger, \ldots, q_n^\dagger, \theta^\dagger) \end{aligned}\right\} \tag{7.20}$$

where $q_j^\dagger$ and $\theta^\dagger$ are the constant functions on $[0, \infty)$ with values q_j and θ. By (5.8), the function $\mathfrak{S}^\circ$, defined in $(7.3)_2$, is an n-tuple of real-valued functions on R^{n+1}:

$$\mathfrak{S}^\circ = (\mathfrak{z}_i^\circ, \ldots, \mathfrak{z}_n^\circ, -\mathfrak{h}^\circ) \tag{7.21}$$

Furthermore, since $\mathbf{\Lambda} = (q_1, \ldots, q_n, \theta)$, for $\nabla \mathfrak{p}^\circ$, defined in (7.15), we have

$$\nabla \mathfrak{p}^\circ = \left(\frac{\partial \mathfrak{p}^\circ}{\partial q_1}, \ldots, \frac{\partial \mathfrak{p}^\circ}{\partial q_n}, \frac{\partial \mathfrak{p}^\circ}{\partial \theta}\right) \tag{7.22}$$

Thus Theorem 7.3 is equivalent to

Remark 7.4. The equilibrium response functions defined in (7.20) obey the classical relations

$$\mathfrak{h}^\circ = -\frac{\partial}{\partial \theta}\mathfrak{p}^\circ \tag{7.23}$$

$$\mathfrak{z}_i^\circ = \frac{\partial}{\partial q_i}\mathfrak{p}^\circ, \qquad i = 1, \ldots, n \tag{7.24}$$

If we let $\mathbf{z}^\circ$ be the R^n-valued function on R^{n+1} with the components

$$\mathbf{z}^\circ = (\mathfrak{z}_1^\circ, \ldots, \mathfrak{z}_n^\circ) \tag{7.25}$$

* Coleman[1] (Remark 11, p. 27); see also Coleman and Mizel[9] (Theorem 4, p. 269). Note that because the proof of (7.19) requires equation (7.10), this theorem, although expected, is not trivial. The passage from (6.8) to (7.19) recently has been discussed from a more general point of view by Gurtin[10] and Coleman and Owen.[11]

then, in an obvious notation, (7.24) becomes

$$\mathfrak{Z}^\circ = \nabla_{\mathbf{q}} \mathfrak{p}^\circ \tag{7.26}$$

This equation is called the *internal force relation for equilibrium*.

As a corollary of Remark 7.4 we have the following classical analogue of Theorem 6.2.

Remark 7.5. If $\mathfrak{Z}$ is continuously Fréchet-differentiable, then

$$\frac{\partial}{\partial q_j} \mathfrak{Z}_i^\circ = \frac{\partial}{\partial q_i} \mathfrak{Z}_j^\circ , \quad \frac{\partial}{\partial \theta} \mathfrak{Z}_i^\circ = -\frac{\partial}{\partial q_i} \mathfrak{h}^\circ , \qquad i, j = 1, \ldots, n \tag{7.27}$$

Thus, as expected, in equiibrium we have Maxwell relations of the following types: $\partial \Xi_i / \partial q_j = \partial \Xi_j / \partial q_i$ and $\partial \Xi_i / \partial \theta = -\partial \eta / \partial q_i$.

VIII. LAGRANGIANS AND HAMILTONIANS

In this section, as well as in Sections IX and X, attention is confined to isothermal processes, and hence it is assumed that the rate of addition of heat is always such that θ is constant in time. By Theorem 5.1, if an admissible process $\mathscr{P}$ is known to be isothermal, and if the temperature θ is specified in advance, then $\mathscr{P}$ is completely determined by giving the coordinate vector $\mathbf{q}$ as a function $\mathbf{q}(\cdot)$ of time. The function $\mathbf{q}(\cdot)$ is called the *motion* of the system.

Here the constitutive equations (2.23) can be written in the simpler forms

$$\psi(t) = \mathfrak{p}(\mathbf{q}^t) \tag{8.1}$$

$$\Xi(t) = \mathfrak{Z}(\mathbf{q}^t) \tag{8.2}$$

$$\eta(t) = \mathfrak{h}(\mathbf{q}^t) \tag{8.3}$$

with $\mathfrak{q}$, $\mathfrak{Z}$, and $\mathfrak{h}$ functionals depending on θ as a parameter. Remark 6.2 here tells us that

$$\Xi_i(t) = D_{q_i} \mathfrak{p}(\mathbf{q}^t) , \qquad i = 1, \ldots, n \tag{8.4}$$

or, as in (6.24),

$$\mathfrak{Z} = D_{\mathbf{q}} \mathfrak{p} \tag{8.5}$$

$D_{\mathbf{q}}\mathfrak{p}$ is defined in (6.13); $D_{q_i}\mathfrak{p}$, the ith component of $D_{\mathbf{q}}\mathfrak{p}$, is called the *instantaneous derivative of* $\mathfrak{p}$ *with respect to the* i*th coordinate* and is defined in (6.16).

For functions of present values alone, the operators $D_{\mathbf{q}}$ and D_{q_i} reduce to the ordinary gradient operator ∇ and partial derivative operator $\partial / \partial q_i$, respectively. Thus, if f maps R^n into R^1, then

$$D_{\mathbf{q}} f(\mathbf{q}) = \nabla f(\mathbf{q})$$

and

$$D_{q_i}f = \frac{\partial}{\partial q_i} f(q_1, \ldots, q_i, \ldots, q_n)$$

By way of example, (2.6) yields

$$\frac{\partial T}{\partial q_i} = \frac{1}{2}\frac{\partial}{\partial q_i}\mathscr{K}(\mathbf{q}, \dot{\mathbf{q}}) = \tfrac{1}{2}D_{q_i}\mathscr{K}(\mathbf{q}, \dot{\mathbf{q}}) \tag{8.6}$$

$$p_i \stackrel{\text{def}}{=} \frac{\partial T}{\partial \dot{q}_i} = \frac{1}{2}\frac{\partial}{\partial \dot{q}_i}\mathscr{K}(\mathbf{q}, \dot{\mathbf{q}}) = \tfrac{1}{2}D_{\dot{q}_i}\mathscr{K}(\mathbf{q}, \dot{\mathbf{q}}) \tag{8.7}$$

The number p_i defined here is called the *momentum associated with the* i*th coordinate*. In view of (8.7) and (2.12), the *momentum vector*, $\mathbf{p} = (p_1, \ldots, p_n)$, obeys the relations

$$\mathbf{p} = \tfrac{1}{2}D_{\dot{\mathbf{q}}}\mathscr{K}(\mathbf{q}, \dot{\mathbf{q}}) = K(\mathbf{q})\dot{\mathbf{q}} \tag{8.8}$$

By (2.7) and (8.8), in a given process,

$$T(t) = \tfrac{1}{2}\mathbf{p}(t) \cdot \mathbf{q}(t) \tag{8.9}$$

It follows from (2.24) that for isothermal processes, (2.9), which, by (2.19), can be written

$$\frac{d}{dt}\frac{\partial T}{\partial \dot{q}_i} - \frac{\partial T}{\partial q_i} = F_i - \Xi_i$$

yields a system of functional-differential equations for the motion $\mathbf{q}(\cdot)$ whenever the applied forces are given as either functions of t or functions of $\mathbf{q}$. Indeed, we have here the following *dynamical equations:*

$$\frac{1}{2}\left[\frac{d}{dt}\frac{\partial}{\partial \dot{q}_i}\mathscr{K}(\mathbf{q}, \dot{\mathbf{q}}) - \frac{\partial}{\partial q_i}\mathscr{K}(\mathbf{q}, \dot{\mathbf{q}})\right] = F_i - \mathfrak{Z}_i(\mathbf{q}^t), \qquad i = 1, \ldots, n \tag{8.10}$$

or, in a more compact notation,

$$\frac{1}{2}\left[\frac{d}{dt}D_{\dot{\mathbf{q}}}\mathscr{K}(\mathbf{q}, \dot{\mathbf{q}}) - D_{\mathbf{q}}\mathscr{K}(\mathbf{q}, \dot{\mathbf{q}})\right] = \mathbf{F} - \mathfrak{Z}(\mathbf{q}^t) \tag{8.11}$$

Problems that occur in physical applications of our theory are often of the following type: Suppose the functional $\mathfrak{Z}$ is specified. Given the history of the coordinates $\mathbf{q}$ up to some time, say, $t = 0$, and the velocity vector at that time, find the motion for future times, assuming that the applied forces are known or are functions of $\mathbf{q}$. That is, given $\mathbf{q}(t)$ for $t \leqslant 0$, given $\lim_{t\to 0+} \dot{\mathbf{q}}(t)$, and given some rule for calculating $\mathbf{F}(t)$ for $t \geqslant 0$, find $\mathbf{q}(t)$ for $t \geqslant 0$. This problem is one of solving the functional-differential equations (8.10).

From this point on, let us assume that for $t \geqslant 0$ the applied forces are functions of $\mathbf{q}$ and have a potential. There then exists a function $\hat{\zeta}$ on R^n such that

$$\text{for } t \geqslant 0, \qquad \mathbf{F}(t) = -\nabla\hat{\zeta}(\mathbf{q})\Big|_{\mathbf{q}=\mathbf{q}(t)} = -D_{\mathbf{q}}\hat{\zeta}(\mathbf{q})\Big|_{\mathbf{q}=\mathbf{q}(t)} \tag{8.12}$$

Of course, the function ζ, defined by

$$\zeta(t) = \hat{\zeta}(\mathbf{q}(t)) \tag{8.13}$$

is, for $t \geqslant 0$, the same as the function defined in (3.8) and there called the *mechanical potential of the applied forces;* that is, (8.12) implies that

$$\dot{\zeta}(t) \stackrel{\text{def}}{=} \frac{d}{dt}\,\zeta(\mathbf{q}(t)) = -\mathbf{F}(t)\cdot\dot{\mathbf{q}}(t)$$

and hence,

$$\zeta(t) = -\int_0^t \mathbf{F}(\tau)\cdot\dot{\mathbf{q}}(\tau)\,d\tau + c$$

For each function $\mathbf{\Psi}$ in $\mathfrak{C}^n$ and each vector $\mathbf{u}$ in R^n, let us put

$$\mathscr{L}(\mathbf{\Psi}, \mathbf{u}) \stackrel{\text{def}}{=} \tfrac{1}{2}\mathscr{K}(\mathbf{\Psi}(0), \mathbf{u}) - \mathfrak{p}(\mathbf{\Psi}) - \hat{\zeta}(\mathbf{\Psi}(0)) \tag{8.14}$$

The function $\mathscr{L}$ so defined maps $\mathfrak{C}^n \times R^n$ into R^1 and may be called the *Lagrangian functional* for the system under consideration. Setting $\mathbf{\Psi} = \mathbf{q}^t$ and $\mathbf{u} = \dot{\mathbf{q}}$ in (8.14), we have

$$\mathscr{L}(\mathbf{q}^t, \dot{\mathbf{q}}) = \tfrac{1}{2}\mathscr{K}(\mathbf{q}, \dot{\mathbf{q}}) - \mathfrak{p}(\mathbf{q}^t) - \hat{\zeta}(\mathbf{q}) \tag{8.15}$$

where $\mathbf{q}$ stands for $\mathbf{q}^t(0)$. It follows from (2.6), (8.1), and (8.13), that in a given admissible process, for each $t \geqslant 0$,

$$\mathscr{L}(\mathbf{q}^t, \dot{\mathbf{q}}(t)) = T(t) - \psi(t) - \zeta(t) \tag{8.16}$$

Furthermore,

$$D_{\dot{\mathbf{q}}}\mathscr{L}(\mathbf{q}^t, \dot{\mathbf{q}}) = \tfrac{1}{2}D_{\dot{\mathbf{q}}}\,\mathscr{K}(\mathbf{q}, \dot{\mathbf{q}}) \tag{8.17}$$

and, by (8.5) and (8.12),

$$D_{\dot{\mathbf{q}}}\,\mathscr{L}(\mathbf{q}^t, \dot{\mathbf{q}}) = \tfrac{1}{2}D_{\mathbf{q}}\mathscr{K}(\mathbf{q}, \dot{\mathbf{q}}) - \mathfrak{Z}(\mathbf{q}^t) + \mathbf{F}(\mathbf{q}) \tag{8.18}$$

From (8.11), (8.17), and (8.18), we may read off the following theorem which gives us an analogue, for systems with memory, of a famous result of Lagrange.

Theorem 8.1. *If (8.12) holds, then for $t \geqslant 0$ the dynamical equations (8.10) can be written*

$$\frac{d}{dt}\,D_{\dot{\mathbf{q}}}\mathscr{L}(\mathbf{q}^t, \dot{\mathbf{q}}(t)) - D_{\mathbf{q}}\mathscr{L}(\mathbf{q}^t, \dot{\mathbf{q}}(t)) = \mathbf{0} \tag{8.19}$$

with $\mathscr{L}$ the "Lagrangian functional" defined in (8.14).

The vector equation (8.19) is equivalent to the n equations

$$\frac{d}{dt}\frac{\partial}{\partial \dot{q}_i}\mathscr{L}(\mathbf{q}^t, \dot{\mathbf{q}}(t)) = D_{q_i}\mathscr{L}(\mathbf{q}^t, \dot{\mathbf{q}}(t)), \qquad i = 1, \ldots, n \tag{8.20}$$

Theorem 8.1 is clearly just a corollary to item 1 of Theorem 6.1.

Let us now assume that the inertia tensor $\mathbf{K}(\mathbf{q})$ is, for each value of $\mathbf{q}$, a nonsingular linear transformation with inverse $\mathbf{K}(\mathbf{q})^{-1}$. This assumption holds when, for example, for each $\mathbf{q}$ the quadratic form $\mathscr{K}(\mathbf{q}, \cdot)$ is positive definite. When $\mathbf{K}(\mathbf{q})$ is nonsingular, the velocity vector $\dot{\mathbf{q}}$ can be expressed as a function of the coordinate vector and the momentum vector $\mathbf{p}$ of (8.8),

$$\dot{\mathbf{q}} = \mathbf{K}(\mathbf{q})^{-1}\mathbf{p} \tag{8.21}$$

and the kinetic energy T is given by a function $\tilde{\mathscr{K}}$ of $\mathbf{q}$ and $\mathbf{p}$:

$$T = \tfrac{1}{2}\tilde{\mathscr{K}}(\mathbf{q}, \mathbf{p}) = \tfrac{1}{2}\mathscr{K}(\mathbf{q}, \mathbf{K}(\mathbf{q})^{-1}\mathbf{p}) \tag{8.22}$$

By (2.5) and (8.21), $\tilde{\mathscr{K}}(\mathbf{q}, \mathbf{p})$ has the form

$$\tilde{\mathscr{K}}(\mathbf{q}, \mathbf{p}) = [\mathbf{K}(\mathbf{q})^{-1}\mathbf{p}] \cdot \mathbf{K}(\mathbf{q})[\mathbf{K}(\mathbf{q})^{-1}\mathbf{p}] = \mathbf{p} \cdot \mathbf{K}(\mathbf{q})^{-1}\mathbf{p} \tag{8.23}$$

Since the linear transformation $\mathbf{K}(\mathbf{q})^{-1}$ is symmetric, (8.23) yields

$$D_{\mathbf{p}}\tilde{\mathscr{K}}(\mathbf{q}, \mathbf{p}) = \nabla_{\mathbf{p}}\tilde{\mathscr{K}}(\mathbf{q}, \mathbf{p}) = 2\mathbf{K}(\mathbf{q})^{-1}\mathbf{p}$$

and therefore, in view of (8.21), we have the important formula

$$\tfrac{1}{2}D_{\mathbf{p}}\tilde{\mathscr{K}}(\mathbf{q}, \mathbf{p}) = \dot{\mathbf{q}} \tag{8.24}$$

Moreover,

$$\begin{aligned}\frac{\partial}{\partial q_i}\tilde{\mathscr{K}}(\mathbf{q}, \mathbf{p}) &= \mathbf{p} \cdot \frac{\partial}{\partial q_i}[\mathbf{K}(\mathbf{q})^{-1}\mathbf{p}] = -\mathbf{p} \cdot \mathbf{K}(\mathbf{q})^{-1}\left[\frac{\partial \mathbf{K}(\mathbf{q})}{\partial q_i}\right]\mathbf{K}(\mathbf{q})^{-1}\mathbf{p} \\ &= -[\mathbf{K}(\mathbf{q})^{-1}\mathbf{p}] \cdot \left[\frac{\partial \mathbf{K}(\mathbf{q})}{\partial q_i}\right]\mathbf{K}(\mathbf{q})^{-1}\mathbf{p} = -\dot{\mathbf{q}} \cdot \frac{\partial}{\partial q_i}\mathbf{K}(\mathbf{q})\dot{\mathbf{q}} \\ &= -\frac{\partial}{\partial q_i}\dot{\mathbf{q}} \cdot \mathbf{K}(\mathbf{q})\dot{\mathbf{q}} = -\frac{\partial}{\partial q_i}\mathscr{K}(\mathbf{q}, \dot{\mathbf{q}})\end{aligned}$$

that is,

$$D_{\mathbf{q}}\tilde{\mathscr{K}}(\mathbf{q}, \mathbf{p}) = -D_{\mathbf{q}}\mathscr{K}(\mathbf{q}, \dot{\mathbf{q}}) \tag{8.25}$$

Let $\mathscr{H}$ be the real-valued function on $\mathfrak{C}^n \times R^n$ defined by

$$\mathscr{H}(\mathbf{\Psi}, \mathbf{u}) = \mathfrak{p}(\mathbf{\Psi}) + \mathfrak{z}(\mathbf{\Psi}(0)) + \tfrac{1}{2}\tilde{\mathscr{K}}(\mathbf{\Psi}(0), \mathbf{u}) \tag{8.26}$$

so that

$$\mathscr{H}(\mathbf{q}^t, \mathbf{p}) = \mathfrak{p}(\mathbf{q}^t) + \mathfrak{z}(\mathbf{q}) + \tfrac{1}{2}\tilde{\mathscr{K}}(\mathbf{q}, \mathbf{p}) \tag{8.27}$$

with $\mathbf{q} = \mathbf{q}^t(0)$. I call $\mathscr{H}$ the *Hamiltonian functional* for the system. In a given admissible process, for $t \geqslant 0$, we have, by (8.1), (8.13), and (8.22),

$$\mathscr{H}(\mathbf{q}^t, \mathbf{p}(t)) = \psi(t) + \zeta(t) + T(t) \tag{8.28}$$

and thus the value of $\mathscr{H}$ is the *canonical free energy* defined in (3.9):

$$\mathscr{H}(\mathbf{q}^t, \mathbf{p}(t)) = \Phi(t) \tag{8.29}$$

Since we are considering isothermal processes, Theorem 2.1 tells us that $\mathscr{H}(\mathbf{q}^t, \mathbf{p}(t))$ is a nonincreasing function of t.

Furthermore, as an immediate consequence of (8.28), (8.16), and (8.9), we have

Remark 8.1. For each admissible process obeying (8.12),

$$\mathscr{H}(\mathbf{q}^t, \mathbf{p}(t)) = \mathbf{q}(t) \cdot \mathbf{p}(t) - \mathscr{L}(\mathbf{q}^t, \dot{\mathbf{q}}(t)) \tag{8.30}$$

for $t \geqslant 0$.

It follows immediately from (8.27) and (8.24) that

$$D_{\mathbf{p}}\mathscr{H}(\mathbf{q}^t, \mathbf{p}) = \tfrac{1}{2} D_{\mathbf{p}}\tilde{\mathscr{K}}(\mathbf{q}, \mathbf{p}) = \dot{\mathbf{q}} \tag{8.31}$$

Let us now calculate $D_{\mathbf{q}}\mathscr{H}$. By (8.27), (8.5), (8.12), and (8.25),

$$\begin{aligned} D_{\mathbf{q}}\mathscr{H}(\mathbf{q}^t, \mathbf{p}) &= D_{\mathbf{q}}\mathfrak{p}(\mathbf{q}^t) + D_{\mathbf{q}}\hat{\zeta}(\mathbf{q}) + \tfrac{1}{2} D_{\mathbf{q}}\tilde{\mathscr{K}}(\mathbf{q}, \mathbf{p}) \\ &= \mathfrak{Z}(\mathbf{q}^t) - \mathbf{F}(\mathbf{q}) - \tfrac{1}{2} D_{\mathbf{q}}\mathscr{K}(\mathbf{q}, \dot{\mathbf{q}}) \end{aligned} \tag{8.32}$$

and therefore (8.11) and (8.8) yield

$$D_{\mathbf{q}}\mathscr{H}(\mathbf{q}^t, \mathbf{p}) = -\frac{d}{dt} D_{\dot{\mathbf{q}}}\mathscr{K}(\mathbf{q}, \dot{\mathbf{q}}) = -\dot{\mathbf{p}} \tag{8.33}$$

This proves the following theorem which is a broad generalization of a well-known theorem of Hamilton.

Theorem 8.2. *If (8.12) holds, then for $t \geqslant 0$ the dynamical equations (8.10) can be written in the following* **canonical form**:

$$\dot{\mathbf{p}}(t) = -D_{\mathbf{q}}\mathscr{H}(\mathbf{q}^t, \mathbf{p}(t)), \qquad \dot{\mathbf{q}}(t) = D_{\mathbf{p}}\mathscr{H}(\mathbf{q}^t, \mathbf{p}(t)) \tag{8.34}$$

Here $\mathscr{H}$ is the "Hamiltonian functional" defined in (8.26). Since $\mathscr{H}(\mathbf{q}^t, \mathbf{p}(t))$ is the canonical free energy at time t, on each solution of the equations (8.33) there holds

$$\frac{d}{dt}\mathscr{H}(\mathbf{q}^t, \mathbf{p}(t)) \leqslant 0 \tag{8.35}$$

Let f and g be two real-valued functions of the histories $\mathbf{q}^t$ and $\mathbf{p}^t$, and suppose that $D_{\mathbf{q}}f(\mathbf{q}^t, \mathbf{p}^t)$, $D_{\mathbf{p}}f(\mathbf{q}^t, \mathbf{p}^t)$, $D_{\mathbf{q}}g(\mathbf{q}^t, \mathbf{p}^t)$, $D_{\mathbf{p}}g(\mathbf{q}^t, \mathbf{p}^t)$ exist throughout the common domain of f and g. We may define the *instantaneous*

Poisson bracket of f and g, $\{f, g\}$, as follows:

$$\{f, g\} \stackrel{\text{def}}{=} D_{\mathbf{q}} f \cdot D_{\mathbf{p}} g - D_{\mathbf{q}} g \cdot D_{\mathbf{p}} f \tag{8.36}$$

Of course, on each solution of (8.34), $\{f, g\}$ is a function of t. In particular, if $g = \mathscr{H}$ and f is a function of only the present values of the coordinates and momenta, we have

$$\{f, \mathscr{H}\} = \{f, \mathscr{H}\}(t) = D_{\mathbf{q}} f(\mathbf{q}, \mathbf{p}) \cdot D_{\mathbf{p}} \mathscr{H}(\mathbf{q}^t, \mathbf{p}) - D_{\mathbf{p}} f(\mathbf{q}, \mathbf{p}) \cdot D_{\mathbf{q}} \mathscr{H}(\mathbf{q}^t, \mathbf{p})$$

with $\mathbf{q} = \mathbf{q}^t(0)$ and $\mathbf{p} = \mathbf{p}(t)$. By (8.34), this expression can be written

$$\{f, \mathscr{H}\} = D_{\mathbf{q}} f(\mathbf{q}, \mathbf{p}) \cdot \dot{\mathbf{q}} + D_{\mathbf{q}} f(\mathbf{q}, \mathbf{p}) \cdot \dot{\mathbf{p}}$$

and since f is a function of $\mathbf{q}$ and $\mathbf{p}$ alone, the right side here is just

$$\dot{f} = \frac{d}{dt} f(\mathbf{q}(t), \mathbf{p}(t))$$

and we have

Theorem 8.3. *If f is a differentiable real-valued function of the present values of the coordinates and momenta, then along each solution of the equations (8.34) the time-derivative of f obeys the formula*

$$\dot{f} = \{f, \mathscr{H}\} \tag{8.37}$$

where $\{\ \}$ is the instantaneous Poisson bracket defined in (8.36).

Remark 8.2. On putting first $f(\mathbf{p}, \mathbf{q}) = p_i$ and then $f(\mathbf{p}, \mathbf{q}) = q_i$ in (8.37), we recover the equations (8.34), which therefore can be written

$$\dot{p}_i = \{p_i, \mathscr{H}\}, \quad \dot{q}_i = \{q_i, \mathscr{H}\}, \qquad i = 1, \ldots, n \tag{8.38}$$

Remark 8.3. If f is a differentiable function of $\mathbf{p}$, $\mathbf{q}$, and t, then in place of (8.37) we have the formula

$$\dot{f} = \{f, \mathscr{H}\} + \partial_t f \tag{8.39}$$

where

$$\dot{f} = \frac{d}{dt} f(\mathbf{p}(t), \mathbf{q}(t), t) \quad \text{and} \quad \partial_t f = \frac{\partial}{\partial \tau} f(\mathbf{p}, \mathbf{q}, \tau)\Big|_{\tau=t}$$

If f is a Fréchet-differentiable function on $\mathfrak{C}^n \times R^n$, then

$$\frac{d}{dt} f(\mathbf{q}^t, \mathbf{p}(t)) = \{f, \mathscr{H}\} + \delta_{\mathbf{q}} f(\mathbf{q}^t, \mathbf{p} \,|\, {}_r\dot{\mathbf{q}}^t) \tag{8.40}$$

whenever $\mathbf{q}^t$ is a tame history in $\mathfrak{C}^n$.

Hamiltonians and Lagrangians for the Dangling Spider

If the filament and the ball, or spider, discussed at the end of Section I have a temperature θ which remains constant in time, then the constitutive equations

(1.17) and (1.18) reduce to

$$\left.\begin{aligned} \psi(t) &= \mathfrak{p}(q^t) \\ \Xi(t) &= \mathfrak{Z}(q^t) \end{aligned}\right\} \tag{8.41}$$

and the dynamical equation (1.20) becomes

$$M\ddot{q}(t) = F(t) - \mathfrak{Z}(q^t) \tag{8.42}$$

Here M is the mass of the ball, q^t is the history up to t of the length q of the filament; $\mathfrak{p}$ and $\mathfrak{Z}$ are constitutive functionals which depend on θ as a parameter; the value Ξ of $\mathfrak{Z}$ is the tension in the filament; and F is an externally applied force acting downward on the ball.

By (8.4),

$$\mathfrak{Z} = D_q\mathfrak{p} \tag{8.43}$$

Let us now assume that for $t \geqslant 0$, F is a function of q. We may put

$$\hat{\zeta}(q) = -\int_0^q F(\sigma)\, d\sigma + c \tag{8.44}$$

By Theorem 3.1, the quantity

$$\Phi(t) \overset{\text{def}}{=} \psi(t) + \tfrac{1}{2}M\dot{q}(t)^2 + \hat{\zeta}(q(t)) \tag{8.45}$$

decreases, for $t \geqslant 0$, on each solution of (8.42), because by (3.9), $\Phi(t)$ is the canonical free energy of the system here considered (ball + filament). The Lagrangian functional (8.14) for this system is given by

$$\mathscr{L}(q^t, \dot{q}) = \tfrac{1}{2}M\dot{q}^2 - \mathfrak{p}(q^t) - \hat{\zeta}(q^t(0)) \tag{8.46}$$

and, in view of (8.43) and (8.44), it is obvious that (8.42) is equivalent to the equation

$$\frac{d}{dt} D_{\dot{q}}\mathscr{L}(q^t, \dot{q}) = D_q\mathscr{L}(q^t, \dot{q}) \tag{8.47}$$

here given to us by Theorem 8.1. For the Hamiltonian functional (8.26) we now have

$$\mathscr{H}(q^t, p) = \mathfrak{p}(q^t) + \hat{\zeta}(q^t(0)) + \frac{p^2}{2M} \tag{8.48}$$

Since $p = p(t) = M\dot{q}(t)$, it is obvious that the value of $\mathscr{H}$ is the quantity $\Phi(t)$ defined in (8.45). In the canonical form (8.34), the dynamical equation (8.42) reads

$$\dot{p}(t) = -\frac{d}{dq}\hat{\zeta}(q(t)) - D_q\mathfrak{p}(q^t), \qquad \dot{q}(t) = \frac{p(t)}{M} \tag{8.49}$$

IX. INFLATION OF A CYLINDRICAL TUBE

In this and the following section we shall consider nontrivial examples from continuum mechanics of motions which can be described by giving the temporal dependence of a single real variable q.

Let $\mathscr{B}$ be a materially homogeneous, incompressible, body which in its reference configuration has the form of a hollow circular tube with inner radius R_I and outer radius R_O. Employing a fixed cylindrical coordinate system with the z-axis along the common axis of the cylinders which bound the tube, let us suppose that $\mathscr{B}$ is undergoing a motion of the type

$$z = Z\,, \qquad r = r(R, t)\,, \vartheta = \Theta \tag{9.1}$$

where z, r, ϑ are the coordinates at time t of the material point which has the coordinates Z, R, Θ in the reference configuration. In such a motion $\mathscr{B}$ remains a circular tube at all times; its inner and outer radii at time t are $r_I(t) = r(R_I, t)$ and $r_Q(t) = r(R_O, t)$. Since the material comprising $\mathscr{B}$ is assumed incompressible, the mass density ρ of $\mathscr{B}$ is not only constant throughout $\mathscr{B}$ but does not change with time. Let us further assume that the temperature is held constant in space throughout $\mathscr{B}$ and in time forever, that is, that the processes we consider are isothermal. It is easily shown that because every motion of $\mathscr{B}$ is isochoric, the function $r(\cdot, \cdot)$ in (9.1) must have the form

$$r(R, t)^2 = R^2 - R_I^2 + q(t) \tag{9.2}$$

Thus each motion of the form (9.1) is completely determined by a scalar function $q(\cdot)$. The value of this function is the square of the inner radius r_I:

$$q(t) = r_I(t)^2 = r(R_I, t)^2 \tag{9.3}$$

In terms of q the outer radius is given by

$$r_O(t)^2 = r(R_O, t)^2 = B + q(t) \tag{9.4}$$

where

$$B = R_O^2 - R_I^2 > 0 \tag{9.5}$$

that is, $\pi B \rho$ is the mass of the tube per unit length. In each process q is subject to the constraint

$$q(t) > 0 \tag{9.6}$$

If we put

$$v \overset{\text{def}}{=} \frac{\partial}{\partial t} r(R, t) = v(r, t)$$

then, by (9.2),

$$v(r, t) = \frac{1}{2}\frac{1}{r}\,\dot{q}(t) \tag{9.7}$$

The kinetic energy of a unit length of the tube is*

$$T(t) = \int_{r_I(t)}^{r_O(t)} \tfrac{1}{2}\rho v^2 2\pi r\, dr = \dot{q}(t)^2 \frac{\pi}{4} \rho \ln \frac{r_O(t)}{r_I(t)}$$

$$= \dot{q}(t)^2 \frac{\pi}{8} \rho \ln \frac{B + q(t)}{q(t)} \tag{9.8}$$

Note that this equation is of the general form (2.7); here the inertia tensor **K**(**q**) of (2.5) reduces to a number $K(q)$, and we have

$$T = \tfrac{1}{2}\mathscr{K}(q, \dot{q}) = \tfrac{1}{2}K(q)\dot{q}^2 \tag{9.9}$$

with

$$K(q) = \frac{\pi}{4} \rho \ln \frac{B + q}{q} > 0 \tag{9.10}$$

As in (8.8), the momentum p associated with q is defined by the equations

$$p \overset{\text{def}}{=} \tfrac{1}{2} D_{\dot{q}} \mathscr{K}(q, \dot{q}) = \frac{1}{2} \frac{\partial}{\partial \dot{q}} \mathscr{K}(q, \dot{q}) = K(q)\dot{q}$$

and hence

$$p = \dot{q} \frac{\pi}{4} \rho \ln \frac{B + q}{q} \tag{9.11}$$

The inertial force Q associated with q is defined as in (2.9), that is,

$$Q \overset{\text{def}}{=} \frac{d}{dt} \frac{\partial T}{\partial \dot{q}} - \frac{\partial T}{\partial q} = \frac{1}{2} \frac{d}{dt} \frac{\partial}{\partial \dot{q}} \mathscr{K}(q, \dot{q}) - \frac{1}{2} \frac{\partial}{\partial q} \mathscr{K}(q, \dot{q}) \tag{9.12}$$

and hence (9.8) yields

$$Q = K(q)\ddot{q} - \Omega(q)\dot{q}^2 \tag{9.13}$$

with K as in (9.10) and

$$\Omega(q) = \frac{\pi}{8} \rho \left[\frac{1}{q} - \frac{1}{B + q} \right] > 0 \tag{9.14}$$

Let us now assume that the material comprising the tube is simple† and transversely isotropic with its axis of symmetry along the z-axis. This implies that for each motion of the form (9.1) the forces on the bounding surfaces of the tube are normal thrusts.‡ Let $P_I(t)$ and $P_O(t)$ be the external pressures (per unit area) acting on the inner and outer bounding cylindrical surfaces. If we assume that body forces are absent, for the rate

* Coleman and Dill[26] [equation (5.27), p. 210].

† In the sense used in continuum mechanics by Noll[31] and in continuum thermodynamics by Coleman.[1,2]

‡ Coleman[37] [§5, particularly equation (5.7)].

w of working of the forces applied to a unit length of the tube, we have

$$w(t) = 2\pi[r_I(t)v(r_I, t)P_I(t) - r_O(t)v(r_O, t)P_O(t)]$$

Upon use of (9.7) this becomes

$$w(t) = \pi\dot{q}(t)\,\Delta P(t) \tag{9.15}$$

with

$$\Delta P = P_I - P_O \tag{9.16}$$

Clearly, w is to be identified with the rate of working of the applied forces, $\mathbf{F}\cdot\dot{\mathbf{q}}$, which appears in the law of balance of energy (2.4). Since $\mathbf{F}\cdot\dot{\mathbf{q}}$ must here reduce to $F\dot{q}$, with F a scalar, (9.15) tells us that this scalar F, which can be called the *applied force associated with* q, is given by the simple formula

$$F = \pi\,\Delta P \tag{9.17}$$

with ΔP the difference in the pressures acting on the inner and outer surfaces of $\mathscr{B}$.

The difference,

$$\Xi = F - Q \tag{9.18}$$

between the applied force F and the inertial force Q is the internal force associated with q. (This force Ξ results from the fact that if the tube is not in equilibrium in an undistorted reference configuration, then the deviatoric part of the stress tensor need not be zero and will give rise to forces which can oppose the applied forces.) E. H. Dill and I* have shown that Ξ is given by a function $\mathfrak{Z}$ of the history of q:

$$\Xi(t) = \mathfrak{Z}(q^t) \tag{9.19}$$

Although the theory of Refs. 26 and 37 relates $\mathfrak{Z}$ to the stress-functional occurring in the full (three-dimensional) theory of isotropic materials with memory, it here suffices for us to know that the functional $\mathfrak{Z}$ is determined when the material comprising the tube, its temperature, the choice of the reference configuration, and the numbers R_I and R_O are specified. We are not at present interested in the form of $\mathfrak{Z}$. By combining (9.12), (9.18), and (9.19), one obtains, as in (8.10), the following general dynamical equation:

$$\frac{1}{2}\frac{d}{dt}\frac{\partial}{\partial\dot{q}}\mathscr{K}(q, \dot{q}) - \frac{1}{2}\frac{\partial}{\partial q}\mathscr{K}(q, \dot{q}) = F - \mathfrak{Z}(q^t) \tag{9.20}$$

* Ref. 26. We there write $-\pi\hbar(\beta^t)$ for the present $\mathfrak{Z}(q^t)$, with $\beta(t) = q(t) - R_I{}^2$.

One may substitute (9.13), (9.17), and (9.19) into (9.18) to obtain the following explicit functional-differential equation for q:*

$$K(q)\ddot{q} - \Omega(q)\dot{q}^2 = \pi\,\Delta P - \mathfrak{Z}(q^t) \tag{9.21}$$

with K and Ω given by (9.10) and (9.14).

So as to be able to cast this equation into the canonical form (8.34), let us assume that from time $t = 0$ onward ΔP is given by a prescribed function of q; then,

$$F = \pi\,\Delta P = \hat{F}(q) \qquad \text{for} \quad t \geqslant 0 \tag{9.22}$$

Of course, situations in which P_I and P_O are held constant for $t \geqslant 0$ fall as special cases of the assumption (9.22). To have another example one may suppose that the tube is filled with an ideal gas obeying Boyle's law, so that in each deformed state

$$P_I = P_{\mathscr{R}} \frac{R_I{}^2}{r_I{}^2} = P_{\mathscr{R}} \frac{R_I{}^2}{q}$$

where $P_{\mathscr{R}}$ is a positive constant equal to the pressure exerted by the enclosed gas when the tube is in its reference configuration. If such a tube is immersed in an atmosphere at constant pressure P_O, we have

$$\hat{F}(q) = \pi\left[\frac{P_{\mathscr{R}} R_I{}^2}{q} - P_O\right] \tag{9.23}$$

The assumption (9.22) implies the existence of a potential function $\hat{\zeta}$ such that

$$\pi\,\Delta P(t) = -\frac{d}{dq}\hat{\zeta}(q)\bigg|_{q=q(t)} \tag{9.24}$$

when t is positive; indeed, we may put

$$\hat{\zeta}(q) \overset{\text{def}}{=} -\int_0^q \hat{F}(\sigma)\,d\sigma \tag{9.25}$$

In the thermodynamics of materials with memory, it is easily shown that there exists a functional $\mathfrak{p}$ which gives the Helmholtz free energy ψ of a

* In 1960, Knowles[39,40] observed that each motion obeying (9.1) and (9.2) is dynamically admissible in every isotropic, incompressible elastic material and found conditions under which such motions can be expected to be periodic. (See also Truesdell[41] and §§61, 62 of the treatise of Truesdell and Noll.[33]) For simple materials with memory, such motions have been discussed recently from various points of view by Carroll,[42] Fosdick,[43] Coleman,[37] Marrucci and Murch,[44] and Coleman and Dill.[26] The present equation (9.21) differs only in notation from equation (5.16) of Coleman and Dill[26] and is shown in Ref. 26 to be equivalent to the full three-dimensional vectorial equation of motion for the transversely isotropic, incompressible, material comprising the tube, provided the motion maintains the form (9.1), (9.2).

unit length of the tube as a function of the history of q:*

$$\psi(t) = \mathfrak{p}(q^t) \tag{9.26}$$

If we assume that the material comprising the tube obeys the principle of fading memory as employed in continuum physics,† then $\mathfrak{p}$ has the smoothness properties discussed in Section IV and Theorem 6.1 holds. In particular, the instantaneous derivative $D_q\mathfrak{p}$ exists and (6.8) [or (8.5)] yields

$$\mathfrak{Z}(q^t) = D_q\mathfrak{p}(q^t) \tag{9.27}$$

for each history q^t in the domain $\mathfrak{C}$ of $\mathfrak{p}$. The Lagrangian functional defined in (8.14), that is,

$$\mathscr{L} = \mathscr{K} - \mathfrak{p} - \hat{\zeta}$$

here has the form

$$\mathscr{L}(q^t, \dot{q}) = \frac{\pi}{8}\rho\dot{q}^2 \ln \frac{B + q^t(0)}{q^t(0)} - \mathfrak{p}(q^t) - \hat{\zeta}(q^t(0)) \tag{9.28}$$

Using (9.24) and (9.27), one easily verifies that the dynamical equation (9.21) can be written

$$\frac{d}{dt} D_{\dot{q}}\mathscr{L}(q^t, \dot{q}) = D_q\mathscr{L}(q^t, \dot{q}) \tag{9.29}$$

as expected from Theorem 8.1. It follows from (9.8), (9.9), and (9.11) that the kinetic energy T, as a function of the coordinate q and its associated momentum p, has the form

$$T = \tfrac{1}{2}\tilde{\mathscr{K}}(p, q) = \frac{2p^2}{\pi\rho}\left[\ln \frac{B + q}{q}\right]^{-1} = \frac{p^2}{2K(q^t(0))} \tag{9.30}$$

and therefore the Hamiltonian function $\mathscr{H}$ of (8.26)–(8.28) is here given by the formula

$$\mathscr{H}(q^t, p) = \mathfrak{p}(q^t) + \hat{\zeta}(q^t(0)) + \frac{2p^2}{\pi\rho}\left[\ln \frac{B + q^t(0)}{q^t(0)}\right]^{-1} \tag{9.31}$$

The canonical form (see Theorem 8.2) of (9.21) is

$$\left.\begin{aligned} \dot{q} &= D_p\mathscr{H}(q^t, p) = \frac{p}{K(q)} \\ \dot{p} &= -D_q\mathscr{H}(q^t, p) = \hat{F}(q) - D_q\mathfrak{p}(q^t) - \frac{\Omega(q)}{K^2(q)}p^2 \end{aligned}\right\} \tag{9.32}$$

* Coleman and Dill[26] [equations (4.10) and (5.13); in the notation employed there, $\mathfrak{p}(q^t) = \ell(\beta^t)$]. Of course, $\mathfrak{p}$, like $\mathfrak{Z}$, depends on the material under consideration, the temperature, the choice of reference configuration, and the values of R_I and R_O,

† That is, the proposal of Coleman and Mizel,[9,23] which generalizes the earlier formulations of Coleman and Noll[18–20] and Coleman.[1,2]

where $q = q(t), p = p(t)$, while K, Ω, and $\hat{F}$ are given by (9.10), (9.14), and (9.22). By (8.35), the quantity $\Phi(t) = \mathscr{H}(q^t, p(t))$ cannot increase on any solution of (9.32).

X. INFLATION OF A SPHERICAL SHELL

Let us now consider a materially homogeneous, incompressible body $\mathscr{B}$ which in its reference configuration has the form of a spherical shell with inner radius R_I and outer radius R_O. It is here assumed that $\mathscr{B}$ is undergoing an isothermal motion which has the form

$$\vartheta = \Theta\,, \qquad \phi = \Phi\,, \qquad r = r(R, t) \tag{10.1}$$

in a fixed spherical coordinate system with its origin at the center of curvature of the shell. In (10.1), ϑ, ϕ, r are the coordinates at time t of the material point which has the coordinates Θ, Φ, R in the reference configuration. In a motion of this type, $\mathscr{B}$ is a spherical shell at each time t, with radii $r_I = r(R_I, t)$, $r_O = r(R_O, t)$. The incompressibility of the material comprising $\mathscr{B}$ requires that the function $r(\cdot, \cdot)$ of (10.1) have the form

$$r(R, t)^3 = R^3 - R_I{}^3 + q(t) \tag{10.2}$$

and hence we again have a class of motions each of which is completely determined by specifying a single scalar function $q(\cdot)$. Here the value of $q(\cdot)$ is the cube of the inner radius r_I of the shell:

$$q(t) = r_I(t)^3 = r(R_I, t)^3 \tag{10.3}$$

and for the outer radius we have

$$r_O(t)^3 = r(R_O, t)^3 = B + q(t) \tag{10.4}$$

where

$$B = R_O{}^3 - R_I{}^3 > 0 \tag{10.5}$$

If we let A be the total solid angle in steradians subtended at the origin by the shell and let ρ be the mass density of the material, then the mass of the shell is ρAB. Once again q is subject to the constraint

$$q(t) > 0 \tag{10.6}$$

For the speed

$$v \stackrel{\text{def}}{=} \frac{\partial}{\partial t}\, r(R, t) = v(r, t)$$

we have here, by (10.3)

$$v(r, t) = \frac{1}{3}\frac{1}{r^2}\,\dot{q}(t) \tag{10.7}$$

and the kinetic energy of the shell is*

$$T(t) = A\int_{r_I(t)}^{r_O(t)} \tfrac{1}{2}\rho v^2 r^2\, dr = \frac{A}{18}\,\rho\left[\frac{1}{r_I} - \frac{1}{r_O}\right]\dot{q}(t)^2$$

$$= \frac{A}{18}\,\rho[q(t)^{-1/3} - (B + q(t))^{-1/3}]\dot{q}(t)^2 = \tfrac{1}{2}\mathscr{K}(q, \dot{q}) \qquad (10.8)$$

Thus the inertia tensor $\mathbf{K}(q)$ of (2.5) is here a number; that is,

$$\mathscr{K}(q, \dot{q}) = K(q)\dot{q}^2 \qquad (10.9)$$

with

$$K(q) = \frac{A}{9}\,\rho[q^{-1/3} - (B + q)^{-1/3}] > 0 \qquad (10.10)$$

For the momentum p associated with q we have

$$p = \frac{1}{2}\frac{\partial}{\partial \dot{q}}\mathscr{K}(q, \dot{q}) = K(q)\dot{q} = \frac{A}{9}\,\rho\dot{q}[q^{-1/3} - (B + q)^{-1/3}] \qquad (10.11)$$

and the corresponding inertial force Q is

$$Q = \frac{1}{2}\frac{d}{dt}\frac{\partial}{\partial \dot{q}}\mathscr{K}(q, \hat{q}) - \frac{1}{2}\frac{\partial}{\partial q}\mathscr{K}(q, \dot{q}) = K(q)\ddot{q} - \Omega(q)\dot{q}^2 \qquad (10.12)$$

with K as in (10.10) and

$$\Omega(q) = \frac{A\rho}{54}\,[q^{-4/3} - (B + q)^{-4/3}] > 0 \qquad (10.13)$$

If we now assume that the material comprising the shell is simple and isotropic and that an undistorted configuration has been chosen as the reference configuration, then the forces on the bounding surfaces $r = r_I$ and $r = r_O$ are normal to these surfaces,† that is, are pressures $P_I(t)$ and $P_O(t)$ (per unit area). Assuming that body forces are absent, for the rate of working of the forces on the shell one obtains

$$w(t) = A[r_I(t)^2 v(r_I, t)P_I(t) - r_O(t)^2 v(r_O, t)P_O(t)] = \frac{A}{3}\,\dot{q}(t)\,\Delta P(t) \qquad (10.14)$$

where

$$\Delta P = P_I - P_O \qquad (10.15)$$

It follows that the applied force associated with q is just

$$F = \frac{A}{3}\Delta P \qquad (10.16)$$

* Coleman and Dill[26] [equation (6.27), p. 217].
† Coleman[37] [§6, particularly equation (6.5)].

with ΔP the difference in the pressures on the inner and outer surfaces of the shell.

The difference

$$\Xi = F - Q \tag{10.17}$$

between the applied forces of (10.16) and inertial force of (10.12) is the internal force associated with q. It follows from an analysis of the present problem in its three-dimensional setting that Ξ is given by a function of q^t,*

$$\Xi(t) = \mathfrak{Z}(q^t) \tag{10.18}$$

where the functional $\mathfrak{Z}$ is proportional to A and depends on the material comprising the tube, the temperature, the choice of reference configuration, and the radii R_I and R_O. By putting (10.16), (10.12), and (10.18) into (10.17), one obtains the following functional-differential equation for q:†

$$K(q)\ddot{q} - \Omega(q)\dot{q}^2 = \tfrac{1}{3}A\,\Delta P - \mathfrak{Z}(q^t) \tag{10.19}$$

with K and Ω given by (10.10) and (10.13).

Let us now assume that from $t = 0$ onward ΔP is a prescribed function of q, so that

$$F = \tfrac{1}{3}A\,\Delta P = \hat{F}(q) \qquad \text{for } t \geqslant 0 \tag{10.20}$$

If, for example, the shell is closed, that is, encloses completely, without leaks, the spherical region $\{r \mid r \leqslant r_I\} = \mathscr{S}$ (instead of just covering a spherical segment), then $A = 4\pi$, and if the region $\mathscr{S}$ is filled with a gas obeying Boyle's law, we have

$$P_I = P_{\mathscr{R}}\frac{{R_I}^3}{{r_I}^3} = P_{\mathscr{R}}\frac{{R_I}^3}{q} \tag{10.21}$$

where $P_{\mathscr{R}}$ is the pressure in the gas when the shell is in its reference configuration. If such a closed shell is immersed in an atmosphere at constant pressure P_O,

$$F(q) = \tfrac{4}{3}\pi\left[\frac{P_{\mathscr{R}}{R_I}^3}{q} - P_O\right] \tag{10.22}$$

For the system considered here the Lagrangian functional $\mathscr{L}$ of (8.14)

* Coleman and Dill[26]; in that reference the present $\mathfrak{Z}(q^t)$ is written $= -A\mathscr{E}(\beta^t)/3$ with $\beta(t) = q(t) - {R_I}^3$.

† For elastic materials, motions obeying (10.1) and (10.2) have been treated by Truesdell,[41] Guo Zhong-Heng and Solecki,[45,46] Wang,[47] and Knowles and Jakub[48]; for a survey see Truesdell and Noll[41] (§62). For simple materials with memory, such motions have been discussed recently by Carroll,[42] Fosdick,[43] Coleman,[37] Marrucci and Murch,[44] and Coleman and Dill.[26] The present equation (10.19) corresponds to equation (6.17) of Coleman and Dill[26] and is equivalent to the three-dimensional vectorial equation of motion of the incompressible isotropic material, provided only that (10.1) and (10.2) hold.

has the form

$$\mathscr{L}(q^t, \dot{q}) = K(q^t(0))\dot{q}^2 - \mathfrak{p}(q^t) - \hat{\zeta}(q^t(0)) \tag{10.23}$$

where K is given by (10.10),

$$\hat{\zeta}(q) = -\int_0^q F(\sigma)\, d\sigma \tag{10.24}$$

and $\mathfrak{p}$ is a functional which gives the Helmholtz free energy of the shell as a function of the history of q:*

$$\psi(t) = \mathfrak{p}(q^t) \tag{10.25}$$

It is easily verified, by direct calculation and the use of (8.5) and (10.23), that the dynamical equation (10.19) can be cast into the form (9.29). By (10.8)–(10.11),

$$T = \frac{p^2}{2K(q)} \tag{10.26}$$

and hence for the Hamiltonian functional $\mathscr{H}$ we have

$$\mathscr{H}(q^t, p) = \mathfrak{p}(q^t) + \zeta(q^t(0)) + \frac{p^2}{2K(q^t(0))} \tag{10.27}$$

It follows from Theorem 8.2 that $\mathscr{H}(q^t, p(t))$ cannot increase on any solution of the dynamical equation (10.19); that equation here has the canonical form (9.32) with K, Ω, and $\hat{F}$ given by (10.10), (10.13), and (10.20).

XI. APPENDIX ON GENERALIZATIONS

In the branch of continuum physics called the thermodynamics of materials with memory, a *simple material* is one for which the free energy density, the stress tensor, the entropy density, and the heat flux are determined at a point X when the history $\mathbf{G}^t$ of the deformation gradient $\mathbf{G}$, the history θ^t of the temperature θ, and the present value of the temperature gradient are given at X. Thus each simple material is characterized by a list of constitutive equations of the general form

$$\mathbf{g}(t) = \mathfrak{g}(\mathbf{G}^t, \theta^t)$$

where $\mathbf{g}(t)$ stands for the value at time t of some quantity, such as the stress tensor or the density of the Helmholtz free energy; the constitutive functional $\mathfrak{g}$ may or may not depend on the temperature gradient. A central problem in the thermodynamics of materials with memory is that of rendering explicit the restrictions which the second law places on constitutive functionals, as is done here in Section VI. For simple materials I first treated the problem[1,2] assuming that the response functionals obey an hypothesis of smoothness introduced in

* In the notation of Ref. 26 [equation (6.10)], $\mathfrak{p}(q^t) = A\ell(\beta^t)$.

work done earlier with Noll[18–20] and mentioned here after Theorem 6.1.* Later work done with Mizel[9] showed that the main conclusions of the earlier papers remain valid under a weaker smoothness assumption of the type discussed here at length in Section IV and called the "principle of fading memory" at the beginning of Section V. Shortly afterward, Owen[13] discussed the problem of thermodynamical restrictions for simple materials employing a markedly different postulate of regularity for constitutive functionals. Although Owen's postulate involved a history-dependent elastic range and was compatible with plastic behavior and rate-independent response, rather than fading memory, among his results are analogues of those which hold under the principle of fading memory. More recently Owen and I[14] studied a class of materials broad enough to contain as special cases both Owen's materials with elastic range and the previously studied materials with fading memory. This recent research shows that the main theorem of the present paper, Theorem 6.1, and its principal consequences, discussed in Section VIII, can be widely generalized. I give below, in terms of the constitutive relations of discrete systems with memory, a brief statement of the hypothesis of smoothness which Owen and I employed in Ref. 14.† The statement starts with a few preliminary definitions.

Let $\mathbf{\Phi}_\delta$ be a one-parameter family of functions mapping $[0, \infty)$ into R^{n+1} with δ varying over $(0, a)$ for some $a > 0$. It is said that $\mathbf{\Phi}_\delta$ is a *regular family of functions into* R^{n+1} *that vanishes quickly with* δ if, for each δ, the function $\mathbf{\Phi}_\delta$ is continuous and piecewise continuously differentiable, *and* there exist positive numbers M and N such that

$$|\mathbf{\Phi}_\delta(s)| \leqslant M\chi_{[0,\delta]}(s)\delta \,, \qquad \left|\frac{d}{ds}\mathbf{\Phi}_\delta(s)\right| \leqslant N\chi_{[0,\delta]}(s)$$

for all s in $[0, \infty)$ and all δ in $(0, a)$; here $\chi_{[0,\delta]}$ is the characteristic function of the interval $[0, \delta]$, and d/ds denotes the right-hand derivative. Given a number $\sigma \geqslant 0$ and a function $\mathbf{\Phi}$ on $[0, \infty)$, one may define a function $L_{(\sigma)}\mathbf{\Phi}$ on $[0, \infty)$ through the formula

$$L_{(\sigma)}\mathbf{\Phi}(s) = \begin{cases} \mathbf{\Phi}(\sigma) \,, & 0 \leqslant s < \sigma \\ \mathbf{\Phi}(s) \,, & \sigma < s < \infty \end{cases}$$

$L_{(\sigma)}\mathbf{\Phi}$ is called the σ-*level* of $\mathbf{\Phi}$. The σ-*section* of $\mathbf{\Phi}$, $\mathbf{\Phi}_{(\sigma)}$, is defined as in (4.5), and the *static continuation*, $\mathbf{\Phi}^{(\sigma)}$, is defined as in (4.4).

Now, let $\mathbf{\Lambda}$ in C^{n+1} and $\mathbf{\Sigma}$ in R^{n+1} be as in (5.1) and (5.6), and consider the constitutive equations (5.7), with $\mathbf{\Lambda}^t$ defined in (5.3). Suppose that there is given a Hausdorff topological vector space $\mathfrak{B}$, formed from functions mapping $[0, \infty)$ into R^{n+1}, with scalar multiplication and addition of functions defined pointwise, as usual. The elements of $\mathfrak{B}$ with range in C^{n+1} form a cone $\mathfrak{C}$ in $\mathfrak{B}$. Let this

* A fresh approach to the problem is given in the work of Day,[11] and his recent monograph[12] surveys much of the literature on the thermodynamics of materials with fading memory.

† For other applications of hypotheses of this type see Coleman and Dill.[16]

cone $\mathfrak{C}$ be the domain of definition of $\mathfrak{p}$ and $\mathfrak{S}$, and let the topology of $\mathfrak{C}$ be that induced by $\mathfrak{V}$.

Postulate I. *The constitutive functionals* $\mathfrak{p}$ *and* $\mathfrak{S}$ *are continuous functions on* $\mathfrak{C}$.

Postulate II. *If* $\mathbf{\Phi}_\delta$ *is a regular family of functions into* R^{n+1} *that vanishes quickly with* δ, *then each function of the family* $\mathbf{\Phi}_\delta$ *is an element of* $\mathfrak{V}$. *Furthermore, with respect to the topology of* $\mathfrak{V}$, $\lim_{\delta\to 0} \mathbf{\Phi}_\delta = \mathbf{0}^\dagger$, *where* $\mathbf{0}^\dagger$, *the zero element of* $\mathfrak{V}$, *is the constant function on* $[0, \infty)$ *with value zero.*

Postulate III. *The functional* $\mathfrak{p}$ *determines functions,* $\mathsf{D}\mathfrak{p}$ *and* $d\mathfrak{p}$, *mapping* $\hat{\mathfrak{C}}$, *a dense subset of* $\mathfrak{C}$, *into* R^{n+1} *and* R^1, *respectively; furthermore, the functions* $\mathsf{D}\mathfrak{p}$ *and* $d\mathfrak{p}$, *and their domain* $\hat{\mathfrak{C}}$, *are such that if* $\mathbf{\Psi}$ *is in* $\hat{\mathfrak{C}}$, *then*

(i) $\mathbf{\Psi}$ *is continuously differentiable on some interval* $[0, \lambda]$, $\lambda > 0$,

(ii) *if* $\mathbf{\Phi}_\delta$ *is a regular family of functions into* R^{n+1} *that vanishes quickly with* δ, *then, for all* δ *in some interval* $(0, b)$, $b > 0$, *the functions* $\mathbf{\Psi} + \mathbf{\Phi}_\delta$ *are in* $\hat{\mathfrak{C}}$, *and, for small* δ,

$$\mathfrak{p}(\mathbf{\Psi} + \mathbf{\Phi}_\delta) = \mathfrak{p}(\mathbf{\Psi}) + \mathsf{D}\mathfrak{p}(\mathbf{\Psi}) \cdot \mathbf{\Phi}_\delta(0) + o(\delta)$$

while

$$\lim_{\delta\to 0} \mathsf{D}\mathfrak{p}(\mathbf{\Phi} + \mathbf{\Psi}_\delta) = \mathsf{D}\mathfrak{p}(\mathbf{\Phi}) \quad \textit{and} \quad \lim_{\delta\to 0} d\mathfrak{p}(\mathbf{\Psi} + \mathbf{\Phi}_\delta) = d\mathfrak{p}(\mathbf{\Psi})$$

(iii) $L_{(\sigma)}\mathbf{\Psi}$ *and* $\mathbf{\Psi}_{(\sigma)}$ *are in* $\hat{\mathfrak{C}}$ *for all* $\sigma \geqslant 0$, *and*

$$\lim_{\sigma\to 0} \frac{1}{\sigma} [\mathfrak{p}(L_{(\sigma)}\mathbf{\Psi}) - \mathfrak{p}(\mathbf{\Psi}_{(\sigma)})] = d\mathfrak{p}(\mathbf{\Psi})$$

For constitutive functionals obeying the postulates just listed, the following theorem plays the role of Theorem 4.7, which gives the chain rule of the theory of fading memory [see also (4.24)].

Theorem 11.1.* *If* $\mathfrak{p}$ *obeys Postulates I–III, and if* $\mathbf{\Psi}$ *is in* $\hat{\mathfrak{C}}$, *then*

$$\dot{\psi} \overset{\text{def}}{=} \lim_{\sigma\to 0+} \frac{1}{\sigma} [\mathfrak{p}(\mathbf{\Psi}) - \mathfrak{p}(\mathbf{\Psi}_{(\sigma)})]$$

exists and obeys the formula

$$\dot{\psi} = \mathsf{D}\mathfrak{p}(\mathbf{\Psi}) \cdot \dot{\mathbf{\Psi}}(0) + d\mathfrak{p}(\mathbf{\Psi})$$

where

$$\dot{\mathbf{\Psi}}(0) = \lim_{s\to 0} \frac{\mathbf{\Psi}(0) - \mathbf{\Psi}(s)}{s}$$

The quantities $d\mathfrak{p}(\mathbf{\Psi})$ and $\mathsf{D}\mathfrak{p}(\mathbf{\Psi})$ play here essentially the role played by $\delta\mathfrak{p}(\mathbf{\Psi} \mid {}_r\dot{\mathbf{\Psi}})$ and $D\mathfrak{p}(\mathbf{\Psi})$ in Sections V and VI. Indeed, the theory of constitutive functionals obeying Postulates I–III generalizes the theory of fading memory in

* Coleman and Owen[14] (Theorem 1).

a precise sense. In viewing the theory of Sections V and VI as a special case of the present, one takes $\mathfrak{C}^{n+1}$ to be $\mathfrak{C}$, and $\hat{\mathfrak{C}}$ to be those tame histories in $\mathfrak{C}^{n+1}$ which are continuously differentiable on a nonempty interval of the type $[0, \lambda]$, and one identifies $\mathscr{d}\mathfrak{p}(\mathbf{\Psi})$ and $\mathsf{D}\mathfrak{p}(\mathbf{\Psi})$ with the restrictions of $\delta\mathfrak{p}(\mathbf{\Psi} \mid {}_r\mathbf{\Psi})$ and $D\mathfrak{p}(\mathbf{\Psi})$ to histories in $\hat{\mathfrak{C}}$.* The functional $\mathsf{D}\mathfrak{p}$, like the functional $D\mathfrak{p}$ in the theory of fading memory, is called the *instantaneous derivative* of $\mathfrak{p}$.

Arguments employed by Coleman and Owen† in their study of the thermodynamics of simple materials with memory can be applied here to prove the following generalization of Theorem 6.1:

Theorem 11.2.‡ *Whenever the constitutive functionals of (5.7) obey Postulates I–III, the second law of thermodynamics requires that for each history* $\mathbf{\Psi}$ *in* $\hat{\mathfrak{C}}$:

$$\mathfrak{S}(\mathbf{\Psi}) = \mathsf{D}\mathfrak{p}(\mathbf{\Psi}) \tag{11.1}$$

and

$$\mathscr{d}\mathfrak{p}(\mathbf{\Psi}) \leqslant 0 \tag{11.2}$$

Although (11.1) holds only in $\hat{\mathfrak{C}}$, and indeed, $\mathsf{D}\mathfrak{p}(\mathbf{\Psi})$ has been defined here only for $\mathbf{\Psi}$ in $\hat{\mathfrak{C}}$, we may observe, as was done on p. 257 of Ref. 14, that the relation (11.1) means that the functional $\mathfrak{p}$ determines the functional $\mathfrak{S}$ on their entire domain $\mathfrak{C}$, because $\mathfrak{S}$ is continuous on all of $\mathfrak{C}$, and $\hat{\mathfrak{C}}$ is dense in $\mathfrak{C}$.

Once Theorem 11.2 is established, it becomes obvious that Theorems 8.1–8.3 and the discussion of Sections IX and X remain valid if the principle of fading memory, stated at the beginning of Section V, is replaced with Postulates I–III.

It appears, however, that the assumptions made so far in this Appendix are not strong enough for the theory of Section VII. To enable a discussion of equilibrium and the approach to equilibrium,§ one may add Postulates IV and V below to the list I, II, and III.

Postulate IV. *If* $\mathbf{\Psi}$ *is in* $\hat{\mathfrak{C}}$, *then so also are the constant function* $\mathbf{\Psi}(0)^\dagger$ *and all static continuations* $\mathbf{\Psi}^{(\sigma)}$ *of* $\mathbf{\Psi}$, *and in the topology of* $\hat{\mathfrak{C}}$, $\lim_{\sigma\to 0} \mathbf{\Psi}^{(\sigma)} = \mathbf{\Psi}(0)^\dagger$. *Furthermore, for each* $\mathbf{\Psi}$ *in* $\hat{\mathfrak{C}}$, *the function* $\sigma \mapsto \mathbf{\Psi}^{(\sigma)}$ *is a continuous mapping of* $[0, \infty)$ *into* $\hat{\mathfrak{C}}$.

* Here $\mathscr{d}\mathfrak{p}$ is defined only on $\hat{\mathfrak{C}}$, while in the theory of fading memory $\delta\mathfrak{p}(\mathbf{\Psi} \mid {}_r\dot{\mathbf{\Psi}})$ has meaning if and only if $\mathbf{\Psi}$ is a tame history. Of course, in the theory of fading memory, $D\mathfrak{p}$, defined in (4.21), has the same domain as $\mathfrak{p}$. For a discussion of the problem of extending $\mathsf{D}\mathfrak{p}$ from $\hat{\mathfrak{C}}$ to all of $\mathfrak{C}$, see Coleman and Owen[14] (p. 257) and the discussion after Theorem 11.2 below.

† Ref. 14 (§5, pp. 255–259).

‡ As it is not assumed here that $\mathfrak{h}$ has the same properties of regularity as $\mathfrak{p}$, one here uses the inequality (3.1) in the form (3.7) and makes certain minor modifications in the definition of a smooth process. Otherwise one should replace $\mathfrak{p}$ with the pair $(\mathfrak{p}, \mathfrak{h})$ in stating Postulates I–III, and add assumptions which permit one to demonstrate not only that both $\psi(\cdot)$ and $\eta(\cdot)$ have left-hand derivatives but also that these functions are continuous. I take the latter route in a forthcoming article.

§ Compare Ref. 14 (pp. 260, 261).

Let $\hat{C}^{n+1}$ be the set of vectors $\mathbf{\Lambda}$ in C^{n+1} for which the corresponding constant functions $\mathbf{\Lambda}^{\dagger}$ are in $\hat{\mathfrak{C}}$. On $\hat{C}^{n+1}$ one defines the equilibrium response functions $\mathfrak{p}^{\circ}$ and $\mathfrak{S}^{\circ}$ through the equations (7.18). One can show* that if $\mathbf{\Lambda}$ is in $\hat{C}^{n+1}$, then for each $\mathbf{\Gamma}$ in R^{n+1}, there exists a $b > 0$ such that $\mathbf{\Lambda}^{\dagger} + \nu\mathbf{\Gamma}$ is in $\hat{C}^{n+1}$ for every ν in $(0, b)$.

Postulate V. *There exists a function* $\nabla\mathfrak{p}^{\circ}$, *mapping* $\hat{\mathfrak{C}}^{n+1}$ *into* R^{n+1}, *such that for each* $\mathbf{\Lambda}$ *in* $\hat{C}^{n+1}$,

$$\lim_{\nu \to 0} \frac{1}{\nu} [\mathfrak{p}^{\circ}(\mathbf{\Lambda} + \nu\mathbf{\Gamma}) - \mathfrak{p}^{\circ}(\mathbf{\Lambda})] = \nabla\mathfrak{p}^{\circ}(\mathbf{\Lambda}) \cdot \mathbf{\Gamma}$$

for every $\mathbf{\Gamma}$ *in* R^{n+1}.

It is clear that Postulates IV and V hold in the theory of fading memory; in that theory the continuity of the function $\sigma \mapsto \mathbf{\Psi}^{(\sigma)}$ is given by Theorem 4.5, and the existence of $\nabla\mathfrak{p}^{\circ}$ follows from the assumed Fréchet-differentiability of $\mathfrak{p}$.

Employing Postulates I–V and arguments given in §5 of Ref. 14, one can prove the following generalizations of Theorems 7.1 and 7.3.

Theorem 11.3† *If* $\mathbf{\Psi}$ *is in* $\hat{\mathfrak{C}}$, *then*

$$\mathfrak{p}(\mathbf{\Psi}) \geqslant \mathfrak{p}^{\circ}(\mathbf{\Psi}(0))$$

Theorem 11.4‡ *For each* $\mathbf{\Lambda}$ *in* $\hat{C}^{n+1}$

$$D\mathfrak{p}(\mathbf{\Lambda}^{\dagger}) = \nabla\mathfrak{p}^{\circ}(\Gamma)$$

and hence (11.1) implies the classical relation

$$\mathfrak{S}^{\circ} = \nabla\mathfrak{p}^{\circ}$$

Acknowledgments

While writing this article, I have received, on various crucial occasions, substantial help and valuable suggestions from the following companions in research: Walter Noll, Victor J. Mizel, Morton E. Gurtin, Ellis H. Dill, Lincoln E. Bragg, and David R. Owen. It is to my collaboration with Professor Noll, which goes back to 1958, with Professor Mizel from 1962 to date, and with Professor Gurtin in the period 1964 to 1967 that I owe the bulk of my understanding of the theory of fading memory. In 1964 Professor Dill helped me to understand, within the framework of the three-dimensional theory of the mechanics of continua with memory, the classes of motions discussed here in Sections IX and X. For several years prior to 1968 I knew that for isothermal processes the general internal force relation (6.24) implies that the time-derivative of any function of $\mathbf{q}$ and $\dot{\mathbf{q}}$ (or $\mathbf{q}$ and $\mathbf{p}$) is determined through instantaneous derivatives once the free-energy functional $\mathfrak{p}$ and the pair $(\mathbf{q}^t, \dot{\mathbf{q}}(t))$ are given, but I had not cast this dependence into the simple form shown in Theorem 8.3. In the summer of 1968, while we

* Coleman and Owen[14] (p. 261).

† Compare Coleman and Owen[14] (Theorem 3).

‡ Compare Coleman and Owen[14] (Theorems 4 and 5), and also the article of Gurtin[10] which influenced the work in Ref. 14.

were discussing the construction of Lyapunov functionals for the functional-differential equations of mechanics, Professor Bragg suggested that one should be able to apply the theory of Poisson brackets and contact transformations to such problems, and this suggestion led immediately to Theorems 8.1–8.3. The generalization of the principle of fading memory discussed in the Appendix of this article was worked out with Professor Owen in the summer of 1969, after he developed his approach to the theory of plastic materials with elastic range.

Parts of the research reported here were supported by the Air Force Office of Scientific Research and the National Science Foundation.

References

1. B. D. Coleman, *Arch. Ration. Mech. Anal.*, **17,** 1 (1964).
2. B. D. Coleman, *Arch. Ration. Mech. Anal.*, **17,** 230 (1964).
3. B. D. Coleman, *Arch. Ration. Mech. Anal.*, **36,** 1 (1970).
4. B. D. Coleman and M. E. Gurtin, *Z. angew. Math. Phys.*, **18,** 199 (1967).
5. B. D. Coleman and M. E. Gurtin, *Arch. Ration. Mech. Anal.*, **19,** 266, 317 (1965).
6. B. D. Coleman and M. E. Gurtin, *Proc. Royal Soc. (London) Ser. A*, **292,** 562 (1966).
7. B. D. Coleman and M. E. Gurtin, *Quart. Appl. Math.*, **24,** 257 (1966).
8. B. D. Coleman and M. E. Gurtin, *Proc. IUTAM Symposia, Vienna, June 22–28, 1966*, Springer-Verlag, Vienna, New York, 1968, p. 54.
9. B. D. Coleman and V. J. Mizel, *Arch. Ration. Mech. Anal.*, **27,** 255 (1967).
10. M. E. Gurtin, *Arch. Ration. Mech. Anal.*, **28,** 40 (1968).
11. W. A. Day, *Arch. Ration. Mech. Anal.*, **31,** 1 (1968).
12. W. A. Day, *The Thermodynamics of Simple Materials with Fading Memory*, Vol. 22 of *Springer Tracts in Natural Philosophy*, Springer-Verlag, Berlin, Heidelberg, New York, 1972.
13. D. R. Owen, *Arch. Ration. Mech. Anal.*, **31,** 91 (1968).
14. B. D. Coleman and D. R. Owen, *Arch. Ration. Mech. Anal.*, **36,** 245 (1970).
15. B. D. Coleman and E. H. Dill, *Z. angew Math Phys* , **22,** 691 (1971)
16 B D Coleman and E. H. Dill, *Arch. Ration. Mech. Anal.*, **41,** 132 (1971).
17. B. D. Coleman and W. Noll, *Arch. Ration. Mech. Anal.*, **13,** 167 (1963).
18. B. D. Coleman and W. Noll, *Arch. Ration. Mech. Anal.*, **6,** 355 (1960).
19. B. D. Coleman and W. Noll, *Rev. Mod. Phys.*, **33,** 239 (1961); errata, **36,** 1103 (1964).
20. B. D. Coleman and W. Noll, *Proc. International Symp. Second-Order Effects, Haifa, April 21–29, 1962*, Pergamon Press, Oxford and Jerusalem Academic Press, Israel, 1964, p. 530.
21. C.-C. Wang, *Arch. Ration. Mech. Anal.*, **18,** 117 (1965).
22. B. D. Coleman and V. J. Mizel, *Arch. Ration. Mech. Anal.*, **23,** 87 (1966).
23. B. D. Coleman and V. J. Mizel, *Arch. Ration. Mech. Anal.*, **29,** 18 (1968).
24. B. D. Coleman and V. J. Mizel, *Arch. Ration. Mech. Anal.*, **29,** 105 (1968).
25. B. D. Coleman and V. J. Mizel, *Arch. Ration. Mech. Anal.*, **30,** 173 (1968).
26. B. D. Coleman and E. H. Dill, *Arch. Ration. Mech. Anal.*, **30,** 197 (1968).
27. E. M. Purcell and R. V. Pound, *Phys. Rev.*, **81,** 279 (1951).
28. N. F. Ramsey, *Phys. Rev.*, **103,** 20 (1956).
29. B. D. Coleman and W. Noll, *Phys. Rev.*, **115,** 262 (1959).
30. J. L. Ericksen, *Int. J. Solids Struct.*, **2,** 573 (1966).
31. W. Noll, *Arch. Ration. Mech. Anal.*, **2,** 197 (1958).
32. P. Perzyna, *Arch. Mech. Stosow.* **19,** 537 (1967).

33. C. Truesdell and W. Noll, *The Non-linear Field Theories of Mechanics*, in *Encyclopedia of Physics*, Vol. III/3, Springer-Verlag, Heidelberg, 1965.
34. W. A. J. Luxemburg, *Indag. Math.*, **27**, 229 (1965).
35. W. A. J. Luxemburg and A. C. Zaanen, *Math. Annal.*, **149**, 150 (1963).
36. V. J. Mizel and C.-C. Wang, *Arch. Ration. Mech. Anal.*, **23**, 124 (1966).
37. B. D. Coleman, *Proc. Roy. Soc. (London) Ser. A*, **306**, 449 (1968).
38. E. T. Whittaker, *A Treatise on the Analytical Dynamics of Particles and Rigid Bodies*, 4th ed., Cambridge University Press, 1937.
39. J. K. Knowles, *Quart. Appl. Math.*, **18**, 71 (1960).
40. J. K. Knowles, *J. Appl. Mech.*, **29**, 283 (1962).
41. C. Truesdell, *Arch. Ration. Mech. Anal.*, **11**, 106 (1962).
42. M. M. Carroll, *Int. J. Eng. Sci.*, **5**, 515 (1967).
43. R. L. Fosdick, *Arch. Ration. Mech. Anal.*, **29**, 272 (1968).
44. G. Marrucci and R. E. Murch, *Ind. Eng. Chem. Fundamentals*, **9**, 498 (1970).
45. Guo Zhong-Heng and R. Solecki, *Arch. Mech. Stosow.*, **15**, 427 (1963).
46. Guo Zhong-Heng and R. Solecki, *Bull. Acad. Polon. Sci. Ser. Sci. Tech.*, **11**, 47 (1963).
47. C.-C. Wang, *Quart. Appl. Math.*, **23**, 270 (1965).
48. J. K. Knowles and M. T. Jakub, *Arch. Ration. Mech. Anal.*, **18**, 367 (1965).

THE TRANSITION FROM ANALYTIC DYNAMICS TO STATISTICAL MECHANICS*

JOSEPH FORD

School of Physics, Georgia Institute of Technology, Atlanta, Georgia

CONTENTS

I. Introduction 155
II. Historical and Mathematical Preliminaries 156
III. *C*-Systems 160
IV. *C*-System Behavior in Nonlinear Oscillator Systems 163
V. Irreversible Rate Equations for Simple *C*-Systems 176
VI. Discussion 183
References 183

I. INTRODUCTION

Although work on ergodic theory in the early thirties by von Neuman, Birkhoff, Hopf, and others[1] provided many of the fundamental concepts required to link analytic dynamics and statistical mechanics, these early efforts foundered on the lack of even one physically interesting system being demonstrably ergodic. This lack, coupled with the mathematical complexities inherent in the theory, has led most workers in statistical mechanics either to ignore ergodic theory or to become openly hostile to it.[2] As an unfortunate consequence, statistical mechanics is presently ill-prepared even to appreciate, much less understand and incorporate, the recent advances[3,4] in ergodic theory of which Sinai's proof[5] of ergodicity and mixing in the hard sphere gas is but one example. Indeed the conceptual gap between ergodic theory and statistical mechanics has grown so large that even mathematical review texts (such as Refs. 3 and 4) are filled with a jargon all but incomprehensible to physicists. Nonetheless these recent mathematical advances offer such promise of yielding a definitive link between analytic dynamics and classical statistical mechanics and of yielding new answers to the old questions on irreversibility that a concerted effort to uproot the apathy and resistance of the past is demanded.

* This work supported in part by the National Science Foundation.

To this end, a number of physical scientists have written review papers[6–12] emphasizing the physical relevance of various aspects of the new mathematical results, and this paper is yet another in that sequence. Our justification for adding to the literature is based on the belief that an extremely simple and clear exposition of the new developments can be made for an audience of physical scientists by subsuming much of the recent work under one easily defined and intuitively appealing mathematical concept—that of *C*-systems.[3] Using the simplest examples and drawing selectively from a wide variety of earlier sources, we seek to achieve three major objectives. After presenting some historical and mathematical preliminaries, we concentrate first on illustrating by example that *C*-systems have precisely those properties intuitively expected of mechanical systems exhibiting statistical mechanical behavior. Indeed these illustrated properties form the underpinning[13] of Sinai's proof of ergodicity and mixing in the hard sphere gas. We then consider several extremely simple mechanical systems, having attractive as well as repulsive forces, and empirically verify the conditions under which a type of *C*-system behavior occurs. In particular we empirically establish the existence of a smooth transition from the non-statistical behavior typical of analytic dynamics to the stochastic (*C*-system) behavior required by statistical mechanics. Finally, again using very simple examples, we illustrate how *C*-system behavior can yield irreversible rate equations governing the approach to equilibrium.

Like most review papers, the following exposition contains little that is new other than the arrangement and interpretation of the topics discussed. However, the interpretive scheme used here, as the author is well aware, has faults accompanying its virtues. By using empirical computer calculations to interpret and extend rigorous mathematical theorems, our discussion gains clarity but loses elegance and perhaps accuracy; by confining our attention to the simplest properties of *C*-systems, which form but one class in a hierarchy of unstable systems, the discourse gains unity but loses precision and comprehensiveness; and by considering only elementary systems as illustrative examples, the presentation gains simplicity but loses generality. Regardless, our central purpose will have been achieved if our intuitive arguments impart to physicists at least some of the flavor and part of the statistical mechanical significance of the recent mathematical advances.

II. HISTORICAL AND MATHEMATICAL PRELIMINARIES

Boltzmann introduced an intuitive notion of ergodicity[14] into statistical mechanics by assuming that the phase space or Γ-space trajectories for an isolated physical system having constant energy pass through every point on the energy surface. Boltzmann then used ergodicity to justify equating the theoretically calculable, microcanonical phase space average of an observable to the intractable, infinite time average of the same observable, which latter was assumed to equal the experimentally measured equilibrium value. Ergodicity also appears in Boltzmann's view of the approach to equilibrium. In essence, he argued[15] that most microstates (q_i, p_i) in Γ-space correspond to macroscopic thermodynamic equilibrium states (P, V, T, or the like). He therefore expected that an isolated system—started in some disequilibrium state and subsequently allowed to follow its assumed ergodic tendency to wander freely over the energy surface—would eventually reach equilibrium. In order to avoid Loschmidt's and Zermelo's paradoxes,[16] he was ultimately forced to admit that the approach to equilibrium might not be monotonic and that even large fluctuations from equilibrium might occur and recur; however, the general trend, as observed in most physical systems, would be for the system to monotonically approach equilibrium and remain there.

In the early thirties, Birkhoff, Khinchin, and von Neumann[1,10,14] constructed a formal and rigorous framework for Boltzmann's intuitive picture. In their formalism, a system is said to be ergodic if and only if it is metrically transitive, that is, provided that the trajectories on the energy surface cannot be decomposed into two invariant sets having positive measure. In less formal terms, ergodicity requires that most phase space trajectories for an isolated physical system densely cover the energy surface. After some technical preliminaries, the precise definition of ergodicity forms the cornerstone for proving the equality of phase space and time averages. Birkhoff, Khinchin, and von Neumann thus reduced Boltzmann's problem concerning phase and time averages to the problem of metric transitivity.

Unfortunately the notion of metric transitivity did not prove to be especially fruitful except in the physically artificial case of geodesic flow on surfaces of negative curvature which, in physical terms, means frictionless motion of a particle on an oddly shaped surface. But aside from a lack of physically realistic examples, the enlightening notion of ergodicity is incomplete, in a physical sense, because it does not ensure that a system "forget" its initial state as required by thermodynamics. For this latter property, two arbitrarily close, but not identical, final states must in a sufficiently remote past have occupied widely separated positions on the energy surface; or continuing into the future, two arbitrarily close initial states must in a sufficiently remote future again be widely separated. Ergodicity alone does not preclude close initial states from remaining close as they densely traverse the energy surface. In short, to ergodicity we must add the concept of mixing.

Interesting enough, Gibbs was formulating an intuitive notion of statistical mechanical mixing[17] at about the time Boltzmann was developing the notion of ergodicity. In order to provide an intuitive picture of the approach to equilibrium, Gibbs considered an ensemble of identical, but isolated, systems whose initial states were uniformly spread over a small region of the energy surface. He then suggested that Newton's equations generated a flow of these initial states leading to a uniform distribution of final states over the energy surface, much as the stirring of a martini leads eventually to a uniform mixture of gin and vermouth. Gibbs's intuitive notion of mixing possesses a special charm for physicists. First, the approach to "equilibrium" generated by stirring a mixture is a commonplace experience, and second, the physicist can easily visualize a nonequilibrium initial state probability distribution evolving through mixing to a uniform, equilibrium distribution on the energy surface, at least in some coarse-grained sense.[17] In addition, Gibbs's model bypasses Zermelo's recurrence paradox since the initial state of an infinite ensemble is not required to recur, whereas Loschmidt's reversibility paradox not only remains, as it must, but is easily demonstrated in the beautifully simple, unmixing experiment of Heller.[18] Indeed reversibility in the mixing problem means only that the system can proceed at the onset through an excursion away from equilibrium, only to return as the inverse stirring continues.

E. Hopf[19] provided a rigorous mathematical definition of mixing, and in so doing, he emphasized not only that mixing is stronger than and hence implies ergodicity but also that ergodicity is insufficient to guarantee the sensitive dependence of final state upon initial state as required,[10] for example, by an honest roulette wheel. The flow generated by Newton's equations on the energy surface for an isolated physical system is said to be mixing if, for every pair of measurable sets A and B on the energy surface E,

$$\lim_{t\to\infty}\left[\frac{\mu(A \cap B_t)}{\mu(A)}\right] = \left[\frac{\mu(B)}{\mu(E)}\right] \tag{1}$$

where μ denotes measure and $(A \cap B_t)$ denotes the intersection of the fixed set A with the set B_t which has evolved under the flow from the original set B. In words, (1) states that, after a sufficiently long time interval, the fractional amount of B_t in every arbitrarily small, but nonzero, set A is the same as the original fractional amount of B in E.

Despite the remarkable intuitive appeal of all these ideas, for decades they lay dormant as ideas whose time had not yet come. During the 1960s, however, the methods developed by Hopf, Hadamard, and others[10] for demonstrating ergodicity and mixing in geodesic flows on surfaces of negative curvature were extended and applied to more general systems by

Anosov, Sinai, Smale, and others.[3] Not only did Sinai establish the hard sphere gas as ergodic and mixing, but Anosov and Sinai[13] also determined the conditions for ergodicity and mixing in smooth dynamical systems, such as systems described by Newton's equations in the presence of smooth force functions. Systems meeting the required conditions are quite naturally called *C*-systems.

In particular Anosov and Sinai noted that orbits near any specified orbit on a surface of negative curvature exhibit a local, exponential instability much as do the solutions of the hyperbolic differential equation

$$\ddot{u} - u = 0 \tag{2}$$

about the solution $\dot{u} \equiv u \equiv 0$. They then identified this local, exponential instability for geodesics on surfaces of negative curvature as the key property to be used in proving ergodicity and mixing for a much wider class of systems. In order to motivate the definition of *C*-systems presented here, let us examine (2), written in the form

$$\dot{u} = v \tag{3a}$$

$$\dot{v} = u \tag{3b}$$

A canonical, 45° rotation of coordinates then allows one to write (3) as

$$\dot{x} = -x \tag{4a}$$

$$\dot{y} = y \tag{4b}$$

Since (4) has solutions $x = x_0e^{-t}$ and $y = y_0e^t$, the Cartesian distance between two close initial points in the (u, v) plane of (3) has an exponentially decreasing x-component, $dx = dx_0e^{-t}$, orthogonal to an exponentially increasing y-component, $dy = dy_0e^t$. In short, for this particular system, each small area-element $(du\, dv)$ in the (u, v) plane exponentially contracts along one direction while exponentially expanding along another in such a way that area is preserved. In physical terms, Anosov and Sinai then generalize and define[20] a *C*-system as any system for which locally each small volume element of Γ-space exponentially contracts in at least one direction while exponentially expanding in at least one other so as to preserve volume; if ζ_1 is the contracting dimension and ζ_2 the expanding one, then locally one requires that

$$\zeta_1(t) \leqslant ae^{-\lambda t}\zeta_1(0) \tag{5a}$$

$$\zeta_2(t) \geqslant be^{\lambda t}\zeta_2(0) \tag{5b}$$

where a, b, and λ are independent of t, $\zeta_1(0)$, and $\zeta_2(0)$. Anosov and Sinai then prove that these *C*-system properties are sufficient to guarantee ergodicity and mixing in smooth dynamical systems.

III. *C*-SYSTEMS

A proof of ergodicity and mixing for either *C*-systems or the hard sphere gas is quite long and mathematically sophisticated. Such proofs occupy scores of journal pages and involve concepts such as diffeomorphisms, transverse foliations, Kolmogorov-Sinai entropy, *K*-systems, and the like. Despite this, one may illustrate that exponential instability leads to ergodic and mixing behavior using a very simple mathematical model and very simple concepts.

Every isolated Hamiltonian system can be regarded as a measure-preserving, that is, area-preserving, transformation of the energy surface onto itself. Thus in order to illustrate the *C*-system behavior expected of many physical systems, let us consider the simple, area-preserving mapping of a unit square onto itself specified by

$$(x_{n+1}, y_{n+1}) = T(x_n, y_n) = (x_n + y_n, x_n + 2y_n)\,, \qquad (\text{mod } 1) \tag{6}$$

where here the continuous time variable is replaced by the discrete integer index n and where (mod 1) means, in essence, periodic boundary conditions. We may examine the local properties of this mapping by considering the differential form of (6) written in the vector-matrix form

$$\begin{pmatrix} dx_{n+1} \\ dy_{n+1} \end{pmatrix} = \begin{pmatrix} 1 & 1 \\ 1 & 2 \end{pmatrix} \begin{pmatrix} dx_n \\ dy_n \end{pmatrix} \tag{7}$$

where we are at liberty to drop the (mod 1), since dx_{n+1} and dy_{n+1} are both less than unity. Let us now rotate the coordinate axes bringing (7) to the form

$$d\zeta_{n+1} = e^{\ln \lambda}\, d\zeta_n, \qquad d\eta_{n+1} = e^{-\ln \lambda}\, d\eta_n \tag{8}$$

where $\lambda = \frac{1}{2}(3 + 5^{1/2})$. Now we may regard $d\zeta_{n+1}$ and $d\eta_{n+1}$ as components of the distance between two points in the unit square initially separated by $d\zeta_n$ and $d\eta_n$, but it is more convenient for our purposes to consider $d\zeta_{n+1}$ and $d\eta_{n+1}$ as the sides of a rectangle having initial values $d\zeta_n$ and $d\eta_n$. After k iterations of (8), an initial rectangle $(d\zeta_n\, d\eta_n)$ is transformed (preserving area) into the rectangle $(d\zeta_{n+k}\, d\eta_{n+k})$ specified by

$$d\zeta_{n+k} = e^{k \ln \lambda}\, d\zeta_n\,, \qquad d\eta_{n+k} = e^{-k \ln \lambda}\, d\eta_n \tag{9}$$

Thus the mapping of (6) is a *C*-system since every small area-element locally exhibits the correct, bounded exponential behavior required by (5), after replacing t by k in that equation.

Moreover, this mapping may be proved to be mixing and hence[21] ergodic. Although a detailed proof of these facts is nontrivial, the central conclusions can at least be made intuitively obvious. As the center of each initial rectangle $(d\zeta_n\, d\eta_n)$ moves according to (6), the $d\zeta$-side of the enclosing rectangle grows and the $d\eta$-side shrinks according to (9). Eventually every original small $(d\zeta_n\, d\eta_n)$ area-element becomes a thin ribbon which "doubles back" across the unit square many times. Moreover, recalling that the unit square (mod 1) is equivalent to a two-dimensional torus and noting that the tangent of the expanding eigenvector direction equals the irrational number $(\lambda - 1)$, we observe that the expansion rate on the torus parallel to the x-axis is incommensurate with the expansion rate parallel to the y-axis. As a consequence, Jacobi's theorem[22] ensures that each initially small area-element $(d\zeta_n\, d\eta_n)$ asymptotically approaches some dense helix on the torus or on the equivalent unit square (mod 1). In short, the (6)-mapping is mixing and hence ergodic. The initial trend toward ergodicity and mixing of the unit square, after only two iterations of the (6)-mapping, is strikingly illustrated in Fig. 1, adapted from the text by Arnold and Avez.[3]

For purposes of comparing the (6)-mapping with others to be discussed later, it is interesting to note that this mapping has a dense set of unstable fixed points throughout the unit square. In order to verify this fact, let us denote k iterations of the mapping by T^k, where $(x_{n+k}, y_{n+k}) = T^k(x_n, y_n)$; then a fixed point of T^k satisfies $(x_n, y_n) = T^k(x_n, y_n)$. One may then verify by direct calculation that the (6)-mapping does have fixed points. For example, the points $(\frac{2}{5}, \frac{1}{5})$ and $(\frac{3}{5}, \frac{4}{5})$ are fixed points of T^2, while $(\frac{1}{4}, \frac{1}{4})$, $(\frac{1}{2}, \frac{3}{4})$, and $(\frac{1}{4}, 0)$ are fixed points of T^3. Indeed every point $(a/b, c/d)$ in the unit square with rational coordinates is a fixed point of T^n for some integer n. Writing the point $(a/b, c/d)$ in common denominator form $(k/m, l/m)$ and noting that (6) has integer coefficients and uses modulo-one arithmetic, one easily establishes that the points $T^Q(a/b, c/d)$, where $Q = 1, 2, 3, \ldots$, are all members of the finite set $(k'/m, l'/m)$, where $0 \leqslant k' \leqslant m$ and $0 \leqslant l' \leqslant m$. Thus for every point $(a/b, c/d) = (k/m, l/m)$ with rotational coordinates, there exists some $Q \leqslant m^2$ for which $T^Q(a/b, c/d) = (a/b, c/d)$; for example, the set of twenty-five points $(k/5, l/5)$, where $0 \leqslant k \leqslant 4$ and $0 \leqslant l \leqslant 4$, contains one fixed point of T itself, four fixed points of T^2, and twenty fixed points of T^{10}. Hence the unit square is everywhere densely covered with fixed points, and because of (9), each of these fixed points is unstable. Finally, then, the ergodic wanderings of iterates of an initial point that is not a fixed point may be intuitively understood as being due to a "scattering" of these iterates off the dense set of unstable fixed points.

In this section we have not only shown that the mapping of (6) provides a simple example[3,13] of a C-system but we have also presented an intuitive picture of how local exponential instability leads to ergodic and mixing behavior. A similar type of exponential instability, though not uniformly

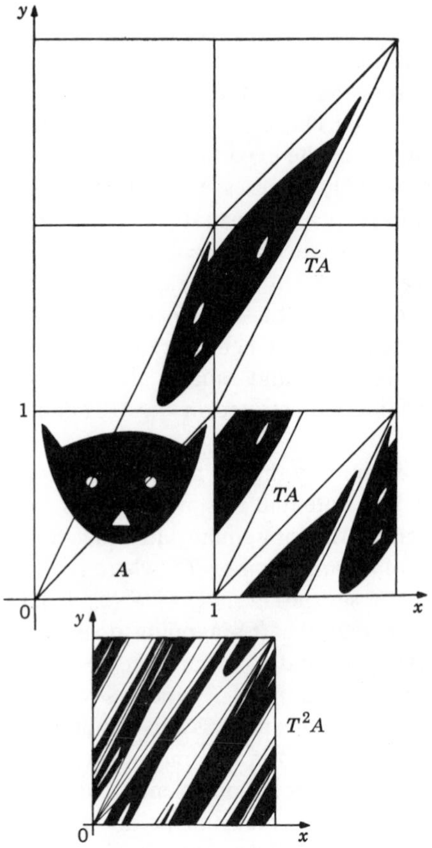

Fig. 1. A drawing of the mapping specified by (6). After only two iterations of the mapping, one already has a crazy, mixed-up cat.

bounded as required by (5), leads to ergodicity and mixing in the hard sphere gas; indeed the same behavior is expected[10] in any many-particle system having only repulsive interparticle forces. One is thus led to inquire[10] whether or not systems with attractive as well as repulsive interparticle forces can exhibit some type of *C*-system behavior. Although a definitive, rigorous answer to this question appears to lie far in the future, in the following section we nonetheless use a combination of rigorous theory and computer-based, empirical evidence to provide a relatively clear outline of this future answer.

IV. C-SYSTEM BEHAVIOR IN NONLINEAR OSCILLATOR SYSTEMS

At sufficiently high energy and low density, classical many-body systems having attractive as well as repulsive forces would be expected to exhibit a type of C-system behavior differing little from that of the hard sphere gas. The success of the Boltzmann equation for dilute gases offers strong support for this conjecture. Severe difficulties are anticipated only when low energy and high density permit bound states and long-range correlations to appear in the system motion. In order to illustrate that physical systems with attractive forces can in general exhibit a type of C-system behavior, we choose to consider the worst case, that is, the lattice vibrations of solids described by nearly linear Hamiltonians of the form

$$H = \frac{1}{2}\sum_{k=1}^{N}\omega_k(P_k^{\,2} + Q_k^{\,2}) + \gamma[V_3 + V_4 + \cdots] \tag{10}$$

where the ω_k are the positive frequencies of the harmonic approximation, N is the number of degrees of freedom, γ is the nonlinear coupling parameter, and $V_3, V_4, \ldots,$ are cubic, quartic, $\ldots,$ polynomials in the Q_k and P_k.

At first glance, the systems described by Hamiltonians (10) appear to be poor candidates for illustrating C-system behavior. When the nonlinear $V_3, V_4, \ldots,$ terms are extremely small, for example, system trajectories can in most cases be accurately calculated using the perturbation theory[23,24] of analytic dynamics. In particular, the perturbing terms $V_3, V_4, \ldots,$ would in general change the unperturbed, harmonic oscillator motion only to the extent of adding small nonlinear harmonics. As a consequence, the small amplitude motion of Hamiltonians (10) would for the most part be no more stochastic or C-systemlike than the unperturbed, harmonic oscillator motion. Even doubts concerning the convergence of such perturbation series have recently been removed[25] by theorems[26] due to Kolmogorov, Arnold, and Moser (hereafter referred to as KAM). KAM rigorously prove that, excepting a set of small measure, most trajectories of Hamiltonians (10) are quasiperiodic orbits lying on smooth N-dimensional integral surfaces embedded in the $2N$-dimensional Γ-space provided, among other things

(i) γ or, equivalently, the total energy is sufficiently small, and

(ii) the harmonic frequencies ω_k do not satisfy low-order resonance conditions of the form $\Sigma_k\, n_k\omega_k \simeq 0$ for integers n_k such that $\Sigma_k\, |n_k| \leqslant 4$.

Now clearly we can choose the γ and the ω_k in Hamiltonians (10) such that the KAM conditions are satisfied, in which event Hamiltonians (10) must for the most part yield the same type of nonstatistical motion that characterizes all the well-known[23] integrable dynamic systems. Thus on the face of it, the KAM theorem makes Hamiltonians (10) appear as an unlikely source of C-system behavior. Looking deeper, however, one finds that the conditions of the KAM theorem, in a sense, delineate the borderline between those γ and ω_k values requiring Hamiltonians (10) to yield nonstatistical behavior and those values allowing them to yield stochastic behavior. In particular in the following paragraphs, we shall first demonstrate using a simple example that varying the system energy (equivalent to varying γ) causes a smooth transition from the integrable system motion of analytic dynamics to a type of C-system motion. We then achieve our central goal regarding lattice vibrations by showing, again by simple example, that allowing the ω_k to violate Condition (ii), as is expected to occur in all solids, also leads to a type of C-system behavior regardless of the size of γ or the system energy.

We begin by considering the nearly linear, two-oscillator Hamiltonian system

$$H = \tfrac{1}{2}(P_1^2 + Q_1^2 + P_2^2 + Q_2^2) + Q_1^2 Q_2 - (\tfrac{1}{3})Q_2^3 \tag{11}$$

first studied by Henon and Heiles.[27] These authors surveyed the allowed motion for Hamiltonian (11) by, in effect, reducing the motion to a plane area-preserving mapping. For fixed energy E, it is possible to generate a plane area-preserving mapping in the (Q_2, P_2) plane using Hamiltonian (11). Starting with a given (Q_2, P_2) point in this plane and setting $Q_1 = 0$, we may solve (11) for $P_1 \geqslant 0$. This provides us with an initial mapping point (Q_2, P_2) and with a set of initial conditions (Q_1, P_1, Q_2, P_2) for integrating a trajectory of Hamiltonian (11). An initial (Q_2, P_2) point then maps into the (Q_2, P_2) coordinates of the next point on this system trajectory for which $Q_1 = 0$ and $P_1 \geqslant 0$. Clearly each trajectory generates a unique sequential set of points in the (Q_2, P_2) plane, and one may use Poincaré's theorem on integral invariants to prove that the mapping is area preserving.

In essence, this area-preserving mapping is a Poincaré surface of section,[28] that is, a two-dimensional cross-section of the three-dimensional energy surface. Since P_1 is determined by Hamiltonian (11) once Q_1, Q_2, and P_2 are given, we may regard the system trajectories as lying in the three-space (Q_1, Q_2, P_2). Plotting those (Q_2, P_2) points on a system trajectory for which $Q_1 = 0$ then provides a picture of the intersection of this trajectory with the (Q_2, P_2) plane. In the event that a trajectory for the Henon-Heiles system lies on one of the smooth, two-dimensional integral surfaces predicted by the KAM theorem, its intersection points with the (Q_2, P_2) plane will lie on a smooth curve.

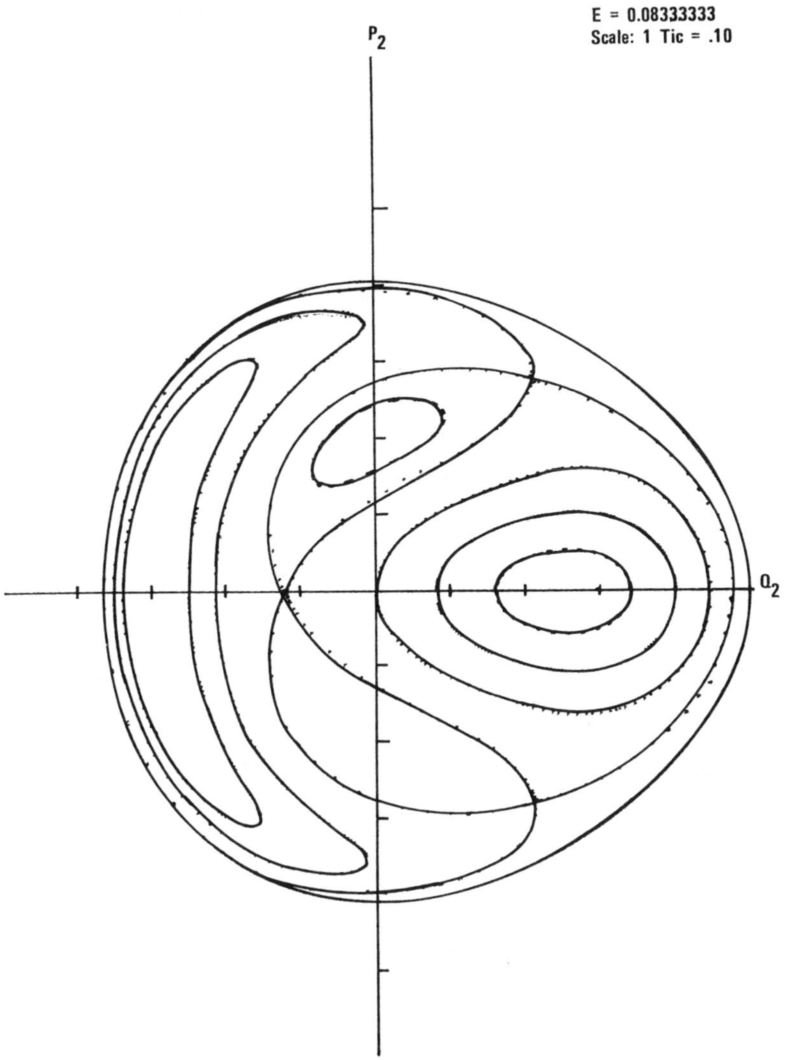

Fig. 2. A surface of section for the Henon-Heiles system at energy $E = \frac{1}{12}$. Here and in the next two figures, whenever the dots representing trajectory intersection points appear to lie on a curve, the curve has been drawn in freehand.

Using a computer, one may obtain the (Q_2, P_2) mapping shown in Fig. 2 which was generated by Hamiltonian (11) at energy $E = \frac{1}{12}$. Here every trajectory investigated yields a smooth curve of intersection points, indicating that the system energy is deep within the region of KAM stability (for reasons to be mentioned later, we may ignore the equality of the unperturbed, harmonic frequencies when considering two-oscillator systems). In Fig. 2, on the Q_2- or P_2-axes at the center of each region of ovals is an elliptic[29] fixed point of the mapping (hereafter called a central fixed point) generated by a stable periodic orbit of Hamiltonian (11). Each oval about a central fixed point is—to computer accuracy—an invariant curve generated by those quasiperiodic orbits of Hamiltonian (11) lying on

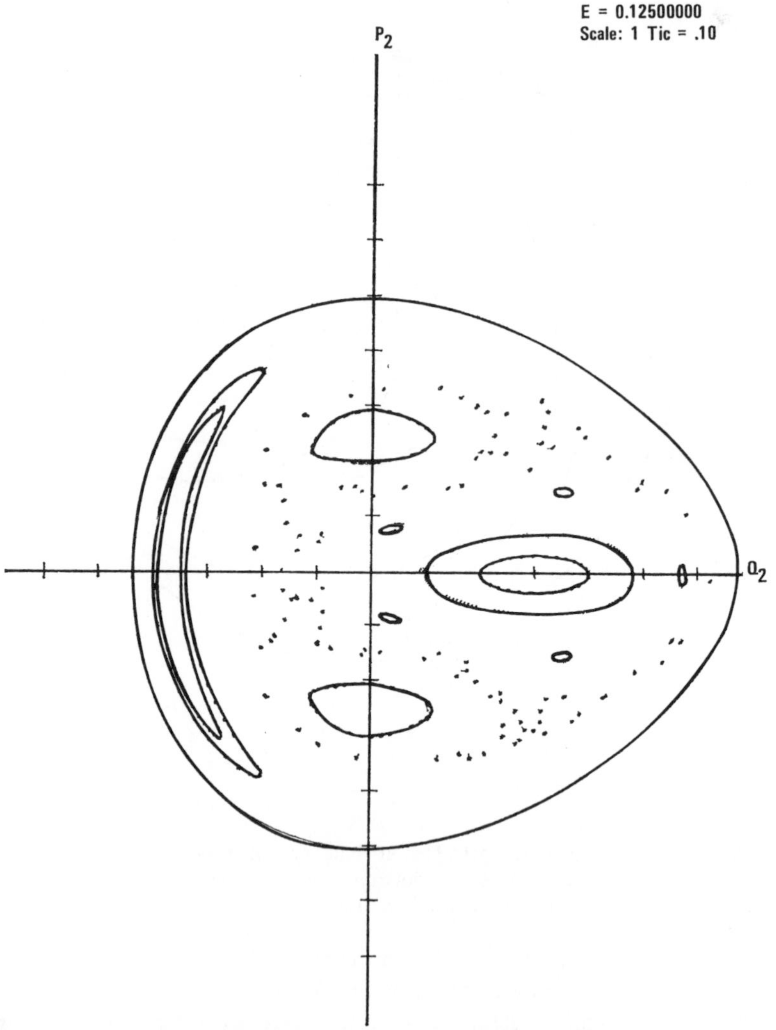

Fig. 3. A surface of section for the Henon-Heiles system at energy $E = \frac{1}{8}$. The unconnected dots are the trajectory intersection points for a single trajectory.

smooth, KAM surfaces. The self-intersection points of the self-intersecting, separatrix curve are hyperbolic[29] fixed points of the mapping generated by unstable periodic orbits. Orbits started elsewhere on this separatrix curve generate mapping iterates which asymptotically approach one or another of the unstable fixed points. Finally all the allowed trajectories for Hamiltonian (11) at $E = \frac{1}{12}$ intersect the (Q_2, P_2) plane at or within the outermost oval in Fig. 2; (Q_2, P_2) points outside this bounding curve yield unphysical negative values for P_1^2.

Increasing the energy to $E = \frac{1}{8}$, we find that Hamiltonian (11) generates the mapping shown in Fig. 3. Although large areas of the allowed (Q_2, P_2) plane in this figure are still covered with invariant curves generated by trajectories lying on smooth KAM surfaces, the remaining area is rather uniformly covered by the intersection points of each (and almost every) orbit initiated in this area. Indeed the somewhat random looking splatter of points in Fig. 3 was generated by a single trajectory; almost any other orbit initiated in the interior of this region would yield a similar pattern of points. Moreover, in these (Q_2, P_2) plane regions of disintegrated invariant curves, initially close points in the plane map apart exponentially,[27] and empirical computer evidence[30] indicates that a dense set of unstable fixed points exists throughout such regions. In short, at $E = \frac{1}{8}$, the Henon-Heiles system exhibits emerging C-system behavior. At $E = \frac{1}{6}$, the mapping shown in Fig. 4 reveals an almost complete transition to C-system behavior. In particular, the dots shown in this figure, which are spread more or less uniformly over the allowed region of the (Q_2, P_2) plane, were generated by a single trajectory.

Thus the Henon-Heiles system is adequate for demonstrating that smooth transition from nonstatistical to stochastic behavior expected of more general oscillator systems. It can also be used to illustrate the pathological behavior responsible for this transition. Let us return to Fig. 3 and consider the region containing ovals surrounding the central fixed point on the positive Q_2-axis. Sequential mapping iterates of an initial (Q_2, P_2) point lying on an oval rotate about the central fixed point and the average angle of rotation (divided by 2π) is called the rotation number ω of the invariant oval curve. The ω associated with each invariant curve varies smoothly as one progresses out from the central fixed point. Moser's statement[31] of the KAM theorem ensures that mathematically exact invariant curves surround the central fixed point provided, among other things, the associated rotation numbers are irrationals satisfying the inequality

$$\left|\omega - \frac{l}{k}\right| > \frac{\epsilon}{k^{5/2}} \tag{12}$$

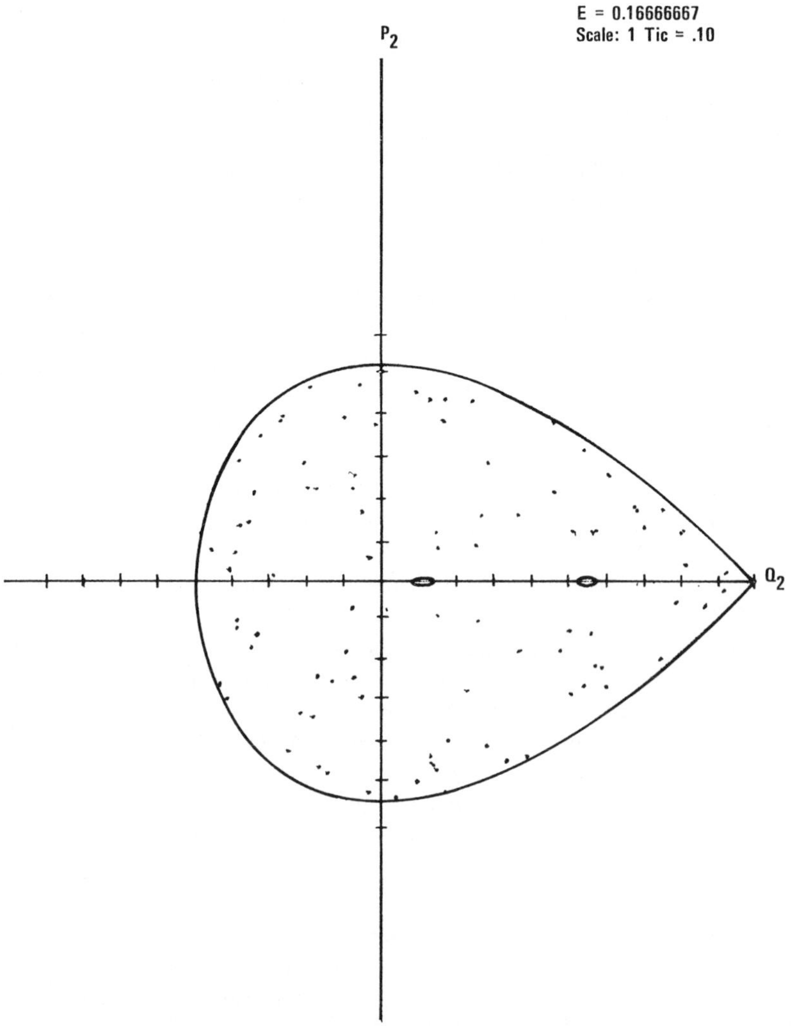

Fig. 4. A surface of section for the Henon-Heiles system at energy $E = \frac{1}{6}$. Again the unconnected dots were generated by a single trajectory.

for all integers l and k, where ϵ is a constant independent of k.

Between these nondense, exact Moser curves, one might expect to find a dense set of invariant curves with rational rotation numbers $\omega = P/Q$ (P and Q relative prime) composed purely of fixed points of Q iterations of the mapping T^Q. Indeed the integrable systems of analytic dynamics do yield such invariant curves. However, a theorem due to Birkhoff[32] assures

Fig. 5. In a neighborhood of the central fixed point on the positive Q_2-axis in Fig. 3, increased computer accuracy might reveal a surface of section looking like this sketch.

us that in general only remnants of such ω-rational invariant curves remain in the form of $2Q$ interleaved fixed points, frequently with half being elliptic and half being hyperbolic. Thus according to Birkhoff, extremely accurate computer calculation of the Fig. 3-mapping in a neighborhood of the central elliptic fixed point on the positive Q_2-axis would reveal a picture similar to that sketched in Fig. 5. An indication of this structure for the Henon-Heiles system at energy $E = \frac{1}{8}$ appears in Fig. 6 showing only a few of the hundreds of empirically calculated fixed points found at this energy.

Returning to Fig. 5, let us emphasize that it is the circles (ovals) surrounding the central fixed point that correspond to the exact Moser invariant curves generated by quasiperiodic orbits on smooth KAM surfaces. Between these invariant curves or the equivalent KAM surfaces, there exists the structure shown in Fig. 5 (some of which can also be predicted by the KAM theorems) and more. The fixed-point families lying between the Moser curves form a dense set; some of these families are alternating elliptic-hyperbolic as shown in Fig. 5 but others are alternating hyperbolic-hyperbolic.[30,33] Moreover, the separatrices emanating from each hyperbolic point have come unglued and no longer smoothly connect

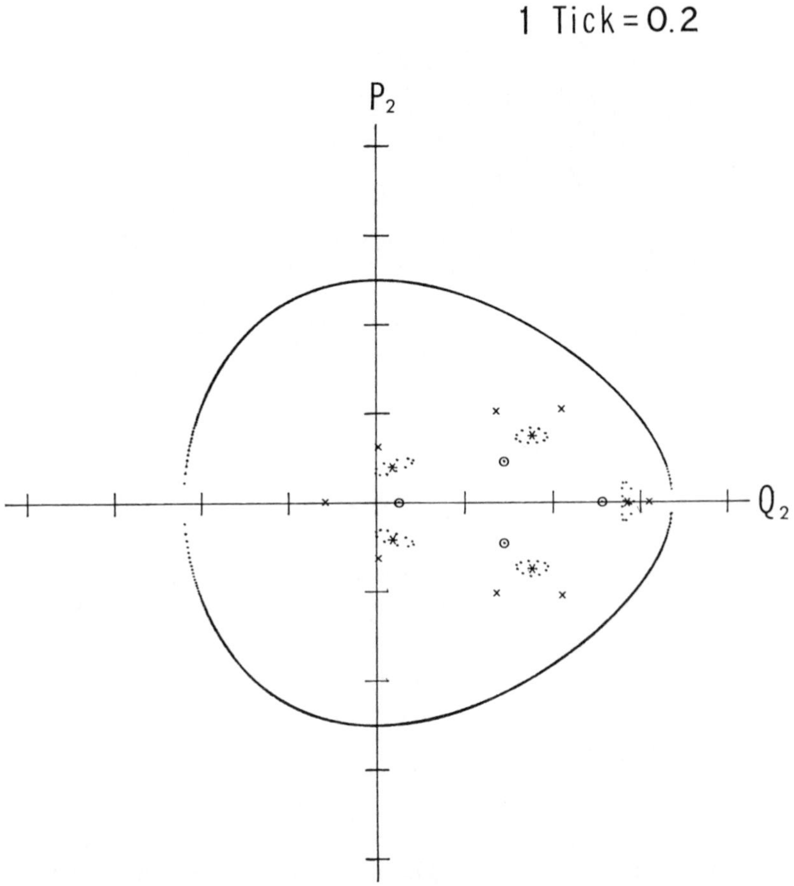

Fig. 6. Selected fixed points of the mapping generated by Hamiltonian (11) at energy $E = \frac{1}{8}$, illustrating that some of the structure sketched in Fig. 5 can actually be observed, The innermost set of four fixed points located at the dots inside small circles are elliptic members of the T^4 family. The fixed points located at the asterisks are elliptic members of the same T^5 family which can be seen in Fig. 3. Surrounding each member of this T^5 family is a set of eleven hyperbolic fixed points belonging to T^{55}. Finally, the eight fixed points located at the ×-symbols are hyperbolic members of T^8. With the exception of the T^8 family, each family of fixed points shown represents half of an alternating elliptic-hyperbolic set. However, all 16 members of the two T^8-families, which lie in the stochastic region of Fig. 3, are hyperbolic.

adjacent hyperbolic fixed points as the separatrix in Fig. 2 appears to do. Not only do separatrices from the same family intersect each other but they also intersect the separatrices belonging to nearby hyperbolic points. In regions containing these many intersecting separatrices, two close initial points locally map exponentially apart. This exponential behavior is especially pronounced as one would expect in those very narrow annular regions containing only hyperbolic fixed points.[33] In short, the phase space for the Henon-Heiles system is always pathologically divided into sets of stochastic and nonstochastic trajectories. At low energy, the Henon-Heiles motion is almost all nonstatistical; as the energy is increased, the originally microscopic regions containing stochastic trajectories[3] also increases in size, eventually covering most of the allowed phase space. Although relatively crude computer calculations indicate a rather abrupt transition at about $E = 0.11$, in actuality, the transition is quite smooth.

Physically, exponential *C*-system behavior in dilute gas systems is to be expected because only a few binary collisions suffice to make the final system state extremely sensitive to the initial state. In nonlinear oscillator systems, Chirikov[8,34] argues that multiple overlap of nonlinear resonances is the source of such behavior. In the Henon-Heiles oscillator system, for example, he would initially argue that the system motion for a specified trajectory involves two frequencies Ω_1 and Ω_2 which depend on initial conditions, since this is an anharmonic system. Moreover, since the unperturbed frequencies are equal, in physical terms he anticipates finding resonant energy exchange between the two oscillators in those initial condition regions for which $\Omega_1 \approx \Omega_2$, just as would occur in coupled harmonic oscillator motion. At $\Omega_1 = \Omega_2$, a periodic orbit would be expected, surrounded by a region of resonant energy exchange; in point of fact, this is precisely the situation observed in Fig. 2 about the central, "$\Omega_1 = \Omega_2$" fixed points. However, this is a nonlinear system which in general must also yield many resonances of the type $n\Omega_1 \approx m\Omega_2$ for suitable initial conditions. Again there would be a periodic orbit for initial conditions such that $n\Omega_1 = m\Omega_2$, surrounded by perhaps small energy exchange regions. Indeed such physical arguments would lead one to expect a surface of section looking very much like Fig. 5, which, as has been noted, is partially verified by Fig. 6. In particular at extremely low energy, one would expect all resonances, except the $\Omega_1 = \Omega_2$ resonance, to have almost undetectable widths; as the energy increases, the width of each resonance in this host would also increase, not only destroying the intervening smooth KAM surfaces but, speaking colloquially, eventually overlapping each other. In such overlap regions containing many resonances, the final system state is extremely sensitive to the initial state because of the numerous intervening trajectory "collisions" with unstable (and perhaps stable) periodic orbits. It is interesting to note that the C-system behavior resulting from microscopic resonance-overlap can be illustrated[7,35] on a macroscopic scale using some simple examples due to Walker and Ford, and the interested reader may consult Refs. 7 and 35 for details.

Let us conclude this section by returning to the question of C-system behavior for lattice vibrations in solids, which include strongly attractive as well as repulsive forces. For solids the oscillator frequency spectrum $\{\omega_k\}$ of the harmonic approximation is almost dense; moreover, there are three such spectra—one longitudinal and two transverse. As a consequence, for solids the cubic and/or quartic terms in Hamiltonian (10) are expected to resonantly couple all degrees of freedom through many low-order resonances, each violating Condition (*ii*) of the KAM theorem. In addition, since the cubic (and sometimes the quartic) terms couple the harmonic modes in first order whereas the major anharmonic frequency shifts are of second order, one anticipates widespread low-order resonance-overlap and violent stochastic behavior regardless of the smallness of the nonlinear coupling parameter γ in Hamiltonian (10).

In order to model this behavior, let us follow Ford and Lunsford[36] and, without loss of significant generality, specialize Hamiltonian (10) to the physically interesting case of resonant "three phonon" interactions for which $\omega_k = k\omega$. We may then write Hamiltonian (10) as

$$H = \sum_{k=1}^{N} k\omega J_k + \gamma \sum A(n_k, n_l, n_m) J_k^{|n_k/2|} J_l^{|n_l/2|} J_m^{|n_m/2|} \times \cos(n_k\varphi_k + n_l\varphi_l + n_m\varphi_m) + \gamma V \tag{13}$$

where $Q_k = (2J_k)^{1/2}\cos\varphi_k$, $P_k = -(2J_k)^{1/2}\sin\varphi_k$, the $A(n_k, n_l, n_m)$ are constants, the second sum is over all resonant V_3 interactions defined here as those V_3 terms having zero Poisson bracket with $\sum_k k\omega J_k$, γV includes all nonresonant V_3 terms as well as all higher-order terms, and the sine terms corresponding to the cosine terms have been omitted for simplicity. The frequency spectrum $\{k\omega\}$ of Hamiltonian (13) is chosen to represent the linear acoustic region of the dispersion curve for solids and the resonant, cubic terms excite the "three phonon" interactions $(\omega_k + \omega_l) \leftrightarrows \omega_m$. If the resonant second sum were absent from Hamiltonian (13), then the KAM theorem could be proved[26] despite the violation of Condition (ii), and for small γ the γV terms would in general only slightly distort the smooth N-dimensional integral surfaces of the $\gamma = 0$ harmonic system. As a consequence when γ is extremely small as it is for a cold solid, any widespread stochastic behavior of Hamiltonian (13) may be determined by investigating the pure resonant Hamiltonian

$$H = \sum_{k=1}^{N} k\omega J_k + \gamma \sum A(n_k, n_l, n_m) J_k^{|n_k/2|} J_l^{|n_l/2|} J_m^{|n_m/2|} \times \cos(n_k\varphi_k + n_l\varphi_l + n_m\varphi_m) \quad (14)$$

However, if we introduce the time-dependent canonical transformation to new variables $(\mathscr{J}_k, \theta_k)$ generated[37] by

$$F = \sum_k \mathscr{J}_k(\varphi_k - k\omega t) \quad (15)$$

into Hamiltonian (14), we obtain

$$\mathscr{H} = \gamma \sum_{k=1}^{N} A(n_k, n_l, n_m) \mathscr{J}_k^{|n_k/2|} \mathscr{J}_l^{|n_l/2|} \mathscr{J}_m^{|n_m/2|} \times \cos(n_k\theta_k + n_l\theta_l + n_m\theta_m) \quad (16)$$

where $H = \sum_k k\omega J_k + \mathscr{H}$. We now observe that γ is only a multiplicative factor in Hamiltonian (16); hence γ primarily determines only the time scale of the motion. As a consequence, any widespread stochastic behavior which occurs for Hamiltonian (16) or the equivalent Hamiltonian (14) will continue to occur no matter how small the value of $\gamma > 0$. Although Hamiltonian (14) has that extensive low-order resonant coupling anticipated to cause wildly erratic trajectories, it nonetheless has one smooth constant of the motion $\sum_k k\omega J_k$ in addition to the Hamiltonian itself. Thus the system trajectories can at best range widely over a $(2N - 2)$-dimensional subspace of the $(2N - 1)$-dimensional energy surface. Therefore resonant oscillator systems with $N = 2$, such as the Henon-Heiles system, are for small γ no more stochastic than systems satisfying both conditions of the KAM theorem. As N becomes greater than two, however, a greater percentage of the energy surface can be reached, first because $[(2N - 2)/(2N - 1)] \to 1$ as $N \to \infty$, and second because the number of overlapping cubic resonant terms increases roughly as N^2.

In order to demonstrate that wildly erratic trajectories can occur in nearly linear oscillator systems for arbitrarily small γ, let us now consider the three-oscillator system

$$H = J_1 + 2J_2 + 3J_3 + \gamma[\alpha J_1 J_2^{1/2} \cos(2\varphi_1 - \varphi_2) + \beta(J_1 J_2 J_3)^{1/2} \cos(\varphi_1 + \varphi_2 + \varphi_3)] \quad (17)$$

This Hamiltonian is especially interesting since it should exhibit the minimum stochasticity to be expected for low-order resonant systems and since the additional constant of the motion $(J_1 + 2J_2 + 3J_3)$ allows us to reduce

Hamiltonian (17) to a two degrees of freedom problem which generates a plane area-preserving mapping. Without entering into the technical details which are discussed elsewhere,[36] we present in Figs. 7 and 8 typical surfaces of section for the reduced problem corresponding to Hamiltonian (17) using $\gamma = 1$, $\alpha = 0.1$, and $\beta = 0.4$. In Fig. 7, $H = 3.0$ and $(J_1 + 2J_2 + 3J_3) = 2.999$, and the surface of section shown reveals that about 70% of the total four-dimensional space available is filled with trajectories lying on smooth, three-dimensional KAM surfaces, and the remaining 30% is rather uniformly covered by each trajectory initiated in this latter region. Setting $H = 3.0$ and $(J_1 + 2J_2 + 3J_3) = 2.901$, we now find in Fig. 8 that only 10% of the allowed four-dimensional space is filled with

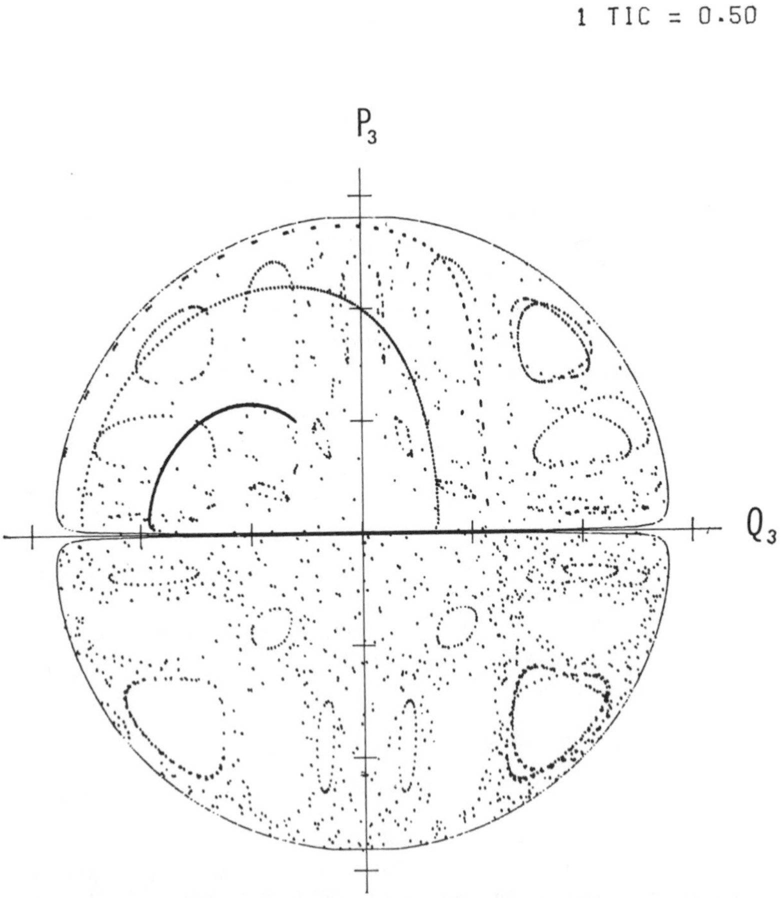

Fig. 7. A typical surface of section for the reduced, two degrees of freedom Hamiltonian equivalent to Hamiltonian (17). Here $H = 3.0$ and $(J_1 + 2J_2 + 3J_3) = 2.999$.

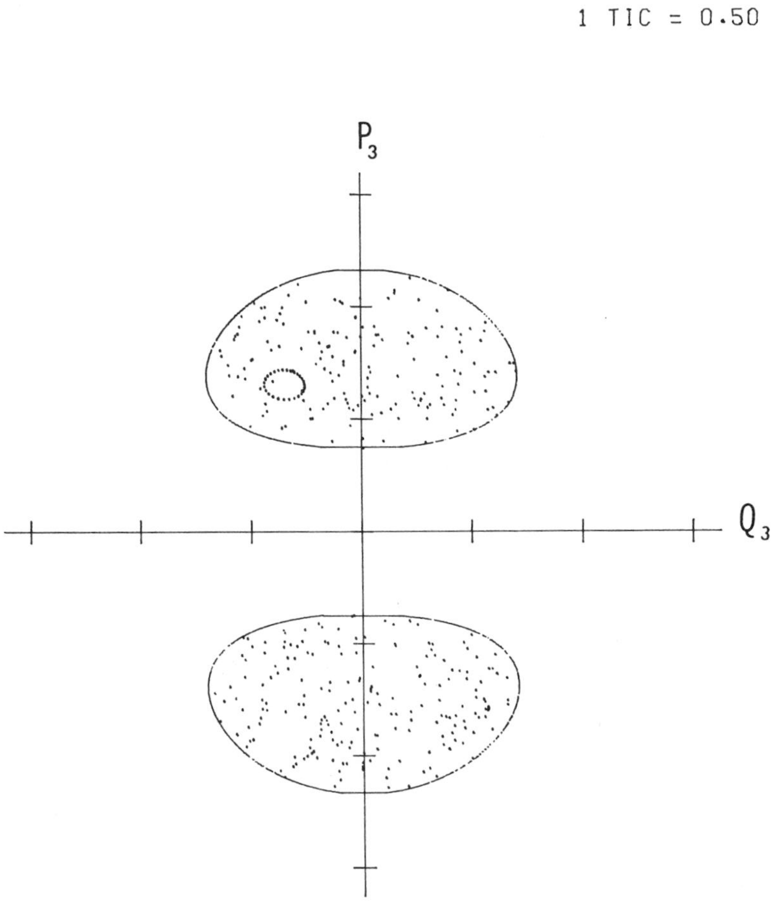

Fig. 8. Another typical surface of section for the reduced version of Hamiltonian (17). Here $H = 3.0$ and $(J_1 + 2J_2 + 3J_3) = 2.901$.

KAM surfaces, whereas 90% contains wild and erratic trajectories. It perhaps should be emphasized that the erratic trajectories shown in these surfaces of section will continue to exist[36] no matter how small the value of $\gamma > 0$. As with the Henon-Heiles system, the stochastic regions exhibit C-system behavior in that initial close orbits locally exponentiate apart and that almost all periodic orbits contained therein are unstable. In Fig. 9, we display the exponential growth of separation distance between orbit pairs initiated in a stochastic region and compare them to the corresponding linear growth of separation distance for orbit pairs started in regions of KAM stability. The time scale in Fig. 9 should be ignored; the

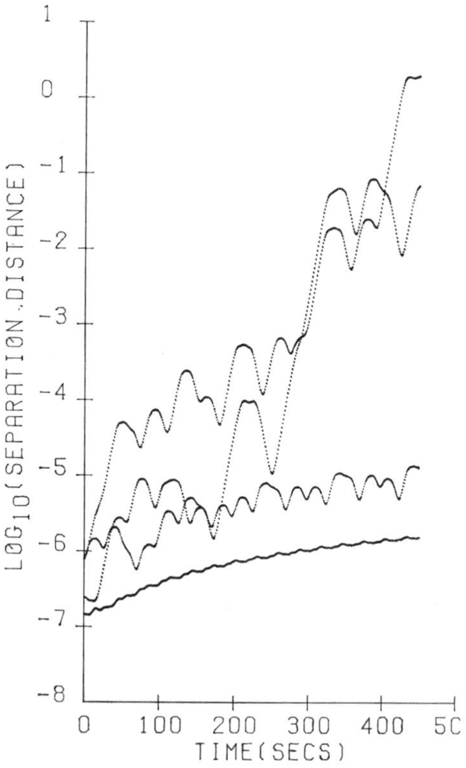

Fig. 9. A log-plot of the separation distance in Γ-space versus time for four distinct orbit-pairs. The two generally upper-lying curves show the typical "exponential" separation for orbits initiated in a stochastic region. The two generally lower-lying curves are for orbit-pairs initiated in a stable KAM region.

significant point is that the exponentiating orbit pairs increase their separation distance by about six powers of ten during the same time interval in which the stable pairs separate by only one power of ten.

Having indicated that *C*-system behavior is relevant to systems having strongly attractive forces as well as to those having only repulsive forces, we now turn to establishing a connection between *C*-system behavior and irreversible rate equations.

V. IRREVERSIBLE RATE EQUATIONS FOR SIMPLE *C*-SYSTEMS

In this section we demonstrate that *C*-system behavior can lead to irreversible rate equations using two distinct examples. The first, previously

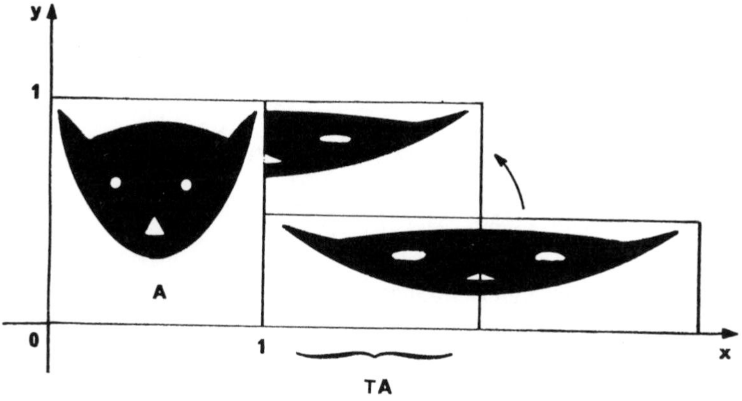

Fig. 10. A drawing of the baker's transformation.

discussed by Penrose[38] and Lebowitz,[39] is exactly solvable and leads to an irreversible rate equation in the absence of widely differing relaxation times. The second, recently studied by Chirikov,[40] Brahic,[41] and Lichtenberg,[42] involves computer calculations and two distinct time scales.

The first system is the so-called baker's transformation, specified by the area-preserving mapping,

$$(x_{n+1}, y_{n+1}) = T(x_n, y_n) = \begin{cases} \left[2x_n, \dfrac{y_n}{2}\right], & 0 \leqslant x_n < \frac{1}{2} \\ \left[(2x_n - 1), \dfrac{(y_n + 1)}{2}\right], & \frac{1}{2} \leqslant x_n < 1 \end{cases} \tag{18}$$

of the unit square onto itself (mod 1) and illustrated in Fig. 10, which was adapted from Ref. 3. In Fig. 10, a unit square of baker's dough is first stretched into a rectangle of twice the original width and one-half the original height. This rectangle is then cut vertically in half and the two pieces rearranged to again form a square. Since this mapping describes the physical mixing of biscuit dough, mathematical mixing and ergodicity are intuitively obvious.[43] Moreover, this system exhibits C-system behavior since each small $(dx\, dy)$ area element exponentially grows in the x-direction, eventually becoming a thin filament which wraps about the unit square in a dense set of horizontal strips.

Since the (18)-mapping preserves area, is reversible, and allows one and only one "trajectory" to pass through each point, let us follow Penrose and Lebowitz and associate a probability density $f(x_n, y_n, n)$ with each point (x_n, y_n) of the unit square, where

$$\iint f(x_0, y_0, 0)\, dx_0\, dy_0 = 1 \tag{19}$$

Using standard physical arguments, we may show that Liouville's equation for this system becomes

$$f(x_n, y_n, n) = f(x_0, y_0, 0) \tag{20}$$

where (x_n, y_n) and (x_0, y_0) are related through the equations of motion given by (18). In particular, we may rewrite (20) as

$$f(x, y, n) = \begin{cases} f\left[\dfrac{x}{2}, 2y, n-1\right], & 0 \leqslant y < \frac{1}{2} \\ f\left[\dfrac{(x+1)}{2}, (2y-1), n-1\right], & \frac{1}{2} \leqslant y < 1 \end{cases} \tag{21}$$

where now the same variables x and y appear on both sides of (21). If we now define the reduced probability density $W(x, n)$ as

$$W(x, n) = \int_0^1 f(x, y, n)\, dy \tag{22}$$

then direct integration of (21) reveals that W satisfies the exact, irreversible rate equation

$$W(x, n) = \frac{1}{2}\left[W\left(\frac{x}{2}, n-1\right) + W\left(\frac{(x+1)}{2}, n-1\right)\right] \tag{23}$$

an equation reminiscent of the difference equation governing the random walk.[44] Although the values of $f(x_0, y_0, 0)$ in every small rectangle $(\Delta x_0\, \Delta y_0)$ become uniformly mixed over the whole unit square, the fine-grained probability density $f(x, y, n)$ does not almost everywhere tend to unity as n tends to infinity. On the other hand, because of mixing, the coarse-grained probability density $W(x, n)$ does almost everywhere tend to unity as n tends to infinity.[39] In the following paragraphs, we empirically obtain a similar result for a physically more interesting system.

Our second example was originally developed by Fermi[45] and Ulam[46] as a highly simplified model of cosmic ray acceleration. Consider a ball bouncing between two infinitely heavy walls, one fixed and one oscillating as shown in Fig. 11. The ball has instantaneous speed v and the moving wall oscillates with amplitude a, period T, and instantaneous speed $V(t)$, where $V(t)$ is a sawtooth function having the maximum value V. The minimum distance between the walls is l. The exact difference equations governing the motion of this system are presented in a paper[40] by Zaslavskii and Chirikov. Following Lieberman and Lichtenberg,[42] we elect to

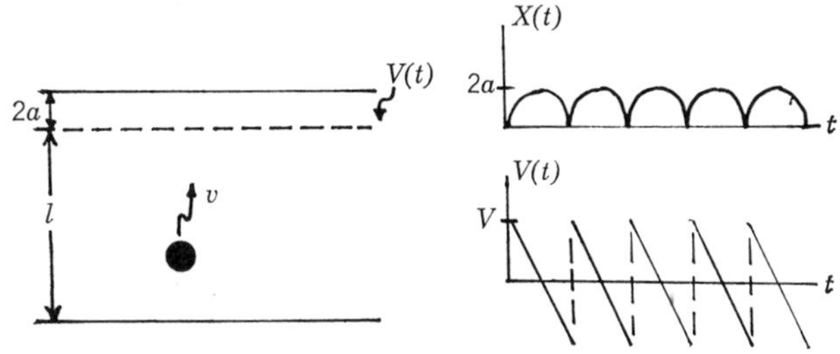

Fig. 11. Diagram of the Fermi-Ulam system used to model the acceleration of cosmic rays.

consider an approximation to these exact difference equations. The approximating equations are

$$u_{n+1} = |u_n + \psi_n - \tfrac{1}{2}| \tag{24a}$$

$$\psi_{n+1} = \left[\psi_n + \left(\frac{M}{u_{n+1}}\right)\right], \quad (\text{mod } 1) \tag{24b}$$

where $u_n = v_n/V$, v_n is the speed of the ball just before the nth collision with the oscillating wall, $\psi_n(0 \leqslant \psi_n \leqslant 1)$ is the phase of the oscillating wall at the nth collision, and $M = (l/16a)$. Equation (24) is a good approximation to the exact equations of motion provided that $M \gg 1$ and $u \gg 1$; however, independent of the goodness of these approximations, (24) yields the same general type of behavior as do the much more complicated exact equations. We therefore confine our attention to (24).

Our first observation, which will by now come as no surprise to the reader, is that (24) reduces our problem to the study of a plane area-preserving mapping. However, this mapping does not exhibit *C*-system behavior throughout the (ψ, u) plane; like the Henon-Heiles mapping, this system exhibits a so-called divided phase space.[8] Indeed, taking differentials of (24), we obtain

$$\begin{aligned} du_{n+1} &= du_n + d\psi_n \\ d\psi_{n+1} &= d\psi_n - \left(\frac{M}{u_{n+1}^2}\right) du_{n+1} \end{aligned} \tag{25}$$

Now for the (ψ, u) region which shall interest us in these calculations, we have $u_n \gg 1$ and $M \gg u_n$; moreover, (24a) shows that $u_n \gg 1$ varies slowly with n. Thus let us approximate in (25) and set (M/u^2) equal to a constant, say, b. Equation (25) may then be written

$$\begin{aligned} du_{n+1} &= du_n + d\psi_n \\ d\psi_{n+1} &= -b\, du_n + (1 - b)\, d\psi_n \end{aligned} \tag{26}$$

where

$$0 < b = \frac{M}{u^2} \tag{27}$$

As with (7), a linear change of variables now permits us to write (26) in the form

$$d\zeta_{n+1} = \lambda\, d\zeta_n\,, \qquad d\eta_{n+1} = \lambda^{-1}\, d\eta_n \tag{28}$$

where

$$\lambda = \tfrac{1}{2}(2 - b) - \tfrac{1}{2}[(2 - b)^2 - 4]^{1/2} \tag{29}$$

From (29), we see that λ is real when $b > 4$ and is imaginary when $0 < b < 4$. Thus referring to (28), we see that iterates of (26) oscillate when $u > M^{1/2}/2$ and they exponentiate when $u < M^{1/2}/2$. We thus expect C-system behavior for the mapping of (24) in that (ψ, u)-plane region for which

$$u < \frac{M^{1/2}}{2} \tag{30}$$

and KAM stability for $u > M^{1/2}/2$. In the stochastic region, each small initial $(d\psi_0\, du_0)$ area-element grows exponentially in the ζ-direction and shrinks exponentially in the η-direction. Further, since the expanding ζ-direction has small but nonzero slope $(\Delta u/\Delta\psi) \approx -(1/b)$ for $b \gg 1$, both variables u and ψ locally spread exponentially, but ψ spreads more rapidly than u.

In Fig. 12, we show a composite sketch of a typical (24)-mapping based on several computer-generated figures presented by Brahic[41] and Lieberman and Lichtenberg[42]; in Brahic's paper especially, some of the mapping pictures represent a striking form of abstract art. In Fig. 12, we note that the boundary of the stochastic behavior occurs at about the predicted value of $u = M^{1/2}/2$. For larger u-values, again as predicted, stable as well as unstable fixed points appear. By direct substitution, one easily finds that the mapping T of (24) has fixed points of T itself at $(\psi, u) = (\frac{1}{2}, M/k)$, where k is a positive integer, and that these fixed points are stable when $u > M^{1/2}/2$. The member of this fixed point set having the largest associated stable region, as seen in Fig. 12, lies at $(\frac{1}{2}, M)$, and physically corresponds to the ball being reflected from the oscillating wall (at $\psi = \frac{1}{2}$ when the moving wall instantaneously has zero speed) and then colliding again with the moving wall after the elapse of precisely one wall period. Fixed points of T^2, T^3, etc., can also be determined through increasingly long and tedious algebraic manipulations of (24).

For motion in the stochastic region where $u < M^{1/2}/2$, Lieberman and Lichtenberg[42] first establish that the relaxation times for the u and ψ motion differ widely. They then use this fact to obtain an irreversible rate

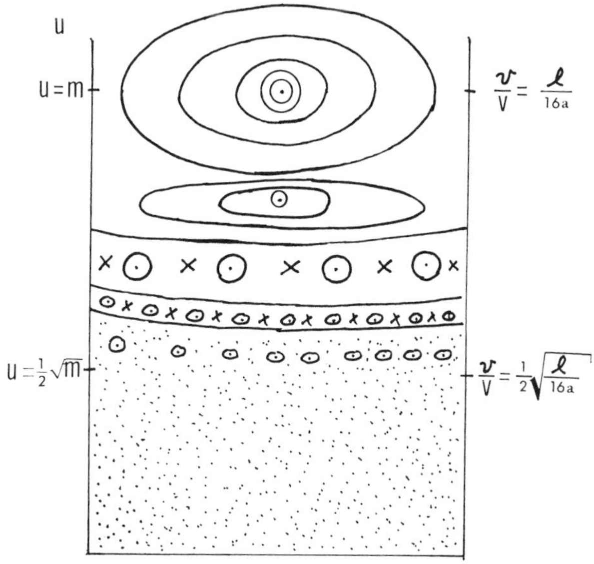

Fig. 12. Sketch of a typical mapping generated by (24).

equation which they validate using a computer. In (24), let us start with a precise initial (ψ_0, u_0) state (a definite, specific state and not a rectangle $d\psi_0\, du_0$) for which $M \gg u \gg 1$. Then as n increases, the sequential iterates of ψ_n will rapidly cover the whole interval $0 \leqslant \psi \leqslant 1$ in a "random" manner much before $(\sum |\Delta u_n|)/u_0$ becomes large. As a consequence, the fractionally small, sequential iterates Δu_n generated by (24a) will be positive or negative with about equal frequency, and u_n will perform a relatively slow "random walk" away from the initial region near u_0. Alternatively, consider an ensemble of systems with initial states spread uniformly over a small rectangle $(d\psi_0\, du_0)$. According to the discussion following (30), this small rectangle will spread exponentially along the ζ-direction into an almost horizontal filament with $\delta\psi \sim 1$ and $(\delta u/u) \ll 1$. Although each small segment of this filament will continue to locally grow exponentially, macroscopically the next iteration of this filament will split into two or more new filaments, each having $\delta\psi \sim 1$. Moreover, since the original filament had $\delta\psi \sim 1$, (24a) ensures that half of the almost horizontal, new filaments lie slightly above (along the u-axis) the original filament and half slightly below. Similarly one more iteration splits each of these new filaments into a newer set, equally split above and below the original, new filament position. Thus in the ensemble, the system phases ψ "randomize" on an exponential time scale followed, on a much longer time scale, by a diffusive spread of the u-values.

On the basis of either of these arguments, one concludes that (24b) causes the fine-grained density $f(\psi, u, n)$ to mix continually along the ψ-direction with exponential rapidity. Equation (24a) then ensures that the reduced probability distribution $W(u, n)$ spreads along the u-direction via a much slower random walk process which is known[44] to lead to a type of diffusion equation. One therefore expects that $W(u, n)$ satisfies the Fokker-Planck equation[44]

$$\frac{\partial W}{\partial n} = -\frac{\partial}{\partial u}(BW) + \frac{1}{2}\frac{\partial^2}{\partial u^2}(DW) \tag{31}$$

In order to verify the use of (31) for this system, B and D are calculated using (24) in the Wang-Uhlenbeck formulas.[44] The results can then be compared with the computer-calculated values for B and D. Starting from $W(u, 0) = \delta(u - u_0)$, (31) predicts that the width of W should grow like $n^{1/2}$, which can also be checked against computer calculations. Finally $W(u, \infty)$ should be a constant over the stochastic region. In all cases, theory and computer experiment agree nicely. For example, in Fig. 13, adapted from Ref. 42, we show a plot of $W(u, \infty)$ versus u obtained by integrating (24) for an ensemble of systems. Here $W(u, \infty)$ is more or less

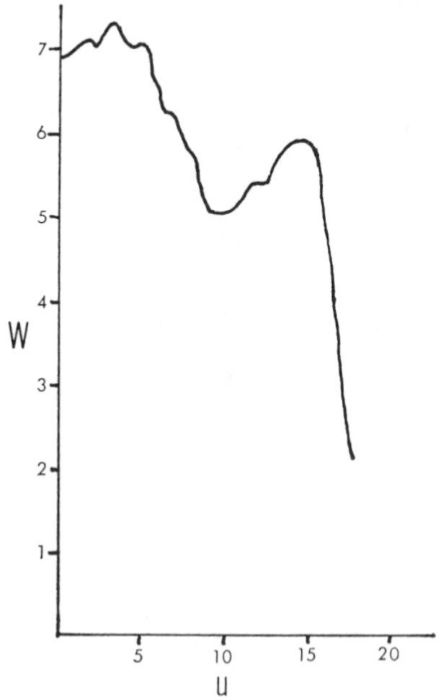

Fig. 13. A graph of the probability density $W(u, \infty)$ as a function of u. W is specified in arbitrary units.

constant up to the stochastic border $u = M^{1/2}/2 = 10^{3/2}/2 \approx 16$, above which W falls off quite rapidly.

A rigourous derivation of (31) from (24), perhaps along the lines used to derive (23) from (18), lies in the future. Nonetheless this example and the previous one point in the direction of future progress[47] in developing rate equations for physical systems exhibiting *C*-system behavior.

VI. DISCUSSION

Ergodic theory, at least as developed during the first half of this century, likely deserves its current reputation among physicists as being irrelevant, useless, and perhaps detrimental to the progress of statistical mechanics. During the last fifteen years, however, the work of Kolmogorov, Moser, Arnold, Anosov, Sinai, Smale, and others has breathed fresh like into the subject, making it vitally significant not only to physics but to other sciences as well. Indeed, Sinai's announcement of ergodicity and mixing in the hard sphere gas might have been thought sufficient by itself to galvanize statistical mechanics into a frenzy of reevaluation. However, attitudes held for half a century change slowly, if at all, and such change becomes even more difficult when the relevant theorems are stated and proved in a mathematical language incomprehensible to most physicists.

In an attempt to make the essentials of the new findings accessible to a wider audience, we have presented an intuitive development of the new results in terms of elementary examples. In addition, we have placed computer experiments on an equal footing with the rigorous mathematical results. In this regard, simplicity of presentation was not our only goal. Our intent was also to emphasize that two- and three-particle systems exhibit most of the complexity of the classical many-body problem discussed in the mathematical theorems while still remaining amenable to detailed computer investigation. Nonetheless, this paper and the several earlier review papers[6–12] form only the first few chapters of a book yet to be written, which would serve as a primer translating into physical terms all the new mathematical results. Most especially, the author would like to learn from the later chapters of this needed, but as yet unwritten, book since he strongly feels with A. S. Wightman[10] that "there is not only much to be done but much that can be done."

References

1. P. R. Halmos, *Lectures on Ergodic Theory*, Chelsea Publishing Co., New York, 1956.
2. For example, see L. Landau and E. Lifshitz, *Statistical Physics*, Addison-Wesley, Reading, Mass., 1958, p. ix.

3. V. I. Arnold and A. Avez, *Ergodic Problems of Classical Mechanics*, Benjamin, New York, 1968.
4. P. Billingsley, *Ergodic Theory and Information*, Wiley, New York, 1965.
5. Ja. G. Sinai, *Russ. Math. Surv.*, **25** (2), 137 (1970).
6. G. Contopoulos, *Bull. Astron.*, **2,** 223 (1967).
7. G. H. Walker and J. Ford, *Phys. Rev.*, **188,** 416 (1969).
8. B. V. Chirikov, "Research Concerning the Theory of Non-Linear Resonance and Stochasticity," Report No. 267, Institute of Nuclear Physics, Novosibirsk, USSR, 1969, unpublished. An English translation of this report is available as *Translation 71-40*, CERN, Geneve, 1971.
9. N. Saito, N. Ooyama, Y. Aizawa, and H. Hirooka, *Progr. Theoret. Phys. Kyoto Suppl.*, **45,** 209 (1970).
10. A. S. Wightman, in *Statistical Mechanics at the Turn of the Decade*, E. G. D. Cohen, Ed., Marcel Dekker, New York, 1971, pp. 1–32.
11. J. L. Lebowitz, in *Proceedings of the I.U.P.A.P. Conference on Statistical Mechanics, Chicago, 1971* (to be published).
12. J. Ford, in *Lectures in Statistical Physics*, Vol. II, W. C. Schieve, Ed., Springer-Verlag, New York, 1973.
13. D. V. Anosov and Ya. G. Sinai, *Russ. Math. Surv.*, **22** (5), 103 (1967). Although strictly speaking the hard sphere gas is not a *C*-system, the two are intimately related, as is discussed on pp. 106–107 of this reference and on p. 78 of Ref. 3.
14. D. ter Haar, *Elements of Statistical Mechanics*, New York, 1954, p. 333.
15. G. E. Uhlenbeck, in *Fundamental Problems in Statistical Mechanics II*, E. G. D. Cohen, Ed., North-Holland, Amsterdam, 1968, p. 4.
16. Ref. 14, p. 341.
17. Ref. 14, p. 364.
18. John P. Heller, *Am. J. Phys.*, **28,** 348 (1960).
19. E. Hopf, *J. Math. Phys.*, **13,** 51 (1934).
20. For a mathematically precise definition, see Ref. 3, p. 55.
21. Ref. 3, p. 20.
22. Ref. 3, p. 7 and Appendix I.
23. E. T. Whittaker, *Analytical Dynamics*, Cambridge University Press, New York, 1965, Chap. 16.
24. G. D. Birkhoff, *Dynamical Systems*, American Mathematical Society, Providence, R.I., 1927, p. 82.
25. J. Moser, *Math. Ann.*, **169,** 136 (1967).
26. See, for example, the excellent review by V. I. Arnold, *Russ. Math. Surv.*, **18** (6), 85 (1963). A good alternative is Chap. 4 of Ref. 3.
27. M. Henon and C. Heiles, *Astron. J.*, **69,** 73 (1964).
28. Ref. 3, Appendix 31.
29. A fixed point is said to be elliptic if invariant curves in its immediate neighborhood are ellipses; it is called hyperbolic if nearby invariant curves are hyperbolas.
30. G. H. Lunsford and J. Ford, *J. Math. Phys.*, **13,** 700 (1972).
31. J. Moser, *Nachr. Akad. Wiss. Goettingen, II. Math. Physik. Kl.*, **1962** (1).
32. G. D. Birkhoff, *Collected Mathematical Papers*, American Mathematical Society, Providence, R.I., 1950, Vol. I, p. 673; Vol. II, p. 252.
33. J. M. Greene, *J. Math. Phys.*, **9,** 760 (1968).
34. F. M. Izrailev and B. V. Chirikov, *Soviet Phys. "Doklady,"* **11,** 30 (1966).
35. J. M. A. Danby in *Periodic Orbits, Stability, and Resonances*, G. E. O. Giacaglia, Ed., D. Reidel Publishing Co., Dordrecht-Holland, 1970, p. 272.

36. J. Ford and G. H. Lunsford, *Phys. Rev.*, *A*, **1**, 59 (1970).
37. H. Goldstein, *Classical Mechanics*, Addison-Wesley, Cambridge, Mass., 1951, p. 241.
38. O. Penrose, *Foundations of Statistical Mechanics*, Pergamon Press, New York, 1970, p. 122.
39. J. L. Lebowitz, "Ergodic Theory and Statistical Mechanics of Non-Equilibrium Processes," lecture given at the 10th Annual Eastern Theoretical Physics Conference, Schenectady, New York (preprint).
40. G. M. Zaslavskii and B. V. Chirikov, *Soviet Phys.* "*Doklady*," **9**, 989 (1965).
41. A. Brahic, *Astron. Astrophys.*, **12,** 98 (1971).
42. M. A. Lieberman and A. J. Lichtenberg, *Phys. Rev.* **A5**, 1852 (1972).
43. This mapping is also isomorphic to the Bernoulli shift $B(\frac{1}{2}, \frac{1}{2})$, and it is therefore intimately related to a random coin-toss game or to a one-dimensional random walk, which is known to lead to a diffusion equation. We refer the reader to Ref. 3 and 4 for additional information.
44. M. C. Wang and G. E. Uhlenbeck, *Rev. Mod. Phys.*, **17,** 323 (1945).
45. E. Fermi, *Phys. Rev.*, **75,** 1169 (1949).
46. S. M. Ulam, in *Proceedings of the Fourth Berkeley Symposium on Mathematical Statistics and Probability*, Vol. III, University of California Press, Berkeley, 1961, p. 315.
47. A. N. Kaufman, *Phys. Rev. Lett.*, **27,** 376 (1971).

VARIATIONAL METHODS IN STATISTICAL MECHANICS*

M. D. GIRARDEAU

Department of Physics and Institute of Theoretical Science, University of Oregon, Eugene, Oregon

R. M. MAZO

Department of Chemistry and Institute of Theoretical Science, University of Oregon, Eugene, Oregon

"It's sure to make *some* change in my size, and as it can't possibly make me larger, it must make me smaller, I suppose."

Alice in Wonderland

CONTENTS

I. Introduction . . . 188
II. Upper Bounds on Thermodynamic Potentials . . . 189
A. The Gibbs-Bogoliubov Inequality . . . 189
B. The Peierls Inequality . . . 193
C. The Jordan-Golden Inequality . . . 194
D. Variation of the Distribution Function . . . 195
III. Relationship Between the Variational Method, Exactly Soluble Models, and the Generalized Hartree-Fock Method . . . 195
A. Quadratic Hamiltonians . . . 197
B. General Hamiltonians . . . 201
IV. Lower Bounds on Thermodynamic Potentials . . . 209
A. The Golden-Thompson Inequality . . . 209
B. The Inverse Gibbs-Bogoliubov Inequality . . . 211
C. Convexity of the Free Energy . . . 212
D. Reduced Density Matrices, the N-Representability Problem, and Lower Bounds . . . 213
V. Bounds on Other Quantities . . . 220
A. Fisher's Bounds on Derivatives . . . 220
B. Bounds on Expectation Values . . . 222

* Supported in part by NSF grants GP-27455 and GP-11728.

VI. Some Applications to Classical Statistical Mechanics 223
A. Examples of General Theorems 223
B. Free Volume Theory 225
C. Pressure of a Hard-Core System Near Close Packing. 226
D. Excluded Volume of Polymer Chains 227
E. Solution Theory 230
VII. Applications to Quantum Statistical Mechanics 232
A. BCS Theory of Superconductivity 233
B. Condensation in an Interacting Fermi Gas 239
C. Localized Ferromagnetism 242
D. Itinerant Ferromagnetism 245
E. Boson Models of Liquid Helium 248
F. Isotope Mixtures 248
VIII. Conclusions 250
Appendix. Convex Functions 251
Acknowledgment 253
References 253

I. INTRODUCTION

The use of variational principles in theoretical physics and chemistry is very widespread indeed. The uses fall into two main groups: (1) the use of an "action principle" of some sort for the derivation of the equations of motion and conservation laws of a theory, and (2) the use of a variational principle for the approximate computation of numerical results. In this article we shall be concerned primarily with the second class of uses.

Uses in the second class have spanned a wide variety of fields, among which are electrostatics[1] (Thomson's principle), quantum mechanical bound state energies,[2] and scattering problems (quantal,[3] electromagnetic,[4] acoustic[5]). Indeed, the variational principle

$$\frac{\langle\psi|\, H \,|\psi\rangle}{\langle\psi \mid \psi\rangle} \geqslant E_0 \tag{1.1}$$

where E_0 is the ground state energy of a quantal system, so pervades molecular quantum mechanics that is is hard to imagine what the subject would be like without it.

On the other hand, for many-body problems, and particularly for statistical mechanical problems, variation principles, although known since the beginning of the subject, have been used relatively seldom in the actual solution of problems. In this article we intend to review some of the applications that have been made in both quantum and classical statistical mechanics. Our hope is that such a review will expose both the power and the limitations of the variational method. We consider here equilibrium statistical mechanics only. Variational methods for nonequilibrium statistical mechnics are virtually noanexistent.

Let us begin by defining what we mean by a variational principle. A variation principle is an inequality of the form

$$\mathscr{F}[\Phi] \geqslant X \tag{1.2}$$

where $\mathscr{F}$ is a functional, Φ is the argument of the functional which may be a Hamiltonian, a wave function, a density matrix, etc., and which we may vary, and X is some physical quantity in which we may be interested. We usually require that the equality sign holds for some choice, Φ_0, of the argument, Φ. Thus we exclude from consideration in this article a number of rigorous inequalities[6] which have been found both interesting and useful in many-body problems, but which do not have the flexibility of varying the left-hand side, as in (1.2).

Although it seems that (1.2) restricts us to functionals which form upper bounds, this is only apparent. For if we have a lower bound $\mathscr{G}[\Phi] \leqslant Y$, then, defining $\mathscr{F} = -\mathscr{G}$, $X = -Y$, we recover (1.2).

The variation principle of bound state quantum mechanics, (1.1), is of course of the form (1.2). Here the argument is the wave function, ψ.

In this review, we attempt to give a fairly thorough treatment of the general theory underlying the use of variational methods in statistical mechanics. The level of rigor attempted in the derivations of the basic equations can generally be described as "physicists' level." Questions of the domain of operators, convergence, and interchange of limits have generally been ignored. The reader who would like to see derivations perhaps more palatable from the mathematically rigorous point of view is advised to consult the article of Huber.[7] Though it is an important article, we do not review Huber's work here, since that would be too much like writing a "review of a review."

One last word about notation: we consistently use the symbol, A, for Helmholtz free energy, not F, as is sometimes done. Furthermore, we use the symbols β and $1/kT$ interchangeably. V stands for volume throughout, not a potential energy, which is generally denoted by U. Lowercase letters u, v, etc., are used for two-body potentials.

II. UPPER BOUNDS ON THERMODYNAMIC POTENTIALS

A. The Gibbs-Bogoliubov Inequality

The statistical-mechanical variational principle that has seen the most use is the principle of Gibbs-Bogoliubov. This principle is also the one most closely related to the ground state energy principle, (1.1).

The history of this principle is rather interesting. Bogoliubov's quantal variational principle is a variational principle for the Helmholtz function. It was apparently never published by Bogoliubov himself, but was given, and attributed to Bogoliubov, by Tolmachev.[8] Girardeau,[9] feeling that the Bogoliubov principle ought to be valid as a theorem in classical statistical mechanics, succeeded in giving a purely classical derivation. It was immediately pointed out by Brown[10] that the classical form of Bogoliubov's principle had already been derived by Gibbs in his fundamental monograph,[11] and the relation between the quantal and classical cases was then further clarified by Girardeau.[12]

In view of this history, we have decided to call the principle in question the Gibbs-Bogoliubov principle. Further historical references are given by Huber.[7] Let us now state the principle.

Let A be the Helmholtz free energy of a system with Hamiltonian H, $A = -kT \ln \text{Tr}\,\{\exp(-\beta H)\}$, and let A_0 be the Helmholtz free energy of a system with Hamiltonian H_0. Then

$$A \leqslant A_0 + \langle H - H_0 \rangle_0 \tag{2.1}$$

Here the symbol $\langle X \rangle_0$ means

$$\langle X \rangle_0 = \begin{cases} \dfrac{\text{Tr}\,\{X \exp(-\beta H_0)\}}{\text{Tr}\,\{\exp(-\beta H_0)\}} & \text{(quantal case)} \quad (2.2\text{a}) \\[2ex] \dfrac{\displaystyle\int X \exp(-\beta H_0)\, d^N p\, d^N q}{\displaystyle\int \exp(-\beta H_0)\, d^N p\, d^N q} & \text{(classical case)} \quad (2.2\text{b}) \end{cases}$$

where X is any dynamical variable.

Let us first give a proof of the variational principle (2.1) in the classical case. The proof is based on the following formula:

$$\phi(z) \equiv z e^z + 1 - e^z \geqslant 0 \tag{2.3}$$

To verify this we note that $\phi'(z) < 0$ for $z < 0$, $\phi'(z) > 0$ for $z > 0$, and $\phi(0) = 0$. Therefore ϕ is incapable of taking on negative values (for real z, of course).

To shorten the formulas, we write $\Delta A = A - A_0$, $\Delta H = H - H_0$, and $\int \cdots d^N p\, d^N q$ as $\int \cdots dp\, dq$. Consider the integral

$$\int \{-\beta(\Delta A - \Delta H) \exp[-\beta(\Delta A - \Delta H)] + 1 - \exp[-\beta(\Delta A - \Delta H)]\} \times \exp[\beta(A - H)]\, dp\, dq \geqslant 0 \tag{2.4}$$

by (2.3). Note that $\exp[-\beta(\Delta A - \Delta H) + \beta(A - H)] = \exp[\beta(A_0 - H_0)]$ and that $\int \exp[\beta(A_0 - H_0)]\,dp\,dq = \int \exp[\beta(A - H)]\,dp\,dq = 1$. Hence (2.4) becomes

$$-\beta(\Delta A - \langle \Delta H \rangle_0) + 1 - 1 \geqslant 0 \tag{2.5}$$

or

$$\Delta A \leqslant \langle \Delta H \rangle_0 \tag{2.6}$$

which is (2.1).

This proof is essentially the same as that originally given by Gibbs, only the notation has been modernized. It is interesting to note that this result appears as Theorem III out of nine theorems in Chapter XI of Gibbs's treatise.[11] This led Brown to suggest[10] that the other eight theorems ought to be investigated for their use as additional variational principles. It seems to us, however, that all nine theorems are, in effect, alternative ways of writing (2.3), and we have been unable to derive any other useful principle, independent of (2.6), from them.

We now turn to the quantal version of (2.1). In deriving the quantal version, we cannot merely replace integrals by traces, since H and H_0 do not necessarily commute; a different approach is needed. We follow here a version given by Koppe.[13]

Let **A** and **B** be two hermitian operators, and define

$$Z(\lambda) = \mathrm{Tr}\,\{\exp(\mathbf{A} + \lambda\mathbf{B})\} \tag{2.7}$$

We expand $Z(\lambda)$ in a Taylor's series (with remainder) about $\lambda = 0$

$$Z(\lambda) = Z(0) + \lambda Z'(0) + \frac{\lambda^2}{2} Z''(\lambda') \tag{2.8}$$

where $0 \leqslant \lambda' \leqslant \lambda$. We shall show below that $Z''(\lambda') \geqslant 0$. Assuming, for the moment, the truth of this assertion, we have

$$Z(1) \leqslant Z(0) + Z'(0) \tag{2.9}$$

Now set

$$\mathbf{A} = -\beta H_0 - \beta\langle H - H_0\rangle_0\,, \qquad \mathbf{B} = -\beta(H - H_0) + \beta\langle H - H_0\rangle_0$$

Therefore

$$Z(1) = \exp(\beta A)\,, \qquad Z(0) = \exp(\beta A_0)\exp(-\beta\langle H - H_0\rangle_0)$$

Furthermore, since $Z'(\lambda) = \mathrm{Tr}\,\{\mathbf{B}\exp(\mathbf{A} + \lambda\mathbf{B})\}$ by the cyclic property of the trace, we have $Z'(0) = 0$ for our choice of **A** and **B**. Hence

$$\exp(\beta A) \leqslant \exp(\beta A_0 + \beta\langle H - H_0\rangle_0) \tag{2.10}$$

Taking logarithms, we recover (2.1).

To complete the proof, we must show that $Z''(\lambda) \geqslant 0$. First of all, differentiation of (2.7) gives

$$Z'(\lambda) = \mathrm{Tr}\,\{\mathbf{B} \exp(\mathbf{A} + \lambda\mathbf{B})\} \tag{2.11}$$

where $\mathbf{B}$ may be placed either to the left or right of $\exp(\mathbf{A} + \lambda\mathbf{B})$ in view of cyclic invariance of the trace. A second differentiation gives

$$Z''(\lambda) = \mathrm{Tr}\left\{\mathbf{B}\frac{d}{d\lambda}\exp(\mathbf{A} + \lambda\mathbf{B})\right\} \tag{2.12}$$

At this stage it is important to preserve the proper order of operator factors in evaluating the derivative of $\exp(\mathbf{A} + \lambda\mathbf{B})$. We shall make use of the general formula[14]

$$\frac{\partial}{\partial\lambda} e^{-\gamma\mathbf{H}(\lambda)} = -\int_0^\gamma e^{-(\gamma-u)\mathbf{H}(\lambda)} \frac{\partial\mathbf{H}}{\partial\lambda} e^{-u\mathbf{H}(\lambda)}\,du \tag{2.13}$$

where $\mathbf{H}(\lambda)$ is any operator depending on the parameter λ. This identity is easily proved[14] by noting that, for each fixed value of λ, both sides of (2.13), denoted by $G(\gamma, \lambda)$, satisfy the differential equation

$$\frac{\partial G(\gamma, \lambda)}{\partial\gamma} + \mathbf{H}(\lambda)G(\gamma, \lambda) = -\frac{\partial\mathbf{H}(\lambda)}{\partial\lambda} e^{-\gamma\mathbf{H}(\lambda)} \tag{2.14}$$

and the initial condition $G(0, \lambda) = 0$. Putting $\gamma = -1$, $\mathbf{H}(\lambda) = \mathbf{A} + \lambda\mathbf{B}$, $u = -x$ in (2.13) we find

$$\frac{d}{d\lambda} e^{\mathbf{A}+\lambda\mathbf{B}} = \int_0^1 e^{(1-x)(\mathbf{A}+\lambda\mathbf{B})}\mathbf{B}e^{x(\mathbf{A}+\lambda\mathbf{B})}\,dx \tag{2.15}$$

Then substitution into (2.12) and use of cyclic invariance of the trace yields

$$Z''(\lambda) = \int_0^1 \mathrm{Tr}\left\{e^{x(\mathbf{A}+\lambda\mathbf{B})/2}\mathbf{B}e^{(1-x)(\mathbf{A}+\lambda\mathbf{B})}\mathbf{B}e^{x(\mathbf{A}+\lambda\mathbf{B})/2}\right\}dx \tag{2.16}$$

which is Koppe's[13] starting point, except that the corresponding formula in Koppe's Anhang 1 contains a spurious factor 2.

That the factor of 2 is, in fact, incorrect, can easily be seen by considering the case where $\mathbf{A}$ and $\mathbf{B}$ commute. Assuming (as is the case in applications) that $\mathbf{A}$ and $\mathbf{B}$ are hermitian and λ real, and defining

$$\mathbf{C}(x) = e^{x(\mathbf{A}+\lambda\mathbf{B})/2}\mathbf{B}e^{(1-x)(\mathbf{A}+\lambda\mathbf{B})/2} \tag{2.17}$$

one has

$$Z''(\lambda) = \int_0^1 \mathrm{Tr}\,\{\mathbf{C}(x)\mathbf{C}^\dagger(x)\}\,dx \geqslant 0 \tag{2.18}$$

which completes the proof of (2.1).

B. The Peierls Inequality

Antedating Bogoliubov's quantal inequality by many years is an inequality due to Peierls.[15] Let the set of states $\{|\mu\rangle\}$ be a complete orthonormal set. Then

$$\operatorname{Tr}\{e^{-\beta H}\} \geqslant \sum_{\mu} e^{-\beta H_{\mu\mu}} \tag{2.19}$$

where $H_{\mu\mu} \equiv \langle\mu| H |\mu\rangle$. The Peierls inequality, (2.19), is completely equivalent to that of Bogoliubov (2.1). Instead of proving (2.19) as Peierls did, we demonstrate this equivalence.

First we show that (2.1) implies (2.19). Given the set of states $\{|\mu\rangle\}$, choose H_0 to be

$$H_0 = \sum_{\mu} |\mu\rangle\langle\mu| H_{\mu\mu} \tag{2.20}$$

that is, H_0 is diagonal in the μ basis, and has the same diagonal matrix elements as H in this basis. (Note that, if $\{|\mu\rangle\}$ happens to be the set of eigenstates of H then the equality sign in (2.19) obviously holds.) For this choice of H_0, $\langle H - H_0\rangle_0 = 0$. Hence (2.1) implies

$$A \leqslant A_0 \tag{2.21}$$

or

$$e^{-\beta A} \geqslant e^{-\beta A_0} = \operatorname{Tr}\{e^{-\beta H_0}\} = \sum_{\mu} e^{-\beta H_{\mu\mu}} \tag{2.22}$$

which is (2.19).

To complete the proof of equivalence, we show that (2.19) implies (2.1). For this, we need some properties of convex functions, which are outlined in Appendix I. Since e^{-x} is a convex function of x, the convex function theorem states that

$$\langle e^{-x}\rangle \geqslant \exp(-\langle x\rangle) \tag{2.23}$$

where the brackets, $\langle\cdots\rangle$, denote the average formed with any nonnegative normalized weight function. We will choose the states $\{|\mu\rangle\}$ of (2.19) to be the eigenstates of H_0 and choose the weight function to be $\exp(\beta[A_0 - H_{0\mu}])$, where $H_{0\mu} = \langle\mu| H_0 |\mu\rangle$. Then

$$\begin{aligned} e^{-\beta A} &\geqslant \sum_{\mu} e^{-\beta H_{\mu\mu}} = e^{-\beta A_0} \sum_{\mu} e^{\beta(A_0 - H_{0\mu})} e^{-\beta\langle\mu|H - H_0|\mu\rangle} \\ &= e^{-\beta A_0} \langle \exp(-\beta[H - H_0]_d)\rangle_0 \geqslant e^{-\beta A_0 - \beta\langle H - H_0\rangle_0} \end{aligned} \tag{2.24}$$

by (2.23). Taking logarithms, we recover (2.1).

The fact of the equivalence of the Bogoliubov and Peierls principles was mentioned to us some years ago by E. Lieb, but we have not previously seen an explicit proof in the literature.

The notation $(H - H_0)_d$ means the diagonal part of $H - H_0$ in the μ representation, for example, $\sum_{\mu} |\mu\rangle\langle\mu| (H_{\mu\mu} - H_{0\mu})$.

C. The Jordan-Golden Inequality

The variational principles discussed so far have been in the form of bounds on A, the Helmholtz function. That is, they are derived by consideration of the canonical ensemble of Gibbs. Jordan and Golden[16] have given a variational principle for the chemical potential of a system, based on the grand canonical ensemble.

In the grand ensemble, one has

$$e^{\beta pV} = \sum_{N\geqslant 0} z^N e^{-\beta A_N} \tag{2.25}$$

where $z = \exp(\beta\mu)$, and A_N is the Helmholtz function for a system of N particles (in a volume V at temperature $T = 1/k\beta$). Let $\{H_N\}$ be the set of Hamiltonians for varying N, and let $\{H_{0,N}\}$ be a set of trial Hamiltonians for varying N. From the Gibbs-Bogoliubov inequality

$$A_N \leqslant A_{0N} + \langle H_N - H_{0N}\rangle_{0,N} \equiv A_N^* \tag{2.26}$$

Hence

$$e^{\beta pV} \leqslant \sum_{N\geqslant 0} z^N e^{-\beta A_N^*} \tag{2.27}$$

Let us *define* μ^* by

$$e^{\beta pV} = \sum_{N\geqslant 0} \zeta^N e^{-\beta A_N^*}, \qquad \zeta = \exp(\beta\mu^*) \tag{2.28}$$

For each p, this defines ζ as a function of V and T, just as (2.25) defines z as a function of V and T for each p. As $V \to \infty$ (p, T fixed) we expect both z and ζ to approach certain definite limits.

From (2.25) and (2.28)

$$\sum_{N\geqslant 0} (z^N - \zeta^N) e^{-\beta A_N^*} \leqslant 0 \tag{2.29}$$

Since $\exp(-\beta A_N^*) > 0$, we must have $\zeta \geqslant z$ or

$$\mu^* \geqslant \mu \tag{2.30}$$

and this gives an upper bound to μ which may be varied by varying the set $\{H_{N,0}\}$.

However, because of the definition of μ^*, (2.28), all comparisons of different sets of trial Hamiltonians $\{H_{N,0}\}$ must be made at the same p, V, and T. On the other hand, there is no requirement that the forms of the Hamiltonians $H_{N,0}$ need be similar for different values of N. This gives a certain freedom of variation which is not present in the canonical ensemble, and this freedom has been exploited by Jordan and Golden in suggesting applications of their principle.

D. Variation of the Distribution Function

In some applications, it is more convenient to vary the probability of occurrence of states, rather than the Hamiltonian. This leads to a variational principle which is equivalent to that of Bogoliubov-Gibbs, but which may be more easily handled in certain circumstances.

Suppose we have a system with energy levels E_i, and probabilities of the various states P_i. Then we assert that

$$\begin{aligned} A &\leqslant \sum_i E_i P_i + kT \sum_i P_i \ln P_i \\ &= \mathrm{Tr}\,\{HP\} + kT\,\mathrm{Tr}\,\{P \ln P\} \end{aligned} \tag{2.31}$$

The operator P is any, not necessarily canonical, density operator, but H is the actual Hamiltonian of the system. To see that (2.31) implies (2.1), we merely define H_0 by $P = \exp(\beta(A_0 - H_0))$; H_0 may, of course, be temperature dependent. Thus (2.31) can be written

$$A \leqslant \langle H \rangle_0 + \frac{1}{\beta} \langle \beta(A_0 + H_0) \rangle_0 = A_0 + \langle H - H_0 \rangle_0 \tag{2.32}$$

which is (2.1).

The inverse implication, (2.1) implies (2.31), can be carried out similarly. Define P by the same formula as above. Then, reversing the steps of the preceding demonstration, (2.1) implies (2.31).

It is obvious from the definition of A that (2.31) reduces to an equality when P is taken to be canonical. See Huber[7] for further references.

III. RELATIONSHIP BETWEEN THE VARIATIONAL METHOD, EXACTLY SOLUBLE MODELS, AND THE GENERALIZED HARTREE-FOCK METHOD

Most quantum-mechanical applications of statistical-mechanical variational methods are formulated in terms of second quantization. The reason is that, in evaluating traces in terms of the many-body Schrödinger representation, the necessity of restricting the trace to symmetric (boson) or antisymmetric (fermion) wave functions introduces complex combinatorics when one enforces such symmetry by use of symmetrizing or antisymmetrizing operators. On the other hand, if all states and observables are expressed in terms of annihilation and creation operators for the given species of identical particles, then the correct symmetry restrictions are an automatic consequence of the commutation (bosons) or anticommutation (fermions) properties of the annihilation and creation operators. The description in this section and much of Section VII will therefore be based on such a representation, in which the Hamiltonian of a system of identical particles with two-body interactions has the general form[17]

$$H = \sum_{ij} \varepsilon_{ij} a_i^{\dagger} a_j + \tfrac{1}{2} \sum_{ijkl} v_{ijkl} a_i^{\dagger} a_j^{\dagger} a_l a_k \tag{3.1}$$

Here the ε_{ij} are single-particle matrix elements of the kinetic energy and any external field (if present), the v_{ijkl} are interparticle interaction matrix elements, and the a_i and $a_i^{\dagger}$ are single-particle annihilation and creation operators satisfying the usual relations

$$[a_i, a_j^{\dagger}]_{\pm} = \delta_{ij} , \qquad [a_i, a_j]_{\pm} = [a_i^{\dagger}, a_j^{\dagger}]_{\pm} = 0 \tag{3.2}$$

where anticommutators $[\ \ ]_+$ are to be taken for the case of Fermi statistics and commutators $[\ \ ]_-$ for Bose statistics. The indices $i, j, \ldots$ are most commonly taken to refer to momentum in the case of Bose particles and momentum plus spin in the case of Fermi particles, but we do not restrict ourselves to this particular choice.

As always in applications of second quantization, it is important to realize that the space in which (3.1) operates includes states with all possible numbers (i.e., 0, 1, 2, . . .) of particles. Thus, in evaluating the Helmholtz free energy A_n of a system with n particles, it is necessary to restrict the trace to the subspace of n-particle states; we shall denote such a restricted trace by Tr_n. Then

$$e^{-\beta A_n} = \mathrm{Tr}_n\, e^{-\beta H} \tag{3.3}$$

The restriction implied in (3.3) is expressed in terms of annihilation and creation operators by the statement that Tr_n stands for the trace over the space of simultaneous eigenstates of H and N, with fixed eigenvalue n for the latter. Here N is the total particle number operator

$$N = \sum_i N_i , \qquad N_i \equiv a_i^{\dagger} a_i \tag{3.4}$$

If, on the other hand, we wish to work with a grand canonical ensemble, then the trace is not restricted. Denoting such an unrestricted trace simply by Tr, one has

$$\Xi = e^{-\beta\Omega} = \mathrm{Tr}\, e^{-\beta(H-\mu N)} \tag{3.5}$$

where Ξ is the grand partition function and Ω is the grand potential, related to the pressure p and volume V by $\Omega = -pV$. One simple choice of a complete orthonormal set of states (but not necessarily the most useful in applications) in terms of which to evaluate such traces is the set of simultaneous occupation number eigenstates ("independent particle states"):

$$|\{n_i\}\rangle = [\Pi_i(n_i!)^{-1/2}(a_i^\dagger)^{n_i}]\,|0\rangle \tag{3.6}$$

where $|0\rangle$ is the normalized vacuum state, satisfying

$$a_i\,|0\rangle = 0\,, \quad \text{all } i; \qquad \langle 0\,|\,0\rangle = 1 \tag{3.7}$$

It is readily verified that the states (3.6) are orthonormal and satisfy

$$N_i\,|\{n_i\}\rangle = n_i\,|\{n_i\}\rangle \tag{3.8}$$

In the Fermi case, each of the n_i can take on only the values 0 or 1, whereas all integral nonnegative eigenvalues are allowed in the Bose case. The restricted trace in (3.3) is over the subspace spanned by those states (3.8) for which $\sum_i n_i = n$.

A. Quadratic Hamiltonians

As motivation for the more realistic cases considered in Section IIIB, we shall first consider the simplified case in which the interaction terms v_{ijkl} in (3.1) are absent:

$$H = \sum_{ij} \varepsilon_{ij} a_i^\dagger a_j \tag{3.9}$$

in which case H is a quadratic form in annihilation and creation operators. We shall see in Section IIIB that a significant part of the interaction Hamiltonian can effectively be thrown into the form of (3.9) by appropriate redefinition of the ε_{ij} (generalized Hartree-Fock method), so that (3.9) is less restrictive than it appears to be.

For the simple case (3.9), there is a very direct connection between exact diagonalization of (3.9) and minimization of a trial free energy based on the Gibbs-Bogoliubov inequality (2.1) or Peierls' inequality (2.19), which is equivalent to the well-known relationship between minimization and diagonalization of a quadratic form. Suppose that H_0 in (2.1) is chosen to have the diagonal form

$$H_0 = \sum_i \omega_i b_i^\dagger b_i \tag{3.10}$$

where the ω_i are variational parameters to be determined, and the b_i and $b_i^\dagger$ are annihilation and creation operators satisfying the same commutation or anticommutation relations as the a_i and $a_i^\dagger$:

$$[b_i, b_j^\dagger]_\pm = \delta_{ij}, \; [b_i, b_j]_\pm = [b_i^\dagger, b_j^\dagger]_\pm = 0 \tag{3.11}$$

The physical interpretation is that the b_i and $b_i^\dagger$ are annihilation and creation operators for a set of noninteracting quasiparticles, with energies ω_i; on the other hand, the a_i and $a_i^\dagger$ annihilate and create "bare" particles.

Equations (3.2) and (3.11) imply that the bare particles and quasiparticles are related by a canonical transformation. If we were working with a *finite*-dimensional Hilbert space $\mathscr{H}$ of states, it would follow[18,19] that the canonical transformation is unitarily implementable, that is, that there exists some *unitary* operator U on $\mathscr{H}$ such that

$$U^{-1}b_iU = a_i \tag{3.12}$$

In applications, however, $\mathscr{H}$ is an infinite-dimensional Hilbert space (even for a finite number of particles), and in that case, existence of such a U is not assured; this is the problem of "inequivalent representations" of the commutation or anticommutation relations.[20,21] In the present case, we shall assume that the explicit algebraic form of the canonical transformation is linear and homogeneous:

$$a_i = \sum_{ij} u_{ij}b_j \tag{3.13}$$

where the u_{ij} are numerical matrix elements (*not* operators). The condition that the transformation be canonical then reduces to the requirement that (u_{ij}) be a unitary matrix:

$$\sum_k u_{ij}u_{jk}{}^* = \delta_{ij} \tag{3.14}$$

The condition (3.14), together with (3.13), is sufficient[22] to ensure that the transformation is not only canonical, but unitarily implementable. On the other hand, if H_0 were expressed in terms of operators b_i and $b_i{}^\dagger$ related to the a_i and $a_i{}^\dagger$ by an improper (not unitarily implementable) canonical transformation, then the Gibbs-Bogoliubov and Peierls inequalities would not necessarily hold, since their derivations assume that H_0 and H are defined on the *same* Hilbert space $\mathscr{H}$. Insufficient attention to this requirement has sometimes led to incorrect results in the literature. In the present simple case of a *linear* transformation (3.13), it can be shown[20,21] that U is of the form e^{iS} where the hermitian operator S is a quadratic form in the b_i and $b_i{}^\dagger$ operators, hence, by (3.13), also in the a_i and $a_i{}^\dagger$ operators:

$$U = e^{iS}, \qquad S = \sum_{ij} s_{ij}b_i{}^\dagger b_j \tag{3.15}$$

The explicit form of the connection between the matrices (u_{ij}) and (s_{ij}) is known,[20] but in applications to equilibrium statistical mechanics the explicit expression is not needed; it is sufficient to know that such a U exists.

Let us now apply the Gibbs-Bogoliubov inequality (2.1), but with H replaced by $H - \mu N$. Then by (3.5)

$$\Omega \leqslant \Omega_0 + \langle H - H_0\rangle_0 \tag{3.16}$$

where

$$\begin{aligned}\Omega &= -\beta^{-1} \ln \operatorname{Tr} e^{-\beta(H-\mu N)} = -\beta^{-1} \ln \Xi \\ \Omega_0 &= -\beta^{-1} \ln \operatorname{Tr} e^{-\beta(H_0-\mu N)} = -\beta^{-1} \ln \Xi_0 \\ \langle \cdots \rangle_0 &= \Xi_0^{-1} \operatorname{Tr} [\cdots e^{-\beta(H_0-\mu N)}]\end{aligned} \tag{3.17}$$

We note from (3.14) that (3.4) can also be written

$$N = \sum_i b_i^\dagger b_i \tag{3.18}$$

so that

$$\Xi_0 = \operatorname{Tr} \exp \left[-\beta \sum_i (\omega_i - \mu) b_i^\dagger b_i \right] \tag{3.19}$$

Evaluating the trace in a basis differing from (3.6) only in replacement of the $a_i^\dagger$ by $b_i^\dagger$, or equivalently, noting that the quasiparticle occupation number operators $b_i^\dagger b_i$ are mutually commuting, with eigenvalues 0, 1 in the Fermi case and 0, 1, 2, . . . in the Bose case, one obtains the familiar ideal Fermi or Bose gas expressions

$$\begin{aligned}\Xi_0 &= \prod_i (1 \pm z e^{-\beta\omega_i})^{\pm 1} \\ \Omega_0 &= \mp \beta^{-1} \sum_i \ln (1 \pm z e^{-\beta\omega_i})\end{aligned} \tag{3.20}$$

Here $z = e^{\beta\mu}$ is the absolute activity, and the upper signs are to be taken for fermions and the lower for bosons. We also have to evaluate the term $\langle H - H_0 \rangle_0$ in (3.16). To evaluate $\langle H_0 \rangle_0$, we make use of the theorem

$$\frac{\partial \Omega_0}{\partial \lambda} = \left\langle \frac{\partial H_0}{\partial \lambda} \right\rangle_0 \tag{3.21}$$

which follows upon differentiation of the expression (3.17) for Ω_0 and use of the cyclic invariance of the trace; here λ is any parameter upon which H_0 depends (this is a generalization of the Hellman-Feynman theorem, due presumably to Pauli). Taking $\lambda = \omega_i$, one has

$$\langle b_i^\dagger b_i \rangle_0 = \frac{\partial \Omega_0}{\partial \omega_i} = (z^{-1} e^{\beta\omega_i} \pm 1)^{-1} \tag{3.22}$$

and hence

$$\langle H_0 \rangle_0 = \sum_i \omega_i \langle b_i^\dagger b_i \rangle_0 = \sum_i \frac{\omega_i}{z^{-1} e^{\beta\omega_i} \pm 1} \tag{3.23}$$

Finally, to evaluate $\langle H_0 \rangle$ we note from (3.9) and (3.13) that

$$H = \sum_{ijkl} \varepsilon_{ij} u_{ik}^* u_{jl} b_k^\dagger b_l \tag{3.24}$$

Imagining the trace evaluated in a basis of simultaneous eigenstates of the $b_i^\dagger b_i$, one sees that only diagonal terms have nonzero averages:

$$\langle b_k^\dagger b_l \rangle_0 = \delta_{kl} \langle b_k^\dagger b_k \rangle_0 \tag{3.25}$$

and hence

$$\langle H \rangle_0 = \sum_{ijk} \varepsilon_{ij} u_{ik}^* u_{jk} \langle b_k^\dagger b_k \rangle_0 \tag{3.26}$$

Then with (3.23) and (3.22)

$$\langle H - H_0 \rangle_0 = \sum_i \left(\sum_{jk} u_{ji}^* \varepsilon_{jk} u_{ki} - \omega_i \right) (z^{-1} e^{\beta\omega_i} \pm 1)^{-1} \tag{3.27}$$

In order to obtain the best approximation to the grand potential Ω, we minimize the right side of (3.16) with respect to variations of the ω_i and u_{ij}, in analogy with the usual quantum-mechanical variational method based on (1.1). First we minimize with respect to the ω_i, for given fixed values of the u_{ij}. Thus we require

$$\frac{\partial \Omega_0}{\partial \omega_i} + \frac{\partial \langle H - H_0 \rangle_0}{\partial \omega_i} = 0 \tag{3.28}$$

By (3.27) and (3.22)

$$\frac{\partial \langle H - H_0 \rangle_0}{\partial \omega_i} = - \frac{\partial \Omega_0}{\partial \omega_i} - (z^{-1} e^{\beta\omega_i} \pm 1)^{-2} z^{-1} \beta e^{\beta\omega_i} \left(\sum_{jk} u_{ji}^* \varepsilon_{jk} u_{ki} - \omega_i \right) \tag{3.29}$$

and hence by (3.28)

$$\sum_{jk} u_{ji}^* \varepsilon_{jk} u_{ki} = \omega_i \tag{3.30}$$

Thus *the upper bound (3.16) is minimized with respect to the ω_i by choosing them to be the diagonal elements of the matrix* $\mathbf{u}^\dagger \boldsymbol{\epsilon} \mathbf{u}$, *where* $\mathbf{u}$ *is the unitary matrix with elements u_{ij} and* $\boldsymbol{\epsilon}$ *is the hermitian matrix with elements ε_{ij}.* It follows from (3.27) that the same choice makes $\langle H - H_0 \rangle_0 = 0$.

It remains to minimize the upper bound with respect to the u_{ij}. The above result strongly suggests that the matrix $\mathbf{u}$ should be chosen so as to diagonalize the matrix $\boldsymbol{\epsilon}$. In fact, if this is done (3.24) reduces simply to

$$H = \sum_i \omega_i b_i^\dagger b_i = H_0$$

on substitution of (3.30), in which case $\Omega = \Omega_0$. *Thus the upper bound (3.16) reduces to the exact free energy Ω if* $\mathbf{u}$ *is chosen so as to diagonalize* $\boldsymbol{\epsilon}$ *and the ω_i are chosen as the corresponding diagonal elements.*

This result is, of course, self-evident from (3.9), (3.10), and (3.13), since the best quadratic approximation H_0 to a quadratic Hamiltonian H

is H itself; $\mathbf{u}$ is merely introduced to facilitate evaluation of the trace by diagonalizing H. However, we have chosen to derive the result in a pedestrian way since it serves to illustrate techniques used in Section IIIB on the full Hamiltonian (3.1), which cannot in general be diagonalized by a linear transformation or indeed by any transformation of finite degree.

B. General Hamiltonians

By "general" here we mean the general Hamiltonian (3.1) for a system of identical bosons or fermions with two-body interactions. In order that the inequality (3.16) can be made the basis of a variational principle for obtaining the best approximation to Ω consistent with a given class of "trial Hamiltonians" H_0, it is important that the right side of (3.16) be evaluated exactly, or at least that the error in its evaluation be known to be either *positive* (so that the sense of the inequality is preserved) or else negligible in the thermodynamic limit. If, on the other hand, the error in the evaluation of the right side of (3.16) is nonnegligible and of unknown sign, it is entirely possible that for some values of the variational parameters, the "approximate bound" may in fact violate the inequality and lie *below* Ω. In such a case *minimization* of the "approximate bound" would not bring it as close to Ω as possible, but would in fact take it *away* from Ω. Furthermore, for a many-body problem, the only type of trial Hamiltonian for which the Gibbs-Bogoliubov bound can be evaluated exactly is an independent particle Hamiltonian of the form (3.10), or some Hamiltonian reducible to this form by a transformation that can be carried out in closed form. Therefore, in this section we shall assume H_0 to be of the form (3.10). Furthermore, in order to obtain a simple closed-form relationship between the a_i and b_i, we shall first restrict ourselves to the form (3.13); some generalizations important in theories of liquid He^4 and superconductivity will be considered later.

The terms Ω_0 and $\langle H_0\rangle_0$ in (3.16) have already been given in (3.20) and (3.23). In order to evaluate $\langle H\rangle_0$, it is necessary to first transform (3.1) into the b_i, $b_i^\dagger$ representation. The transform of the single-particle part has already been given in (3.24), which is the special case $v_{ijkl} = 0$. Transforming the interaction part similarly, one finds

$$H = \sum_{ijkl} \varepsilon_{ij} u_{ik}{}^* u_{jl} b_k{}^\dagger b_l + \tfrac{1}{2} \sum_{ijklmnop} v_{ijkl} u_{im}{}^* u_{jn}{}^* u_{ko} u_{lp} b_m{}^\dagger b_n{}^\dagger b_p b_o \qquad (3.31)$$

In order to evaluate the thermal averages $\langle\cdots\rangle_0$ [Eq. (3.17)] of the various terms in (3.31), one can make use of Matsubara's theorem[23,24] (a nonzero temperature generalization of Wick's theorem) according to which the thermal average of any product of annihilation and creation operators in an ensemble appropriate to independent fermions and/or bosons can be expressed in terms of products of *contractions*, each such contraction being the thermal average of a product of only two annihilation and/or creation operators. According to (3.25) and the fact that H_0 commutes with N [Eq. (3.18)], one has

$$\langle b_i^\dagger b_j\rangle_0 = 0\,, \quad i \neq j; \qquad \langle b_i b_j\rangle_0 = \langle b_i^\dagger b_j^\dagger\rangle_0 = 0 \tag{3.32}$$

that is, the only nonzero contractions contributing in the expression for $\langle H\rangle_0$ according to Matsubara's theorem are the thermal averages (3.22) of the quasiparticle occupation number operators $b_i^\dagger b_i$. Thus one has[23,24]

$$\langle H\rangle_0 = \sum_i w_i f_i + \tfrac{1}{2}\sum_{ij} w_{ij} f_i f_j \tag{3.33}$$

where

$$\begin{aligned} f_i &= \langle b_i^\dagger b_i\rangle_0 \\ w_i &= \sum_{jk} u_{ji}{}^* \varepsilon_{jk} u_{ki} \\ w_{ij} &= \sum_{klmn} u_{ki}{}^* u_{lj}{}^* v_{klmn}(u_{mi}u_{nj} \mp u_{mj}u_{ni}) \end{aligned} \tag{3.34}$$

and the upper or lower signs are to be taken in the expression for w_{ij} depending on whether the particles are fermions or bosons. Combining (3.23) and (3.33), one has

$$\langle H - H_0\rangle_0 = \sum_i (w_i - \omega_i) f_i + \tfrac{1}{2}\sum_{ij} w_{ij} f_i f_j \tag{3.35}$$

Minimizing the upper bound (3.16) with respect to variation of the ω_i for given values of the u_{ij}, as in Section IIIA, one obtains the necessary condition (3.28). Now clearly

$$\frac{\partial\langle H - H_0\rangle_0}{\partial\omega_i} = -f_i + (w_i - \omega_i)\frac{\partial f_i}{\partial\omega_i} + \sum_j w_{ij} f_j \frac{\partial f_i}{\partial\omega_i} \tag{3.36}$$

where it has been assumed, without loss of generality, that w_{ij} is symmetric, $w_{ji} = w_{ij}$ [it follows from (3.1) and the commutation properties of the annihilation and creation operators that one may assume $v_{lknm} = v_{klmn}$ without loss of generality]. Noting (3.22) and (3.34), one sees that substitution of (3.36) into (3.28) yields

$$\left(w_i - \omega_i + \sum_j w_{ij} f_j\right)\frac{\partial f_i}{\partial\omega_i} = 0 \tag{3.37}$$

It follows from (3.34) and (3.22) that

$$\frac{\partial f_i}{\partial\omega} = -(z^{-1}e^{\beta\omega_i} \pm 1)^{-2} z^{-1}\beta e^{\beta\omega_i} \neq 0 \tag{3.38}$$

so that the only solution of (3.37) is

$$\omega_i = w_i + \sum_j w_{ij} f_j \tag{3.39}$$

The physical interpretation is that the best (in the variational sense) choice of ω_i includes not only the contribution $(\mathbf{u}^{-1}\boldsymbol{\epsilon}\mathbf{u})_{ii}$ from the single-quasiparticle part of H, but also the contribution $\sum_i w_{ij} f_j$ representing the thermally averaged interaction energy of a quasiparticle of index i due to its interaction with the other quasiparticles, the average being taken in the grand ensemble appropriate to H_0. This is a generalization of the Hartree-Fock method to nonzero temperature. In view of (3.34) and (3.22), which can be combined as

$$f_i = (z^{-1} e^{\beta\omega_i} \pm 1)^{-1} \tag{3.40}$$

equation (3.39) is not an explicit solution for the ω_i but rather a system of coupled nonlinear equations for these parameters, as is also the case in the usual Hartree-Fock method. At the minimum, one has by (3.35) and (3.39)

$$\langle H - H_0 \rangle_0 = -\tfrac{1}{2} \sum_{ij} w_{ij} f_i f_j \tag{3.41}$$

and hence, with (3.20) and (3.40)

$$\Omega_{\min} = \pm \beta^{-1} \sum_i \ln (1 \mp f_i) - \tfrac{1}{2} \sum_{ij} w_{ij} f_i f_j \tag{3.42}$$

where $\Omega_{\min}$ denotes the minimum of the upper bound (3.16) with respect to variation of the ω_i. The first term in (3.42) represents the usual expression for the thermodynamic potential in terms of the Fermi-Dirac or Bose-Einstein distribution function f_i of an "ideal gas of quasiparticles," whereas the second term represents the self-consistent shift due to quasiparticle interactions. In view of the nonlinearity of the coupled equations (3.39) and (3.40), the solution need not be unique; in such a case, the solution giving the lowest value of the expression (3.42) should be chosen. In fact, in applications to phase transitions (condensation, ferromagnetism, superconductivity, etc.) such multiple solutions do occur below the transition temperature, the one of lowest thermodynamic potential representing the condensed phase and the other representing the analytic continuation of the "normal" phase. Some examples of this will be discussed in Section VII.

The minimization with respect to the parameters u_{ij} specifying the real particle–quasiparticle transformation (3.13) remains to be carried out. In the case of a purely quadratic Hamiltonian studied in Section IIIA, we found that the minimum of the upper bound (3.16) with respect to variation of the u_{ij} is achieved by choosing the matrix $\mathbf{u} = (u_{ij})$ so that $\mathbf{u}^\dagger \boldsymbol{\epsilon} \mathbf{u}$ is diagonal, and that the minimum gives the exact Ω inasmuch as the diagonalized form of (3.9) is precisely H_0. In the present case of the more general Hamiltonian (3.1), we certainly cannot expect that the best choice of the *quadratic* Hamiltonian H_0 will reproduce the exact Ω; the many-body problem is not that simple! However, it is possible, by an appropriate decomposition of the diagonal part of H, to cast the minimization problem into a form in which the connection between the best choice of the u_{ij} and the best choice of H_0 is clarified, and the problem becomes similar to that of the purely quadratic Hamiltonian (3.9). We note first from (3.32) and Matsubara's theorem that

$$\langle H \rangle_0 = \langle H_{\text{diag}} \rangle_0 \tag{3.43}$$

where H_{diag} is the diagonal part of H in the b_i, $b_i^\dagger$ representation, that is, the part expressible solely in terms of the quasiparticle occupation numbers $b_i^\dagger b_i$:

$$H_{\text{diag}} = \sum_i W_i b_i^\dagger b_i + \tfrac{1}{2} \sum_{ij} w_{ij} b_i^\dagger b_i b_j^\dagger b_j \tag{3.44}$$

Here

$$\begin{aligned} W_i &= w_i - w_{ii} \\ &= w_i - \delta_B \sum_{jklm} u_{ji}{}^* u_{ki}{}^* v_{jklm} u_{li} u_{mi} \end{aligned} \tag{3.45}$$

and δ_B is zero for a Fermi system and unity for a Bose system. In view of (3.43) and (3.44), a variational treatment of (3.1) based on Peierls' inequality leads to the same result as one based on the Gibbs-Bogoliubov inequality; this is a special case of the general equivalence of these two inequalities pointed out in Section II. The previous treatment[16] was based on Peierls' inequality.

To proceed we separate H_{diag} into an "independent quasiparticle" part H_{quas} and a "fluctuation part" H_{fluc}. Inserting the identity

$$b_i^\dagger b_i b_j^\dagger b_j = (b_i^\dagger b_i - f_i)(b_j^\dagger b_j - f_j) + (f_i b_j^\dagger b_j + f_j b_i^\dagger b_i) - f_i f_j \tag{3.46}$$

into (3.44), one finds

$$\begin{aligned} H_{\text{diag}} &= H_{\text{quas}} + H_{\text{fluc}} \\ H_{\text{quas}} &= W_0 + \sum_i \omega_i b_i^\dagger b_i \\ H_{\text{fluc}} &= \tfrac{1}{2} \sum_{ij} w_{ij} (b_i^\dagger b_i - f_i)(b_j^\dagger b_j - f_j) \end{aligned} \tag{3.47}$$

with

$$\begin{aligned} W_0 &= -\tfrac{1}{2} \sum_{ij} w_{ij} f_i f_j \\ \omega_i &= W_i + \sum_j w_{ij} f_j \end{aligned} \tag{3.48}$$

The justification for using the same symbols f_i and ω_i previously used in a different context will presently become apparent.

Since H_{quas} is quadratic and in fact, apart from the c-number term W_0 which gives a trivial contribution W_0 to Ω, of the same form as the H_0 previously used, it is natural to try a variational treatment based on (3.16), with $H_0 = H_{\text{quas}}$. The variational parameters are then the f_i and the u_{ij}. One has clearly

$$\Omega_0 = W_0 \mp \beta^{-1} \sum_i \ln(1 \mp z e^{-\beta\omega_i}) \tag{3.49}$$

differing from (3.20) only in the additive term W_0 and the differing definition of ω_i. In order to evaluate

$$\langle H - H_0 \rangle_0 = \langle H_{\text{fluc}} \rangle_0 = \tfrac{1}{2} \sum_{ij} w_{ij} \langle (b_i^\dagger b_i - f_i)(b_j^\dagger b_j - f_j) \rangle_0 \tag{3.50}$$

one can either use Matsubara's theorem as was done previously, or else express the thermal averages in (3.50) in terms of appropriate derivatives of Ω_0. In either case, one finds

$$\begin{aligned}\langle H_{\text{fluc}} \rangle_0 &= \tfrac{1}{2} \sum_{ij} w_{ij} f_i f_j - \sum_{ij} w_{ij} f_i \langle b_j^\dagger b_j \rangle_0 \\ &\quad + \tfrac{1}{2} \sum_{ij} w_{ij} \langle b_i^\dagger b_i \rangle_0 \langle b_j^\dagger b_j \rangle_0 + \tfrac{1}{2} \sum_i w_{ii} \langle b_i^\dagger b_i \rangle_0 (1 \mp \langle b_i^\dagger b_i \rangle_0)\end{aligned} \tag{3.51}$$

where $\langle b_i^\dagger b_i \rangle_0$ is given by (3.22). Since, by (3.34), the w_{ij} do not depend on the f_i, the dependence of (3.51) on the f_i enters only through its explicit dependence and through the implicit dependence of the ω_i in (3.22) on the f_i. By (3.48) and (3.45), one has

$$\frac{\partial \omega_j}{\partial f_i} = w_{ji} \tag{3.52}$$

and hence with (3.22)

$$\frac{\partial \langle b_j^\dagger b_j \rangle_0}{\partial f_i} = -(z^{-1} e^{\beta\omega_j} \pm 1)^{-2} z^{-1} \beta e^{\beta\omega_j} w_{ji} \tag{3.53}$$

Then by straightforward algebra

$$\begin{aligned}\frac{\partial \langle H_{\text{fluc}} \rangle_0}{\partial f_i} &= \sum_j w_{ij}(f_j - \langle b_j^\dagger b_j \rangle_0) \\ &\quad + \sum_{jk} w_{kj}(f_k - \langle b_k^\dagger b_k \rangle_0)(z^{-1} e^{\beta\omega_j} \pm 1)^{-2} z^{-1} \beta e^{\beta\omega_j} w_{ji} \\ &\quad - \tfrac{1}{2} \sum_j w_{jj}(z^{-1} e^{\beta\omega_j} \pm 1)^{-2} z^{-1} \beta e^{\beta\omega_j} w_{ji}(1 \mp 2\langle b_j^\dagger b_j \rangle_0)\end{aligned} \tag{3.54}$$

Similarly, with (3.48), (3.49), (3.52), and (3.22) one finds

$$\frac{\partial \Omega_0}{\partial f_i} = -\sum_j w_{ij}(f_j - \langle b_j^\dagger b_j \rangle_0) \tag{3.55}$$

Then denoting the right side of (3.16) by Ω_{var}, one has

$$\begin{aligned}\frac{\partial\Omega_{\text{var}}}{\partial f_i} &= \frac{\partial\Omega_0}{\partial f_i} + \frac{\partial\langle H_{\text{fluc}}\rangle_0}{\partial f_i} \\ &= \sum_{jk} w_{kj}(f_k - \langle b_k{}^\dagger b_k\rangle_0)(z^{-1}e^{\beta\omega_j} \pm 1)^{-2}z^{-1}\beta e^{\beta\omega_j}w_{ji} \\ &\quad - \tfrac{1}{2}\sum_j w_{jj}(z^{-1}e^{\beta\omega_j} \pm 1)^{-2}z^{-1}\beta e^{\beta\omega_j}w_{ji}(1 \mp 2\langle b_j{}^\dagger b_j\rangle_0) \end{aligned} \quad (3.56)$$

For a macroscopic system, the double-summation term in (3.56) is independent of the volume of the system in the thermodynamic limit, whereas the single-summation term is inversely proportional to the volume and hence vanishes in the same limit (this will be clear from the examples studied in Section VII). Thus, neglecting the single-summation term, one sees that the necessary condition for a minimum of Ω_{var} with respect to variation of the f_i will be satisfied if f_i is taken as the thermal average (3.22) of the quasiparticle occupation number operator, that is, if (3.40) is satisfied. This justifies our use of the same notation f_i as previously, and suggests that use of the Gibbs-Bogoliubov variational method with $H_0 = H_{\text{quas}}$ is equivalent to the former calculation with an H_0 "inserted from the outside." To see that this is in fact the case, it is only necessary to show that the expression for Ω_{var} at its minimum with respect to variation of the f_i is the same as the previous expression (3.42) obtained by minimization of a superficially different expression with respect to variation of the ω_i. Denoting the minimum of Ω_{var} by Ω_{min} as in (3.42), one finds easily

$$\Omega_{\text{min}} = \pm\beta^{-1}\sum_i \ln(1 \mp f_i) - \tfrac{1}{2}\sum_{ij} w_{ij}f_i f_j + \tfrac{1}{2}\sum_i w_{ii}f_i(1 \mp f_i) \quad (3.57)$$

which differs from (3.42) only in the presence of the terms involving the w_{ii}, and similarly in that by (3.48) and (3.45)

$$\omega_i = w_i - w_{ii} + \sum_j w_{ij}f_j \quad (3.58)$$

whereas in (3.39) the term $-w_{ii}$ is absent. But by (3.34) w_{ii} vanishes identically for a Fermi system (upper signs). Furthermore, in many cases (e.g., if the indices $i, j, \ldots$ stand for momenta) w_{ij} is inversely proportional to the volume and hence w_{ii} is negligible even for a Bose system. Finally, even if (as in the case of the Ising model) the w_{ij} are volume-independent, it is usually possible, by appropriate redefinition of the single-particle part of H, to arrange that $w_{ii} = 0$ (in fact, this is true automatically for the Ising model). Hence we conclude that *if the system is composed of fermions, or more generally if the w_{ii} either vanish, or else have a vanishing thermodynamic limit, then the variational method based on the decomposition*

(3.47) is equivalent to the previous method based on a trial Hamiltonian H_0 inserted "from the outside."

Finally, we consider, in this light, the problem of minimizing $\Omega_{\min}$ with respect to variation of the u_{ij}. We have noted that only the diagonal part of H in the b_i, $b_i^\dagger$ representation contributes to the upper bound (3.16) and hence also to (3.57), and that the expression (3.57) (with neglect of the w_{ii}) or (3.42) differs from the thermodynamic potential of an "ideal quasiparticle gas" only in the presence of the quasiparticle interaction contribution $-\frac{1}{2}\sum_{ij} w_{ij} f_i f_j$ and in that the quasiparticle energies ω_i are to be determined self-consistently by (3.39), and are therefore temperature-dependent. The minimization of the bound (3.16) with respect to variation of the u_{ij} then is equivalent to finding the best choice of quasiparticles so as to bring this bound as close to Ω as possible, subject to the constraint that the quasiparticles are related to the actual particles by the simple linear canonical transformation (3.13). Actually, it turns out that (3.13) can be generalized to include inhomogeneous terms and terms involving the $b_j^\dagger$ on the right side, and that such generalizations are important in theories of liquid He^4 and superconductivity. The variational treatment goes through with no essential changes, except for the more general definition of the b_i and $b_i^\dagger$. The details will be discussed in some special cases in Section VII. Due to their extreme nonlinearity the variational equations obtained by variation of the u_{ij} are analytically tractable only in certain simplified cases; otherwise they have to be handled numerically, for example, by iteration as in the usual Hartree-Fock method. We shall therefore not attempt any general discussion of the solutions here.

Corrections to the approximation to Ω provided by the variational method can be investigated by statistical-mechanical perturbation theory. Suppose, for example, that in the appropriate b_i, $b_i^\dagger$ representation, the Hamiltonian is purely diagonal, hence of the form (3.44). Then one can expand the grand partition function in powers of H_{fluc} as follows:

$$\Xi = e^{-\beta\Omega} = \operatorname{Tr} e^{-\beta(H_{\text{quas}}+H_{\text{fluc}}-\mu N)}$$

$$= \Xi_0 \sum_{j=0}^{\infty} \frac{(-\beta)^j}{j!} \langle (H_{\text{fluc}})^j \rangle_0 \tag{3.59}$$

and hence

$$\Omega = \Omega_0 - \beta^{-1} \ln \sum_{j=0}^{\infty} \frac{(-\beta)^j}{j!} \langle (H_{\text{fluc}})^j \rangle_0 \tag{3.60}$$

where Ω_0 is given by (3.49) and

$$\langle \cdots \rangle_0 = \Xi_0^{-1} \operatorname{Tr} [\cdots e^{-\beta(H_{\text{quas}}-\mu N)}] \tag{3.61}$$

a special case of (3.17). It follows from (3.51), with neglect of the w_{ii}, that $\langle H_{\text{fluc}}\rangle_0 = 0$ with the correct self-consistent choice (3.22), (3.40) of the f_i, that is, that the $j = 1$ term in (3.60) vanishes. In general, however, the higher-order terms do not vanish. Thus, for example, for $j = 2$, one has

$$\langle(H_{\text{fluc}})^2\rangle_0 = \tfrac{1}{2}\sum_{ijkl} w_{ij}w_{kl}\langle(b_i^\dagger b_i - f_i)(b_j^\dagger b_j - f_j)(b_k^\dagger b_k - f_k)(b_l^\dagger b_l - f_l)\rangle_0 \tag{3.62}$$

The various contributions to (3.62) can be evaluated by forming all possible contractions according to Matsubara's theorem. Terms in which each $b_i^\dagger$ (or $b_j^\dagger$, $b_k^\dagger$, or $b_l^\dagger$) is contracted with its neighbor b_i (or b_j, b_k, or b_l) give no net contribution, such contractions being canceled by the corresponding f_i (or f_j, f_k, or f_l) since $f_i = \langle b_i^\dagger b_i\rangle_0$. On the other hand, terms in which each $b_i^\dagger$ or b_i is contracted with some other such operator other than its neighbor do not in general vanish. However, because of such contractions, there remain only two free summations in (3.62), giving two factors of the volume of the system (or the number of spins in the case of, e.g., the Ising model). In some cases, the w_{ij} are inversely proportional to the volume; in fact, this is quite generally true if the indices $i, j, \ldots$ stand for momenta. *In such a case, the total contribution* (3.62) *will be volume-independent*. The same argument can be extended to arbitrary order, with the conclusion that *if the w_{ij} are inversely proportional to the volume, then every term in the argument of the logarithm in* (3.60) *will be volume-independent*, and hence, assuming convergence, *the perturbation H_{fluc} will cause only a negligible volume-independent shift*, *whereas Ω_0, and hence Ω, is volume-proportional* (*extensive*). *In such a case, the variational bound Ω_0 becomes equal to the exact Ω in the thermodynamic limit.* This situation was first noted by Bogoliubov, Zubarev, and Tserkovnikov[25] in the case of the BCS model of superconductivity, and then generalized by Wentzel.[26] It is important to notice that, in addition to the assumption that the w_{ij} are inversely proportional to the volume, it has also been assumed that H is purely diagonal in an appropriate quasiparticle representation; thus the argument is restricted to certain simplified models. The assumption of complete diagonality can be weakened slightly (in fact, the BCS case[25] and those studied by Wentzel[26] are somewhat more general), but the restrictions are still so strong as to limit the applicability of the theorem to Hartree-Fock-like and BCS-like models. Nevertheless, it is interesting that for such a model, the variational treatment in fact gives the exact equilibrium thermodynamic functions. Furthermore, the thermodynamic formulas of Landau's Fermi-liquid theory (or the corresponding Bose-liquid theory) can be shown[27] to hold for such a model.

IV. LOWER BOUNDS ON THERMODYNAMIC POTENTIALS

In addition to upper bounds on the free energy, lower bounds are very useful. In conjunction with upper bounds, they can be used to bracket the free energy. In some cases, they may be easier to use for computation and thus provide worthwhile approximations in themselves; generally, however, this circumstance does not appear to hold.

A. The Golden-Thompson Inequality

A useful lower bound is due to Golden[28] and Thompson[29] (independently). We give here only Golden's development.

The Golden-Thompson principle can be expressed as

$$A \geqslant A_0 - \frac{1}{\beta} \ln \langle e^{-\beta(H-H_0)} \rangle_0 \tag{4.1}$$

Before deriving (4.1), let us note that the equality sign always holds in the classical case, so that (4.1) is not very useful for classical statistical mechanics.

Golden's proof of the desired inequality (4.1) depends on two lemmas.

Lemma 1. Let **A** and **B** be two bounded, positive definite hermitian operators. Then, for p an integer $\geqslant 0$

$$\mathrm{Tr}\,\{(\mathbf{AB}^{2^{p+1}})\} \leqslant \mathrm{Tr}\,\{(\mathbf{A}^2\mathbf{B}^2)^{2^p}\}$$

The basic tool for the proof is the Cauchy-Schwarz inequality for operators which states that

$$\mathrm{Tr}\,\{UV^\dagger\} \leqslant [\mathrm{Tr}\,\{UU^\dagger\}\,\mathrm{Tr}\,\{VV^\dagger\}]^{1/2} \tag{4.2}$$

where † denotes hermitian conjugate. We first consider the case $p = 0$. Take $U = \mathbf{AB}$, $V = \mathbf{BA}$ in (4.2):

$$\begin{aligned}\mathrm{Tr}\,\{\mathbf{ABAB}\} &\leqslant [\mathrm{Tr}\,\{\mathbf{AB}(\mathbf{AB})^\dagger\}\,\mathrm{Tr}\,\{\mathbf{BA}(\mathbf{BA})^\dagger\}]^{1/2} \\ &= [\mathrm{Tr}\,\{\mathbf{A}^2\mathbf{B}^2\}\,\mathrm{Tr}\,\{\mathbf{A}^2\mathbf{B}^2\}]^{1/2} = \mathrm{Tr}\,\{\mathbf{A}^2\mathbf{B}^2\}\end{aligned} \tag{4.3}$$

so the lemma holds for $p = 0$.

Let us suppose the lemma holds for all $m < p$. In general we have

$$\mathrm{Tr}\,\{\mathbf{AB}\}^{2^{p+1}}\} \leqslant \mathrm{Tr}\,\{(\mathbf{AB})^{2^p}(\mathbf{BA})^{2^p}\} \tag{4.4}$$

by (4.2). Now define

$$\begin{aligned}X_n &= (\mathbf{AB})^{2^{(p-n)}}(\mathbf{BA})^{2^{(p-n)}} \\ Y_n &= (\mathbf{BA})^{2^{(p-n)}}(\mathbf{AB})^{2^{(p-n)}}\end{aligned} \tag{4.5}$$

Clearly $X_n^\dagger = X_n$, $Y_n^\dagger = Y_n$. Then, from (4.4)

$$\mathrm{Tr}\,\{(\mathbf{AB})^{2^{p+1}}\} \leqslant \mathrm{Tr}\,\{X_1 Y_1\} \leqslant \mathrm{Tr}\,\{X_1^2\} \tag{4.6}$$

where the second inequality follows from (4.2). Now

$$\mathrm{Tr}\,\{X_1^2\} = \mathrm{Tr}\,\{(X_2 Y_2)^2\} \tag{4.7}$$

by the cyclic invariance of the trace. In fact

$$\mathrm{Tr}\,\{X_n^N\} = \mathrm{Tr}\,\{(X_{n+1} Y_{n+1})^N\} \tag{4.8}$$

for arbitrary integer N, and $p > n$.

Now from the case $p = 0$ [first inequality in (4.9)] and Eq. (4.2) [second inequality in (4.9)]

$$\mathrm{Tr}\,\{(X_2 Y_2)^2\} \leqslant \mathrm{Tr}\,\{X_2^2 Y_2^2\} \leqslant \mathrm{Tr}\,\{X_2^4\} \tag{4.9}$$

By (4.8)

$$\mathrm{Tr}\,\{X_2^4\} = \mathrm{Tr}\,\{(X_3 Y_3)^4\} \tag{4.10}$$

Using the lemma for $m = 1$ and $m = 0$

$$\mathrm{Tr}\,\{(X_3 Y_3)^4\} \leqslant \mathrm{Tr}\,\{(X_3^2 Y_3^2)^2\} \leqslant \mathrm{Tr}\,\{X_3^4 Y_3^4\} \leqslant \mathrm{Tr}\,\{X_3^8\} \tag{4.11}$$

This process can clearly be continued, using the inductive hypothesis, until, finally

$$\mathrm{Tr}\,\{(\mathbf{AB})^{2^{p+1}}\} \leqslant \mathrm{Tr}\,\{X_p^{\,2^p}\} = \mathrm{Tr}\,\{(\mathbf{ABBA})^{2^p}\} = \mathrm{Tr}\,\{(\mathbf{A}^2\mathbf{B}^2)^{2^p}\} \tag{4.12}$$

which is the lemma.

Lemma 2.

$$e^{(\mathbf{A}+\mathbf{B})} = \lim_{N\to\infty} (e^{\mathbf{A}/N} e^{\mathbf{B}/N})^N \tag{4.13}$$

We write

$$\lim_{N\to\infty} N \ln (e^{\mathbf{A}/N} e^{\mathbf{B}/N}) = \lim_{x\to 0} \frac{1}{x} \ln (e^{\mathbf{A}x} e^{\mathbf{B}x}) \tag{4.14}$$

Expanding the logarithm in Taylor's series

$$\lim_{x\to 0} \frac{1}{x} \ln (e^{\mathbf{A}x} e^{\mathbf{B}x}) = \lim_{x\to 0} \sum \frac{(-1)^m}{m} \frac{(1 - e^{\mathbf{A}x} e^{\mathbf{B}x})^m}{x} \tag{4.15}$$

Interchanging the sum and limit, we get

$$\lim_{N\to\infty} N \ln (e^{\mathbf{A}/N} e^{\mathbf{B}/N}) = \mathbf{A} + \mathbf{B} \tag{4.16}$$

Of course, to make this rigorous, one must justify the interchange of limit and sum. A rigorous proof has been given by Trotter[30] (see Nelson[31] for an outline and references) but we content ourselves with Golden's proof above.

We can now demonstrate (4.1). Consider

$$Z_N = \mathrm{Tr}\,\{[e^{-\beta H_0/N} e^{-\beta(H-H_0)/N}]^N\} \tag{4.17}$$

We can, of course, take N to be of the form $N = 2^{p+1}$. Then, using Lemma 1 $p + 1$ times

$$\begin{aligned} Z_N &\leqslant \mathrm{Tr}\,\{e^{-\beta H_0} e^{-\beta(H-H_0)}\} \\ &= e^{-\beta A_0}\langle e^{-\beta(H-H_0)}\rangle_0 \end{aligned} \tag{4.18}$$

taking the limit as $N \to \infty$, $Z_N \to e^{-\beta A}$ (assuming the trace and limit operations are commutable). Finally, taking logarithms, one recovers (4.1).

B. The Inverse Gibbs-Bogoliubov Inequality

Let us note at this point that another lower bound can be obtained directly from the Gibbs-Bogoliubov inequality. One need only interchange the roles of H and H_0, obtaining

$$A_0 \leqslant A + \langle H_0 - H\rangle \tag{4.19}$$

or

$$A \geqslant A_0 + \langle H - H_0\rangle \tag{4.20}$$

We call this the "inverse Gibbs-Bogoliubov" inequality.

Of course, to use (4.20), one must average the "perturbation" $H - H_0$ over the ensemble appropriate to H, rather than that appropriate to H_0 (note the lack of a zero subscript on the carets). Such averages are difficult to evaluate; the usual use of variational methods is to avoid them. This suggests that (4.20) is not likely to be of great utility in practice. In the Golden-Thompson principle, one must average $\exp(-\beta[H - H_0])$ over the H_0 ensemble, to be sure. Nevertheless, this exponential is usually a relatively complicated operator, and the Golden-Thompson principle is likely to be less useful in practice than the Gibbs-Bogoliubov principle.

Still another lower bound on A has been given by Yeh.[32] It reads

$$\begin{aligned} A &\geqslant -kT \ln \sum_{\mu} \exp(-\beta\langle\mu|\, H\, |\mu\rangle a) \\ a &= \exp(-\beta\langle\mu|\, H^2\, |\mu\rangle/\langle\mu|\, H\, |\mu\rangle) \end{aligned} \tag{4.21}$$

where $\{|\mu\rangle\}$ is any complete orthonormal set.

We do not discuss the derivation of this formula, since it seems most unlikely that it will be of any use in practice. The reason for this is that $\langle\mu|\, H^2\, |\mu\rangle/\langle\mu|\, H\, |\mu\rangle$ will be of order N for an N-body system. Thus, a will be of order e^{-N}, and each exponential in the sum in (4.21) of order 1.

Hence what (4.21) will tell us is that $A \geqslant$ some very large negative number, a lower bound indeed, but not a very useful one.

C. Convexity of the Free Energy

The Helmholtz free energy has a very general convexity property, of which the Gibbs-Bogoliubov and Peierls inequalities and a number of other useful inequalities can be regarded as special cases.[33] In order to formulate it, we regard the free energy A as a functional of the Hamiltonian operator H, $A = A[H]$, in the sense that given any sufficiently well-behaved Hamiltonian defined on a given state space, a numerical free energy $A[H]$ is determined by the usual formula

$$A[H] = -\beta^{-1} \ln \operatorname{Tr} \{\exp(-\beta H)\}$$

We use the notation $A[H]$, rather than $A(H)$, to emphasize that A is a functional, rather than a function, of H. The general convexity property is then

$$A\left[\sum_\nu \lambda_\nu H_\nu\right] \geqslant \sum_\nu \lambda_\nu A[H_\nu] \tag{4.22}$$

where the λ_ν are weights, that is,

$$\lambda_\nu \geqslant 0\,, \quad \text{all } \nu; \qquad \sum_\nu \lambda_\nu = 1 \tag{4.23}$$

For general values of the λ_ν and general partial Hamiltonians H_ν, the argument $\sum_\nu \lambda_\nu H_\nu$ on the left of (4.22) is in general not equal to the actual Hamiltonian H of the system of interest. However, with special choices of the λ_ν and H_ν, one can arrange that $\sum_\nu \lambda_\nu H_\nu$ is, in fact, equal to H or some other closely related Hamiltonian, in which case (4.22) yields a lower bound on the corresponding free energy.

We shall prove (4.22) by a nonrigorous version of a proof due to Ruelle.[34] The lack of rigor arises from the fact that, as in our previous proofs, we assume that all operators are well enough behaved that all the traces and thermal averages involved exist; Ruelle states conditions sufficient to ensure this.

Let $\{|\mu\rangle\}$ be any complete orthonormal set. Then

$$\begin{aligned}\sum_\mu \exp\left(-\beta\langle\mu|\sum_\nu \lambda_\nu H_\nu|\mu\rangle\right) &= \sum_\mu \prod\nolimits_\nu \exp(-\beta\lambda_\nu\langle\mu|\,H_\nu\,|\mu\rangle)\\ &= \sum_\mu \prod\nolimits_\nu [\exp(-\beta\langle\mu|\,H_\nu\,|\mu\rangle)]^{\lambda_\nu} \end{aligned}\tag{4.24}$$

Now Hölder's inequality[35] states that

$$\sum_{\mu} \prod_{\nu} [f_\nu(\mu)]^{\lambda_\nu} \leqslant \prod_{\nu} \left[\sum_{\mu} f_\nu(\mu) \right]^{\lambda_\nu} \tag{4.25}$$

where the λ_ν are nonnegative weights [Eq. (4.23)] and the $f_\nu(\mu)$ are nonnegative functions of the summation index μ. Taking

$$f_\nu(\mu) = \exp(-\beta \langle \mu | H_\nu | \mu \rangle) \tag{4.26}$$

and the set $\{|\mu\rangle\}$ to be the complete orthonormal set of eigenstates of $\sum_\nu \lambda_\nu H_\nu$, one concludes from (4.24) that

$$\mathrm{Tr}\left\{\exp\left(-\beta \sum_\nu \lambda_\nu H_\nu\right)\right\} \leqslant \prod_\nu \left[\sum_\mu \exp(-\beta \langle \mu | H_\nu | \mu \rangle) \right]^{\lambda_\nu} \tag{4.27}$$

Now by Peierls' theorem (2.19)

$$\sum_\mu \exp(-\beta \langle \mu | H_\nu | \mu \rangle) \leqslant \mathrm{Tr}\{\exp(-\beta H_\nu)\} \tag{4.28}$$

which remains an inequality (of the same sense) if the λ_ν power of both sides is taken; hence

$$\mathrm{Tr}\left\{\exp\left(-\beta \sum_\nu \lambda_\nu H_\nu\right)\right\} \leqslant \prod_\nu [\mathrm{Tr}\{\exp(-\beta H_\nu)\}]^{\lambda_\nu} \tag{4.29}$$

Thus, since the logarithm is a monotonic function,

$$\begin{aligned} A\left[\sum_\nu \lambda_\nu H_\nu\right] &= -\beta^{-1} \ln \mathrm{Tr}\left\{\exp\left(-\beta \sum_\nu \lambda_\nu H_\nu\right)\right\} \\ &\geqslant -\beta^{-1} \sum_\nu \lambda_\nu \ln \mathrm{Tr}\{\exp(-\beta H_\nu)\} = \sum_\nu \lambda_\nu A[H_\nu] \end{aligned} \tag{4.30}$$

D. Reduced Density Matrices, the N-Representability Problem, and Lower Bounds

Let $\psi(\mathbf{r}_1, \ldots, \mathbf{r}_N)$ be any normalized wave function of a system of identical particles. If the particles have spin or other internal degrees of freedom, then the corresponding discrete variables should be included as arguments of the wave function; however, we shall suppress them, since they play no essential role in our general arguments. The *N-particle density matrix* associated with the many-body wave function ψ is defined as

$$\rho_N(\mathbf{r}_1, \ldots, \mathbf{r}_N, \mathbf{r}_1', \ldots, \mathbf{r}_N') = \psi(\mathbf{r}_1, \ldots, \mathbf{r}_N)\psi^*(\mathbf{r}_1', \ldots, \mathbf{r}_N') \tag{4.31}$$

The "diagonal elements" $\rho_N(\mathbf{r}_1, \ldots, \mathbf{r}_N, \mathbf{r}_1, \ldots, \mathbf{r}_N)$ of this density "matrix" define the configurational probability density $|\psi(\mathbf{r}_1, \ldots, \mathbf{r}_N)|^2$.

One can define *reduced density matrices or j-particle density matrices* $(1 \leqslant j \leqslant N-1)$ by

$$\rho_j(\mathbf{r}_1, \ldots, \mathbf{r}_j, \mathbf{r}_1', \ldots, \mathbf{r}_j') = \int \rho_N(\mathbf{r}_1, \ldots, \mathbf{r}_N, \mathbf{r}_1', \ldots, \mathbf{r}_j', \mathbf{r}_{j+1}, \ldots, \mathbf{r}_N)\, d^3r_{j+1}, \ldots, d^3r_N \quad (4.32)$$

It is important to note that ρ_N and the ρ_j are completely symmetric or antisymmetric under all permutations of either their unprimed or primed arguments, since the wave functions ψ are either completely symmetric (Bose) or completely antisymmetric (Fermi). In view of the fact that ψ is normalized and the ρ_j are related by (4.31), one has the normalization and recursion relations

$$\int \rho_j(\mathbf{r}_1, \ldots, \mathbf{r}_j, \mathbf{r}_1, \ldots, \mathbf{r}_j)\, d^3r_1, \ldots, d^3r_j = 1\,, \qquad 1 \leqslant j \leqslant N$$

$$\int \rho_{j+1}(\mathbf{r}_1, \ldots, \mathbf{r}_{j+1}, \mathbf{r}_1', \ldots, \mathbf{r}_j'\mathbf{r}_{j+1})\, d^3r_{j+1} = \rho_j(\mathbf{r}_1, \ldots, \mathbf{r}_j, \mathbf{r}_1', \ldots, \mathbf{r}_j')\,, \qquad 1 \leqslant j \leqslant N-1 \quad (4.33)$$

In addition, if one regards ρ_j as an integral operator with kernel $\rho_j(\mathbf{r}_1, \ldots, \mathbf{r}_j, \mathbf{r}_1', \ldots, \mathbf{r}_j')$, in the sense

$$\rho_j f(\mathbf{r}_1, \ldots, \mathbf{r}_j) \equiv \int \rho_j(\mathbf{r}_1, \ldots, \mathbf{r}_j, \mathbf{r}_1', \ldots, \mathbf{r}_j') f(\mathbf{r}_1', \ldots, \mathbf{r}_j')\, d^3r_1' \cdots d^3r_j' \quad (4.34)$$

then each ρ_j is positive semidefinite:

$$\begin{aligned}(f, \rho_j f) &\equiv \int f^*(\mathbf{r}_1, \ldots, \mathbf{r}_j)\rho_j f(\mathbf{r}_1, \ldots, \mathbf{r}_j)\, d^3r_1, \ldots, d^3r_j \\ &= \int f^*(\mathbf{r}_1, \ldots, \mathbf{r}_j)\, \rho_j(\mathbf{r}_1, \ldots, \mathbf{r}_j, \mathbf{r}_1', \ldots, \mathbf{r}_j') \\ &\quad \times f(\mathbf{r}_1', \ldots, \mathbf{r}_j')\, d^3r_1 \cdots d^3r_j\, d^3r_1' \cdots d^3r_j' \geqslant 0\,, \\ &\qquad 1 \leqslant j \leqslant N \quad (4.35)\end{aligned}$$

One simple way of proving (4.35) is to note that

$$(f, \rho_j f) = \int |I(\mathbf{r}_{j+1}, \ldots, \mathbf{r}_N)|^2\, d^3r_{j+1} \ldots d^3r_N \geqslant 0 \quad (4.36)$$

where I is the partial inner product

$$I(\mathbf{r}_{j+1}, \ldots, \mathbf{r}_N) \equiv \int f^*(\mathbf{r}_1, \ldots, \mathbf{r}_j)\psi(\mathbf{r}_1, \ldots, \mathbf{r}_N)\, d^3r_1, \cdots, d^3r_j \quad (4.37)$$

The definitions (4.31) and (4.32) apply to the special case of a "pure state" ψ. However, the generalization to statistical mechanics is immediate. Let $\{\psi_i\}$ be any statistical ensemble of orthonormal N-particle states with associated statistical weights P_i $\left(P_i \geqslant 0, \sum_i P_i = 1\right)$. Then the N-particle density matrix of this ensemble is defined as

$$\rho_N(\mathbf{r}_1, \ldots, \mathbf{r}_N, \mathbf{r}_1', \ldots, \mathbf{r}_N') = \sum_i P_i \psi_i(\mathbf{r}_1, \ldots, \mathbf{r}_N)\psi_i^*(\mathbf{r}_1', \ldots, \mathbf{r}_N') \quad (4.38)$$

and the reduced density matrices of the ensemble are defined recursively as before [Eq. (4.32)]. It is easy to show that Eqs. (4.33) and (4.35) remain valid. The density matrices of the canonical ensemble are obtained by the special choice $P_i = \exp(-\beta E_i)/\sum_j \exp(-\beta E_j)$ where ψ_i and E_i are the N-particle energy eigenstates and eigenvalues.

The importance of reduced density matrices in many-body quantum and statistical mechanics arises from the fact that most physically important expectation values and thermal averages can be expressed in terms of reduced density matrices of low order. For example, the mean kinetic energy $\langle K \rangle$ is

$$\langle K \rangle = \sum_k \frac{\hbar^2 k^2}{2m} n(\mathbf{k}) \quad (4.39)$$

where $n(\mathbf{k})$, the expectation value or ensemble average of the number of particles with momentum $\hbar\mathbf{k}$, can be shown to be

$$n(\mathbf{k}) = \rho \int \rho_1(\mathbf{r}, \mathbf{r}')e^{-i\mathbf{k}\cdot(\mathbf{r}-\mathbf{r}')} \, d^3r \, d^3r' \quad (4.40)$$

Here $\rho = N/V$, the mean particle number density, and we have assumed that the wave functions satisfy periodic boundary conditions with macroscopic periodicity cube of volume V, in which case the allowed wave vectors have components which are integral multiples of $2\pi V^{-1/3}$; however, for a macroscopic system (4.39 and 4.40) are correct to within negligible surface terms for any reasonable boundary conditions. Similarly, for a system with only pair interactions $v(\mathbf{r}_{ij}) = v(\mathbf{r}_i - \mathbf{r}_j)$ the mean potential energy $\langle U \rangle$ is

$$\langle U \rangle = \tfrac{1}{2}N(N-1)\int \rho_2(\mathbf{r}_1, \mathbf{r}_2, \mathbf{r}_1, \mathbf{r}_2)V(\mathbf{r}_{12}) \, d^3r_1 \, d^3r_2 \quad (4.41)$$

For an energy eigenstate in the absence of any external forces, or more generally for a canonical ensemble under the same conditions, ρ_1 in (4.40) depends only on $\mathbf{r} - \mathbf{r}'$, and ρ_2 in (4.41) only on $\mathbf{r}_{12} = \mathbf{r}_1 - \mathbf{r}_2$, in which case one integration in each of these expressions becomes trivial.

However, (4.38) is valid even for a nonequilibrium ensemble, which is not necessarily translationally invariant.

The energy expectation value E of a pure state ψ or the internal energy E of an ensemble (e.g., canonical) is related to (4.39)–(4.41) by

$$E = \langle K \rangle + \langle U \rangle \tag{4.42}$$

and can thus be expressed only in terms of the one- and two-particle reduced density matrices (in the presence of an external potential, the expression for $\langle U \rangle$ includes a term involving ρ_1). This immediately suggests a variational approach to determination of the ground state energy or the internal energy, based on minimizing the expression (4.42) with respect to variations of ρ_1 and ρ_2 subject to the constraints (4.33) and (4.35). However, a little more thought shows that there must be something seriously wrong with such an approach, since neither the expressions (4.39)–(4.41) nor the constraints (4.33), (4.35) contain any explicit reference to the space of allowed wave functions of the system, or to the temperature, or indeed to any other property of the statistical ensemble. These misgivings are substantiated by the fact that one can show by counterexamples that the expression (4.42) can lie *below* the true ground state energy for certain choices of functions ρ_1 and ρ_2 satisfying (4.33) and (4.35). This shows that variation of ρ_1 and ρ_2 subject to these constraints is *not* equivalent to variation of the wave function ψ subject to its normalization, since the familiar quantum-mechanical variational principle (1.1) ensures that the minimum value attainable by such variation is the true ground-state energy E_0. This manifest incorrectness of a minimum principle based on variation of ρ_1 and/or ρ_2 has not prevented publication of a number of papers based on just such a "variational principle."

The logical error, made in supposing that the minimum of (4.42) with respect to ρ_1 and ρ_2 subject to (4.33) and (4.35) is E_0, is simply that the conditions (4.33) and (4.35), although *necessary*, are *not sufficient* to ensure the existence of *any* N-particle wave function ψ *of the proper (Bose or Fermi) symmetry* from which the ρ_j are derivable according to (4.31) and (4.32). In fact, they are not even sufficient to justify the existence of any ensemble (let alone the canonical one) from which the ρ_j are derivable according to (4.38) and (4.32), again restricting the allowed ψ to the proper (Bose or Fermi) symmetry. The problem of trying to find a set of both necessary and sufficient conditions on the low-order ρ_j (not referring explicitly to ψ or the high-order ρ_j) which ensure their derivability from allowed ψ's or ensembles of such ψ's is called the *N-representability problem*,[36–38] and is unsolved. In fact, there is little hope that it ever will be solved, since this would essentially reduce the N-body problem to a few-body problem.

However, as soon as one realizes the nature of the error in the invalid *upper*-bound principle based on ρ_1 and ρ_2, one can turn it into a correct *lower*-bound principle. Suppose that there exists some (unknown) set of *both necessary and sufficient* conditions which include the known necessary conditions (4.33) and (4.35). Then certainly minimization of (4.42) with respect to ρ_1 and ρ_2 subject to these conditions will yield the true ground state energy E_0. Suppose that the constraints expressed by the *unknown* conditions are relaxed, that is, discarded, leaving only the constraints expressed by the known necessary conditions. Then since relaxation of constraints always *lowers* (or at least never raises) the lowest eigenvalue, one concludes that[37]

$$E_0 \geqslant \min E[\rho_1, \rho_2] \tag{4.43}$$

where $E[\rho_1, \rho_2]$ denotes the functional of ρ_1 and ρ_2 defined by (4.39)–(4.42), and the minimization is to be carried out subject to (4.33) and (4.35). In this way one obtains a *lower* bound on the true E_0. It is important to realize that (4.43) is valid only if a *complete functional* minimization is carried out; on the other hand, any restriction of ρ_1 and ρ_2 implied by assumed functional forms may raise a *parametric* minimum *above* E_0, so that the bounding property is lost, and one does not know whether the result thus obtained is below or above E_0.

The lower bound (4.43) can be put into a more general and explicit form, applicable to any N-body Hamiltonian H expressible as a sum of single-particle Hamiltonians H_i and pair-interaction Hamiltonians H_{ij}:

$$H = \sum_{i=1}^{N} H_i + \sum_{1 \leqslant i < j \leqslant N} H_{ij} \tag{4.44}$$

Here H_i is the sum of the kinetic energy operator $(-\hbar^2/2m)\nabla^2$ and any external potential $v(\mathbf{r}_i)$ acting on the ith particle, and H_{ij} is the pair interaction potential $v(\mathbf{r}_i, \mathbf{r}_j)$, which need not be restricted to translationally invariant form $v(\mathbf{r}_{ij})$. Define

$$\operatorname{tr}(H_{12}\rho_2) \equiv \int \{[H_{12}\rho_2(\mathbf{r}_1\mathbf{r}_2, \mathbf{r}_1'\mathbf{r}_2')]_{\mathbf{r}_1'=\mathbf{r}_1, \mathbf{r}_2'=\mathbf{r}_2}\}\, d^3r_1\, d^3r_2 \tag{4.45}$$

where H_{12} acts only on the dependence of ρ_2 on $\mathbf{r}_1$ and $\mathbf{r}_2$, not on its dependence on $\mathbf{r}_1'$ and $\mathbf{r}_2'$, and the subscripts on the square bracket imply that the replacement $\mathbf{r}_1' \to \mathbf{r}_1$, $\mathbf{r}_2' \to \mathbf{r}_2$ is to be carried out *after* H_{12} acts on ρ_2. Then it is not difficult to show from the symmetry or antisymmetry of ψ and the definitions (4.31) and (4.32) that[36]

$$(\psi, H\psi) = \tfrac{1}{2}N \operatorname{tr}(H^{(2)}\rho_2) \tag{4.46}$$

with

$$H^{(2)} = H_1 + H_2 + (N-1)H_{12} \tag{4.47}$$

Then the lower bound (4.43) assumes the more explicit and general form

$$E_0 \geqslant \tfrac{1}{2}N \min_{\rho_2} \operatorname{tr}\,(H^{(2)}\rho_2) \tag{4.48}$$

where the minimization is to be carried out over the class of all functions $\rho_2(\mathbf{r}_1\mathbf{r}_2, \mathbf{r}_1', \mathbf{r}_2')$ satisfying the normalization and positivity conditions

$$\operatorname{tr}\rho_2 = 1\,; \qquad (f, \rho_2 f) \geqslant 0\,, \quad \text{all } f \tag{4.49}$$

with $(f, \rho_2 f)$ defined by (4.35), plus any other known[39,40] necessary conditions. Again it is important that the minimization be a *functional* one; assuming some known functional form involving parameters may well destroy the lower-bound property (4.48). Carrying out such a functional minimization is not necessarily a trivial problem, since the positivity constraint and other less trivial ones[39,40] are inequalities and hence cannot be handled by the Lagrange multiplier method. Nevertheless, some success has been achieved by such approaches.[37,38]

This method can be generalized to nonzero temperature as follows.[38] First, defining the two-particle density matrix associated with any given statistical ensemble by (4.38) and (4.32), one readily verifies that (4.46) generalizes to

$$\langle H\rangle = \tfrac{1}{2}N \operatorname{tr}\,(H^{(2)}\rho_2) \tag{4.50}$$

where $\langle H\rangle$ is the ensemble average of H, reducing to the canonical average $\operatorname{Tr}\{H \exp(\beta(A - H))\}$ in the case of the canonical ensemble. Thus the inverse Gibbs-Bogoliubov inequality (4.20) can be written in the form

$$A \geqslant A_0 + \tfrac{1}{2}N \operatorname{tr}\,[(H^{(2)} - H_0^{(2)})\rho_2] \tag{4.51}$$

where H_0 is assumed decomposable into single-particle Hamiltonians H_{0i} and interaction Hamiltonians H_{0ij} (possibly zero) in the same manner (4.44) as H is, and $H_0^{(2)}$ is defined by replacing H by H_0 in (4.47). Evaluation of ρ_2 in (4.51) would involve first finding the N-particle energy eigenstates ψ_i and eigenvalues E_i, constructing ρ_N by (4.38) with $P_i = \exp(-\beta E_i)/\sum_i \exp(-\beta E_i)$, and finally integrating over $N - 2$ variables to obtain ρ_2 by (4.32). Such calculations obviously cannot in general be carried out for $N \sim 10^{23}$. However, in analogy with the situation in the zero-temperature case, we note that the minimum of the right side of (4.51) with respect to functional variations of ρ_2 subject only to the necessary conditions (4.49) (plus any other known necessary conditions) is surely lower than the actual value that the right-hand side would assume if ρ_2 were actually evaluated from the N-body energy eigenstates. Thus

$$A \geqslant A_0 + \tfrac{1}{2}N \min_{\rho_2} \operatorname{tr}\,[(H^{(2)} - H_0^{(2)})\rho_2] \tag{4.52}$$

where the minimization is with respect to functional variations of ρ_2 subject only to (4.49) and any other known necessary conditions on ρ_2 which one wishes to employ. Finally, since this equality is true for any H_0 decomposable in the form (4.44), the equality will remain valid if this *minimum* is *maximized* with respect to any parameters or undetermined functions contained in H_0:

$$A \geqslant \max_{H_0} \left\{ A_0 + \tfrac{1}{2}N \min_{\rho_2} \operatorname{tr} [(H^{(2)} - H_0^{(2)})\rho_2] \right\} \tag{4.53}$$

In contrast to the minimization with respect to ρ_2, the maximization with respect to H_0 can be a parametric rather than a complete functional maximization. This is fortunate, since the functional maximization cannot be carried out in practice, inasmuch as A_0 can be evaluated only for certain simple choices of H_0 (e.g., an effective field Hamiltonian with no explicit interaction terms). If the only necessary conditions imposed on ρ_2 are the obvious ones (4.49) and the requirement that $\rho_2(\mathbf{r}_1, \mathbf{r}_2, \mathbf{r}_1', \mathbf{r}_2')$ be symmetric (Bose) or antisymmetric (Fermi, in which case spin variables should also be included), then the minimization with respect to ρ_2 yields the ground state energy $E_0[H^{(2)} - H_0^{(2)}]$ of the Hamiltonian $H^{(2)} - H_0^{(2)}$ for the given (Bose or Fermi) symmetry, and (4.53) reduces to

$$A \geqslant \max_{H_0} \{A_0 + \tfrac{1}{2}NE_0[H^{(2)} - H_0^{(2)}]\} \tag{4.54}$$

A better (higher) lower bound can be obtained[38] by starting not with the inverse Gibbs-Bogoliubov inequality (4.20), but instead with the modified Gibbs-Bogoliubov inequality (2.31). Noting that the internal energy E is equal to the expression (4.50) and that the inequality (2.31) becomes an equality when P is the canonical density operator, one has

$$A = \min_{P} \{\tfrac{1}{2}N \operatorname{tr} (H^{(2)}\rho_2) - TS[P]\} \tag{4.55}$$

where $S[P]$ is derived from P in the usual way, $S[P] = -k \operatorname{Tr} \{P \ln P\}$, the minimization is over the class of N-body density operators P which have unit trace and are nonnegative, and ρ_2 is the reduced density matrix derived from P according to (4.38) and (4.32) (P_i is the expansion coefficient in the spectral representation $P = \sum_i P_i \, |\psi_i\rangle\langle\psi_i|$). Now the inequality (2.31) is also valid if H is replaced by any "trial Hamiltonian" H_0; hence, assuming that H_0 can be decomposed in the same manner (4.44) as H, one has

$$A_0 \leqslant \tfrac{1}{2}N \operatorname{tr} (H_0^{(2)}\rho_2) - TS[P] \tag{4.56}$$

or

$$-TS[P] \geqslant A_0 - \tfrac{1}{2}N \operatorname{tr} (H_0^{(2)}\rho_2) \tag{4.57}$$

Since this is true for all H_0 of the form (4.44), one can maximize this lower bound with respect to undetermined parameters or functions in H_0:

$$-TS[P] \geqslant \max_{H_0} [A_0 - \tfrac{1}{2}N \operatorname{tr}(H_0^{(2)}\rho_2)] \tag{4.58}$$

Substituting into (4.55), one concludes that

$$A \geqslant \min_{P} \max_{H_0} \{A_0 + \tfrac{1}{2}N \operatorname{tr}[(H^{(2)} - H_0^{(2)})\rho_2]\} \tag{4.59}$$

This is valid under the assumption that ρ_2 is derived from P according to (4.38) and (4.32). If this constraint is relaxed, then the minimum will be lower, preserving the sense of the inequality:

$$A \geqslant \min_{\rho_2} \max_{H_0} \{A_0 + \tfrac{1}{2}N \operatorname{tr}[(H^{(2)} - H_0^{(2)})\rho_2]\} \tag{4.60}$$

Here, as in (4.53), the minimization is with respect to functional variations of ρ_2 subject only to (4.49) and any other known necessary conditions on ρ_2 which one wishes to employ. Even if only the obvious necessary conditions (4.49) are employed, this expression does not simplify into one involving the ground state of $H^{(2)} - H_0^{(2)}$, in contrast with (4.54), and hence is in general more difficult to evaluate then (4.54). Since the lower bound (4.60) is in general higher (hence better) than (4.53) or (4.54), one gains additional accuracy for the extra effort in evaluation.[38]

Not many applications of these density-matrix variational principles have been made as yet (not counting incorrect applications based on the *minimum* principle). However, the results so far obtained[37,38] on simple test cases are rather encouraging, and suggest that significant results may be obtained by application of such methods to more realistic cases. Since one thus obtains optimal lower bounds, there is the prospect of combining such lower bounds with optimal upper bounds obtained from the Gibbs-Bogoliubov variational principle so as to narrowly circumscribe the free energy, and even to obtain good bounds on other thermodynamic quantities by the methods described in Section V.

V. BOUNDS ON OTHER QUANTITIES

A. Fisher's Bounds on Derivatives

In thermodynamics, we are not interested solely in the free energy, A, but also in its derivatives, $p = -(\partial A/\partial V)_T$, $S = -(\partial A/\partial T)_V$, and of course, higher derivatives. Even if we have good variational bounds on A, it is not clear that differentiating the bound gives a bound on the derivative. The bound may, for example, have a small amplitude but high frequency "ripple." Fisher[41] has shown how one may obtain upper and lower bounds to the derivative of a *convex* function, $f(x)$, from upper and lower bounds on f itself. It is not even necessary to assume that the bounds on f are differentiable.

Suppose f is a function which is convex upwards (or concave, in the usual mathematical terminology). Then

$$f(\tfrac{1}{2}(x_1 + x_2)) \geqslant \tfrac{1}{2}f(x_1) + \tfrac{1}{2}f(x_2) \tag{5.1}$$

that is, the curve lies above its chords. It safely can be assumed (see Appendix) that f must be continuous, and differentiable everywhere except at denumerably many points where the derivative may have jumps. Furthermore, df/dx is monotone nonincreasing and hence f lies below any of its tangents.

Now suppose that we have somehow found an upper bound for f at a particular x_0. Call this bound f_0^+,

$$f(x_0) \leqslant f_0^+ \tag{5.2}$$

and, correspondingly, a lower bound at $x = x_1 > x_0$. Call this f_1^-:

$$f(x_1) \geqslant f_1^- \tag{5.3}$$

Then

$$m_- = \frac{(f_1^- - f_0^+)}{(x_1 - x_0)} \tag{5.4}$$

is a lower bound to df/dx at x_0. For, suppose the contrary, $m_- > f'(x_0)$. Then the tangent to $f(x)$ at x_0 would lie below the line joining the upper and lower bounds at x_0 and x_1, respectively. Since $f(x)$ lies below this tangent, this would imply that $f(x_1) < f_1^-$, which is contrary to hypothesis.

Similarly if we have a lower bound to f at $x_2 < x_0$

$$f(x_2) \geqslant f_2^- \tag{5.5}$$

then

$$m_+ = \frac{(f_0^+ - f_2^-)}{(x_0 - x_2)} \tag{5.6}$$

is an upper bound to $f'(x)$ at x_0. The proof follows the same lines as the preceding paragraph. The geometrical aspects of the problem can be understood most clearly by consideration of Fig. 1, which is taken from Fisher's paper[41] (with permission).

Now suppose that, instead of having lower bounds f_1^- and f_2^- at isolated points, one has a function $f^-(x)$, with continuous derivative, which is a lower bound on $f(x)$. Which values of x_1 and x_2 should one pick to give the optimal bounds, m_+ and m_-, on the derivative $f'(x)$?

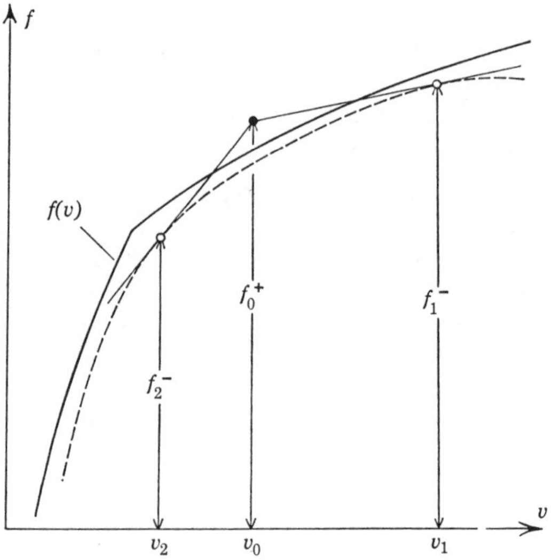

Fig. 1. A convex function, $f(v)$, shown by the heavy solid curve, with upper and lower bounds as indicated. The dashed line is a continuous lower bound.

Fisher remarks that the optimal bounds are obtained by constructing tangents to $f^{-}(x)$ from $f(x_0)$, that is, by solving the equation

$$f_0^+ + \left(\frac{df^-}{dx}\right)(x - x_0) = f^-(x) \tag{5.7}$$

for x_1 and x_2 with $x_1 < x_0 < x_2$.

The application of these results to the free energy derives from the fact that A is a convex function of the temperature, T, and volume V. Rigorous statistical mechanical proofs have been given by Ruelle[34] and Fisher.[42] A thermodynamic argument is the following: a sufficient condition for a twice-differentiable function to be convex is that its second derivative be negative. Since $(\partial^2 A/\partial T^2)_V = -C_V/T$, $(\partial^2 A/\partial V^2)_T = -(\partial P/\partial V)_T$, thermodynamic stability ($C_V > 0$, $\kappa_T > 0$) implies convexity.

B. Bounds on Expectation Values

Falk[43] has shown that the Gibbs-Bogoliubov, Golden-Thompson, and Fisher inequalities can be combined to give upper and lower bounds on any hermitian operator, Q. Let us set

$$A(\gamma) = kT \ln \mathrm{Tr}\,\mathrm{p}\,(-\beta[H + \gamma Q])\} \tag{5.8}$$

Clearly, $A(0)$ is the Helmholtz function of a system with Hamiltonian H. Furthermore

$$\left(\frac{\partial A}{\partial \gamma}\right)_{\gamma=0} = \langle Q \rangle \tag{5.9}$$

and $A(\gamma)$ is a convex function, by the argument used in Section II. A to show that $Z''(\lambda) > 0$.

If we write H as $H = H_0 + H_1$, then an upper bound for $A(0)$ is

$$A(0) \leqslant A^+(0) \equiv A_0 + \langle H_1 \rangle_0 \tag{5.10}$$

by the Gibbs-Bogoliubov inequality. Similarly, a lower bound for $A(\gamma)$ is

$$A(\gamma) \geqslant A^-(\gamma) \equiv A_0 - kT \ln \langle \exp(-\beta[H_1 + \gamma Q]) \rangle_0 \tag{5.11}$$

by the Golden-Thompson principle. Finally, from Fisher's bounds, we get

$$\frac{-kT \ln \langle \exp(-\beta[H_1 + \gamma_2 Q]\rangle_0 - \langle H_1 \rangle_0}{\gamma_2} \geqslant \langle Q \rangle \geqslant \frac{-kT \ln \langle \exp(-\beta[H_1 + \gamma_1 Q])\rangle_0 - \langle H_1 \rangle_0}{\gamma_1} \tag{5.12}$$

where $\gamma_2 > 0 > \gamma_1$.

Just as for the Golden-Thompson principle, the Falk inequality (5.12) is likely to be difficult to use because of the occurrence of Q and H_1 in the exponential.

Yeh[44] has given bounds for the average value of a function in terms of averages of moments of its argument. However, there does not appear to be much one can vary in his results, and so we do not discuss them here.

VI. SOME APPLICATIONS TO CLASSICAL STATISTICAL MECHANICS

Applications of variational methods, both for classical and quantum statistics can be grouped, for convenience, into two categories: (1) the proving of theorems, usually involving the establishment of a bound, and (2) the actual calculation of thermodynamic quantities by, hopefully, clever choice of variational parameters. In the following sections we shall give examples to show what has been done along these lines already. We shall not, however, go into the full details of the various calculations, the reader being urged to consult the original papers if full detail is desired.

A. Examples of General Theorems

The first category of applications is not terribly extensive to date. We give just two examples. Our first example will be to establish the proposition that the second virial coefficient of a gas with an angle-dependent two-body potential, $v(r, \Omega)$ is less than that for a gas with a corresponding potential, averaged over angles, $v_0(r) = \int d\Omega\, v(r, \Omega)$.

It is known[45] that the second virial coefficient, B, for a gas can be written

$$B = -\frac{1}{2V}(Z_2 - Z_1^{2}) \tag{6.1}$$

where Z_2 is the partition function for two particles in a box of volume V, whereas Z_1 is that for a single particle. That is, $Z_1 = V$. Let Z_2 be the two-body partition function for the aspherical potential $v(r, \Omega)$ and let $Z_2{}^0$ be the two-body partition function for $v_0(r)$. Then we have, from the Bogoliubov-Gibbs inequality

$$Z_2 \geqslant Z_2{}^0 \exp(-\beta\langle v(r, \Omega) - v_0(r)\rangle_0) \tag{6.2}$$

Now, we assert that $\langle v - v_0\rangle_0$ vanishes, since an angular average is part of $\langle\cdots\rangle_0$ and the angular average of v is v_0. Hence $Z_2 \geqslant Z_2{}^0$, and, because of the minus sign in (6.1),

$$B \leqslant B_0 \tag{6.3}$$

where B_0 is the second virial coefficient for a gas with potential v_0. This argument is similar to that previously used in a note on the dissociation pressure of gas hydrates,[46] but has not previously been published.

As our second example, we give Leff's argument[47] that the entropy of a classical imperfect gas is less than that of a classical ideal gas at the same temperature and density and same molecular mass. The free energies and entropies of the imperfect and perfect gases are A and S, and A_0 and S_0, respectively. The Hamiltonian for the imperfect gas is $K + U$ (kinetic and potential energy) and that for the ideal gas is K.

We use the Bogoliubov-Gibbs inequality in the form (4.20)

$$A_0 \leqslant A - \langle U\rangle \tag{6.4}$$

Since $A = E - TS$, we may rewrite this as

$$\begin{aligned} S - S_0 &\leqslant \frac{1}{T}(\langle K\rangle + \langle U\rangle - \langle K\rangle_0 - \langle U\rangle) \\ &= \frac{1}{T}(\langle K\rangle - \langle K\rangle_0) \end{aligned} \tag{6.5}$$

In classical statistical mechanics $\langle K\rangle = \langle K\rangle_0$, so that $S \leqslant S_0$. This accords with the intuitive feeling that the ideal gas is, in some sense, more disordered than the imperfect gas.

Leff also derives a lower bound for $S - S_0$ which is $\langle U\rangle - \langle U\rangle_0$. But for any physical potential, $\langle U\rangle_0$ is ∞, so that this bound is not very illuminating. He also points out that (6.5) holds for quantal systems also, but that $\langle K\rangle - \langle K\rangle_0$ does not vanish for quantum statistics, and its sign is unknown in general.

B. Free Volume Theory

Now let us begin to discuss applications in the second category. Our first example is the free volume theory of liquids. This theory was first put forward by Lennard-Jones and Devonshire;[48] some years later it was shown by Kirkwood[49] how, on the basis of a minimum principle, the Lennard-Jones-Devonshire theory could be derived and sys[illegible]tically improved. The minimum principle in question was ess[illegible]). Kirkwood's method was then extended somewhat by Maye[illegible].[50] Since Kirkwood's derivation is reasonably well known, we shall not repeat it here, but instead, show how the integral equation for the free volume can be obtained directly from (2.1), the Gibbs-Bogoliubov principle. Of course, this is completely equivalent to what Kirkwood did.

As usual in classical problems, the contribution of the kinetic energy to the partition function is trivial to evaluate, and we assume that this has already been done. Thus the Hamiltonian for our system is

$$H = \tfrac{1}{2}\sum_{i\neq j} v(\mathbf{r}_i - \mathbf{r}_j) \tag{6.6}$$

To define H_0, we divide up our system into N cells (N is the number of molecules present), all identical and, in toto, spanning the volume, and define

$$u_i(\mathbf{r}) = \begin{cases} \text{to be determined;} & \mathbf{r} \text{ in cell } i \\ \infty \quad ; & \mathbf{r} \text{ outside of cell } i \end{cases} \tag{6.7}$$

and

$$H_0 = \sum_i u_i(r_i) \tag{6.8}$$

Then

$$\begin{aligned} A_0 + \langle H - H_0\rangle_0 = &-kT \ln\left[\int_\Delta e^{-\beta u(r)}\, d^3r\right]^N \\ &+ \frac{\tfrac{1}{2}\sum_{i\neq j}\int_{\Delta_i} d^3r_i \int_{\Delta_j} d^3r_j v(\mathbf{r}_i - \mathbf{r}_j) e^{-\beta(u_i(r_i)+u_j(r_j))}}{\left[\int_\Delta u(r) e^{-\beta u(r)}\, d^3r\right]^2} \\ &- \frac{N\int_\Delta u(r)e^{-\beta u(r)}\, d^3r}{\int_\Delta e^{-\beta u(r)}\, d^3r} \end{aligned} \tag{6.9}$$

where Δ is the volume of integration, that is, one cell. One now attempts to minimize $A_0 + \langle H - H_0 \rangle_0$ with respect to $u(r)$. Setting the variational derivative of $A_0 + \langle H - H_0 \rangle_0$ equal to zero, one gets the nonlinear integral equation

$$u(r) - \langle u \rangle_0 = \frac{\sum_i \int_{\Delta_i} v(\mathbf{r} - \mathbf{r}_i) e^{-\beta u(r_i)} \, d^3 r_i}{\int_{\Delta} e^{-\beta u(r)} \, d^3 r} - \langle H \rangle_0 \qquad (6.10)$$

The details of the calculation are very similar to those of Kirkwood's calculation, so we omit them. The result is exactly that of Kirkwood, if we identify his $\psi(r)$ with our $u(r) - \langle u \rangle_0$. Indeed, this identification must be made, since, according to Kirkwood's results $\langle \psi \rangle_0 = 0$. We stop here with the basic integral equation, the ensuing development of free volume theory being outside the realm of variational approximation.

C. Pressure of a Hard-Core System Near Close Packing

There is, however, one further point in the theory which has been investigated variationally. If we choose the unit of length such that the minimum volume per molecule of the system is 1, then, for systems with hard cores in d dimensions, free volume theory predicts that

$$\frac{A_N}{NkT} = d \ln (v - 1) + 0(1) \qquad (6.11)$$

as $v \to 1$.

If it is permissible to use the derivative of the approximate A to get the pressure

$$\frac{p}{kT} = \frac{d}{v - 1} + 0(1) \qquad (6.12)$$

where $0(1)$ denotes a quantity which remains bounded as $v \to 1$.

Since, intuitively, one would expect free volume theory to work best near close packing, these formulas may give important information on the equation of state of a real fluid near close packing.

Fisher[41] has applied his formulas for bounds on the derivative of a convex function to this problem. The final conclusions were rather inconclusive because of rather poor bound on A [recall from (5.4) and (5.6) that bounds on the function f are ingredients of the bounds on df/dx]. Hoover[51] had shown that, for a system of oriented hard cubes

$$d \ln (v - 1) + c^- \leqslant \frac{A}{NkT} \leqslant d \ln (v - 1) + c^+ \qquad (6.13)$$

where c^- and c^+ are fixed numbers. Oriented hard cubes are hard cubes for which edges of all molecules are parallel, and for which the three orthogonal side directions remain fixed, for all configurations. The proof of these bounds involves rather special geometrical properties of the hard cube system, and so we refer the reader to Ref. 51 for details. It follows after some algebra from Fisher's bound that

$$\frac{\xi_1 d}{v-1} \leqslant \frac{p}{kT} \leqslant \frac{\xi_2 d}{v-1}$$
$$0 \leqslant \xi_1 \leqslant 1 \leqslant \xi_2 < \infty \tag{6.14}$$

ξ_1 and ξ_2 are, in fact, well-defined numbers, given in terms of c^- and c^+. Thus for the oriented hard cube system (6.12) holds in the logarithmic sense

$$\frac{\ln p}{kT} = \ln\left(\frac{d}{(v-1)}\right) + 0(1) \tag{6.15}$$

For a system with a less special core shape, for example, hard spheres, bounds such as (6.13) are not known rigorously. More precisely, the lower bound of the form (6.13) still holds, but the upper has not been proved. The best bounds which Fisher has been able to obtain are

$$\eta_1 \ln (v-1) \leqslant \frac{p}{kT} \leqslant \eta_2 d(v-1)^{-1} \ln (v-1)^{-1} \tag{6.16}$$

for any η_1, η_2 obeying $\eta_1 < 1 < \eta_2$. This shows that the pressure indeed diverges as $v \to 1$, but is not very specific about the rate of growth.

This example, the details of which we have omitted, shows clearly the weakness of the variational estimation of derivatives. The weakness is that one is limited by the goodness of the upper and lower bounds one has for the function itself.

D. Excluded Volume of Polymer Chains

Let us now turn to a problem for which the variational approach appears to be successful, the problem of the "excluded volume" of polymer chains. The problem, simply stated, is what is the effect of the finite size of monomer elements (or alternatively stated, of the mutual forces exerted by the monomer elements on each other) on the configurations of a polymer chain? This problem has been recognized for a long time, but up till fairly recently has resisted any systematic theory. The reason is that it is not only a many-body problem, but one in which the forces are of a long-range nature. Although the segment–segment forces are assumed to be of short range in ordinary three-dimensional space (except for polyelectrolytes), in polymer statistics, one customarily indexes distances *along the chain*, that is, by the number of intervening segments. As the chain folds around in space, segments far removed along the chain can interact, and this is the long-range aspect of the problem.

The quantity most often discussed is the mean square end-to-end distance in ordinary space $\langle R_N{}^2\rangle$ for a chain of N elements. This is the mean square distance between the first and last chain elements, averaged over all chain configurations. One expects for sufficiently long chains

$$\langle R_N{}^2\rangle \sim CN^{1+\epsilon} \tag{6.17}$$

For a freely jointed chain with no excluded volume restriction, it is known that $\langle R_N{}^2\rangle_0 = Nl^2$ where l is the length of one segment. Restrictions of fixed bond angle keep the N dependence the same but change the coefficient C. The excluded volume problem, or at least one aspect of it, is to ask what is C and what is ϵ? The question of ϵ, that is, the rate of growth of the chain dimensions with N, is of most interest. Numerical work, consisting of the actual generation of chains on a lattice,[52] gives very strong indication that $\epsilon = \frac{1}{5}$.

The approach to the excluded volume problem which we wish to discuss is a self-consistent field theory, pioneered by Edwards[53] (see also de Gennes[54]). Reiss[55] has shown how this theory can be derived, and systematically improved, by the use of the variational principle (2.31).

We can investigate the configurational properties of the chain if we know the distribution function of the end-to-end distance, and we get a variational approximation to this as follows. Let $P_N(\mathbf{R}^N)$ be the distribution function for the positions of all N segments of the chain, the zeroth segment being supposed attached to the origin of the coordinate system. $P_N(\mathbf{R}^N)$ has the form

$$P_N(\mathbf{R}^N) = \frac{1}{Z}\Psi(\mathbf{R}^N)\exp\left(-\frac{\beta}{2}\sum_{i\neq j} v_{ij}\right) \tag{6.18}$$

Here Ψ is a function expressing the constraints on bond length and bond angle inherent in any chain model; its form does not concern us here. The exponential factor in (6.18) contains the excluded volume effect; v_{ij} is the interaction between segment i and segment j.

The idea of Reiss was to substitute for P_N, the function $P_N{}^0$

$$P_N{}^0 = \frac{1}{Z_0}\Psi(\mathbf{R}^N)\prod_i \exp\left(-\beta\phi_i(\mathbf{R}_i)\right) \tag{6.19}$$

That is, to replace the actual intersegment forces by a self-consistent field ϕ_i. It is not assumed that the self-consistent field is the same for all segments; that is why ϕ_i has a subscript. The reason for this is that we have tied one end of the molecule down to the origin, and thus destroyed translational invariance. Note that we have not tampered with the function Ψ. It is this function which ensures that we have a chain molecule, and not a gas, and we alter it at our peril. We have only decoupled that part of the problem which concerns the excluded volume, leaving the problem of the "unperturbed chain" alone.

The procedure is now as follows: insert (6.19) in (2.31) and vary the resulting functional with respect to the potentials ϕ_i, setting the coefficient of $\delta\phi_i$ equal to zero. The algebra is rather involved, sufficiently so that Reiss missed one term in giving his original result. (This was pointed out by Yeh and Isihara;[56] see also Yamakawa.[57]) Nevertheless, since there are no tricky points involved we omit the algebraic details.

If we define

$$P_N^{(s)}(\mathbf{R}^s) = \int P_N(\mathbf{R}^N)\, d^{N-s}R \tag{6.20}$$

where the integration is over the coordinates of $N - s$ specific particles, then the self-consistent field equation is

$$\begin{aligned}
&\phi_l(\mathbf{R}_l)P_N^{(1)}(\mathbf{R}_l) - \sum_{i\neq l}\int v_{li}P_N^{(2)}(\mathbf{R}_l, \mathbf{R}_i)\, d^3R_i \\
&\quad - P_N^{(1)}(\mathbf{R}_l)\int\left[\phi_k(\mathbf{R}_k)P_N^{(1)}(\mathbf{R}_k) - \sum_{i\neq k}\int v_{ki}P_N^{(2)}(\mathbf{R}_k, \mathbf{R}_i)\, d^3R_i\right] d^3R_k \\
&\quad + \sum_{i\neq l}\int\Big[\phi_i\{P_N^{(2)}(\mathbf{R}_i, \mathbf{R}_l) - P_N^{(1)}(\mathbf{R}_i)P_N^{(1)}(\mathbf{R}_l)\} \\
&\quad - \tfrac{1}{2}\sum_{j\neq i,\neq l}\int v_{ij}\{P_N^{(3)}(\mathbf{R}_i, \mathbf{R}_j, \mathbf{R}_l) - P_N^{(1)}(\mathbf{R}_l)P_N^{(2)}(\mathbf{R}_i, \mathbf{R}_j)\}\, d^3R_j\Big]\, d^3R_i = 0
\end{aligned} \tag{6.21}$$

Obviously, this is a rather complicated equation; the $P_N^{(s)}$ functions are complicated functionals of the ϕ's. The question arises of how to solve it. To date, no one has attempted a fully self-consistent solution. Various approximations have been discussed, by Yamakawa,[57] but it would take us too far away from variational methods proper to consider them here. Suffice it to say that all reasonable approximations so far suggested, except the original one of Reiss, yield $\epsilon = \frac{1}{5}$, but differ with respect to predictions of C, which in all cases, however, appears to be of order unity. The original approximate solution of Reiss gave $\epsilon = \frac{1}{3}$, but this is presumably due to the missing term previously mentioned.

The self-consistent field theory, based on a variational principle, seems to be a most promising *analytical* method for studying excluded volume effects. It seems that when a self-consistent field approximation is physically reasonable, a variational principle can be used to advantage to derive it.

E. Solution Theory

The last classical example which we shall discuss is the statistical mechanical theory of mixtures, of course confining ourselves to variational approaches. The aim of a theory of the thermodynamics of mixtures is to compute the thermodynamic excess functions of the mixture, that is, the difference between the property of the mixture, and the value of the property which would obtain if the mixture were ideal. Most workers are content to take the properties of the pure components as given data, not requiring them to be computed a priori.

In a paper which appeared in 1956, a very active period in solution theory, Scott[58] proposed a "one-fluid" model for a mixture. The potentials between pairs of molecules of type α and β, respectively, in the real mixture is supposed to be of the form

$$v_{\alpha\beta}(r) = \epsilon_{\alpha\beta} f\left(\frac{r}{\sigma_{\alpha\beta}}\right) \tag{6.22}$$

the function f being the same for all pairs. f is dimensionless, $\epsilon_{\alpha\beta}$ is an energy parameter, and $\sigma_{\alpha\beta}$ a distance parameter. Such solutions are called "conformal." Scott's "one-fluid" model consists of replacing the mixture by a single fluid with potential

$$v(r) = \epsilon f\left(\frac{r}{\sigma}\right) \tag{6.23}$$

and gave a prescription [see (6.25) below] for choosing ϵ and σ when f is of the Lennard-Jones form, $f(x) = 4(x^{-12} - x^{-6})$. Finally, he noted that the Helmholtz function, computed with the potential v, is an upper bound to the Helmholtz function computed when the $v_{\alpha\beta}$'s are used in the Hamiltonian. We should mention that Scott's paper considers other models also, which will not concern us here.

Some years later, Mazo[59] showed that Scott's choice of parameters, ϵ and σ, was the best choice that can be made, in the sense of the Gibbs-Bogoliubov principle, for the Lennard-Jones fluid. In outline the demonstration runs as follows, $\langle H - H_0\rangle_0$ is given by

$$\langle H - H_0\rangle_0 = \tfrac{1}{2}N(N-1)\left[\sum_{\alpha\beta} 4x_\alpha x_\beta \epsilon_{\alpha\beta}({\sigma_{\alpha\beta}}^{12}B_{12} - {\sigma_{\alpha\beta}}^{6}B_6) - 4\epsilon(\sigma^{12}B_{12} - \sigma^6 B_6)\right] \tag{6.24}$$

where $B_n = \langle r^{-n}\rangle_0$, and x_γ is the mole fraction of species γ present.

If one now differentiates $A_0 + \langle H - H_0 \rangle_0$ with respect to ϵ and σ, and sets the derivatives equal to zero, to minimize the right-hand side of the Gibbs-Bogoliubov inequality, one gets

$$\left[\sum_{\alpha\beta} x_\alpha x_\beta \epsilon_{\alpha\beta} \sigma_{\alpha\beta}{}^{12} - \epsilon\sigma^{12}\right]\frac{\partial B_{12}}{\partial \epsilon} - \left[\sum_{\alpha\beta} x_\alpha x_\beta \epsilon_{\alpha\beta} \sigma_{\alpha\beta}{}^{6} - \epsilon\sigma^{6}\right]\frac{\partial B_6}{\partial \epsilon} = 0 \quad (6.25)$$

and a similar equation containing $\partial B_{12}/\partial\sigma$, $\partial B_6/\partial\sigma$. Temperature- and density-independent solutions are obtained by setting the quantities in square brackets in (6.25) separately equal to zero. This determines ϵ and σ, and these are exactly the rules for ϵ and σ given by Scott.

Several years later, apparently unaware of this earlier work, Mansoori and co-workers took up again the question of a variational formulation for liquid theory, rederiving the Gibbs-Bogoliubov principle in the process.[60] In fact, they derived a second, incorrect inequality also, but this was subsequently corrected.[61] Their first paper[60] considered pure liquids only, and used a hard-sphere fluid for the reference system, described by the Hamiltonian H_0. In a later paper, this technique was extended to mixtures, using a mixture of hard spheres as the reference fluid.[62] The variational principle is used to determine the reference hard-sphere diameters by minimizing the right hand side of the Gibbs-Bogoliubov inequality with respect to these parameters. We should add that Henderson and Barker[63] point out certain difficulties which may arise if one uses the hard sphere H_0 for systems with a perhaps steep, but nevertheless soft, potential.

What are the numerical consequences of these variational theories of mixtures? In short, they are disappointing. The free energies come out too large (as they must, from their variational genesis), but the magnitude of the error is disappointingly big. One might think that the work of Mansoori and Leland[62] to be an exception to this statement, since there, $G^{(\text{ex})}$(variational) $<$ $G^{(\text{ex})}$(experimental), where $G^{(\text{ex})}$ is the excess Gibbs free energy which, at low pressures, is substantially identical to $A^{(\text{ex})}$. The reason for this is that Mansoori and Leland calculated both G(mixture) and G(pure components) variationally. After taking the difference, there is no assurance that one is left with an upper bound.

On the other hand, perturbation theories of simple fluid mixtures can now give quite reasonable agreement with experiment. Although this article does not concern itself, in general, with perturbation theory, a few remarks about the relationship of perturbation theory and variational methods may serve to finish this section.

Suppose we have a system with a Hamiltonian, $H(\boldsymbol{\lambda})$, depending on a set of numerical parameters $\boldsymbol{\lambda} = \{\lambda_1, \lambda_2, \ldots, \lambda_n\}$ (in addition to its dynamical variables). In perturbation theory, one tries to expand the free energy about some set $\boldsymbol{\lambda}^0$ as

$$A(\boldsymbol{\lambda}) = A(\boldsymbol{\lambda}^0) + \sum_i (\lambda_i - \lambda_i^\circ)\left(\frac{\partial A}{\partial \lambda_i}\right)_{\lambda_i=\lambda_i^\circ} + \text{higher order terms} \quad (6.26)$$

It usually turns out that one is powerless to evaluate the higher-order terms.

Now suppose we are given a Hamiltonian H and set

$$H(\lambda) = H_0 + \lambda(H - H_0) \quad (6.27)$$

Then perturbation theory gives us, setting $\lambda = 1$, $\lambda^0 = 0$

$$A = A_0 + \langle H - H_0\rangle_0 + \text{higher-order terms} \quad (6.28)$$

That is, *for this particular type of perturbation*, first-order perturbation theory gives an upper bound on the free energy. On the other hand, if the parameters, $\boldsymbol{\lambda}$, are intrinsic in the Hamiltonian (force constants, radii, etc.) and not artifically introduced as in (6.27), then first-order perturbation theory and a variational calculation may have little relation to each other. In perturbation theory, one even has a wide choice of functions of the λ_i which one may use as expansion parameters (pointed out to us by J. S. Rowlinson). In particular, in solution theory at least, judicious choices of the reference perturbation parameters $\boldsymbol{\lambda}^0$ appear to give much better results than a variational treatment.[64] We feel that it ought to be possible to use the variational method to reduce the amount of intuition necessary for this judicious choice, but do not know how to accomplish this aim at present.

All in all, the application of variational principles to problems in classical statistical mechanics has so far been disappointing. Some interesting general inequalities have been established, but when it comes to computation, the results have not been encouraging; an exception may be the excluded volume problem. The reason for this, we think, is that one is limited in practice to trial Hamiltonians which one can handle analytically or numerically. These apparently do not provide the flexibility to simulate the correlations which dominate the actual phenomena. We see no reason why this situation should be permanent, and we expect it to improve as our experience grows and our analytical abilities sharpen.

VII. APPLICATIONS TO QUANTUM STATISTICAL MECHANICS

As is quite generally the case in approximate solutions of physical problems, the same approximation can often be derived and interpreted in superficially quite different ways. Thus, for example, the Weiss molecular field theory of ferromagnetism can be derived and interpreted as a typical self-consistent field theory, but like the Hartree-Fock self-consistent field theory of atomic structure, it can also be derived variationally. The same is true of several of the applications to classical statistical mechanics discussed in Section VI. The general relationship between the variational method, self-consistent field methods (generalized Hartree-Fock method), and soluble models in quantum statistical mechanics has already been discussed in Section III, and most of the applications which we shall discuss in this section furnish examples of this general relationship. Thus some of these applications, for example, to the theory of magnetism, were first derived by physical reasoning based on the idea of self-consistent fields, whereas others, for example, the BCS theory of superconductivity, were initially formulated as variational theories, but subsequently shown to be interpretable in terms of soluble models and generalized self-consistent fields ("order parameters"). The approach via variational principles has the advantage that the nature of the approximations involved is clearly exhibited, with the concomitant disadvantage that the results obtained are limited by one's physical intuition in choosing a "trial Hamiltonian" and by the necessity of assuming the statistical mechanics of this trial Hamiltonian to be known. Also, some knowledge of the error (if only its sign) is always provided by the variational method.

A. BCS Theory of Superconductivity

Probably the first and also most successful explicitly variational calculation in the quantum statistical mechanics of interacting particles is the BCS theory of superconductivity.[65] By arguments which we shall not go into here, Bardeen, Cooper, and Schrieffer motivated the following "reduced Hamiltonian" H_{red} as an approximate model of a system of electrons in a solid, interacting directly by the Coulomb interaction and indirectly by exchange of phonons (lattice vibrations):

$$H_{\text{red}} = \sum_k \varepsilon_k a_k^\dagger a_k + \tfrac{1}{2} \sum_{kk'} v_{kk'} a_k^\dagger a_{-k}^\dagger a_{-k'} a_{k'} \tag{7.1}$$

Here k stands for the set $(\mathbf{k}, \sigma)$ composed of the wave vector $\mathbf{k}$ and spin variable ($= \uparrow$ or $\downarrow$) of a single electron, $-k$ stands for $(-\mathbf{k}, -\sigma)$, and a_k and $a_k^\dagger$ are electron (fermion) annihilation and creation operators; the notation in BCS[65] has been changed in order to conform with Section III. BCS discuss the ground state and statistical mechanics of this model separately; we shall limit our discussion here to the statistical mechanics, which is in fact essentially no more complicated than the ground state when approached via the Gibbs-Bogoliubov variational principle.

The variational treatment of the statistical mechanics of the model (7.1) given by BCS was based on an expression for the free energy [Eq. (3.16) of BCS][65] which they justified only by rather heuristic physical arguments, but can in fact be shown to be equivalent to the upper bound (2.31), with P taken as the density operator of an ideal Fermi gas of quasiparticles with single-particle energies to be determined variationally. From examination of low-orders terms in perturbation theory, BCS conjectured (Appendix A of BCS)[65] that their variational approximation to the ground-state energy of H_{red} is in fact exact in the thermodynamic limit. This conjecture was later verified by Bogoliubov, Zubarev, and Tserkovnikov,[25] who showed that the BCS variational solution is exact (for the given approximate Hamiltonian H_{red}) in the thermodynamic limit even at nonzero temperature. The BZT proof,[25] based on the decomposition (3.46)–(3.48), has already been discussed in Section III, and showed that H_{red} is a "soluble model" as far as equilibrium statistical mechanics is concerned (we remark parenthetically that some, but by no means all or even the most interesting, nonequilibrium properties of H_{red} are also soluble by essentially the same method).[66] The reader is referred to the voluminous literature on the theory of superconductivity for discussions of the details of the BCS variational solution and its rather successful application to the understanding of experimental observations. Our discussion here will be limited mainly to a derivation as a specific example of the general method discussed in Section III.

Suppose that the indices i, j, k, l in (3.1) are taken to be wave vector and spin pairs $k_i = (\mathbf{k}_i, \sigma_i)$, as in (7.1):

$$H = \sum_k \varepsilon_k a_k{}^\dagger a_k + \tfrac{1}{2} \sum_{k_1k_2k_3k_4} v_{k_1k_2k_3k_4} a_{k_1}{}^\dagger a_{k_2}{}^\dagger a_{k_4} a_{k_3} \tag{7.2}$$

We have assumed that $\varepsilon_{kk'}$ is already diagonal, $\varepsilon_{kk'} = \varepsilon_k \delta_{kk'}$, as will be the case for a translationally invariant system (following BCS, we neglect the lattice structure). Under the same assumption, $v_{k_1k_2k_3k_4}$ will satisfy the total linear momentum conservation selection rule of vanishing unless $\mathbf{k}_1 + \mathbf{k}_2 = \mathbf{k}_3 + \mathbf{k}_4$. As a guide to the choice of the trial Hamiltonian H_0, we note that the BCS reduced Hamiltonian (7.1) implies a strong correlation between electrons labeled by $\pm k$, that is, electrons with equal and opposite momenta and opposite spins. This suggests that not only the occupation-number operators $a_k{}^\dagger a_k$, but also the pair creation and annihilation operators $a_k{}^\dagger a_{-k}{}^\dagger$ and $a_{-k} a_k$, ought to have nonzero thermal averages in the superconducting phase (in fact, this is the essential physical content of the BCS theory). Thus we select not only the diagonal (in a_k, $a_k{}^\dagger$ representation) interaction terms in (7.2), but also the pairing interaction terms with $k_2 = -k_1$, $k_4 = -k_3$, for special attention, and decompose H by a generalization of (3.47):

$$
\begin{aligned}
H - \mu N &= H_{\text{quas}} + H_{\text{fluc}} + U' \\
H_{\text{quas}} &= -\tfrac{1}{2}\sum_{kk'} (u_{kk'}\xi_k\xi_{k'} + v_{kk'}\eta_k{}^*\eta_{k'}) \\
&\quad + \sum_k \left(\varepsilon_k - \mu + \sum_{k'} u_{kk'}\xi_{k'}\right) a_k{}^\dagger a_k \\
&\quad + \tfrac{1}{2}\sum_{kk'} (v_{kk'}\eta_{k'}a_k{}^\dagger a_{-k}{}^\dagger + v_{kk'}^*\eta_{k'}{}^* a_{-k}a_k) \\
H_{\text{fluc}} &= \tfrac{1}{2}\sum_{kk'} [u_{kk'}(a_k{}^\dagger a_k - \xi_k)(a_{k'}{}^\dagger a_{k'} - \xi_{k'}) \\
&\quad + v_{kk'}(a_k{}^\dagger a_{-k}{}^\dagger - \eta_k{}^*)(a_{-k'}a_{k'} - \eta_{k'})] \qquad (7.3) \\
U' &= \tfrac{1}{2}\sum_{k_1k_2k_3k_4}{}' v_{k_1k_2k_3k_4} a_{k_1}{}^\dagger a_{k_2}{}^\dagger a_{k_4} a_{k_3}
\end{aligned}
$$

Here

$$
\begin{aligned}
u_{kk'} &\equiv v_{kk'kk'} - v_{kk'k'k} \\
v_{kk'} &\equiv v_{k,-k,k',-k'} \qquad (7.4)
\end{aligned}
$$

the ξ_k and η_k are variational parameters to be determined, and the prime in the definition of U' implies exclusion of interaction terms already included in $H_{\text{quas}} + H_{\text{fluc}}$, that is, those with $(k_3 = k_1, k_4 = k_2)$ or $(k_3 = k_2, k_4 = k_1)$ or $(k_2 = -k_1, k_4 = -k_3)$.

We now apply the inequality (3.16), (3.17) with $H_0 - \mu N = H_{\text{quas}}$. Since H_{quas} is a quadratic form in annihilation and creation operators, it can be reduced to diagonal form by a linear canonical transformation, and it is clear from the structure of H_{quas} that this transformation must mix the operators a_k and $a_{-k}{}^\dagger$. The most general form of such a transformation, subject to the requirement that it be canonical (i.e., the anticommutation relations are preserved) is[67,68]

$$
\begin{aligned}
a_k &= U^{-1}b_kU = \frac{b_k + \varphi_k b_{-k}{}^\dagger}{\sqrt{1 + |\varphi_k|^2}} \\
b_k &= Ua_kU^{-1} = \frac{a_k - \varphi_k a_{-k}{}^\dagger}{\sqrt{1 + |\varphi_k|^2}}
\end{aligned}
\qquad (7.5)
$$

where the complex parameters φ_k are arbitrary except for the requirement of oddness:

$$\varphi_{-k} = -\varphi_k \qquad (7.6)$$

We see that the quasiparticles annihilated and created by b_k and $b_k{}^\dagger$ are linear combinations of electrons and holes with the same k [note that since a hole is the absence of an electron, it actually carries the same momentum (plus spin) $+k$ as does the electron, in spite of the fact that the index of $a^\dagger$ is $-k$].

The parameters φ_k can be determined by substituting (7.5) into the expression (7.3) for H_{quas} and requiring that it take the diagonal form (3.47), that is, that the coefficients of the terms $b_k{}^\dagger b_{-k}{}^\dagger$ and $b_{-k}b_k$ vanish. This requirement implies that

$$\tfrac{1}{2}\Delta_k{}^*\varphi_k{}^2 - (\varepsilon_k - \mu + s_k)\varphi_k - \tfrac{1}{2}\Delta_k = 0 \tag{7.7}$$

where the "Hartree-Fock shift" s_k and "energy gap function" Δ_k are defined as

$$s_k = \sum_{k'} u_{kk'}\xi_{k'}, \qquad \Delta_k = \sum_{k'} v_{kk'}\eta_{k'} \tag{7.8}$$

Either solution of the quadratic equation (7.7) diagonalizes H_{quas}, and hence both result in the same free energy; arbitrarily choosing the negative square root, one has

$$\varphi_k = \frac{\varepsilon_k - \mu + s_k - \omega_k}{\Delta_k{}^*} \tag{7.9}$$

where

$$\omega_k = [(\varepsilon_k - \mu + s_k)^2 + |\Delta_k|^2]^{1/2} \tag{7.10}$$

The corresponding diagonal expression for H_{quas} can be shown, assuming ε_k and s_k to be even function of k, to be

$$H_{\text{quas}} = W_0 + \sum_k \omega_k b_k{}^\dagger b_k \tag{7.11}$$

with

$$W_0 = -\tfrac{1}{2}\sum_{kk'}(u_{kk'}\xi_k\xi_{k'} + v_{kk'}\eta_k{}^*\eta_{k'}) + \sum_k(\varepsilon_k - \mu + s_k - \omega_k) \tag{7.12}$$

The corresponding thermodynamic potential is

$$\Omega_0 = \text{Tr}\, e^{-\beta H_{\text{quas}}} = W_0 - \beta^{-1}\sum_k \ln(1 + e^{-\beta\omega_k}) \tag{7.13}$$

Since the chemical potential μ is not merely an additive term, its contribution cannot be separated out explicitly in terms of $z = e^{\beta\mu}$, as was done in (3.49); instead, μ occurs implicitly in ω_k.

The bound (3.16) takes the form

$$\Omega \leqslant \Omega_{\text{var}} = \Omega_0 + \langle H_{\text{fluc}} + U'\rangle_0 \tag{7.14}$$

where $\langle\cdots\rangle_0$ denotes the thermal average in the ensemble with (grand canonical) density operator $\Omega_0{}^{-1}e^{-\beta H_{\text{quas}}}$. Note, again, that $-\mu N$ should not be added to H_{quas}, but is included in its definition. It follows from Matsubara's theorem, as in Section III, that $\langle U'\rangle_0$ vanishes, since

U' is completely off-diagonal in the b_k, $b_k^\dagger$ representation. Furthermore, $\langle H_{\text{fluc}} \rangle_0$ can be evaluated in analogy with the simpler case (3.50), and the minimization of Ω_{var} with respect to the real parameters ξ_k and complex parameters η_k can be carried through similarly. The result is that the minimum occurs when ξ_k and η_k are chosen as

$$\xi_k = \langle a_k^\dagger a_k \rangle_0 \,, \qquad \eta_k = \langle a_{-k} a_k \rangle_0 \,, \tag{7.15}$$

so that H_{fluc} [Eq. (7.3)] is indeed a fluctuation Hamiltonian. This is analogous to the top equation (3.34). The thermal averages in (7.15) can be expressed in terms of the quasiparticle Fermi distribution function

$$f_k = \langle b_k^\dagger b_k \rangle_0 = (e^{\beta \omega_k} + 1)^{-1} \tag{7.16}$$

since only the diagonal parts of the corresponding operators contribute, with the result

$$\begin{aligned} \xi_k &= \frac{|\varphi_k|^2}{1 + |\varphi_k|^2} + \left(\frac{1 - |\varphi_k|^2}{1 + |\varphi_k|^2} \right) f_k \\ \eta_k &= \frac{\varphi_k}{1 + |\varphi_k|^2} - 2\left(\frac{\varphi_k}{1 + |\varphi_k|^2} \right) f_k \end{aligned} \tag{7.17}$$

With this choice, the contribution $\langle H_{\text{fluc}} \rangle_0$ vanishes apart from a thermodynamically negligible (volume-independent) term, in analogy with the situation in Section III, and the bound (7.14) reduces to Ω_0.

Even after substitution of (7.9), the expressions (7.17) for ξ_k and η_k are not really explicit, since f_k depends (highly nonlinearly) on ω_k; φ_k depends on ω_k, s_k, and Δ_k; ω_k depends (nonlinearly) on s_k and Δ_k; and s_k and Δ_k depend on ξ_k and η_k. We are really dealing with a system of coupled nonlinear integral equations, as is to be expected for a generalized Hartree-Fock treatment. The qualitative features of the solution have been discussed by Wentzel.[69] Rather than repeating such a discussion here, we shall, following BCS, simplify the remaining discussion by neglecting the Hartree-Fock terms $u_{kk'}$ in (7.3), that is, we suppose that the corresponding matrix elements vanish. Then s_k is to be put equal to zero in (7.9), (7.10), and (7.12), and $u_{kk'}$ is also to be put equal to zero in (7.12). Then we have a single nonlinear integral equation to solve, namely, the equation (7.8) for Δ_k [we call it an integral equation since it becomes one in the thermodynamic limit, when $\sum_k \rightarrow (2\pi)^{-3} V \int d^3k$]. According to (7.17), (7.16), and (7.9) one has

$$\eta_k = -\left(\frac{\Delta_k}{2\omega_k} \right) \tanh\left(\tfrac{1}{2}\beta\omega_k\right) \tag{7.18}$$

so that the integral equation (7.8) becomes

$$\Delta_k = -\sum_{k'} v_{kk'}\left(\frac{\Delta_{k'}}{2\omega_{k'}}\right) \tanh\left(\tfrac{1}{2}\beta\omega_{k'}\right) \tag{7.19}$$

The nonlinearity arises because, by (7.10),

$$\omega_k = [(\varepsilon_k - \mu)^2 + |\Delta_k|^2]^{1/2} \tag{7.20}$$

If the pairing interaction $v_{kk'}$ is real (not merely hermitian), then the solution Δ_k of (7.19) may be assumed real without loss of generality. In the contrary case (e.g., for pairing in nonspherically symmetric states) Δ_k will be complex.

Obviously (7.19) admits a trivial solution $\Delta_k = 0$, for which also $\eta_k = 0$, $\omega_k = \varepsilon_k - \mu$, and the trial thermodynamic potential reduces to only the contribution from the kinetic energy part of the Hamiltonian. However, under certain conditions, there is also a nontrivial solution at sufficiently low temperatures. In the original BCS paper[65] it is shown that this is the case, for example, if $v_{kk'}$ is negative (attractive) and separable ($v_{kk'} = -v_k v_{k'}$). The separability assumption violates translational invariance and is only introduced for the purpose of reducing the integral equation to a transcendental algebraic equation; more generally, it is well known that a solution exists at sufficiently low temperatures T if $v_{kk'}$ is mainly negative. For details of the solution, the reader is referred to the literature; we wish here only to present an argument, due to Bogoliubov et al.,[70] for determining the transition temperature T_c below which the nontrivial solution exists. Suppose that for $T > T_c$ there is no nontrivial solution, and that as T drops through T_c a nontrivial solution appears, representing a transition to the superconducting state (it can be shown that the nontrivial solution, when it exists, has lower free energy than the trivial solution). Then $T = T_c$ is a "point of bifurcation" at which the nontrivial solution splits off the trivial one. For T only slightly less than T_c, Δ_k will be very small (it vanishes at T_c), so that the integral equation can be linearized by neglecting the term $|\Delta_k|^2$ in the expression (7.20) for ω_k:

$$\Delta_k = -\sum_{k'} v_{kk'}\left[\frac{\Delta_{k'}}{2(\varepsilon_{k'} - \mu)}\right] \tanh\left[\tfrac{1}{2}\beta_c(\varepsilon_{k'} - \mu)\right] \tag{7.21}$$

where β has been replaced by β_c since $T \approx T_c$. This is now a homogeneous, linear integral equation, and has nontrivial solutions only for certain values of the eigenvalue β_c. Since β_c occurs nonlinearly in the kernel, this is not a Fredholm equation; nevertheless, under reasonable conditions on the kernel, eigenvalues and eigenfunctions still exist. If $v_{kk'}$ is *negative*, the eigenvalues will be *positive* (note that the dominant region of k' is near the Fermi surface, that is, $\varepsilon_{k'} \approx \mu$). Then the *lowest* eigenvalue β_c corresponds to the *highest* temperature T_c for which the linear equation (7.21), hence the nonlinear equation (7.19), has a nontrivial solution. The gap function Δ_k and the pairing amplitude $\eta_k = \langle a_{-k}a_k \rangle_0$ are then nonzero for $T < T_c$, and may be interpreted as order parameters.

A rigorous justification of the identification of the transition temperature $T_c = (k\beta_c)^{-1}$ with the lowest nonnegative eigenvalue β_c of the linear integral equation (7.21) could probably be made on the basis of some theorems of Hammerstein[71] on the existence and multiplicity of solutions of certain classes of nonlinear integral equations, but to our knowledge, the details of such a proof have not been carried out. The application of the Hammerstein theorems to the closely related problem of condensation in an interacting Fermi gas will be discussed in Section VIIB (immediately following).

B. Condensation in an Interacting Fermi Gas

Gartenhaus, Stranahan, and Andersen[72,73] have shown that a model of the gas–liquid condensation in an interacting Fermi gas can be obtained by application of the Peierls (or equivalently, Gibbs-Bogoliubov) inequality to the Hamiltonian (7.2) of a system of interacting fermions, choosing a trial Hamiltonian such that only the Hartree-Fock terms, that is, those with $(k_1 = k_3, k_2 = k_4)$ or $(k_1 = k_4, k_2 = k_3)$, contribute to the variational thermodynamic potential Ω_{var}. We thus decompose $H - \mu N$ as follows:

$$
\begin{aligned}
H - \mu N &= H_{\text{quas}} + H_{\text{fluc}} + U' \\
H_{\text{quas}} &= -\tfrac{1}{2} \sum_{kk'} u_{kk'} \xi_k \xi_{k'} + \sum_k \left(\varepsilon_k - \mu + \sum_{k'} u_{kk'} \xi_{k'} \right) a_k^\dagger a_k \\
H_{\text{fluc}} &= \tfrac{1}{2} \sum_{kk'} u_{kk'} (a_k^\dagger a_k - \xi_k)(a_{k'}^\dagger a_{k'} - \xi_{k'}) \\
U' &= \tfrac{1}{2} \sum_{k_1k_2k_3k_4}{}' v_{k_1k_2k_3k_4} a_{k_1}^\dagger a_{k_2}^\dagger a_{k_4} a_{k_3}
\end{aligned}
\tag{7.22}
$$

where $u_{kk'}$ is defined by (7.4), and the prime in the definition of U' implies that it does not contain the Hartree-Fock interaction terms already included in $H_{\text{quas}} + H_{\text{fluc}}$.

Clearly (7.22) differs from (7.3) only in that η_k is taken to be identically zero and the superconductive "pairing interaction" terms are included in U'. Thus the derivation goes through as in Section III, or by simplifying the derivation for the superconductive case, by putting φ_k identically zero. The minimum of the expression (7.14) with respect to the variational parameters ξ_k occurs when

$$\xi_k = \langle a_k{}^\dagger a_k \rangle_0 = f_k \equiv (e^{\beta\omega_k} + 1)^{-1} \tag{7.23}$$

where

$$\begin{aligned} \omega_k &= \varepsilon_k - \mu + s_k \\ s_k &= \sum_{k'} u_{kk'} f_{k'} \end{aligned} \tag{7.24}$$

As the basic nonlinear integral equation of the theory one may take the one for s_k obtained, in analogy with the derivation of (7.18), by substitution of (7.23) into (7.24):

$$s_k = \sum_{k'} \frac{u_{kk'}}{e^{\beta\omega_{k'}} + 1} \tag{7.25}$$

with the top equation (7.24) substituted for $\omega_{k'}$. Alternatively, following Gartenhaus and Stranahan,[72] one may regard

$$f_k = \left\{ \exp\left[\beta\left(\varepsilon_k - \mu + \sum_{k'} u_{kk'} f_{k'}\right)\right] + 1 \right\}^{-1} \tag{7.26}$$

as the basic nonlinear integral equation for f_k. A third possibility is to take the top equation (7.24), with s_k expressed in terms of the $\omega_{k'}$ via (7.23) and the second equation (7.24), as the basic nonlinear integral equation for ω_k.

Regardless of which of these three descriptions is chosen, the description of possible phase transitions is more complicated than in the case of the BCS superconductive transition, since trivial possibilities such as $s_k = 0$ or $\omega_k = \varepsilon_k - \mu$ or $f_k = 0$ are easily seen *not* to satisfy the integral equations, so that the Bogoliubov argument[70] leading to (7.21) cannot be applied without modification. Thus, for example, $s_k = 0$ is in general compatible with (7.25) only if $u_{kk'} = 0$ (*ideal* Fermi gas). Nevertheless, considerable progress in analytically characterizing the multiplicity of solutions was made by Gartenhaus and Stranahan[72] by use of the Hammerstein theorems.[71] Assuming the hermitian matrix $u_{kk'}$ negative definite, a condition (attractive interaction) which clearly favors a gas–liquid transition, these authors showed that if the temperature T satisfies

$$kT > \tfrac{1}{4}\lambda_1^{-1} \tag{7.27}$$

where $\lambda_1 > 0$ is the lowest eigenvalue of the linear integral equation

$$\psi_k = -\lambda \sum_{k'} u_{kk'} \psi_{k'} \tag{7.28}$$

(recall that $u_{kk'}$ is assumed *negative* definite), then (7.26) has exactly one solution f_k. The physical interpretation is that the temperature region (7.27) certainly is in the gaseous phase, and the value of s_k ($\neq 0$) simply gives the self-consistent shift of single-particle energies due to interactions. For the very special (and rather trivial) case $\varepsilon_k = 0$, $u_{kk'} = -\text{Const}\ \delta_{kk'}$ (Mermin model[74]), the system does in fact have a gas–liquid phase transition exactly at the critical temperature T_c given by (7.27), that is, $kT_c = \frac{1}{4}\lambda_1^{-1}$. However, in general (7.27) only serves to *exclude* the possibility of a phase transition for $kT > \frac{1}{4}\lambda_1^{-1}$, that is, *if a phase transition exists*, it will surely be at a temperature T_c satisfying $kT_c \leqslant \frac{1}{4}\lambda_1^{-1}$.

Denote the *unique* solution of (7.26) for the case $kT > \frac{1}{4}\lambda_1^{-1}$ by $f_k^{(\text{gas})}$. Then, by numerical solution if necessary, one can analytically continue this solution down into the region $kT < \frac{1}{4}\lambda_1^{-1}$. At temperatures such that a liquid phase actually exists, the solution $f_k^{(\text{gas})}$ then represents the analytic continuation of the gaseous phase into the liquid region. Using the Hammerstein theorems, Gartenhaus and Stranahan succeeded in establishing a very useful sufficient condition for the existence of multiple solutions of (7.26) [see (20), (27), and (28) of Ref. 71], as follows. Consider the eigenfunctions ψ_k and eigenvalues ε of the homogeneous, *linear* integral equation

$$\psi_k = \varepsilon\beta \sum_{k'} u_{kk'} f_{k'}^{\text{gas}}(1 - f_{k'}^{\text{gas}})\psi_{k'} \tag{7.29}$$

in which $f_k^{(\text{gas})}$ is regarded as known, and the ψ_k and ε, which depend on β, are to be determined. Then if, at the temperature $T = (k\beta)^{-1}$, (7.29) has one or more eigenvalues ε satisfying $0 > \varepsilon > -1$, the nonlinear integral equation (7.26) *must* have *at least two* solutions f_k *in addition to* the "gaseous solution" $f_k^{(\text{gas})}$ [strictly speaking, $f_k^{(\text{gas})}$ in (7.29) can be *any* known solution, but in applications one will generally want to choose it to be the analytic continuation of the gaseous solution]. Furthermore, the thermodynamic potential Ω_{var} is necessarily *lower* for at least one of these "nongaseous" solutions than it is for $f_k^{(\text{gas})}$. This *ensures* the existence of a phase transition under suitable conditions. For the special case of the Mermin model, Gartenhaus and Stranahan show that for $kT < \frac{1}{4}\lambda_1^{-1}$ there are precisely three solutions, one ($f_k^{(\text{gas})}$) representing the analytic continuation of the gaseous phase, another ($f_k^{(\text{liquid})}$, the one of lowest free energy) representing the liquid phase, and the third being thermodynamically unstable in the sense of having negative compressibility. Quite generally, Gartenhaus and Stranahan show that the limiting eigenvalue $\varepsilon = -1$ of (7.29) (which can occur only at one value of β and μ, if at all) corresponds to the critical point, in that the density fluctuations are *necessarily* infinite at that point.

In further, more detailed numerical work with a simple two-parameter choice of $u_{kk'}$, Andersen, Gartenhaus, and Stranahan[73] verified these general conclusions and derived explicit thermodynamic functions and phase diagrams. They were even able to obtain reasonable agreement with the experimental data on the gas–liquid transition of He^3 by suitable choice of the parameters. The reader is referred to Refs. 72 and 73 for further details of the work, particularly with regard to the representation and interpretation of the multiple solutions and the connection with Maxwell's equal-area rule.

Since the criterion for multiple solutions of (7.26) based on (7.29) is only a *sufficient* condition, it does not tell us directly the critical temperature T_c at which multiple solutions first appear, which represents physically the gas–liquid condensation temperature (for given chemical potential μ). One might be tempted, by refining the bifurcation argument[70] by which (7.21) was derived, to try to derive a linear, homogeneous integral equation of which $\beta_c = (kT_c)^{-1}$ is the lowest eigenvalue. However, such an attempt is doomed to failure. The reason is that, in contrast to the superconducting transition, which is of second order, the gas–liquid transition is of first order. Hence the gaseous solution $f_k^{(\text{gas})}$ and liquid solution $f_k^{(\text{liquid})}$ are *not* in general equal at the transition temperature (for given μ), and the bifurcation argument fails. Only at the critical point (which occurs only at a *particular* value, μ_{crit}, of the chemical potential) do $f_k^{(\text{gas})}$ and $f_k^{(\text{liquid})}$ become equal. The bifurcation argument can indeed be applied to obtain a criterion for the critical temperature in terms of the lowest eigenvalue of a homogeneous, linear integral equation, but this criterion for the critical point is equivalent to that already given in terms of (7.29) by Gartenhaus and Stranahan.

C. Localized Ferromagnetism

The simplest model of a ferromagnet is the Weiss molecular field model. It is well known that the results of the Weiss theory can in fact be derived variationally. Our description here will therefore be very brief, intended merely to establish the connection with our general treatment.

Consider an Ising model (arbitrary lattice, arbitrary dimensionality) in zero external field. The Hamiltonian is

$$H = -J \sum_{\langle ij \rangle} \mu_i \mu_j \,, \qquad \mu_i = \pm 1 \tag{7.30}$$

The sum is over nearest-neighbor pairs. Let us choose

$$H_0 = -\mathscr{H} \sum_i \mu_i \tag{7.31}$$

where $\mathscr{H}$ has the significance of some effective magnetic field times an appropriate magneton. If there are N sites, each with z neighbors, we easily find

$$A_0 = -\frac{N}{\beta}[\ln z + \ln \cosh(\beta\mathscr{H})]$$
$$\langle H - H_0\rangle_0 = -\frac{zNJ}{2}\tanh^2(\beta\mathscr{H}) + N\mathscr{H}\tanh(\beta\mathscr{H}) \tag{7.32}$$

If we now minimize $A_0 + \langle H - H_0\rangle_0$ with respect to $\mathscr{H}$ we find

$$-N\tanh(\beta\mathscr{H}) - NzJ\beta\tanh(\beta\mathscr{H})\,\text{sech}^2(\beta\mathscr{H}) + N\tanh(\beta\mathscr{H}) + N\beta\mathscr{H}\,\text{sech}^2(\beta\mathscr{H}) = 0 \tag{7.33}$$

or

$$\tanh(\beta\mathscr{H}) = \frac{\mathscr{H}}{zJ} \tag{7.34}$$

which determines $\mathscr{H}$. We can put this in another form by noting that $\langle\mu\rangle_0 = \tanh(\beta\mathscr{H})$. Hence $\mathscr{H} = zJ\langle\mu\rangle_0$ and

$$\tanh(z\beta J\langle\mu\rangle_0) = \langle\mu\rangle_0 \tag{7.35}$$

which will immediately be recognized as the basic equation of the Bragg-Williams approximation. These considerations may be easily generalized, with little extra effort, to the case of an Ising model in an external field, or to a Heisenberg model, but it would not be particularly useful to do so here. The thing to be learned from this exercise is that the Bragg-Williams approximation is the best effective field approximation one can make, in the sense of approximating the free energy by a system of independent spins. Note that, in this case, $\langle H - H_0\rangle_0 \neq 0$, so that A_0, by itself, is not necessarily a bound on A. This is also a feature of the conventional derivations of the Bragg-Williams theory.

One generalization of (7.30) that is instructive to consider, from the point of view of relating the Curie temperature T_c to the lowest eigenvalue of a linear integral equation, is

$$H = -\tfrac{1}{2}\sum_{ij} J_{ij}\mu_i\mu_j \tag{7.36}$$

Then (7.30) is the special case $J_{ij} = J$, i and j nearest neighbors; $J_{ij} = 0$, otherwise. The obvious generalization of (7.31) is then

$$H_0 = -\sum_i \mathscr{H}_i\mu_i \tag{7.37}$$

One finds easily, assuming $J_{ii} = 0$ and $J_{ji} = J_{ij}$

$$A_0 = -\beta^{-1}\Big[N \ln z + \sum_i \ln \cosh (\beta\mathscr{H}_i)\Big]$$
$$\langle H - H_0\rangle_0 = -\sum_i \tanh (\beta\mathscr{H}_i)\Big[\mathscr{H}_i - \tfrac{1}{2}\sum_j J_{ij} \tanh (\beta\mathscr{H}_j)\Big] \tag{7.38}$$

The generalization of the variational equation (7.34) is then easily found to be

$$\mathscr{H}_i = \sum_j J_{ij} \tanh (\beta\mathscr{H}_j) \tag{7.39}$$

which is analogous to the nonlinear integral equation (7.19) of the theory of superconductivity and that (7.26) of the Gartenhaus-Stranahan theory of condensation. Since there is no mean magnetization for $T > T_c$ in the absence of an external field (i.e., the transition is of second order), the $\mathscr{H}_i$ vanish at $T = T_c$ so that the previously discussed bifurcation argument[70] applies here, with the conclusion that $T_c = (k\beta_c)^{-1}$ where β_c is the lowest nonnegative eigenvalue of the linear, homogeneous "integral equation"

$$\mathscr{H}_i = \beta_c \sum_j J_{ij}\mathscr{H}_j \tag{7.40}$$

In the ferromagnetic and translationally invariant case (J_{ij} = function of $i - j$), the lowest eigenfunction is $\mathscr{H}_i = \mathscr{H}$ (independent of i), but for suitable J_{ij} the lowest solution may be antiferromagnetic or with even more complicated magnetic ordering. Equation (7.34) is a degenerate case of the nonlinear "integral equation" (7.39), and the corresponding linearized equation $\mathscr{H} = \beta_c zJ\mathscr{H}$ is a degenerate case of (7.40), with the single eigenvalue $\beta_c = (zJ)^{-1}$ which is in fact the Curie β in the Bragg-Williams approximation. One (well-known) defect of this approximation is that a phase transition is predicted even in one dimension, in which case it is known that the exact free energy is an analytic function of temperature, at least for J_{ij} of finite range. Finally, we note that the nonvanishing of $\langle H - H_0\rangle_0$ (more precisely, the fact that it is of order N and hence nonnegligible) is related to the fact that the indices i, j refer to lattice sites, and hence J_{ij} is independent of N. In contrast, the corresponding itinerant model (in which i, j refer to momenta) has J_{ij} inversely proportional to N, hence $\langle H - H_0\rangle_0$ independent of N and thus thermodynamically negligible, in analogy with the situation in the theory of superconductivity and the Gartenhaus-Stranahan theory of condensation.

The Gibbs-Bogoliubov inequality has also been applied[75] to investigation of the thermodynamics of the anisotropic Heisenberg linear chain

$$H = -2J \sum_{j=1}^{N-1} [(1 + \gamma)S_j^x S_{j+1}^x + (1 - \gamma)(S_j^y S_{j+1}^y + S_j^z S_{j+1}^z)] \tag{7.41}$$

taking the anisotropic linear XY model, which is exactly soluble,[76,77] as the trial Hamiltonian H_0:

$$H_0 = -2J' \sum_{j=1}^{N-1} [(1 + \gamma')S_j^x S_{j+1}^x + (1 - \gamma')S_j^y S_{j+1}^y] \qquad (7.42)$$

The variational parameters are then the exchange coupling J' and anisotropy γ' of the XY model. Since the ground state energy of H is known both in the ferromagnetic and antiferromagnetic cases, one can compare the variational approximation with the exact result at $T = 0$, and it is found that the error is less than 8% for all γ in the interval $-1 \leqslant \gamma \leqslant 1$ (the error vanishes identically in γ in the interval $0 \leqslant \gamma \leqslant 1$ for the ferromagnetic case). However, since exact results are not known for $T > 0$ and lower bounds were not derived, the error is not known for general values of the temperature. The variational results show no phase transition for $\gamma \neq 0$ but a second-order phase transition at $kT = 0.452\,J$ in the case $\gamma = 0$. Since, however, the one-dimensional Heisenberg model has no phase transition, this result at $\gamma = 0$ is a spurious one analogous to the spurious phase transition of the Bragg-Williams approximation in the one-dimensional case.

D. Itinerant Ferromagnetism

The simplest itinerant ferromagnetic model is that of the Stoner theory.[78] Suppose that in (7.2) k stands for both the wave vector $\mathbf{k}$ and spin index $\sigma(= \uparrow \text{ or } \downarrow)$ of an electron, and write (7.2) in the more explicit form

$$H = \sum_{\mathbf{k}\sigma} \varepsilon_{\mathbf{k}} a_{\mathbf{k}\sigma}{}^\dagger a_{\mathbf{k}\sigma} + \tfrac{1}{2} \sum_{\mathbf{k}_1\mathbf{k}_2\mathbf{k}_3\mathbf{k}_4,\sigma\sigma'} v_{\mathbf{k}_1\mathbf{k}_2\mathbf{k}_3\mathbf{k}_4}\, a_{\mathbf{k}_1\sigma}{}^\dagger a_{\mathbf{k}_2\sigma'}{}^\dagger a_{\mathbf{k}_4\sigma'}\, a_{\mathbf{k}_3\sigma} \qquad (7.43)$$

where it is assumed that the single-particle energies $\varepsilon_{\mathbf{k}}$ and interaction matrix elements $v_{k_1k_2k_3k_4}$ are spin-independent. The Stoner theory is essentially a temperature-dependent Hartree-Fock treatment of (7.43) based on taking into account only the diagonal interaction terms, which consist of direct terms ($\mathbf{k}_1 = \mathbf{k}_3$, $\mathbf{k}_2 = \mathbf{k}_4$) and exchange terms ($\mathbf{k}_1 = \mathbf{k}_4$, $\mathbf{k}_2 = \mathbf{k}_3$, $\sigma = \sigma'$). The corresponding diagonal part, H_{diag}, of H can be rearranged into the form

$$H_{\text{diag}} = \sum_{\mathbf{k}} \varepsilon_{\mathbf{k}} N_{\mathbf{k}} + \tfrac{1}{2} \sum_{\mathbf{k}\mathbf{k}'} I_{\mathbf{k}\mathbf{k}'} N_{\mathbf{k}} N_{\mathbf{k}'} - \tfrac{1}{2} \sum_{\mathbf{k}\mathbf{k}'} J_{\mathbf{k}\mathbf{k}'} (\tfrac{1}{2} N_{\mathbf{k}} N_{\mathbf{k}'} + 2S_{\mathbf{k}}{}^z S_{\mathbf{k}'}{}^z) \qquad (7.44)$$

where

$$N_{\mathbf{k}} = a_{\mathbf{k}\uparrow}{}^\dagger a_{\mathbf{k}\uparrow} + a_{\mathbf{k}\downarrow}{}^\dagger a_{\mathbf{k}\downarrow}$$
$$S_{\mathbf{k}}{}^z = \tfrac{1}{2}(a_{\mathbf{k}\uparrow}^\dagger a_{\mathbf{k}\uparrow} - a_{\mathbf{k}\downarrow}^\dagger a_{\mathbf{k}\downarrow}) \qquad (7.45)$$

and the direct coupling $I_{\mathbf{k}\mathbf{k}'}$ and exchange coupling $J_{\mathbf{k}\mathbf{k}'}$ are

$$I_{\mathbf{k}\mathbf{k}'} = v_{\mathbf{k}\mathbf{k}'\mathbf{k}\mathbf{k}'}, \qquad J_{\mathbf{k}\mathbf{k}'} = v_{\mathbf{k}\mathbf{k}'\mathbf{k}'\mathbf{k}} \qquad (7.46)$$

One can then decompose (7.43) in analogy with (7.3) and (7.22):

$$H - \mu N = H_{\text{quas}} + H_{\text{fluc}} + U'$$

$$H_{\text{quas}} = \sum_{\mathbf{k}} (\omega_{\mathbf{k}} N_{\mathbf{k}} + w_{\mathbf{k}} S_{\mathbf{k}}{}^{z})$$

$$H_{\text{fluc}} = \tfrac{1}{2} \sum_{\mathbf{kk}'} I_{\mathbf{kk}'}(N_{\mathbf{k}} - \xi_{\mathbf{k}})(N_{\mathbf{k}'} - \xi_{\mathbf{k}'}) - \tfrac{1}{2} \sum_{\mathbf{kk}'} J_{\mathbf{kk}'}[\tfrac{1}{2}(N_{\mathbf{k}} - \xi_{\mathbf{k}})(N_{\mathbf{k}'} - \xi_{\mathbf{k}'}) + 2(S_{\mathbf{k}}{}^{z} - \eta_{\mathbf{k}})(S_{\mathbf{k}'}{}^{z} - \eta_{\mathbf{k}'})] \tag{7.47}$$

where $U'(= H - H_{\text{diag}})$ is the off-diagonal part of the interaction terms in (7.43) (those not included in H_{diag}),

$$\begin{aligned} \omega_{\mathbf{k}} &= \varepsilon_{\mathbf{k}} - \mu + \sum_{\mathbf{k}'} (I_{\mathbf{kk}'} - \tfrac{1}{2} J_{\mathbf{kk}'})\xi_{\mathbf{k}'} \\ w_{\mathbf{k}} &= -2 \sum_{\mathbf{k}'} J_{\mathbf{kk}'} \eta_{\mathbf{k}'} \end{aligned} \tag{7.48}$$

and the $\xi_{\mathbf{k}}$ and $\eta_{\mathbf{k}}$ are real variational parameters to be determined by minimization of the bound (7.14), in which, as before, $\langle U' \rangle_0$ vanishes. The thermodynamic potential, Ω_0, of H_{quas} is easily found to be

$$\Omega_0 = -\beta^{-1} \sum_{\mathbf{k}} \ln [1 + e^{-2\beta\omega_{\mathbf{k}}} + 2e^{-\beta\omega_{\mathbf{k}}} \cosh (\tfrac{1}{2}\beta w_{\mathbf{k}})] \tag{7.49}$$

As before, the minimum condition is that

$$\xi_{\mathbf{k}} = \langle N_{\mathbf{k}} \rangle_0 , \qquad \eta_{\mathbf{k}} = \langle S_{\mathbf{k}}{}^{z} \rangle_0 \tag{7.50}$$

Substituting the values of these thermal averages evaluated by differentiation of Ω, one derives the nonlinear integral equations

$$\begin{aligned} \omega_{\mathbf{k}} &= \varepsilon_{\mathbf{k}} - \mu + \sum_{\mathbf{k}'} (I_{\mathbf{kk}'} - \tfrac{1}{2} J_{\mathbf{kk}'}) \left[\frac{e^{-\beta\omega_{\mathbf{k}}'} + \cosh (\tfrac{1}{2}\beta w_{\mathbf{k}'})}{\cosh (\beta\omega_{\mathbf{k}'}) + \cosh (\tfrac{1}{2}\beta w_{\mathbf{k}'})} \right] \\ w_{\mathbf{k}} &= \sum_{\mathbf{k}'} J_{\mathbf{kk}'} \left[\frac{\sinh (\tfrac{1}{2}\beta w_{\mathbf{k}'})}{\cosh (\beta\omega_{\mathbf{k}'}) + \cosh (\tfrac{1}{2}\beta w_{\mathbf{k}'})} \right] \end{aligned} \tag{7.51}$$

For suitable values of the $I_{\mathbf{kk}'}$ and $J_{\mathbf{kk}'}$, the system will undergo a ferromagnetic transition at some temperature T_c, with $w_{\mathbf{k}} = 0$ for $T > T_c$ but $w_k \neq 0$ for $T < T_c$. The bifurcation argument can then be applied as before, with the conclusion that β_c, the Curie β, is the lowest (nonnegative) eigenvalue of the linear integral equation

$$w_{\mathbf{k}} = \tfrac{1}{2}\beta_c \sum_{\mathbf{k}'} J_{\mathbf{kk}'}[1 + \cosh (\beta_c \omega_{\mathbf{k}'})]^{-1} w_{\mathbf{k}'} \tag{7.52}$$

[note that the eigenvalue β_c also occurs in cosh $(\beta_c\omega_{\mathbf{k}'})$]. Strictly speaking, since the nonlinear equations (7.51) are coupled, one should really solve (7.52) simultaneously with the top equation (7.51) (with $\beta \to \beta_c$ and $w_{\mathbf{k}} \to 0$). However, if one is interested only in qualitative features of the transition, one can make customary approximations (note that only the behavior of $\omega_{\mathbf{k}}$ near the Fermi surface is relevant), in which case $\omega_{\mathbf{k}}$ can be regarded as known. For further details of the Hartree-Fock-Stoner theory, the reader is referred to the voluminous literature; our purpose here was only to establish the connection with the Gibbs-Bogoliubov variational method and the other closely related applications already discussed.

In view of the analogy of the interaction terms $J_{\mathbf{kk}'}S_{\mathbf{k}}^zS_{\mathbf{k}'}^z$ in (7.44) with the Ising model, one is tempted to generalize these terms to obtain an "itinerant Heisenberg model," by including not only those interaction terms in (7.43) expressible in terms of $S_{\mathbf{k}}^zS_{\mathbf{k}'}^z$, but also those expressible in terms of $S_{\mathbf{k}}^xS_{\mathbf{k}'}^x$ and $S_{\mathbf{k}}^yS_{\mathbf{k}'}^y$, where

$$\begin{aligned} S_{\mathbf{k}}^+ &= S_{\mathbf{k}}^x + iS_{\mathbf{k}}^y = a_{\mathbf{k}\uparrow}^\dagger a_{\mathbf{k}\downarrow} \\ S_{\mathbf{k}}^- &= S_{\mathbf{k}}^x - iS_{\mathbf{k}}^y = a_{\mathbf{k}\downarrow}^\dagger a_{\mathbf{k}\uparrow} \end{aligned} \tag{7.53}$$

[Note that $S_{\mathbf{k}}^x$, $S_{\mathbf{k}}^y$, and $S_{\mathbf{k}}^z$ satisfy spin commutation relations, but, since $S_{\mathbf{k}}^z$ can take on the value 0 as well as $\pm\frac{1}{2}$, they do not satisfy $\mathbf{S}_{\mathbf{k}}^2 = \frac{1}{2}(\frac{1}{2}+1)$]. Such a generalization has been carried out.[79] The resultant interaction terms differ from those in (7.44) only in the replacement of $S_{\mathbf{k}}^zS_{\mathbf{k}'}^z$ by $\mathbf{S}_{\mathbf{k}} \cdot \mathbf{S}_{\mathbf{k}'}$, where $\mathbf{S}_{\mathbf{k}}$ is the vector operator with components $S_{\mathbf{k}}^x$, $S_{\mathbf{k}}^y$, and $S_{\mathbf{k}}^z$. The variational theory then goes through with comparatively minor modifications, the main differences being that $w_{\mathbf{k}}$ is replaced by a vector function $\mathbf{w}_{\mathbf{k}}$ and that H_{quas} is not diagonal in the $a_{\mathbf{k}\sigma}$, $a_{\mathbf{k}\sigma}^\dagger$ representation, but has to be diagonalized by a canonical transformation analogous to the Bogoliubov-Valatin transformation (7.5) of superconductivity theory. It is found (not surprisingly) that in the absence of an external field the solution is highly degenerate, in that the vector $\mathbf{w}_{\mathbf{k}}$ has a direction which is independent of $\mathbf{k}$, with all possible choices of this direction giving the same value of the thermodynamic potential. If one chooses $\mathbf{w}_{\mathbf{k}}$ in the z-direction, then it is found that the theory reduces to that based on only the $z - z$ interactions [Eq. (7.44)], so that the generalization to $\mathbf{S}_{\mathbf{k}} \cdot \mathbf{S}_{\mathbf{k}'}$ does not in fact produce a lower thermodynamic potential. The reason for this behavior is not completely understood, but may be related to the fact that in an effective-field theory in $\mathbf{k}$ space, quantum fluctuations of the operators $\mathbf{S}_{\mathbf{k}}$ are negligible (recall that $\langle H_{\text{fluc}}\rangle_0 = 0$ at the minimum). However, just such quantum fluctuations are known to be responsible for the difference between the Ising and Heisenberg models.

E. Boson Models of Liquid Helium

Suppose now that the indices $k, k', \ldots$ in (7.2) stand merely for wave vectors $\mathbf{k}, \mathbf{k}', \ldots$, and that $a_\mathbf{k}$ and $a_\mathbf{k}^\dagger$ are annihilation and creation operators for bosons of wave vector $\mathbf{k}$. By selecting out just the same interaction terms (except that spins are now absent) as in the BCS theory, one obtains a "pair model" of liquid He^4, based on the idea that both Hartree-Fock terms and pairing terms are important.[80] The treatment of the thermodynamics of this model is in one–one correspondence with our previous description of the treatment of the BCS model based on (7.3), except for the replacement of Fermi functions by Bose functions and the (very important) fact that in the Bose system a Bose-Einstein condensation occurs, necessitating a special treatment of the $\mathbf{k} = 0$ mode. The thermal properties of this model were investigated, by this method, by Wentzel,[26] and more detailed investigations were made by Luban[81] and a number of subsequent investigators. Although the physical interpretation is quite different from that of the BCS model, the mathematics of the derivation of the nonlinear integral equations is almost identical, so we shall not discuss the details of this model here. It is found that the system does exhibit a phase transition (Bose-Einstein condensation modified by interaction effects), with both the Bose-Einstein condensate density $N^{-1}\langle a_0^\dagger a_0\rangle_0$ and the "pairing order parameter" $\langle a_\mathbf{k}^\dagger a_{-\mathbf{k}}^\dagger\rangle_0$ vanishing for $T > T_c$, and both nonzero for $T < T_c$. The details of the phase transition (logarithmic specific heat singularity for real liquid He^4) are not given correctly by the variational theory. This is again typical of effective field theories, as, for example, in the case of Onsager's exact solution of the two-dimensional Ising model vs. the Bragg-Williams approximation.

F. Isotope Mixtures

The quantum-mechanical applications so far discussed are based on various assumptions (depending on the application) regarding the important interparticle interaction terms, with the physical effects of these terms then incorporated into the trial Hamiltonian H_0 through the use of generalized self-consistent fields. A rather different approach, which does not depend at all on details of the interparticle interactions, has been applied to the statistical mechanics of isotope mixtures by Byrns and Mazo.[82] Consider a mixture of ν different species. The Hamiltonian is then of the form

$$H = \sum_{\alpha=1}^{\nu} \sum_{i=1}^{n_\alpha} \frac{p_i^2}{2m_\alpha} + U \tag{7.54}$$

where n_α is the number of molecules of species α, and p_i is the quantum-mechanical momentum operator of the ith particle. The only property of the intermolecular interaction U which is essential to the argument is that the interspecies and intraspecies interactions are all the same; this restricts the theory to isotopic mixtures. Then it is natural to choose a trial Hamiltonian H_0 describing $N = \sum_{\alpha} n_\alpha$ molecules of equal mass m_0, with the same intermolecular potential as for the actual system:

$$H_0 = \sum_{i=1}^{N} \frac{p_i^{\,2}}{2m_0} + U \tag{7.55}$$

One can then apply the Gibbs-Bogoliubov inequality (2.1), and the potential energy U cancels between H and H_0:

$$\begin{aligned} A \leqslant A_{\text{var}} &= A_0 + \langle H - H_0 \rangle_0 \\ &= A_0 + \sum_{\alpha=1}^{\nu} \sum_{i=1}^{n_\alpha} \left(\frac{m_0}{m_\alpha} - 1 \right) \left\langle \frac{p_i^{\,2}}{2m_0} \right\rangle_0 \end{aligned} \tag{7.56}$$

For the real system, the trace should be taken with respect to a set of states having the permutation symmetry appropriate to the particular mixture being considered. In principle, the traces in (7.56) should be taken with respect to states having this symmetry, also. Byrns and Mazo assume, however, that the symmetry properties of the wave functions are unimportant; effects due to particle statistics are known to be small for pure liquids (except helium) even when $\hbar$ cannot be treated as small.

Making this assumption, one has

$$\langle p_i^{\,2} \rangle_0 = \langle p^2 \rangle_0 \quad \text{(independent of } i\text{)} \tag{7.57}$$

Granting this, one can easily show, with the aid of the relation

$$\frac{\partial A_0}{\partial m_0} = \left\langle \frac{\partial H_0}{\partial m_0} \right\rangle_0 \tag{7.58}$$

that the necessary condition for a minimum of A_{var} with respect to variation of m_0 is

$$\frac{\partial A_{\text{var}}}{\partial m_0} = N m_0 \left(\sum_{\alpha} \frac{x_\alpha}{m_\alpha} - \frac{1}{m_0} \right) \left(\frac{1}{m_0} + \frac{\partial}{\partial m_0} \right) \left\langle \frac{p^2}{m_0} \right\rangle_0 = 0 \tag{7.59}$$

If the factor $[(1/m_0) + (\partial/\partial m_0)]\langle p^2/2m_0 \rangle_0$ vanishes for any value of m_0, this is almost certainly not the solution of minimum A_{var}, since $\langle p^2 \rangle_0$ is independent of the m_α and x_α and would then yield an m_0 independent of these parameters. Thus the desired solution for m_0 is

$$\frac{1}{m_0} = \sum_{\alpha} \frac{x_\alpha}{m_\alpha} \tag{7.60}$$

Since

$$\langle H - H_0\rangle_0 = Nm_0\left\langle \frac{p^2}{2m_0}\right\rangle_0 \left(\sum_\alpha \frac{x_\alpha}{m_\alpha} - \frac{1}{m_0}\right) \tag{7.61}$$

the trial free energy A_{var} reduces simply to that A_0 of the fictitious pure component at the value of m_0 given by (7.60).

The results of the theory can be compared with experiment using the law of corresponding states which follows from the assumptions already made. The comparison for H_2–D_2 mixtures showed[82] poor agreement. Since the assumption (7.57) is presumably not a significant source of error at the temperature where the comparison was made, the error must be a reflection of the fact that a trial Hamiltonian of the form (7.55) is not capable of simulating the actual Hamiltonian well for any choice of m_0, at least in the sense that no choice of m_0 brings A_{var} close to the actual free energy A of the H_2–D_2 mixtures considered. It is possible, however, that some bound other than the Gibbs-Bogoliubov one (2.1) may in fact yield a good approximation to A for the *same* trial Hamiltonian H_0 and a *different* choice of m_0. In fact, Byrns and Mazo give some arguments in justification of the hypothesis that the Golden-Thompson *lower* bound (4.1), with the following different choice of m_0:

$$m_0 = \prod_\alpha m_\alpha^{x_\alpha} \tag{7.62}$$

also implies $A_{\text{var}} \approx A_0$, and for this choice of m_0 they find (Table I of Ref. 82) reasonable agreement with experiment. A rigorous justification of the neglect of the term $\langle \exp[-\beta(H - H_0)]\rangle_0$ in (4.1) could not, however, be made, since this term cannot be evaluated in closed form.

This completes our discussion of quantum-mechanical applications. Our discussion should not in any sense be regarded as an exhaustive review of such applications. Instead, we have chosen a few examples with the intent of illustrating techniques of more general applicability. To date, very little use has been made of trial Hamiltonians H_0 which include realistic interparticle forces. It is to be expected that as sufficiently accurate computer data on thermodynamic properties of certain model interacting systems (e.g., hard spheres) is accumulated, better variational bounds on real systems can be obtained by use of such interacting model systems as trial Hamiltonians H_0. Such applications should not be subject to the inherent defects (neglect of correlations and fluctuations) of effective-field theories, of which most of the applications we have discussed are examples.

VIII. CONCLUSIONS

In this article we have reviewed the variational bounds which are available for the approximate computation of several thermodynamic potentials, and have discussed some of the applications which have been made of these methods. In several places in the text we have already alluded to some general conclusions, but let us now state our conclusions explicitly.

Suppose the physical situation which is under discussion is dominated by a self-consistent field mechanism. That is, suppose the state of a given sub-system depends primarily on the average interaction with its environment. Then the variational method seems to be a most valuable and useful tool for finding the equations describing this interaction.

On the other hand, when the physical situation is such that correlations dominate the phenomena of interest, then the variational method, *with the trial Hamiltonians that one knows how to handle at the present time*, seems to be much less useful. This is a relative statement, of course. For example, in the BCS theory, the correlations between electrons of opposite momenta and spin, which lie at the heart of superconductivity, are taken into account in the variational ansatz. That is why it works! But we do not yet know how to do this in many other cases. In particular, we are fairly helpless when the important correlations are in coordinate space, rather than momentum space.

But this is not a time-invariant state of affairs. Theoretical physicists and chemists are learning how to handle more complex correlations as time goes on. As we become more expert and more experienced, we can expect the variational method to become an ever more useful tool. Perhaps some day it may even be as important for the N-body problem as it has been for years in the few-body problem.

APPENDIX. CONVEX FUNCTIONS

The theory of convex functions is a very important tool for proving inequalities. Although the theory is reasonably well known, we thought it would be useful to give a compendium of the principal results without proofs. The reader is referred to Hardy, Littlewood, and Polya[35] for further details.

Definition. *A function, $f(x)$, defined in the interval $a < x < b$ is said to be convex in the interval if for any x_1, x_2 in the interval, and for any λ, $0 \leqslant \lambda \leqslant 1$*

$$f(\lambda x_1 + (1 - \lambda)x_2) \leqslant \lambda f(x_1) + (1 - \lambda)f(x_2) \tag{A.1}$$

Geometrically, this means that the chord of the graph of f joining $f(x_1)$ and $f(x_2)$ always lies *above* the graph itself.

Definition. *f is said to be concave if $-f$ is convex.*

Note that the "convex" functions discussed in Section VA were really concave by the above definition. We kept the original wording of Fisher (who also noted that his usage was anomalous) in order not to confuse readers who wish to consult the original paper.

Convex functions are either continuous or very wild indeed. In fact, a discontinuous convex function is necessarily unbounded in every interval. No discontinuous convex function has ever been constructed or exhibited. Their existence has only been inferred from some rather deep results in set theory (axiom of choice). Hence for purposes of physics, we can safely assume that a convex function is continuous, and in the following, when we say "convex" we mean, senso strictu, "continuous convex."

Theorem. *Let $f(x)$ be convex in an open interval. Then at every interior point f has a right-hand and a left-hand derivative. The right-hand derivative is greater than or equal to the left-hand derivative, and both derivatives increase with X.*

This says nothing, of course, about an ordinary derivative; for such to exist, the right and left derivatives must be equal. But we have the

Theorem. *A convex function, f, has a derivative, f', except at perhaps an enumerable set of values of x.*

Perhaps the most important result on convex functions is the

Theorem. *Let $f(x)$ be convex in $a < x < b$. Let $p_1, \ldots, p_n$ be positive numbers such that $\sum p_i = 1$. Then*

$$f\left(\sum_{i=1}^{n} p_i x_i\right) \leqslant \sum_{i=1}^{n} p_i f(x_i)$$
$$a < x_i < b \tag{A.2}$$

This is sometimes called the *convex function theorem*.

Another important result is

Theorem. *Let $f(x)$ be continuous and possess a second derivative $f''(x)$ in an open interval. Then a necessary and sufficient condition for the convexity of f is $f''(x) \geqslant 0$.*

Thus, for example, e^x is convex in any interval.

These definitions and theorems obviously apply to functions of several variables, if one considers convexity in each variable separately. To consider convexity generally, in all variables simultaneously, we use the following definitions. Let $\mathbf{x} = \{x_1, \ldots, x_N\}$ be a point in N-dimensional Euclidean space.

Definition. *A domain D, in N-dimensional Euclidean space is convex if* $\mathbf{x} \in D$, $\mathbf{y} \in D$ *implies* $\lambda\mathbf{x} + (1 - \lambda)\mathbf{y} \in D$, $0 \leqslant \lambda \leqslant 1$.

Geometrically, a convex domain contains the whole of the straight line joining any two of its points.

Definition. *A function* $g(\mathbf{x}) = g(x_1, \ldots, x_N)$ *defined on a convex domain, D, is convex if*

$$g(\lambda\mathbf{x} + (1 - \lambda)\mathbf{y}) \leqslant \lambda g(\mathbf{x}) + (1 - \lambda)g(\mathbf{y}) \tag{A.3}$$

This says that any chord lies above the surface.

The analogue of the convex function theorem for this case is

Theorem. *If* $g(\mathbf{x})$ *is convex and continuous and* $p_i \geqslant 0$, $\sum p_i = 1$, *then*

$$g\left(\sum_{i=1}^{n} p_i\mathbf{x}^i\right) \leqslant \sum_{i=1}^{n} p_i g(\mathbf{x}^i) \tag{A.4}$$

where $\mathbf{x}^i = \{x_1{}^i, x_2{}^i, \ldots, x_N{}^i\}$.

The analogue of the second derivative theorem is

Theorem. *If* $g(\mathbf{x})$ *has second derivatives (with respect to all variables) in an open convex domain, D, then the necessary and sufficient condition that g be convex is that the quadratic form*

$$Q = \sum_{i,j=1}^{N} g_{ij}u_iu_j$$

be positive semidefinite at every point $\mathbf{x}$ *of D. Here* $g_{ij} = \partial^2 g/\partial x_i\partial x_j$.

Acknowledgment

We should like to thank Professor M. E. Fisher for permission to reproduce Fig. 1, and especially for sending his original drawing.

References

1. O. D. Kellog, *Foundations of Potential Theory*, Dover, New York, 1953, p. 277 (reprint).
2. See any textbook on quantum mechanics.
3. N. F. Mott and H. S. W. Massey, *The Theory of Atomic Collisions*, 3rd ed., Oxford University Press, Oxford, 1965.
4. J. Schwinger and D. S. Saxon, *Discontinuities in Wave Guides; Notes on Lectures by J. Schwinger*, Gordon and Breach, New York, 1968.
5. E. Gerjouy and D. S. Saxon, *Phys. Rev.*, **94,** 1445 (1954).
6. S. Okubo, *J. Math. Phys.*, **12,** 1123 (1971), and earlier work cited therein.
7. A. Huber, in *Methods and Problems of Theoretical Physics*, J. E. Bowcock, Ed., North-Holland, Amsterdam, 1970.

8. V. V. Tolmachev, *Dokl. Akad. Nauk. SSSR*, **134,** 1324 (1960) [English transl.: *Soviet Phys.* "*Doklady*," **5,** 984 (1961)].
9. M. D. Girardeau, *J. Chem. Phys.*, **40,** 899 (1964).
10. W. B. Brown, *J. Chem. Phys.*, **41,** 2945 (1964).
11. J. W. Gibbs, *Elementary Principles in Statistical Mechanics*, in *The Collected Works of J. Willard Gibbs*, Vol. II, Yale University Press, New Haven, 1948.
12. M. D. Girardeau, *J. Chem. Phys.*, **41,** 2945 (1964).
13. H. Koppe, in *Werner Heisenberg und die Physik Unserer Zeit*, F. Bopp, Ed., Vieweg, Braunsweig, 1961.
14. R. M. Wilcox, *J. Math. Phys.*, **8,** 962 (1967).
15. R. E. Peierls, *Phys. Rev.*, **54,** 918 (1938).
16. P. C. Jordan and S. Golden, *Phys. Rev.*, **169,** 215 (1968).
17. Further details of the approach described here can be found in M. Girardeau, *J. Math. Phys.*, **3,** 131 (1962).
18. J. von Neumann, *Math. Ann.*, **104,** 570 (1931).
19. P. Jordan and E. P. Wigner, *Z. Physik*, **47,** 631 (1928).
20. F. A. Berezin, *The Method of Second Quantization*, Academic Press, New York, 1966, Chaps. II and III.
21. K. O. Friedricks, *Mathematical Aspects of the Quantum Theory of Fields*, Wiley, New York, 1953.
22. See, for example, Berezin, Ref. 20, pp. 92 and 119.
23. T. Matsubara, *Progr. Theoret. Phys.* (*Kyoto*), **14,** 351 (1955).
24. C. Bloch and C. de Dominicis, *Nucl. Phys.*, **7,** 459 (1958).
25. N. N. Bogoliubov, D. B. Zubarev, and Iu. A. Tserkovnikov, *Soviet Phys.* "*Doklady*," **2,** 535 (1957).
26. G. Wentzel, *Phys. Rev.*, **120,** 1572 (1960).
27. M. D. Girardeau, *Phys. Rev.*, **140,** A1139 (1965).
28. S. Golden, *Phys. Rev.* B, **137,** 1127 (1965).
29. C. J. Thompson, *J. Math. Phys.*, **6,** 1812 (1965).
30. H. Trotter, *Pacific J. Math.*, **8,** 887 (1958).
31. E. Nelson, *J. Math. Phys.*, **5,** 332 (Appendix) (1964).
32. R. H. T. Yeh, *J. Math. Phys.*, **11,** 1521 (1970).
33. R. B. Griffiths, *J. Math. Phys.*, **5,** 1215 (1964).
34. D. Ruelle, *Helv. Phys. Acta*, **36,** 789 (1963).
35. G. H. Hardy, J. E. Littlewood, and G. Polya, *Inequalities*, Cambridge University Press, 1952, p. 22.
36. A. J. Coleman, *Rev. Mod. Phys.*, **35,** 668 (1963).
37. C. Garrod and J. K. Percus, *J. Math. Phys.*, **5,** 1756 (1964).
38. L. J. Kijewski and J. K. Percus, *Phys. Rev.*, **164,** 228 (1967); *J. Math. Phys.*, **8,** 2184 (1967).
39. L. J. Kijewski and J. K. Percus, *Phys. Rev.*, **179,** 45 (1969).
40. M. Rosina, J. K. Percus, L. J. Kijewski, and C. Garrod, *J. Math. Phys.*, **10,** 1761 (1969).
41. M. E. Fisher, *J. Chem. Phys.*, **42,** 3852 (1965).
42. M. E. Fisher, *Arch. Ration. Mech. Anal.*, **17,** 377 (1964).
43. H. Falk, *J. Math. Phys.*, **7,** 977 (1966).
44. R. H. T. Yeh, *J. Math. Phys.*, **12,** 2397 (1971).
45. J. E. Kilpatrick, *J. Chem. Phys.*, **21,** 274 (1953).
46. R. M. Mazo, *Mol. Phys.*, **8,** 515 (1964).
47. H. Leff, *Am. J. Phys.*, **37,** 548 (1969).

48. J. E. Lennard-Jones and A. F. Devonshire, *Proc. Roy. Soc. (London) Ser. A*, **169**, 317 (1939).
49. J. G. Kirkwood, *J. Chem. Phys.*, **18**, 380 (1950).
50. J. E. Mayer and G. Careri, *J. Chem. Phys.*, **20**, 1001 (1952).
51. W. G. Hoover, *J. Chem. Phys.*, **43**, 371 (1965).
52. C. Domb, *Advances in Chemical Physics*, K. E. Shuler, Ed., Wiley, New York, Vol. 15.
53. S. F. Edwards, *Proc. Phys. Soc. (London)*, **85**, 613 (1965).
54. P. G. De Gennes, *Rept. Progr. Phys.*, **32**, 187 (1969).
55. H. Reiss, *J. Chem. Phys.*, **47**, 186 (1967).
56. R. Yeh and A. Isihara, *J. Chem. Phys.*, **51**, 1215 (1969).
57. H. Yamakawa, *J. Chem. Phys.*, **54**, 2484 (1971).
58. R. L. Scott, *J. Chem. Phys.*, **25**, 193 (1956).
59. R. M. Mazo, *J. Chem. Phys.*, **40**, 1454 (1964).
60. G. A. Mansoori and F. B. Canfield, *J. Chem. Phys.*, **51**, 4958 (1969).
61. G. A. Mansoori and F. B. Canfield, *J. Chem. Phys.*, **53**, 1618 (1970).
62. G. A. Mansoori and T. W. Leland, Jr., *J. Chem. Phys.*, **53**, 1931 (1970).
63. D. Henderson and J. A. Barker, *J. Chem. Phys.*, **52**, 2315 (1970).
64. T. W. Leland, J. S. Rowlinson, and G. A. Sather, *Trans. Faraday Soc.*, **64**, 1447 (1968).
65. J. Bardeen, L. N. Cooper, and J. R. Schrieffer, *Phys. Rev.*, **108**, 1175 (1957).
66. M. D. Girardeau, *J. Math. Phys.*, **8**, 389 (1967).
67. N. N. Bogoliubov, *Soviet Phys. JETP*, **7**, 41 (1958).
68. J. G. Valatin, *Nuovo Cimento*, **7**, 843 (1958).
69. G. Wentzel, *Helv. Phys. Acta*, **33**, 859 (1960).
70. N. N. Bogoliubov, V. V. Tolmachev, and D. N. Shirkov, *A New Method in the Theory of Superconductivity*, Consultants Bureau, Inc., New York, 1959, p. 89, Eqs. (7.8)ff.
71. A. Hammerstein, *Acta Math.*, **54**, 117 (1930).
72. S. Gartenhaus and G. Stranahan, *Phys. Rev.*, **138**, A1346 (1965).
73. C. M. Andersen, S. Gartenhaus, and G. Stranahan, *Phys. Rev.*, **146**, 101 (1966).
74. N. D. Mermin, *Ann. Phys. (N.Y.)*, **21**, 99 (1963).
75. J. T. Hunt and M. D. Girardeau, *Phys. Rev.*, **160**, 455 (1967).
76. E. Lieb, T. Schultz, and D. Mattis, *Ann. Phys. (N.Y.)*, **16**, 409 (1961).
77. S. Katsura, *Phys. Rev.*, **127**, 1508 (1962).
78. E. C. Stoner, *Proc. Roy. Soc. (London) Ser. A*, **165**, 372 (1938).
79. M. D. Girardeau and D. Wright, *Physica*, **38**, 464 (1968).
80. M. Girardeau and R. Arnowitt, *Phys. Rev.*, **113**, 755 (1959).
81. M. Luban, *Phys. Rev.*, **128**, 965 (1962).
82. F. L. Byrns and R. M. Mazo, *J. Chem. Phys.*, **47**, 2007 (1967).

KINETIC THEORY OF DENSE FLUIDS AND LIQUIDS REVISITED

H. TED DAVIS

Departments of Chemical Engineering and Chemistry, University of Minnesota, Minneapolis, Minnesota

CONTENTS

I. Introduction . . . 258
II. An Effective Potential Theory . . . 259
A. The Kinetic Equation . . . 259
B. Chapman-Enskog Solution to Kinetic Equation . . . 264
C. The Transport Fluxes and Transport Coefficients . . . 271
III. The Rice-Allnatt Theory . . . 275
A. The Kinetic Equations . . . 275
B. The Transport Coefficients . . . 276
IV. The Prigogine-Nicolis-Misguich Theory . . . 278
A. The Model . . . 278
B. The Transport Coefficients . . . 283
C. Comparison of the RA and PNM Theories with Experiment . . . 285
V. A Generalized Rice-Allnatt Theory . . . 289
A. The Kinetic Equation . . . 289
B. The Transport Coefficients . . . 291
VI. Summation of Diagrams . . . 292
A. Some Numerical Studies . . . 292
B. Foster and Cole's Theory . . . 299
VII. Some General Aspects of Kinetic Equations . . . 300
A. Master Equations . . . 300
B. Homogeneous Solutions to General Linear Collision Operator of Singlet Kinetic Equation . . . 304
VIII. Time Correlation Functions and Memory Functions . . . 308
A. Time Correlation Formulas for the Transport Coefficients . . . 308
B. Gaussian Approximation of Time Correlation Functions . . . 311
C. Memory Functions . . . 318
D. Long-Time Decay of Time Correlation Functions . . . 325
IX. Enskog Theory of Fluids Composed of Molecules Interacting with Impulsive Forces . . . 328
A. Hard Sphere Fluids . . . 328
B. Square-Well Fluids . . . 332
C. Loaded Sphere Fluids . . . 336
D. Rough Sphere Fluids . . . 338
E. Self-Diffusion Coefficients for Smooth, Loaded, and Rough Spheres Interacting with a Square-Well Potential . . . 339
References . . . 340

I. INTRODUCTION

Over the last fifteen years or so irreversible statistical mechanics has seen two very important developments. One of these has been the derivation, from the Liouville or von Neumann equation, of general master equations[1,2] from which the states of thermodynamic equilibrium, hydrodynamical equations, transport coefficients fluctuation, and relaxation phenomena may be deduced as appropriate solutions to the master equations. The other important advancement in statistical mechanics has been the derivation of time correlation function expressions for transport coefficients and the delineation of the role of time correlation functions in relaxation and hydrodynamical processes.[3,4] For example, the time correlation formulas for the transport coefficients provide a rigorous microscopic starting point for computing transport coefficients in the same sense that the partition function provides a starting point for computing thermodynamic functions.

Although the master equations and time correlation formulas provide solutions in principle to kinetic and transport properties of many-body systems, the solutions involve N-body operators which are generally not tractable. Thus potential models, mathematical approximations, diagram techniques, etc., have to be introduced in order to obtain numerical results for particular systems. It is this aspect of the problem that we review here for simple dense fluids. In particular, we shall concentrate on approximate theories which yield tractable expressions for transport coefficients. For the sake of brevity mixtures are not discussed, although for most of the models we discuss the extension to mixtures either has been done or would be immediately obvious.

Polyatomic systems are discussed only in the final section where Enskog's theory of molecules interacting with impulsive forces is surveyed briefly. Very little work has been done on the kinetic theory of realistic polyatomic systems.[5] Even most of the general theoretical treatments neglect internal degrees of freedom early in the development. Since the real world of fluids is primarily polyatomic, these fluids should receive a great deal of attention in the future.

We do not discuss the dynamical theory of critical phenomena in this article because of lack of space and because the kinetic theoretical understanding of dynamical critical behavior in fluids is far from complete and is undergoing rapid changes currently. The semimicroscopic theory of fluids and spin systems of Kadanoff and Swift[6] and the microscopie model of spin systems of Resibois and De Leener[7] are important advances in this area. A series of lectures by Resibois[8] provides a useful survey of the state of the art of the dynamical theory critical phenomena as of 1971. It is interesting that some of the models (see Sections II, IV, and V) we consider in this article do predict singular behavior for the thermal conductivity and bulk viscosity near the critical point. However, it has not been established that the singularities predicted by these models are those observed experimentally. For example, the critical behavior of long-range hydrodynamical fluctuations may dominate contributions of the kinetic type considered here.

On the basis of the models to be presented it is perhaps reasonable to claim that transport coefficients for simple dense fluids away from the critical point can be predicted with no more than about 30% error. However, the lack of accurate theoretical or experimental radial distribution functions for dense fluids remains the number one problem in testing and deciding which models are most meaningful.

Experimental work on kinetic processes in dense fluids has been treated lightly in this article, being invoked only to compare with some of the theoretical models. Rice, Boon, and the author[9] included in a 1968 article a rather extensive review of experimental work on simple fluid transport phenomena up to that time. The recent text by Stanley[10] and an NBS publication[11] of a symposium on critical phenomena are good starting points for anyone wanting to construct a bibliography of publications on transport processes in the critical region.

II. AN EFFECTIVE POTENTIAL THEORY

A. The Kinetic Equation

Several years ago Rice and Allnatt[12] (RA) developed a promising kinetic theory of liquids based on the following idea: a molecule moving through a dense fluid will undergo a motion in which it experiences a "Brownian" motion through the average potential field of the neighboring molecules with intermittent binary hard corelike collisions occurring when the molecule moves within the strongly repulsive force field of a neighboring molecule. More recently, Prigogine, Nicolis, and Misguich[13,15] (PNM) presented a local equilibrium model which follows the Rice-Allnatt decomposition of the motion into hard sphere binary collisions plus continuous field interactions. The PNM model neglects distortion of the singlet distribution function from local equilibrium and, consequently, cannot be used for prediction of diffusional transport properties. However, for thermal conductivity and viscosity the PNM theory is as good as or better than the RA theory in the low-temperature, high-density region where neglect of the singlet distortions is reasonable.[15]

The RA theory has the shortcomings that two different kinetic equations must be solved for the singlet and doublet distribution functions, respectively, and that the RA kinetic equations generate incorrect lowest-order hydrodynamic equations. The PNM theory also does not generate the correct lowest-order hydrodynamic equations and, furthermore, neglects diffusional phenomena and the kinetic contributions to the transport coefficients by neglecting the distortion of the singlet distribution function from local equilibrium.

What we wish to do in this section is to develop a theory which incorporates the desirable features of the RA and PNM theories, eliminates the shortcomings mentioned above, and, not least by any means, takes as a starting point a very simple kinetic equation. A similar approach is the generalized Rice-Allnatt theory discussed in a later section.

We shall consider a system of N identical structureless particles interacting with pairwise additive, centrally symmetric forces. The pair potential will be decomposed as follows:

$$\begin{aligned} V(r) &= V^H(r) + V^S(r) \\ V^H(r) &= 0 \qquad r \geqslant \sigma \\ V^S(r) &= 0 \qquad r \leqslant \sigma \end{aligned} \tag{2.1}$$

V^H denotes the strongly repulsive, short-range part of the potential energy and V^S the softer, longer-range part. The hierarchy equation obeyed by the singlet distribution function, $f(\mathbf{x}, \mathbf{v}_1, t)$, for Particle 1 of the system is

$$\begin{aligned} \partial_t f + \mathbf{v}_1 \cdot \nabla_1 f &= + \frac{1}{m} \int \nabla_1 V(\mathbf{x}_{12}) \cdot \boldsymbol{\partial}_1 f_2(\mathbf{x}_1, \mathbf{x}_2, \mathbf{v}_1, \mathbf{v}_2, t)\, d\mathbf{x}_2\, d\mathbf{v}_2 \\ &= \left(\frac{\partial_c f}{\partial t}\right)^H + \frac{1}{m} \int \nabla_1 V^S(\mathbf{x}_{12}) \cdot \boldsymbol{\partial}_1 f_2(\mathbf{x}_1, \mathbf{x}_2, \mathbf{v}_1, \mathbf{v}_2, t)\, d\mathbf{x}_2\, d\mathbf{v}_2 \end{aligned} \tag{2.2}$$

where

$$\left(\frac{\partial_c f}{\partial t}\right)^H = \frac{1}{m} \int \nabla_1 V^H(\mathbf{x}_{12}) \cdot \boldsymbol{\partial}_1 f_2(\mathbf{x}_1, \mathbf{x}_2, \mathbf{v}_1, \mathbf{v}_2, t)\, d\mathbf{x}_2\, d\mathbf{v}_2 \tag{2.3}$$

$$\partial_t \equiv \frac{\partial}{\partial t} \tag{2.4}$$

$$\nabla_i = \frac{\partial}{\partial \mathbf{x}_i} \tag{2.5}$$

$$\boldsymbol{\partial}_i = \frac{\partial}{\partial \mathbf{v}_i} \tag{2.6}$$

f_2 is the doublet distribution function, and $\mathbf{x}_{12} = \mathbf{x}_1 - \mathbf{x}_2$.

Equation (2.2) is a continuity equation for f. The terms on the right-hand side of (2.2) represent the rate of change of f due to collisions of Particle 1 with the background particles. The first assumption of our model is that the contribution from the strongly repulsive interactions, $(\partial_c f/\partial t)^H$, can be approximated by the binary hard core collision operator. Thus, we have *Assumption 1*:

$$\left(\frac{\partial_c f}{\partial t}\right)^H = \sigma^2 \int_{\mathbf{v}_{21}\cdot\mathbf{k}>0} [f_2(\mathbf{x}_1, \mathbf{x}_1 + \sigma\mathbf{k}, \mathbf{v}_1', \mathbf{v}_2', t) - f_2(\mathbf{x}_1, \mathbf{x}_1 - \sigma\mathbf{k}, \mathbf{v}_1, \mathbf{v}_2, t)]\mathbf{v}_{21} \cdot \mathbf{k}\, d\mathbf{k}\, d\mathbf{v}_2 \quad (2.7)$$

$\mathbf{k}$ is a unit vector directed from the center of Molecule 2 to the center of Molecule 1, $d\mathbf{k}$ is the incremental solid angle associated with $\mathbf{k}$, and $\mathbf{v}_1'$, $\mathbf{v}_2'$ are the resulting velocities of a pair of molecules with initial velocities $\mathbf{v}_1$, $\mathbf{v}_2$ which undergo a hard sphere collision. Note that we do not make the chaos assumption of Enskog[16] that f_2 may be factorized into the product of the singlet distribution functions and the local equilibrium radial distribution function. Although we shall not do so in any computations presented here, the fact that V^H is not the hard sphere potential for real systems can be compensated for by assuming the cutoff diameter σ to be temperature dependent. This is done with a good deal of success, for example, in the scaled particle theory of equilibrium properties of liquids.[17]

To obtain a solvable equation for f we must relate f_2 to f. The so-called "master equations" give a formal solution to this problem by relating f_2 to the initial correlations in the system and to f through a nonlinear non-Markovian collision operator. In an important paper evolved, from the "Brussels school," Severne[2] has shown that in the transport or long-time limit the contribution of the initial correlations decays to zero for fluids with initial correlations of finite range, so that f_2 (and f_n for that matter) is a functional of f.

At this point it is useful to introduce the doublet correlation function, g_2, defined by the expression

$$g_2(\mathbf{x}_1, \mathbf{x}_2, \mathbf{v}_1, \mathbf{v}_2, t) = f_2(\mathbf{x}_1, \mathbf{x}_2, \mathbf{v}_1, \mathbf{v}_2, t) - f(\mathbf{x}_1, \mathbf{v}_2, t) f(\mathbf{x}_2, \mathbf{v}_2, t) \quad (2.8)$$

and its Fourier transform

$$g_2(\mathbf{l}, \mathbf{v}_1, \mathbf{v}_2; \mathbf{x}_1, t) = \frac{1}{8\pi^3} \int d\mathbf{x}_{21} e^{i\mathbf{l}\cdot\mathbf{x}_{21}} g_2(\mathbf{x}_1, \mathbf{x}_1 + \mathbf{x}_{21}, \mathbf{v}_1, \mathbf{v}_2, t) \quad (2.9)$$

To lowest order in the interaction potential (i.e., to the weak coupling approximation) Severne's exact long-time form for g_2 is

$$g_2(\mathbf{l}, \mathbf{v}_1, \mathbf{v}_2; \mathbf{x}_1, t) = \frac{1}{m} \int d\mathbf{x}_2\, \delta(\mathbf{x}_1 - \mathbf{x}_2) \times e^{-i\nabla_2\cdot(\partial/\partial\mathbf{l})} \frac{1}{\mathbf{l}\cdot\mathbf{v}_{12} - i(\mathbf{v}_1\cdot\nabla_1 + \mathbf{v}_2\cdot\nabla_2) - i\partial_t - io} \times V_l \mathbf{l}\cdot\boldsymbol{\partial}_{12} f(\mathbf{x}_1, \mathbf{v}_1, t) f(\mathbf{x}_2, \mathbf{v}_2, t) \quad (2.10)$$

where $\partial_{12} = \partial_1 - \partial_2$ and V_l is the Fourier transform of $V(\mathbf{x}_{12})$ defined by the relation

$$V_l = \frac{1}{8\pi^3} \int d\mathbf{r} e^{-i\mathbf{l}\cdot\mathbf{r}} V(\mathbf{r}) \tag{2.11}$$

and where the abbreviation

$$i\pi\delta(A) + \mathfrak{p}\frac{1}{A} \equiv \frac{1}{A - io} \tag{2.12}$$

with $A = \mathbf{l} \cdot \mathbf{v}_{12} - i(\mathbf{v}_1 \cdot \nabla_1 + \mathbf{v}_2 \cdot \nabla_2) - i\partial_t$, has been used. δ is the Dirac delta function operator and $\mathfrak{p}$ is the principal part operator. The operators $\mathbf{v}_i \cdot \nabla_i$ and $i\partial_t$ in A arise, respectively, from the time evolution of spatial delocalization and from the time evolution (memory) of the distribution functions of colliding particles. These "temporal" effects contribute to the nondissipative part of the hydrodynamic equations as well as to the transport fluxes. Dowling and the author[18,19] have presented a detailed study of the temporal contributions in the weak coupling kinetic equation.

Severne's full result for g_2 involves a complicated operator expansion in powers of the interaction potential V. The integrand of (2.10) is the first term of the expansion. According to the diagram theory of Prigogine and co-workers, one tries to pick out a dominant and tractable class of terms from the exact expansion of g_2. The PNM theory, to be discussed in a later section, is based on summing a certain infinite class of terms which, when the f's are approximated by their local equilibrium values, leads to a result of the same basic form as (2.10) but with V_l replaced by a combination of the local equilibrium radial distribution function. Thus, roughly speaking, the effect of the many-body interactions in the fluid is to replace the bare two-body potential in (2.10) by an effective many-body potential. On the other hand, one obtains immediately the RA result if, in (2.10), (1) the spatial and temporal delocalization operators are neglected; that is,

$$e^{-i\nabla_2\cdot(\partial/\partial\mathbf{l})} \simeq 1 \qquad \text{and} \qquad A \simeq \mathbf{l} \cdot \mathbf{v}_{12} - io$$

and (2) V_l is replaced by the effective potential $-kTG_l$, where G_l is the Fourier transform of $g(r) - 1$, $g(r)$ being the local equilibrium radial distribution function. The resulting g_2 combined with (2.2) and (2.7) gives the RA singlet kinetic equation.

What we propose to do now is introduce an ansatz, motivated by the experience of the RA and PNM theories, that incorporates the basic features of these theories and eliminates some of their shortcomings discussed above. Accordingly, our second and last approximation is

Assumption 2: The transport doublet distribution function is of the form

$$g_2(\mathbf{l}, \mathbf{v}_1, \mathbf{v}_2; \mathbf{x}_1, t) = \frac{1}{m}\int d\mathbf{x}_2\, \delta(\mathbf{x}_1 - \mathbf{x}_2)e^{-i\nabla_2\cdot(\partial/\partial\mathbf{l})}$$
$$\times \frac{1}{\mathbf{l}\cdot\mathbf{v}_{12} - i(\mathbf{v}_1\cdot\nabla_1 + \mathbf{v}_2\cdot\nabla_2) - i\partial_t - io}$$
$$\times\, \bar{V}_l\left(\frac{\mathbf{x}_1 + \mathbf{x}_2}{2}, t\right)\mathbf{l}\cdot\boldsymbol{\partial}_{12} f(\mathbf{x}_1, \mathbf{v}_1, t) f(\mathbf{x}_2, \mathbf{v}_2, t) \quad (2.13)$$

where $\bar{V}_l((\mathbf{x}_1 + \mathbf{x}_2)/2, t)$ is an effective potential chosen to satisfy the following conditions:

(i) Equation (2.13) is exact in the weak coupling limit (i.e., $\bar{V}_l$ reduces to V_l to first order in V).
(ii) Equation (2.13) is exact in the equilibrium limit.
(iii) Equation (2.13) when inserted in the kinetic equation (2.2) leads to the exact nondissipative hydrodynamic equations which occur in the Chapman-Enskog solution of the kinetic equation.

The choice of $\bar{V}_l$ which satisfies these three conditions is

$$\bar{V}_l = -kT\left(\frac{\mathbf{x}_1 + \mathbf{x}_2}{2}, t\right)G_l\left(\frac{\mathbf{x}_1 + \mathbf{x}_2}{2}, t\right) \quad (2.14)$$

where

$$G_l = \frac{1}{8\pi^3}\int d\mathbf{r}e^{-i\mathbf{l}\cdot\mathbf{r}}(g(r) - 1) \quad (2.15)$$

$T[(\mathbf{x}_1 + \mathbf{x}_2)/2, t]$ is the kinetic temperature evaluated at the center of mass of the colliding particles and $g(r)$ is the equilibrium radial distribution function evaluated for the density and kinetic temperature at the center of mass of the colliding particles. This choice of $\bar{V}_l$ also has the advantages that in the limit that the f's have their local equilibrium values (2.13) is very similar to the PNM result for g_2 and in the limit that spatial and time delocalizations in the integrand of (2.13) are neglected the RA kinetic equation results except that Enskog's chaos assumption has not been used in (2.7). Because of the neglect of the spatial and time delocalizations in (2.13) the RA theory does not satisfy Condition (iii) and, therefore, one has to go to a Schmoluchowski equation to calculate part of the transport coefficients.

Thus, in summary, our basic kinetic equation is

$$\begin{aligned}
\partial_t f + \mathbf{v}_1 \cdot \nabla_1 f + m^{-1}\mathbf{F}^S(\mathbf{x}_1, t) \cdot \boldsymbol{\partial}_1 f
&= \sigma^2 \int_{\mathbf{v}_{21}\mathbf{k}>0} [f_2(\mathbf{x}_1, \mathbf{x}_1 + \sigma\mathbf{k}, \mathbf{v}_1', \mathbf{v}_2', t) \\
&\quad - f_2(\mathbf{x}_1, \mathbf{x}_1 - \sigma\mathbf{k}, \mathbf{v}_1, \mathbf{v}_2, t)]\mathbf{v}_{21} \cdot \mathbf{k}\, d\mathbf{k}\, d\mathbf{v}_2 \\
&\quad - i\frac{8\pi^3}{m} \int V_l^S \mathbf{l} \cdot \boldsymbol{\partial}_1 g_2(\mathbf{l}, \mathbf{v}_1, \mathbf{v}_2; \mathbf{x}_1, t)\, d\mathbf{l}\, d\mathbf{v}_2 \\
&= \sigma^2 \int_{\mathbf{v}_{21}\cdot\mathbf{k}>0} [f(\mathbf{x}_1, \mathbf{v}_1', t) f(\mathbf{x}_1 + \sigma\mathbf{k}, \mathbf{v}_2', t) \\
&\quad - f(\mathbf{x}_1, \mathbf{v}_1, t) f(\mathbf{x}_1 - \sigma\mathbf{k}, \mathbf{v}_2, t)]\mathbf{v}_{21} \cdot \mathbf{k}\, d\mathbf{k}\, d\mathbf{v}_2 \\
&\quad + \sigma^2 \int_{\mathbf{v}_{21}\cdot\mathbf{k}>0} [e^{-i\mathbf{l}\cdot\mathbf{k}\sigma} g_2(\mathbf{l}, \mathbf{v}_1', \mathbf{v}_2'; \mathbf{x}_1, t) \\
&\quad - e^{i\mathbf{l}\cdot\mathbf{k}\sigma} g_2(\mathbf{l}, \mathbf{v}_1, \mathbf{v}_2; \mathbf{x}_1, t)]\mathbf{v}_{21} \cdot \mathbf{k}\, d\mathbf{k}\, d\mathbf{v}_2\, d\mathbf{l} \\
&\quad - i\frac{8\pi^3}{m} \int V_l^S \mathbf{l} \cdot \boldsymbol{\partial}_1 g_2(\mathbf{l}, \mathbf{v}_1, \mathbf{v}_2; \mathbf{x}_1, t)\, d\mathbf{v}_2\, d\mathbf{l}
\end{aligned} \tag{2.16}$$

where $\mathbf{F}^S(\mathbf{x}_1, t)$ is the Vlassov force

$$\mathbf{F}_1^S(\mathbf{x}_1, t) = \nabla_1 \cdot \int V^S(\mathbf{x}_{12}) f(\mathbf{x}_2, \mathbf{v}_2, t)\, d\mathbf{x}_2\, d\mathbf{v}_2 \tag{2.17}$$

and $g_2(\mathbf{l}, \mathbf{v}_1, \mathbf{v}_2; \mathbf{x}_1, t)$ is given by (2.13). The generalization of (2.16) to multicomponent systems is immediately obvious and, perhaps more importantly, the generalization to include the rotational degrees of freedom of polyatomic systems should be straightforward. These generalizations are currently being studied.

B. Chapman-Enskog Solution to Kinetic Equation

Since we are interested only in linear transport phenomena here, it is useful to expand the spatial and time operators in the integrand of (2.13) and keep terms only through first order in ∇_i and ∂_t; that is,

$$e^{-i\nabla_2\cdot(\partial/\partial\mathbf{l})} = 1 - i\nabla_2 \cdot \frac{\partial}{\partial \mathbf{l}} + \cdots \tag{2.18}$$

$$\frac{1}{\mathbf{l}\cdot\mathbf{v}_{12} - i(\mathbf{v}_1\cdot\nabla_1 + \mathbf{v}_2\cdot\nabla_2 + \partial_t) - io} = \frac{1}{\mathbf{l}\cdot\mathbf{v}_{12} - io} + \frac{i(\mathbf{v}_1\cdot\nabla_1 + \mathbf{v}_2\cdot\nabla_2 + \partial_t)}{(\mathbf{l}\cdot\mathbf{v}_{12} - io)^2} + \cdots \tag{2.19}$$

To first order, then,

$$g_2(\mathbf{l}, \mathbf{v}_1, \mathbf{v}_2; \mathbf{x}_1, t) = g_2^{(1)} + g_2^{(2)} + g_2^{(3)} + g_2^{(4)} \tag{2.20}$$

where

$$g_2^{(1)} = \frac{m^{-1}\bar{V}_l(\mathbf{x}_1, t)}{\mathbf{l}\cdot\mathbf{v}_{12} - io}\,\mathbf{l}\cdot\boldsymbol{\partial}_{12} f(\mathbf{x}_1, \mathbf{v}_1, t) f(\mathbf{x}_1, \mathbf{v}_2, t) \tag{2.21}$$

$$g_2^{(2)} = -\frac{i}{m}\int d\mathbf{x}_2\, \delta(\mathbf{x}_1 - \mathbf{x}_2)\nabla_2 \cdot \frac{\partial}{\partial \mathbf{l}} \frac{\bar{V}_l\left(\dfrac{\mathbf{x}_1 + \mathbf{x}_2}{2}, t\right)}{\mathbf{l}\cdot\mathbf{v}_{12} - io}\,\mathbf{l} \cdot \boldsymbol{\partial}_{12} f(\mathbf{x}_1, \mathbf{v}_1, t) f(\mathbf{x}_2, \mathbf{v}_2, t) \tag{2.22}$$

$$g_2^{(3)} = im^{-1}\partial_t \frac{\bar{V}_l(\mathbf{x}_1, t)}{(\mathbf{l}\cdot\mathbf{v}_{12} - io)^2}\,\mathbf{l}\cdot\boldsymbol{\partial}_{12} f(\mathbf{x}_1, \mathbf{v}_1, t) f(\mathbf{x}_1, \mathbf{v}_2, t) \tag{2.23}$$

$$g_2^{(4)} = im^{-1}\int d\mathbf{x}_2\, \delta(\mathbf{x}_1 - \mathbf{x}_2)\frac{(\mathbf{v}_1\cdot\nabla_1 + \mathbf{v}_2\cdot\nabla_2)}{(\mathbf{l}\cdot\mathbf{v}_{12} - io)^2}\,\bar{V}_l\left(\frac{\mathbf{x}_1 + \mathbf{x}_2}{2}, t\right)\mathbf{l} \cdot \boldsymbol{\partial}_{12} f(\mathbf{x}_1, \mathbf{v}_1, t) f(\mathbf{x}_2, \mathbf{v}_2, t) \tag{2.24}$$

The term $g_2^{(2)}$ arises from the spatial delocalization of colliding particles, $g_2^{(3)}$ from the memory or time dependence of the distribution functions during a collision, and $g_2^{(4)}$ from the evolution of spatial delocalization during a collision.

We shall also introduce in the Boltzmann-like term in (2.16) the expansion

$$f(\mathbf{x}_1 \pm \sigma\mathbf{k}, \mathbf{v}_2, t) = f(\mathbf{x}_1, \mathbf{v}_2, t) \pm \sigma\mathbf{k}\cdot\nabla_1 f(\mathbf{x}_1, \mathbf{v}_2, t) + \cdots \tag{2.25}$$

Introducing the notations

$$J_{B1} = \sigma^2 \int_{\mathbf{v}_{21}\cdot\mathbf{k}>0} [f(\mathbf{x}_1, \mathbf{v}_1', t) f(\mathbf{x}_1, \mathbf{v}_2', t) - f(\mathbf{x}_1, \mathbf{v}_1, t) f(\mathbf{x}_1, \mathbf{v}_2, t)]\mathbf{v}_{21}\cdot\mathbf{k}\, d\mathbf{k}\, d\mathbf{v}_2 \tag{2.26}$$

$$J_{B2} = \sigma^3 \int_{\mathbf{v}_{21}\cdot\mathbf{k}>0} \mathbf{k}\cdot[f(\mathbf{x}_1, \mathbf{v}_1', t)\nabla_1 f(\mathbf{x}_1, \mathbf{v}_2', t) + f(\mathbf{x}_1, \mathbf{v}_1, t)\nabla_1 f(\mathbf{x}_1, \mathbf{v}_2, t)]\mathbf{v}_{21}\cdot\mathbf{k}\, d\mathbf{k}\, d\mathbf{v}_2 \tag{2.27}$$

$$J(g_2^{(i)}) = \sigma^2 \int_{\mathbf{v}_{21}\cdot\mathbf{k}>0} [e^{-i\mathbf{l}\cdot\mathbf{k}\sigma} g^{(i)}(\mathbf{l}, \mathbf{v}_1', \mathbf{v}_2', \mathbf{x}_1, t) - e^{i\mathbf{l}\cdot\mathbf{k}\sigma} g_2^{(i)}(\mathbf{l}, \mathbf{v}_1, \mathbf{v}_2; \mathbf{x}_1, t)]\mathbf{v}_{21}\cdot\mathbf{k}\, d\mathbf{k}\, d\mathbf{v}_2\, d\mathbf{l} \tag{2.28}$$

$$\mathscr{B}(g_2^{(i)}) = -i\,\frac{8\pi^3}{m}\int V_l^S \mathbf{l}\cdot\boldsymbol{\partial}_1 g_2^{(i)}(\mathbf{l}, \mathbf{v}_1, \mathbf{v}_2; \mathbf{x}_1, t)\, d\mathbf{v}_2\, d\mathbf{l} \tag{2.29}$$

we obtain the following form for the kinetic equation appropriate for the computation of linear transport properties:

$$\partial_t f + \mathbf{v}_1 \cdot \nabla_1 f + m^{-1}\mathbf{F}^S(\mathbf{x}_1, t) \cdot \boldsymbol{\partial}_1 f = J_{B1} + J_{B2} + \sum_{i=1}^{4} J(g_2^{(i)}) + \sum_{i=1}^{4} \mathscr{B}(g_2^{(i)}) \quad (2.30)$$

Equation (2.30) is still nonlinear in the singlet distribution function f. Linearizing by the Chapman-Enskog method,[16] we obtain

$$h(f^0) = \mathscr{A}_H\phi + \mathscr{A}_S\phi \quad (2.31)$$

with

$$h(f^0) \equiv \partial_t^0 f^0 + \mathbf{v}_1 \cdot \nabla_1 f^0 + m^{-1}\mathbf{F}^{S0}(\mathbf{x}_1, t) \cdot \boldsymbol{\partial}_1 f^0 - J_{B2}{}^0 - \sum_{i=2}^{4} J^0(g_2^{(i)}) - \sum_{i=2}^{4} \mathscr{B}^0(g_2^{(i)}) \quad (2.32)$$

where

$$\begin{aligned}\mathscr{A}_H\phi = {} & g(\sigma)\sigma^2 \int_{\mathbf{v}_{21}\cdot\mathbf{k}>0} f^0(1)f^0(2) \\ & \times [\phi'(1) + \phi'(2) - \phi(1) - \phi(2)]\mathbf{v}_{21} \cdot \mathbf{k}\, d\mathbf{k}\, d\mathbf{v}_2 \\ & - m^{-1}kT\sigma^2 \int_{\mathbf{v}_{21}\cdot\mathbf{k}>0} \left[\frac{e^{-i\mathbf{l}\cdot\mathbf{k}\sigma}}{\mathbf{l}\cdot\mathbf{v}_{12}' - io} \mathbf{l} \cdot (\boldsymbol{\partial}_1'\phi(1)' - \boldsymbol{\partial}_2'\phi(2)') \right. \\ & \left. - \frac{e^{i\mathbf{l}\cdot\mathbf{k}\sigma}}{\mathbf{l}\cdot\mathbf{v}_{12}' - io} \mathbf{l} \cdot (\boldsymbol{\partial}_1\phi(1) - \boldsymbol{\partial}_2\phi(2)) \right] G_l f^0(1) f^0(2)\mathbf{v}_{21} \cdot \mathbf{k}\, d\mathbf{k}\, d\mathbf{v}_2\, d\mathbf{l}\end{aligned} \quad (2.33)$$

and

$$\mathscr{A}_S\phi = -\frac{8\pi^4}{m^2} kT \int G_l V_l^S \mathbf{l} \cdot \boldsymbol{\partial}_1\, \delta(\mathbf{l}\cdot\mathbf{v}_{12}) f^0(1) f^0(2) \mathbf{l} \cdot \boldsymbol{\partial}_{12}(\phi(1) + \phi(2)) \quad (2.34)$$

where the superscript "0" on $\mathbf{F}^{S0}$, $J_{B2}{}^0$, $J^0(g_2^{(i)})$ and $\mathscr{B}^0(g_2^{(i)})$ means that in these expressions we have set $f = f^0$, and $\partial_t{}^0$ denotes the zeroth-order time derivative of hydrodynamic variables. The quantity $f^0(i) = f^0(\mathbf{x}_1, \mathbf{v}_i, t)$ is the local equilibrium distribution function defined as

$$f(i)^0 = n(\mathbf{x}_1, t)\left(\frac{m}{2\pi kT(\mathbf{x}_1, t)}\right)^{3/2} e^{-[m/2kT(\mathbf{x}_1,t)](\mathbf{v}_i - \mathbf{u}(\mathbf{x}_1,t))^2} \quad (2.35)$$

The quantities n, T, and $\mathbf{u}$ are the local density, temperature, and mass average velocity, respectively. The quantity $\phi(i)$ represents the deviation of f from local equilibrium and is defined by the relation

$$f(i) = f(i)^0(1 + \phi(i)) \quad (2.36)$$

$\mathscr{A}_S\phi$ is identical to the Fokker-Planck part of the Rice-Allnatt collision operator, but $\mathscr{A}_H\phi$ differs from the Enskog operator employed in the Rice-Allnatt theory since we did not use the chaos assumption of Enskog. The added contribution to the hard sphere collision operator is real since it does not vanish in the weak coupling limit, $-kTG_l \rightarrow V_l$, where our Assumption 2 is exact. The contribution represents dynamic correlations neglected by Enskog's chaos assumption. PNM obtained similar contributions in their local equilibrium theory.[14]

The operators $\mathscr{A}_H$ and $\mathscr{A}_S$ are symmetric and admit the solutions $\psi_h = m$, $\mathbf{v}$, $\mathbf{v}^2$ to the homogeneous equation

$$\mathscr{A}_H\psi_h + \mathscr{A}_S\psi_h = 0 \tag{2.37}$$

Thus the Fredholm solvability condition requires that the left-hand side of (2.31) be orthogonal to m, $\mathbf{v}$, $\mathbf{v}^2$; that is, the left-hand side when multiplied by m, $\mathbf{v}_1$, and $\mathbf{v}_1^2$, respectively, and integrated over $d\mathbf{v}_1$ must equal zero. Performing these integrations, we obtain the following solvability conditions:

$$\partial_t^0 n + \nabla_1 \cdot (n\mathbf{u}) = 0 \tag{2.38}$$

$$\partial_t^0 \mathbf{u} + \mathbf{u} \cdot \nabla_1 \mathbf{u} + \frac{1}{mn} \nabla_1 P = 0 \tag{2.39}$$

$$\partial_t^0 T + \mathbf{u} \cdot \nabla_1 T + \frac{T}{n\tilde{C}_v}\left(\frac{\partial P}{\partial T}\right)_v \nabla_1 \cdot \mathbf{u} = 0 \tag{2.40}$$

These equations (which define the zeroth-order time derivative ∂_t^0) are the exact nondissipative hydrodynamic equations as demanded by Condition (iii) of our model. If, as Rice and Allnatt did, we neglect the contributions $g_2^{(2)}$, $g_2^{(3)}$, and $g_2^{(4)}$, the ideal specific heat $\tilde{C}_v^{ID} = \frac{3}{2}k$ and the hard sphere pressure, P^{HS}, appear in these equations instead of the correct specific heat and pressure. The correct specific heat is obtained only if the memory and time displacement terms $g_2^{(3)}$ and $g_2^{(4)}$ are retained. The reason for this is that neglect of $g_2^{(3)}$ and $g_2^{(4)}$ is equivalent to assuming the collisions are instantaneous and, therefore, occur too quickly to allow kinetic energy to be stored as potential energy. Therefore, only the ideal heat capacity associated with the kinetic energy will be effective.

Using (2.38)–(2.40), one can eliminate the time derivatives $\partial_t^0 n$, $\partial_t^0 \mathbf{u}$, and $\partial_t^0 T$ from $h(f^0)$. However, it is not convenient to do this explicitly. In the calculations that follow, we evaluate integrals over $d\mathbf{v}_1$ involving products of $h(f^0)$ with certain functions of $\mathbf{v}_1$. It is generally easier to evaluate the integrals first and then eliminate $\partial_t^0 n$, $\partial_t^0 \mathbf{u}$, and $\partial_t^0 T$ from the results using (2.38)–(2.40).

On the basis of the kinetic theory of gases, it is expected that a good approximation to ϕ is[16,20]

$$\phi = b_0\left(\frac{m}{2kT}\right)\mathfrak{C}^0\mathfrak{C}:\nabla_1\mathbf{u} + a_1 m\left(\frac{5}{2} - \frac{m\mathfrak{C}^2}{2kT}\right)\mathfrak{C}\cdot\nabla_1 \ln T + b_2\left(\frac{15}{8} - \frac{5m\mathfrak{C}^2}{4kT} + \frac{m^2\mathfrak{C}^4}{8(kT)^2}\right)\nabla_1\cdot\mathbf{u} \quad (2.41)$$

where $\mathfrak{C}^0\mathfrak{C}$ is a traceless dyadic,

$$\mathfrak{C}^0\mathfrak{C} = \mathfrak{C}\mathfrak{C} - \tfrac{1}{3}\mathfrak{C}^2\mathbf{1} \quad (2.42)$$

and

$$\mathfrak{C} = \mathbf{v} - \mathbf{u} \quad (2.43)$$

and where the quantities b_0, b_2, and a_1 are determined from the equations

$$\int d\mathbf{v}_1\mathfrak{C}_1{}^0\mathfrak{C}_1\left\{h(f^0) - b_0\left(\frac{m}{2kT}\right)(\mathscr{A}_H + \mathscr{A}_S)(\mathfrak{C}^0\mathfrak{C}:\nabla_1\mathbf{u})\right\} = 0 \quad (2.44)$$

$$\int d\mathbf{v}_1\left(\frac{15}{8} - \frac{5m\mathfrak{C}_1{}^2}{4kT} + \frac{5m^2\mathfrak{C}_1{}^4}{8(kT)^2}\right) \times\left\{h(f^0) - b_2(\mathscr{A}_H + \mathscr{A}_S)\left(\frac{15}{8} - \frac{5m\mathfrak{C}^2}{4kT} + \frac{5m^2\mathfrak{C}^4}{8(kT)^2}\right)\nabla_1\cdot\mathbf{u}\right\} = 0 \quad (2.45)$$

$$\int d\mathbf{v}_1\mathfrak{C}_1\left(\frac{5}{2} - \frac{m\mathfrak{C}_1{}^2}{2kT}\right) \times\left\{h(f^0) - a_1 m(\mathscr{A}_H + \mathscr{A}_S)\left(\frac{5}{2} - \frac{m\mathfrak{C}^2}{2kT}\right)\cdot\nabla_1 \ln T\right\} = 0 \quad (2.46)$$

with $\mathbf{u} = 0$ in Eq. (2.46).

Performing the lengthy integrations indicated in (2.44)–(2.46) and eliminating $\partial_t{}^0 n$, $\partial_t{}^0\mathbf{u}$, and $\partial_t{}^0 T$ from the results, we obtain

$$b_0 = \frac{-5m}{3(\zeta_{(1)}^H + \zeta_{(2)}^H + \zeta^S)}\left[1 + \frac{2(e^S + P^S + P^H)}{5nkT} - \frac{\pi n\sigma^3}{5}G(\sigma)(1 - \chi_3)\right] \quad (2.47)$$

$$b_2 = -\frac{m}{3(\zeta_{(1)}^H + \zeta_{(2)}^H + \zeta^S)}\left[1 + \frac{P^H}{nkT} - \frac{4}{n\tilde{C}_v}\left(\frac{\partial P}{\partial T}\right)_v\left(\frac{e^S}{nkT} + \frac{3}{8}\right) + \frac{12e^S}{nkT} - \frac{\pi}{3}n\sigma^3 G(\sigma)(\chi_5 + \tfrac{7}{8}\chi_3)\right] \quad (2.48)$$

and

$$a_1 = \frac{5}{4(\zeta^H_{(1)} + 3\zeta^H_{(2)} + \zeta^S)}\left[1 + \frac{3}{5nkT}(e^S + P^S + P^H) + \frac{1}{3nk}\left(\frac{\partial e^S}{\partial T}\right)_P + \frac{2\pi n\sigma^3}{15}G(\sigma)(8\xi_3 - 12\xi_1 - \chi_3 - 4)\right] \tag{2.49}$$

where

$$\zeta^H_{(1)} = \tfrac{8}{3}(\pi mkT)^{1/2}n\sigma^2 g(\sigma) \tag{2.50}$$

$$\zeta^H_{(2)} = -\frac{2\pi}{3}\sigma^2 G(\sigma)n(mkT)^{1/2}\left[\frac{2}{\pi^{1/2}} + 3\chi_2 - 2\chi_4\right] \tag{2.51}$$

$$\zeta^S = -\frac{16\pi^4}{3}n\left(\frac{\pi m}{kT}\right)^{1/2}\int_0^\infty dl\, l^3 V_l^S G_l \tag{2.52}$$

$$P^H = \frac{2\pi}{3}n\sigma^3 g(\sigma)nkT \tag{2.53}$$

$$P^S = -\frac{2\pi}{3}n^2\int_\sigma^\infty dr\, r^3\frac{dV^S(r)}{dr}G(r) \tag{2.54}$$

$$e^S = 2\pi n^2\int_\sigma^\infty dr\, r^2 V^S(r)G(r) \tag{2.55}$$

$$\xi_n = -\frac{(-1)^n I_n}{8\pi^{3/2}G(\sigma)}\int_1^\infty dx\, xG(x\sigma)P_n(x) \tag{2.56}$$

$$I_n = \int dy e^{-y^2}y^{n-1} \tag{2.57}$$

$$P_n(x) = 16\int_{-1}^0 d\mu\,\frac{\mu^n}{\sqrt{x^2 - 1 + \mu^2}} \tag{2.58}$$

$$\chi_n = -\xi_n + \frac{2I_n}{\pi^{3/2}n} + \frac{(-1)^n I_n}{8\pi^{3/2}G(\sigma)}\int_1^\infty dx\, x^2 G(x\sigma)\frac{dP_n(x)}{dx} \tag{2.59}$$

The quantities $P_n(x)$ may be evaluated explicitly. For $x \geqslant 0$, the first five P_n's are

$$P_1(x) = -16(x - \sqrt{x^2 - 1}) \tag{2.60}$$

$$P_2(x) = 8x - 4(x^2 - 1)\ln\left|\frac{x+1}{x-1}\right| \tag{2.61}$$

$$P_3(x) = -16x + \tfrac{32}{3}[x^3 - (x^2 - 1)^{3/2}] \tag{2.62}$$

$$P_4(x) = 2x(5 - 3x^2) + 3(x^2 - 1)^2\ln\left|\frac{x+1}{x-1}\right| \tag{2.63}$$

$$P_5(x) = -16x + \tfrac{64}{3}x^3 - \tfrac{128}{15}[x^5 - (x^2 - 1)^{5/2}] \tag{2.64}$$

TABLE I

Self-Diffusivity (in units of cm^2/sec) of Liquids Argon, Krypton, and Xenon along the Vapor Pressure Curve. Theoretical Predictions Based on (1.3.27) and Experimental Values Taken from Ref. 24

Substance	T (°K)	D(calc.) $\times 10^5$	D(obs.) $\times 10^5$
Argon	90	3.3	2.4
	110	4.8	4.8
	130	6.9	7.3
	140	8.6	8.6
	148	11.5	9.8
Krypton	120	2.5	1.7
	140	3.2	2.7
	160	4.3	3.7
	180	5.7	4.5
	200	8.2	5.6
Xenon	220	3.8	4.4
	240	4.7	5.2
	260	6.0	6.0
	280	8.0	6.2
	285	9.2	6.2

For tracer self-diffusion with $\nabla_1 T = \mathbf{u} = 0$, we can solve the kinetic equation by setting $\phi(2) \equiv 0$ and taking

$$\phi(1) = d\mathbf{v}_1 \cdot \nabla_1 \ln n^* \tag{2.65}$$

where n^* is the tracer element concentration, and where d is obtained from the equation

$$\int dv_1\, \mathbf{v}_1\{h(f^0) - d(\mathscr{A}_H{}^* + \mathscr{A}_S{}^*)\mathbf{v}_1 \cdot \nabla_1 \ln n^*\} = 0 \tag{2.66}$$

The asterisk on $\mathscr{A}_H{}^*$ and $\mathscr{A}_S{}^*$ means that $\phi(2) \equiv 0$ for the self-diffusion case. The result is

$$d = -\frac{1}{\zeta^H_{(1)} + \zeta^H_{(3)} + \zeta^S} \tag{2.67}$$

$$\zeta^H_{(3)} = -\frac{2\pi}{3} n\sigma^2 G(\sigma)(mkT)^{1/2}\chi_2 \tag{2.68}$$

Finally, the doublet distribution function, to first order in gradients of $\mathbf{u}$ and T, is of the form

$$g_2(\mathbf{l}, \mathbf{v}_1, \mathbf{v}_2; \mathbf{x}_1, t) = -\frac{m^{-1}kTG_l}{\mathbf{l}\cdot\mathbf{v}_{12} - io}\,\mathbf{l}\cdot\boldsymbol{\partial}_{12}[f^0(1)f^0(2)(\phi(1) + \phi(2))] + \sum_{i=1}^{4} g_2^{(i)0} \tag{2.69}$$

where the superscript "0" on $g_2^{(i)0}$ indicates that we have set $f = f^0$ in the expressions defining these quantities.

C. The Transport Fluxes and Transport Coefficients

The kinetic parts of the heat flux $\mathbf{q}$ and momentum flux $\mathbf{J}$ are by definition

$$\mathbf{q}_K = \frac{m}{2}\int f(\mathbf{x}_1, \mathbf{v}_1, t)(\mathbf{v}_1 - \mathbf{u})^2(\mathbf{v}_1 - \mathbf{u})\, d\mathbf{v}_1 \tag{2.70}$$

and

$$\mathbf{J}_K = m\int f(\mathbf{x}_1, \mathbf{v}_1, t)(\mathbf{v}_1 - \mathbf{u})(\mathbf{v}_1 - \mathbf{u})\, d\mathbf{v}_1 \tag{2.71}$$

The potential energy contributions to $\mathbf{q}$ and $\mathbf{J}$ from binary hard sphere collisions are[12,16]

$$\mathbf{q}_v{}^H = -\frac{m\sigma^3}{4}\int_{\mathbf{v}_{21}\cdot\mathbf{k}>0} (\mathbf{v}_{21}\cdot\mathbf{k})^2\mathbf{k}\mathbf{k}\cdot(\mathbf{v}_1 + \mathbf{v}_2 - 2\mathbf{u}) \times f_2(\mathbf{x}_1, \mathbf{x}_1 - \sigma\mathbf{k}, \mathbf{v}_1, \mathbf{v}_2, t)\, d\mathbf{k}\, d\mathbf{v}_1\, d\mathbf{v}_2 \tag{2.72}$$

and

$$\mathbf{J}_v{}^H = \frac{m\sigma^3}{2}\int_{\mathbf{v}_{21}\cdot\mathbf{k}>0} (\mathbf{v}_{21}\cdot\mathbf{k})^2\mathbf{k}\mathbf{k} f_2(\mathbf{x}_1, \mathbf{x}_1 - \sigma\mathbf{k}, \mathbf{v}_1, \mathbf{v}_2, t)\, d\mathbf{k}\, d\mathbf{v}_1\, d\mathbf{v}_2 \tag{2.73}$$

The potential energy contributions from the soft part of the potential energy are[21]

$$\begin{aligned}\mathbf{q}_v{}^S &= \frac{1}{4}\int (\mathbf{v}_1 + \mathbf{v}_2 - 2\mathbf{u})\cdot\left[\mathbf{1}V^S(r) - \mathbf{r}\frac{\partial V^S(r)}{\partial \mathbf{r}}\right] \\ &\quad \times f_2(\mathbf{x}_1, \mathbf{x}_1 + \mathbf{r}, \mathbf{v}_1, \mathbf{v}_2, t)\, d\mathbf{r}\, d\mathbf{v}_1\, d\mathbf{v}_2 \\ &= 2\pi^3\int (\mathbf{v}_1 + \mathbf{v}_2 - 2\mathbf{u})\cdot\left[\mathbf{1}V_l^S + \left(\frac{\partial}{\partial \mathbf{l}}\mathbf{l}V_l^S\right)\right] \\ &\quad \times g_2(\mathbf{l}, \mathbf{v}_1, \mathbf{v}_2; \mathbf{x}_1, t)\, d\mathbf{l}\, d\mathbf{v}_1\, d\mathbf{v}_2\end{aligned} \tag{2.74}$$

$$\begin{aligned}\mathbf{J}_v{}^S &= -\frac{1}{2}\int \mathbf{r}\frac{\partial V^S(r)}{\partial \mathbf{r}} f_2(\mathbf{x}_1, \mathbf{x}_1 + \mathbf{r}, \mathbf{v}_1, \mathbf{v}_2, t)\, d\mathbf{r}\, d\mathbf{v}_1\, d\mathbf{v}_2 \\ &= -\frac{n^2}{2}\int \mathbf{r}\frac{\partial V^S(r)}{\partial \mathbf{r}}\, d\mathbf{r} \\ &\quad + 4\pi^3\int \left(\frac{\partial}{\partial \mathbf{l}}\mathbf{l}V_l^S\right) g_2(\mathbf{l}, \mathbf{v}_1, \mathbf{v}_2; \mathbf{x}_1, t)\, d\mathbf{l}\, d\mathbf{v}_1\, d\mathbf{v}_2\end{aligned} \tag{2.75}$$

The term involving $f(\mathbf{x}_1, \mathbf{v}_1, t)f(\mathbf{x}_1 + \mathbf{r}, \mathbf{v}_2, t)$, obtained by setting $f_2 \equiv g_2 + f(1)f(2)$, vanishes in (2.74) through first order in gradients of n, $\mathbf{u}$, and T and to the same order the corresponding term in (2.75) gives the contribution

$$-\frac{n^2}{2}\int \mathbf{r}\frac{\partial V^S(r)}{\partial \mathbf{r}}\, d\mathbf{r} = -\mathbf{1}\frac{n^2}{6}\int r\frac{\partial V^S(r)}{\partial r}\, d\mathbf{r} \tag{2.76}$$

Using the values of f, f_2, and g_2 obtained in the preceding sections, we can compute the various contributions to $\mathbf{q}$ and $\mathbf{J}$. Then, comparing the molecular expressions to Fourier's and Newton's laws,

$$\mathbf{q} = -\kappa \nabla T \tag{2.77}$$

and

$$\mathbf{J} = P\mathbf{1} - \Phi\nabla_1 \cdot \mathbf{u}\mathbf{1} - \eta[\nabla_1\mathbf{u} + \nabla_1\mathbf{u}^T - \tfrac{2}{3}\nabla_1 \cdot \mathbf{u}\mathbf{1}] \tag{2.78}$$

we obtain the various contributions to the thermal conductivity κ, the shear viscosity η, and the bulk viscosity Φ. The results are as follows.

Thermal Conductivity

$$\kappa_K = \frac{\kappa^*}{(\zeta_{(1)}^H + 3\zeta_{(2)}^H + \zeta^S)/\zeta^*}\left[1 + \frac{3}{5nkT}(e^S + P^S + P^H) + \frac{1}{3nk}\left(\frac{\partial e^S}{\partial T}\right)_P + \frac{2\pi n\sigma^3}{15}G(\sigma)(8\xi_3 - 12\xi_1 - \chi_3 - 4)\right] \tag{2.79}$$

$$\kappa_v{}^H = \kappa_K\left[\frac{2}{5}\pi n\sigma^3 g(\sigma) + \frac{\pi}{15}n\sigma^3 G(\sigma)(1 + 2\chi_3)\right] + \kappa^*\left\{\frac{128}{225}\pi n^2\sigma^6\left[g(\sigma) - \frac{\pi^{1/2}}{4}G(\sigma)(\xi_2 - \xi_4)\right] + \frac{32\pi}{225}\left(\frac{\pi m}{kT}\right)^{1/2} T\left[\frac{\partial}{\partial T}\left(n^2\sigma^6 G(\sigma)\left(\frac{kT}{m}\right)^{1/2}\xi_2\right)\right]_P\right\} \tag{2.80}$$

$$\kappa_v{}^S = -\frac{(e^S + 9P^S)}{15nkT}\kappa_K + \frac{256\pi^5}{225}\kappa^*\left\{\frac{n^2\sigma^2}{kT}\left(5W_1 + \frac{9}{2}W_2 + W_3\right) + \frac{1}{k}\left[\frac{\partial}{\partial T}(6n^2\sigma^2 W_1 + n^2\sigma^2 W_2)\right]_P\right\} \tag{2.81}$$

where

$$\kappa^* = \frac{75}{64\sigma^2}\left(\frac{k^3 T}{\pi m}\right)^{1/2} \tag{2.82}$$

and

$$\zeta^* = \tfrac{8}{3}n\sigma^2(\pi mkT)^{1/2} \tag{2.83}$$

Shear Viscosity

$$\eta_K = \frac{\eta^*}{(\zeta_{(1)}^H + \zeta_{(2)}^H + \zeta^S)/\zeta^*}\left[1 + \frac{2(e^S + P^S + P^H)}{5nkT} - \frac{\pi n\sigma^3}{5}G(\sigma)(1 - \chi_3)\right] \tag{2.84}$$

$$\eta_V{}^H = \eta_K\left[\frac{4\pi}{15}n\sigma^3 g(\sigma) + \frac{2\pi}{15}n\sigma^3 G(\sigma)(1 - 3\chi_2)\right] + \frac{64\pi}{75}n^2\sigma^6\eta^*\left[g(\sigma) + \frac{\pi^{1/2}}{4}G(\sigma)\chi_2 - \frac{\pi^{1/2}}{2}G(\sigma)(3\xi_4 - 8\xi_2)\right] \tag{2.85}$$

$$\eta_V{}^S = \frac{2(e^S - P^S)}{5nkT}\eta_K + \frac{128}{75}\frac{\pi^5 n^2\sigma^2}{kT}\eta^*(2W_2 + W_3) \tag{2.86}$$

where

$$\eta^* = \frac{5}{16\sigma^2}\left(\frac{mkT}{\pi}\right)^{1/2} \tag{2.87}$$

Bulk Viscosity

$$\Phi_K = 0 \tag{2.88}$$

$$\begin{aligned}\Phi_V{}^H &= b_2{}^*\pi n\sigma^3 G(\sigma)\eta^* + \frac{64\pi}{45} n^2\sigma^6\eta^*[g(\sigma) - \tfrac{6}{5}G(\sigma)(\xi_4 - \tfrac{7}{3}\xi_2)] \\ &\quad + \frac{16}{15}\left(\frac{\pi}{kT}\right)^{1/2}\eta^*\left\{\left[\left(\frac{\partial}{\partial \ln n}\right)_T + \frac{1}{n\tilde{C}_v}\left(\frac{\partial P}{\partial T}\right)_n\left(\frac{\partial}{\partial \ln T}\right)_n\right]\right. \\ &\quad \left.\times\, [\pi n^2\sigma^6 G(\sigma)(kT)^{1/2}\xi_2]\right\}\end{aligned} \tag{2.89}$$

$$\begin{aligned}\Phi_V{}^S &= -\tfrac{9}{5}b_2{}^*\frac{P^S}{nkT}\eta^* + \frac{128\pi^5}{45}\frac{n^2\sigma^2}{kT}\eta^*(8W_2 + \tfrac{4}{3}W_1 + W_3) \\ &\quad - \frac{512\eta^*}{15(kT)^{1/2}}\left\{\left[\left(\frac{\partial}{\partial \ln n}\right)_T + \frac{1}{n\tilde{C}_v}\left(\frac{\partial P}{\partial T}\right)_n\left(\frac{\partial}{\partial \ln T}\right)_n\right]\left[\frac{\pi^5 n^2\sigma^2}{(kT)^{1/2}}(W_1 + W_2)\right]\right\}\end{aligned} \tag{2.90}$$

$$\begin{aligned}b_2{}^* = \frac{3\zeta^*}{m}b_2 &= \frac{-\zeta^*}{\zeta_{(1)}^H + \zeta_{(2)}^H + \zeta^S}\left[1 + \frac{P^H}{nkT} - \frac{4}{n\tilde{C}_v}\left(\frac{\partial P}{\partial T}\right)_n\left(\frac{e^S}{nkT} + \frac{3}{8}\right)\right. \\ &\quad \left. + \frac{12e^S}{nkT} - \frac{\pi}{3}n\sigma^3 G(\sigma)(\chi_5 + \tfrac{7}{8}\chi_3)\right]\end{aligned} \tag{2.91}$$

The quantities W_n are by definition

$$W_n = \int_0^\infty dl\, G_l l^n \frac{\partial^{n-1}}{\partial l^{n-1}} V_l^S \tag{2.92}$$

The other quantities in (2.79)–(2.92) were defined in (2.47)–(2.59). The total values of κ, η, and Φ are, of course, obtained by summing the three parts given for each.

In the self-diffusion case, the diffusion flux is

$$\mathbf{J}_d = m\int \mathbf{v}_1 f\, d\mathbf{v}_1 = dkT\nabla_1 n^* \tag{2.93}$$

which, when compared to Fick's law,

$$\mathbf{J}_d = -D\nabla_1 n^* \tag{2.94}$$

gives for the self-diffusion coefficient the result

$$D = -dkT \tag{2.95}$$

Taking d from (2.68), we obtain

$$D = \frac{kT}{\zeta^H_{(1)} + \zeta^H_{(3)} + \zeta^S} \tag{2.96}$$

The last terms in (2.81) and (2.90) arise from the memory and temporal delocalization contributions $g_2^{(3)\circ}$ and $g_2^{(4)\circ}$. These terms are singular, and hence of great importance, at the critical point. Also a_1 contains a singular term arising from $g_2^{(3)\circ}$ and $g_2^{(4)\circ}$. Thus the neglect of the memory and time evolution of delocalization would have serious consequences, at least near the critical point. Using van Kampen's[22] approximation to G_l near the critical point, we can conclude from the above results that κ and Φ are singular at the critical point whereas η and D are not. This is in qualitative agreement with experiment. However, whether the quantitative nature of the singularities predicted for κ and Φ agree with experiment has not yet been established. Work on this point is underway at present.

The terms in κ, η, Φ, and D involving the χ_n and ξ_n functions arise from the dynamic correlations mentioned earlier. If Enskog's chaos assumption were used in the hard sphere contributions these terms would not appear.

Using the values for the radial distribution function computed for the superposition approximation by Kirkwood, Lewinson, and Alder[23] for the truncated Lennard-Jones model, we have computed the self-diffusion coefficient for liquids argon, krypton, and xenon. The quantity $\zeta^H_{(3)}$ is much smaller than $\zeta^H_{(1)} + \zeta^S$ so that (1.3.27) gives essentially the same numerical results as the original Rice-Allnatt approximation, $D = kT/(\zeta^H_{(1)} + \zeta^S)$. In Table I, theory and experiment[24] are compared. The theoretical values disagree with experiment by only 20% for temperatures below $0.9T_c$, where T_c is the critical temperature. For higher temperatures the disagreement becomes increasingly greater. Numerical predictions of the viscosities and the thermal conductivity have not been made, although in the dense fluid range they should be similar to the predictions of the PNM theory which are compared to experiment in Section III. Calculations are presently underway for the dense gas range where the kinetic contributions included in this theory and neglected by the PNM theory will be important in the viscosity and thermal conductivity. One of the most severe problems in numerical work at the moment is the lack of accurate radial distribution functions for the truncated Lennard-Jones potential. The calculations of Kirkwood, Lewinson, and Alder are all that are available and are, unfortunately, based on the superposition approximation.

III. The Rice-Allnatt Theory

A. The Kinetic Equations

As stated in Section IIA, the basic physical assumption of the Rice-Allnatt theory is that the motion of a fluid molecule may be viewed as a Brownian motion through the "soft" potential field with intermittent hard core collisions. Under this assumption RA derived a singlet kinetic equation of the form[12,25]

$$\begin{aligned} &\partial_t f + \mathbf{v}_1 \cdot \nabla_1 f + m^{-1}\mathbf{F}_1{}^* \cdot \boldsymbol{\partial}_1 f \\ &= \sigma^2 \int_{\mathbf{v}_{21}\cdot\mathbf{k}>0} [f(\mathbf{x}_1, \mathbf{v}_{1'}, t) f(\mathbf{x}_1 + \sigma\mathbf{k}, \mathbf{v}_{2'}, t) g(\mathbf{x}_1, \mathbf{x}_1 + \sigma\mathbf{k}) \\ &\quad - f(\mathbf{x}_1, \mathbf{v}_1, t) f(\mathbf{x}_1 - \sigma\mathbf{k}, \mathbf{v}_2, t) g(\mathbf{x}_1, \mathbf{x}_1 - \sigma\mathbf{k})] \mathbf{v}_{21} \cdot \mathbf{k}\, d\mathbf{k}\, d\mathbf{v}_2 + \mathscr{B}(g_2^{(1)}) \end{aligned} \tag{3.1}$$

where $\mathbf{F}_1{}^*$ is a mean force which accounts in part for the spatial delocalizations accounted for by the term $\mathscr{B}(g_2^{(2)})$ of the effective potential theory. Baleiko and the author[25] found that if $\mathbf{F}_1{}^*$ is chosen to be

$$\mathbf{F}_1{}^* = -\frac{1}{n(\mathbf{x}_1, t)} \int \nabla_1 V^S(\mathbf{x}_{12}) n(\mathbf{x}_2, t) g(\mathbf{x}_1, \mathbf{x}_2; t)\, d\mathbf{x}_2 \tag{3.2}$$

then (3.1) yields the right momentum equation. However, since temporal terms such as $\mathscr{B}(g_2^{(3)})$ and $\mathscr{B}(g_2^{(4)})$ have been neglected in (3.1), it does not yield the correct hydrodynamic energy equation. The $\mathscr{B}(g_2^{(i)})$ referred to here as those defined by (2.29). In treating the hard sphere collision integral RA used Enskog's chaos assumption that $f_2 = gff$, where g is the local equilibrium radial distribution function. Equation (3.2) is also the result of using Enskog's chaos assumption in the "soft" part of the collision integral.

Using arguments similar to those used in deriving (3.1) Rice and Allnatt derive the following similar equation for the doublet distribution function:

$$\partial_t f_2 + \sum_{i=1,2} (\mathbf{v}_i \cdot \nabla_i + m^{-1}\mathbf{F}_i^{(2)*} \cdot \boldsymbol{\partial}_i) f_2 = \mathscr{J}^H + \mathscr{J}^S \tag{3.3}$$

where $\mathscr{J}^H$ accounts for hard sphere collisions of Particles 1 and 2 with a third particle and $\mathscr{J}^S$ is a Fokker-Planck kind of operator accounting for the Brownian motion of a pair of molecules through the soft field of their neighbors. $\mathscr{J}^H$ and $\mathscr{J}^S$ will not be reproduced here. They are more complicated than the collision operators of (3.1) but similar in content: effects of memory and the time evolution of spatial delocalization have been neglected and an Enskog-like chaos assumption has been used in obtaining $\mathscr{J}^H$ and $\mathscr{J}^S$.

By integrating (3.3) over $d\mathbf{v}_1\, d\mathbf{v}_2$ and making some simplifying assumptions, Rice and Allnatt obtain for the nonequilibrium spatial distribution function, $\rho^{(2)}$, defined by the expression

$$\rho^{(2)}(\mathbf{x}_1, \mathbf{x}_2, t) = \int f_2(\mathbf{x}_1, \mathbf{x}_2, \mathbf{v}_1, \mathbf{v}_2, t)\, d\mathbf{v}_1\, d\mathbf{v}_2 \tag{3.4}$$

a Smoluchowski equation of the form

$$\frac{\partial \rho^{(2)}}{\partial t} = \sum_{i=1,2} \left[\nabla_i \cdot \left(\frac{kT}{\zeta^S} \nabla_i \rho^{(2)} - \frac{\mathbf{F}_i^{(2)}}{\zeta^S} \rho^{(2)} \right) - \nabla_i \cdot (\mathbf{u}(\mathbf{x}_i, t) \rho^{(2)}) \right] \tag{3.5}$$

where

$$\mathbf{F}_i^{(2)} = kT \nabla_i \ln g(\mathbf{x}_{12}) \tag{3.6}$$

and ζ^S is given by (2.52). The part of the momentum flux denoted $\mathbf{J}_v{}^S$ may be computed from $\rho^{(2)}$.

B. The Transport Coefficients

Equations (3.1), (3.3), and (3.5) must be solved in the linear transport limit and then the definitions given in Section IIC used to obtain the RA transport coefficients. We shall not reproduce the solutions here. The reader is referred to Ref. 12 for such details. However, it is interesting to compare the values of the coefficients b_0, a_1, and b_2 of the trial function for ϕ, (2.41) determined from (3.1) with those obtained from the effective potential theory. The RA values are

$$b_0 = -\frac{5}{3(\zeta_{(1)}^H + \zeta^S)} \left[1 + \frac{2P^H}{5nkT} \right] \tag{3.7}$$

$$b_2 = 0 \tag{3.8}$$

and

$$a_1 = \frac{5}{4(\zeta_{(1)}^H + \zeta^S)} \left[1 + \frac{3P^H}{5nkT} \right] \tag{3.9}$$

These results are to be compared with (2.47) and (2.48). The zero value of b_2 is obtained in the RA theory by neglecting $g_2^{(3)}$ and $g_2^{(4)}$ and by eliminating under Enskog's chaos assumption some dynamic correlations in the hard sphere collision operator. Similar effects have been neglected in the RA values of b_0 and a_1. As mentioned earlier, the effective potential result for a_1 contains a term, $(\partial e^S/\partial T)_P$, arising from the temporal contributions $g_2^{(3)}$ and $g_2^{(4)}$, which is singular at the critical point and, therefore, could constitute a serious shortcoming of the RA theory in the critical region.

We shall list here a summary of the RA transport coefficients in a convenient form presented recently by Schrodt, Ku, and Luks.[26] In the notation of Section IIC, these are as follows:

$$\eta = \eta_K + \eta_V \tag{3.10}$$

$$\eta_K = \eta^*\left\{\frac{1 + \frac{8}{5}yg(\sigma)}{(\zeta^H + \zeta^S)/\zeta^*}\right\} \tag{3.11}$$

$$\eta_V = \eta^*\left\{[4yg(\sigma)]\left[\frac{2\eta_K}{5\eta^*} + \frac{192}{25\pi}\,y - \frac{576}{25\pi}\left(\frac{\zeta^S}{\zeta^*}\right)y^2\Psi^{\cdot(2)}(\sigma)\right]\right.$$
$$\left. + \frac{9216}{25\pi}\frac{\zeta^S}{\zeta^*}\,\theta y^3 I_\eta\right\} \tag{3.12}$$

$$\kappa = \kappa \quad + \kappa_V \tag{3.13}$$

$$\kappa_K = \kappa^*\left\{\frac{1 + \frac{12}{5}yg(\sigma)}{(\zeta^H + \zeta^S)/\zeta^*}\right\} \tag{3.14}$$

$$\kappa_V = \kappa^*\left\{[4yg(\sigma)]\left[\frac{3\kappa_K}{5\kappa^*} + \frac{128}{25\pi}\,y\right] - \frac{256}{25}\frac{\zeta^*}{\zeta^S}\,\theta y I_\kappa\right\} \tag{3.15}$$

$$\Phi = \Phi_K + \Phi_V \tag{3.16}$$

$$\Phi_K = 0 \tag{3.17}$$

$$\Phi_V = \eta^*\left\{\frac{256}{5\pi}\,y^2 g(\sigma) + \frac{9216}{15\pi}\frac{\zeta^S}{\zeta^*}\,\theta y^3 I_\Phi\right\} \tag{3.18}$$

$$D = \frac{kT}{\zeta^H_{(1)} + \zeta^S} \tag{3.19}$$

with the definitions

$$\theta = \frac{\varepsilon}{kT} \tag{3.20}$$

$$y = \frac{\pi}{6}\,n\sigma^3 \tag{3.21}$$

$$\eta^* = \frac{5}{16\sigma^2}\left(\frac{mkT}{\pi}\right)^{1/2} \tag{3.22}$$

$$\kappa^* = \frac{75}{64\sigma^2}\left(\frac{k^3T}{\pi m}\right)^{1/2} \tag{3.23}$$

$$\zeta^* = \tfrac{8}{5}n\sigma^2(\pi mkT)^{1/2} \tag{3.24}$$

$$I_\eta = \varepsilon^{-1}\int_1^\infty x^3\,\frac{dV^S}{dx}\,(x\sigma)\Psi^{\cdot(2)}(x\sigma)g(x\sigma)\,dx \tag{3.25}$$

$$I_\kappa = \varepsilon^{-1}\int_1^\infty x^2 g(x\sigma)\left\{V^S(x\sigma) - \frac{x}{3}\frac{dV^S(x\sigma)}{dx} + \frac{x^3}{3}\left[\left(\frac{dV^S(x\sigma)/x}{dx}\right)\right]\frac{d}{dx}\right\}$$
$$\times\left(\frac{\partial}{\partial\theta}\ln g(x\sigma)\right)_p dx \tag{3.26}$$

$$I_\Phi = \varepsilon^{-1}\int_1^\infty x^3\,\frac{dV^S(x\sigma)}{dx}\,\Psi^{\cdot(0)}(x\sigma)g(x\sigma)\,dx \tag{3.27}$$

The quantity ϵ is an arbitrary scale factor with dimensions of energy, x is distance in units of σ, $g(x\sigma)$ is the equilibrium radial distribution function, and $\zeta_{(1)}^{H}$ and ζ^{S} are defined by (2.50) and (2.52). The quantities $\Psi^{\prime(0)}$ and $\Psi^{\prime(2)}$ satisfy the differential equations

$$\frac{d}{dx}\left[x^2 g(x\sigma)\frac{d\Psi^{\prime(2)}(x\sigma)}{dx}\right] - 6g(x\sigma)\Psi^{\prime(2)}(x\sigma) - x^3\frac{dg(x\sigma)}{dx} = 0 \quad (3.28)$$

$$\begin{aligned} \Psi^{\prime(2)} &\to 0\,, & x &\to \infty \\ \frac{d\Psi^{\prime(2)}}{dx} &= 0\,, & x &\to 1+0 \end{aligned} \quad (3.29)$$

and

$$\frac{d}{dx}\left[x^2 g(x\sigma)\frac{d\Psi^{\prime(0)}(x\sigma)}{dx}\right] - x^3\frac{dg(x\sigma)}{dx} + 3x^2\left(\frac{\partial g(x\sigma)}{\partial \ln n}\right)_T = 0 \quad (3.30)$$

$$\begin{aligned} \Psi^{\prime(0)} &\to 0\,, & x &\to \infty \\ \frac{d\Psi^{\prime(0)}}{dx} &= 0\,, & x &\to 1+0 \end{aligned} \quad (3.31)$$

Although the RA formulas look simpler in form than the formulas obtained from the effective potential theory, for viscosity computations the former are not easier to use than the latter because one has to solve the associated differential equations, (3.28) and (3.30), for each desired thermodynamic state. These differential equations, incidentally, are obtained by solving the Smoluchowski equation for viscous transport.

Some comparisons of the RA theory with experiment and with the PNM theory are presented in Section IVC. Other comparisons are available elsewhere. Generally speaking, the RA predictions are lower than experiment, though not more than 30–40% lower, and the qualitative trends and density are correctly predicted.[12]

Wei and the author[27] have extended the RA theory to binary mixtures.

IV. THE PRIGOGINE-NICOLAS-MISGUICH THEORY

A. The Model

In the long-time (or transport) limit, the Fourier transform of the doublet correlation function has been shown by Severne[2] to be of the form

$$g_2(\mathbf{l}, \mathbf{v}_1, \mathbf{v}_2; \mathbf{x}_1, t) = \sum_{s=2}^{N} \int \prod_{i=3}^{s} d\mathbf{v}_i \prod_{j=2}^{N} d\mathbf{x}_j \delta(\mathbf{x}_1 - \mathbf{x}_j) C^{(s)}(\mathbf{l}, \{\nabla\}; +io + i\partial_t) \prod_{i=1}^{s} f(\mathbf{x}_i, \mathbf{v}_i, t) \quad (4.1)$$

where the "creation operator" $C^{(s)}$ is defined by the relation

$$C^{(s)}(\mathbf{l}, \{\nabla\}; +io + i\partial_t)$$
$$= -\frac{\Omega}{8\pi^3} C^{-s+2}$$
$$\times \sum_{n=1}^{\infty} \left\langle \mathbf{l} + \sum_{j=1}^{N} \mathbf{q}_j, -\mathbf{l} \right| [(L_0 - z)^{-1}(-\delta L)]^n \left| \mathbf{q}_1, \ldots, \mathbf{q}_N \right\rangle_s^{ir} \Bigg|_{z \to +io + i\partial_t}^{\mathbf{q}_j \to -i\nabla_j} \tag{4.2}$$

The pointed brackets in (4.2) denote a Fourier matrix element of the form

$$\langle \mathbf{k}_1', \ldots, \mathbf{k}_N' | A | \mathbf{k}_1, \ldots, \mathbf{k}_N \rangle$$
$$\equiv \frac{1}{\Omega^N} \int d\mathbf{x}_1 \cdots d\mathbf{x}_N \exp\left(-i \sum_{i=1}^{N} \mathbf{k}_i' \cdot \mathbf{x}_i\right) A(\mathbf{x}_1, \ldots, \mathbf{x}_N)$$
$$\times \exp\left(+i \sum_{i=1}^{N} \mathbf{k}_i \cdot \mathbf{x}_i\right) \tag{4.3}$$

Ω is the volume of the system containing N particles. The indices in the bra $\left\langle \mathbf{l} + \sum_{j=1}^{N} \mathbf{q}_j, -\mathbf{l} \right|$ mean that $\mathbf{k}_1' = \mathbf{l} + \sum_{i=1}^{N} \mathbf{q}_j$, $\mathbf{k}_2' = -\mathbf{l}$ and $\mathbf{k}_i' = 0$, $i = 3, \ldots, N$.

The quantities L_0 and δL are the free particle and interaction parts of the Liouville operator,

$$L_0 = -i \sum_{j=1}^{N} \mathbf{v}_j \cdot \nabla_j \tag{4.4}$$

and

$$\delta L = i \sum_{i,j}^{N} \nabla_i V(\mathbf{x}_{ij}) \cdot \boldsymbol{\partial}_{ij} \tag{4.5}$$

whose matrix elements are of the form

$$\langle \{\mathbf{k}'\} | L_0 | \{\mathbf{k}\} \rangle = \sum_j \mathbf{k}_j \cdot \mathbf{v}_j \prod_{j=1}^{N} \delta_{\mathbf{k}_j', \mathbf{k}_j}^{Kr} \tag{4.6}$$

and

$$\langle \{k'\} | \delta L | \{k\} \rangle = \frac{8\pi^3}{\Omega} \sum_{i,j} V_{|\mathbf{k}_j' - \mathbf{k}_j|}(\mathbf{k}_j' - \mathbf{k}_j) \cdot \boldsymbol{\partial}_{ji} \delta_{\mathbf{k}_j' + \mathbf{k}_i', \mathbf{k}_j + \mathbf{k}_i}^{Kr} \prod_{l \neq i,j}^{N} \delta_{\mathbf{k}_l', \mathbf{k}_l}^{Kr} \tag{4.7}$$

The subscript "s" on the ket in (4.2) indicates that exactly s distinct particles will be involved in the interaction operators δL defining $C^{(s)}$. The superscript "ir" on the same ket means that, after all the Fourier matrix elements are written out explicitly, no intermediate state $\{\mathbf{k}''\}$ will exist such that all the vectors $\{\mathbf{k}''\} = \{\mathbf{k}_1'', \mathbf{k}_2'', \ldots\}$ can be written exclusively in terms of the vectors $\mathbf{q}_i$. The notation $\mathbf{q}_j \to -i\nabla_j$ means to replace the vectors $\mathbf{q}_j$ by the operators $-i\nabla_j$ after the matrix elements have been fully explicited. And finally, the notation $z \to +io + i\partial_t$ means that the matrix elements of $(L_0 - z)^{-1}$, that is, $(\sum_j \mathbf{v}_j \cdot \mathbf{k}_j - z)^{-1}$, are to be replaced by the operator

$$\frac{1}{\sum_j \mathbf{k}_j \cdot \mathbf{v}_j - i\partial_t - io} \equiv i\pi\delta\left(\sum_j \mathbf{k}_j \cdot \mathbf{v}_j - i\partial_t\right) + \mathfrak{p}\frac{1}{\sum_j \mathbf{k}_j \cdot \mathbf{v}_j - i\partial_t} \tag{4.8}$$

where δ and $\mathfrak{p}$ denote the Dirac delta function and principal part operators, respectively.

The operators ∇_j and ∂_t account for the role of spatial inhomogeneities and delocalizations and memory (or nonstationary) effects in the collisional processes giving rise to the transport limit of the doublet correlation function.

The expression for $C^{(s)}$ can be rewritten in a form more suggestive of the approximations made by PNM, namely,

$$\begin{aligned} C^{(s)}&(\mathbf{l}, \{\nabla\}; +io + i\partial_t) \\ &= -\frac{\Omega}{8\pi^3} C^{-s+2} \sum_{n=1}^{\infty} \Bigg[\exp\left(+\mathbf{q}_2 \cdot \frac{\partial}{\partial \mathbf{l}} + \sum_{j=\nu+1}^{N} \mathbf{q}_i \cdot \frac{\partial}{\partial \mathbf{l}}\right) \Big\langle \mathbf{l} + \mathbf{q}_1 + \sum_{j=3}^{\nu} \mathbf{q}_j, \\ &\quad -\mathbf{l} + \mathbf{q}_2 + \sum_{j=\nu+1}^{N} \mathbf{q}_j \Big| [(L_0 - z)^{-1}(-\delta L)]^n \Big| \mathbf{q}_1, \ldots, \mathbf{q}_N \Big\rangle_s^{ir} \Bigg] \Bigg|_{z \to io + ij}^{\mathbf{q}_1 \to i\nabla_j} \end{aligned} \tag{4.9}$$

where ν is some integer greater than or equal to 3. It turns out that it is in the nature of the PNM approximations that ν does not have to be chosen explicitly.

The *first approximation* of the Prigogine, Nicolis, and Misquich[13] is to neglect the deviation of f from local equilibrium; that is, they approximate f by

$$f^0(\mathbf{x}_i, \mathbf{v}_i, t) = n(\mathbf{x}_i, t)\left(\frac{m}{2\pi kT(\mathbf{x}_i, t)}\right)^{3/2} \exp\left[-\frac{m}{2kT(\mathbf{x}_i, t)}(\mathbf{v}_i - \mathbf{u}(\mathbf{x}_i, t))^2\right] \tag{4.10}$$

where n, T, and $\mathbf{u}$ are the local hydrodynamic density, temperature, and mass average velocity. This approximation eliminates diffusional transport phenomena and the kinetic contributions to the transport coefficients such as viscosity and thermal conductivity. For liquids at low temperatures and high densities the kinetic parts of η, Φ, and κ are usually small compared to the potential parts.

The ordering of the factors $(L_0 - z)^{-1}$ in the operator $C^{(s)}$ actually corresponds to the evolution in time of the interactions among the particles of the system. The time is read from right to left, that is, the leftmost operator

$$(L_0 - z)^{-1} \rightarrow (\mathbf{l} \cdot \mathbf{v}_{12} - i(\mathbf{v}_1 \cdot \nabla_1 + \mathbf{v}_2 \cdot \nabla_2 + \partial_t) - io)^{-1} \qquad (4.11)$$

refers to the most recent effects of inhomogeneities and nonstationarities on the behavior of interacting particles. The *second approximation* of PNM is to neglect the effects of inhomogeneities and nonstationarities except for those appearing most recently, that is, those appearing in the leftmost matrix elements of $(L_0 - z)^{-1}$ and those included in the operator

$$\exp\left(\mathbf{q}_2 \cdot \frac{\partial}{\partial \mathbf{l}} + \sum_{j=n+1}^{N} \mathbf{q}_j \cdot \frac{\partial}{\partial \mathbf{l}}\right)$$

To help see the motivation for projecting out the inhomogeneities included in the latter operator, the reader is advised to examine the explicit form of the weak coupling (first order in V) limit of g_2. The factor $e^{+\mathbf{q}_2 \cdot (\partial/\partial \mathbf{l})}$ is necessary, for example, to account for the delocalization of a pair of interacting particles and to obtain the correct nondissipative hydrodynamic equation for $\mathbf{u}$ from the hierarchy equation for f.

Under the two assumptions described above, the PNM doublet correlation function becomes

$$g_2^{PNM}(\mathbf{l}, \mathbf{v}_1, \mathbf{v}_2; \mathbf{x}_1, t)$$

$$= \sum_{s=2}^{N} \int \prod_{j=3}^{s} d\mathbf{v}_j \prod_{j=2}^{N} d\mathbf{x}_i \delta(\mathbf{x}_1 - \mathbf{x}_j) \exp\left(-i\nabla_2 \cdot \frac{\partial}{\partial \mathbf{l}} - i \sum_{j=\nu+1}^{N} \nabla_j \cdot \frac{\partial}{\partial \mathbf{l}}\right)$$

$$\times \frac{1}{\mathbf{l} \cdot \mathbf{v}_{12} - i\left(\mathbf{v}_1 \cdot \nabla_1 + \mathbf{v}_2 \cdot \nabla_2 + \sum_{i=3}^{\nu} \mathbf{v}_1 \cdot \nabla_j + \sum_{j=\nu+1}^{N} \mathbf{v}_2 \cdot \nabla_j\right) - i\partial_t - io} G_2^{PNM} \qquad (4.12)$$

where

$$G_2^{PNM} = \frac{\Omega}{8\pi^3} c^{-s+2} \sum_{n=1}^{\infty} \langle \mathbf{l}, -\mathbf{l} | \, \delta L[(L_0 - io)^{-1}(-\delta L)]^n \, |0\rangle_s^{ir} \prod_{j=1}^{s} f^0(\mathbf{x}_i, \mathbf{v}_i, t) \qquad (4.13)$$

The *third approximation* of the PNM theory is that one can neglect deviations of G_2^{PNM} from its corresponding local equilibrium value (that is, its value for an ensemble in which each particle is in equilibrium with respect to the density, temperature and hydrodynamic velocity at its location). Using this approximation in (4.13) and linearizing G_2^{PNM} in ∇_i and ∂_t, Misguich, in a rather lengthy work,[14] derives the following expression for g_2^{PNM}:

$$g_2^{PNM}(\mathbf{l}, \mathbf{v}_1, \mathbf{v}_2; \mathbf{x}_1, t) = G_l f^0(1) f^0(2) + g_{2s}^{(1)} + g_{2p}^{(1)}(st.) + g_{2p}^{(1)}(n.st.) + g_{2s}^{COR} \tag{4.14}$$

$$g_{2s}^{(1)} = -i \frac{\partial}{\partial \mathbf{l}} \cdot \left\{ \frac{1}{\mathbf{l} \cdot \mathbf{v}_{12} - io} (\mathbf{Q}\nabla_1 \ln T + \nabla_1 \mathbf{u}^T) \cdot \left(\mathbf{1} - \frac{m}{2kT} \mathbf{v}_{12} \cdot \mathbf{l}\mathbf{v}_{12}\right) G_l f^0(1) f^0(2) \right\} \tag{4.15}$$

$$g_{2p}^{(1)}(st.) = \frac{i}{\mathbf{l} \cdot \mathbf{v}_{12} - io} G_l f^{\circ}(1) f^0(2) \left\{ \left[\frac{m}{2kT} [2(\mathbf{Q} + \mathbf{u})(Q^2 + \tfrac{1}{4}v_{12}^2) + \mathbf{v}_{12} \cdot \mathbf{Q}(\mathbf{v}_{12} + \mathbf{u})] + \left(T \frac{\partial}{\partial T} \ln G_l\right)(\mathbf{Q} + \mathbf{u}) \right] \cdot \nabla_1 \ln T + \frac{m}{2kT} (4\mathbf{Q}\mathbf{Q} + \mathbf{v}_{12}\mathbf{v}_{12}) : \nabla \mathbf{u} + (\mathbf{Q} + \mathbf{u}) \cdot \nabla_1 \ln n \right\} \tag{4.16}$$

$$g_{2p}^{(1)}(n.st.) = \partial_t \frac{i}{\mathbf{l} \cdot \mathbf{v}_{12} - io} G_l f^0(1) f^0(1) \tag{4.17}$$

$$g_{2s}^{COR} = + \frac{i}{2} \nabla_1 \ln T \cdot \frac{\partial}{\partial \mathbf{l}} \left\{ \frac{1}{\mathbf{l} \cdot \mathbf{v}_{12} - io} \left[G_l \mathbf{l} + \frac{1}{kT} \int d\mathbf{l}' V_{\mathbf{l}-\mathbf{l}'} (\mathbf{l} - \mathbf{l}') G_{l'} \right] \cdot \mathbf{Q} f^0(1) f^0(2) \tag{4.18}$$

Recall that the quantity G_l [defined in (2.15)] is the Fourier transform of the excess equilibrium radial distribution function evaluated at $T(\mathbf{x}_1, t)$ and $n(\mathbf{x}_1, t)$. The definition

$$\mathbf{Q} = \tfrac{1}{2}(\mathbf{v}_1 + \mathbf{v}_2) - \mathbf{u}(\mathbf{x}_1, t) \tag{4.19}$$

has been used.

The quantity $G_l f^0 f^0$ corresponds to the local equilibrium ($\phi \equiv 0$) limit of the $g_2^{(1)}$ of the effective potential model. The quantities $g_{2s}^{(1)} + g_{2s}^{COR}$, $g_{2p}^{(1)}(n.st.)$ and $g_{2p}^{(1)}(st.)$ have the same physical origin as $g_2^{(2)}$, $g_2^{(3)}$, and $g_2^{(4)}$, respectively, of the effective potential model, and are quite similar in structure. In fact, $g_2^{(3)}$ is identical to $g_{2p}^{(1)}(n.st.)$.

The quantity g_{2s}^{COR} did not appear in the original PNM theory. However, under the three assumptions outlined above, this term, affecting only the thermal conductivity, must be kept. Misguich[28] has recently derived the corrections g_2^{COR} gives to the original PNM formula for thermal conductivity.

B. The Transport Coefficients[14]

The PNM transport coefficients are obtained by inserting g_2^{PNM} into the flux equations, (2.72)–(2.75), performing the integrations, and comparing the resulting expressions with the phenomenological laws, (2.77) and (2.78).

The PNM *thermal conductivity* is

$$\kappa^{PNM} = \kappa^E + \tilde{\kappa}_s + \kappa_p + \kappa_s^{COR} \tag{4.20}$$

κ^E is an Enskog-like hard core contribution (coming from a contribution of $g_{2s}^{(1)}$ to $\mathbf{q}_v{}^H$)

$$\kappa^E = \kappa^0 g(\sigma) \tag{4.21}$$

$$\kappa^0 = \frac{2}{3} n^2 \sigma^4 \frac{k}{m} \sqrt{\pi m k T} = \frac{128\pi}{225} n^2 \sigma^6 \kappa^* \tag{4.22}$$

$g_{2s}^{(1)}$ gives the further contribution

$$\tilde{\kappa}_s = \kappa_s{}^H + \kappa_s{}^S \tag{4.23}$$

where $\kappa_s{}^H$ comes from $\mathbf{q}_v{}^H$

$$\kappa_s{}^H = -\frac{\kappa^0}{2}\left\{G(\sigma) + \tfrac{1}{8}\int_0^\infty dx G(\sigma x) \frac{d}{dx}[P_2(x) - P_4(x)]\right\} \tag{4.24}$$

and from $\mathbf{q}_v{}^S$

$$\kappa_s{}^S = \kappa^* \left[\frac{256\pi^5}{225}\left(\frac{n^2\sigma^2}{kT}\right)(5W_2 + W_3)\right] \tag{4.25}$$

κ_p comes from $g_{2p}^{(1)}$ and is composed of two parts, $\kappa_p{}^H$ and $\kappa_p{}^S$, coming from $\mathbf{q}_v{}^H$ and $\mathbf{q}_v{}^S$, respectively.

$$\kappa_p{}^H = \frac{\kappa^0}{32}\int_0^\infty dx\, x\left\{[g(\sigma x) - 1][8P_4(x) - P_2(x)] + 2P_2(x)\left(T\frac{\partial g(\sigma x)}{\partial T}\right)_n\right\} \tag{4.26}$$

where the P_n are defined by (2.60)–(2.64).

$$\kappa_p{}^S = \kappa^* \left\{\frac{256\pi^5}{225}\frac{n^2\sigma^2}{kT}\left[\left(T\frac{\partial}{\partial T}[6W_1 + W_2]\right)_n - 5W_1 - \frac{3}{2}W_2\right]\right\} \tag{4.27}$$

where W_n are defined by (2.92).

Finally, the contributions from g_{2s}^{COR} are[28]

$$\kappa_s^{COR} = \kappa_N - \tfrac{1}{2}\tilde{\kappa}_s \tag{4.28}$$

where

$$\kappa_N = \kappa^E\left\{\frac{1}{2} + \frac{1}{32kT}\int_0^\infty dx\left[V^S(\sigma x) - \frac{\partial}{\partial x}V^S(\sigma x)\right][P_2(x) - P_4(x) - 4]\right\}$$
$$+ \frac{\kappa^0}{32kT}\int_0^\infty dx\, g(\sigma x)\frac{dV^S(\sigma x)}{dx}[P_2 - P_4 - 4]$$
$$+ \frac{\kappa^0}{8(kT)^2}\int_1^\infty dx\, x\left[V^S - x\frac{dV^S}{dx}\right]\int_1^\infty dx'\frac{dV^S}{dx'}g(\sigma x')$$
$$\times\left[xx' - \tfrac{1}{2}(x^2 + x'^2)\ln\left|\frac{x + x'}{x - x'}\right|\right] \tag{4.29}$$

By setting $\kappa_s^{COR} \equiv 0$ in (4.20), one recovers the original version of the *PNM* theory. The original κ^{PNM} was sometimes larger than experiment, so that if κ_s^{COR} turns out to be negative it will be an improvement of the theory, especially since the kinetic contributions neglected by PNM are positive and would, therefore, give even larger predicted thermal conductivities if included.

If the kinetic terms in the effective potential theory are ignored, we obtain formulas similar to (but not identical to) the PNM results.

The PNM formulas for the shear and the bulk viscosities are not affected by the term g_{2s}^{COR}. These are, respectively,

$$\eta^{PNM} = \eta_s^E + \eta_s^H + \eta_p \tag{4.30}$$

and

$$\Phi^{PNM} = \tfrac{5}{3}\eta^E + \tfrac{5}{3}\eta_s^H + \Phi_p \tag{4.31}$$

Again η^E is the Enskog hard core term (a contribution of $g_{2s}^{(1)}$ to $\mathbf{J}_V^H$).

$$\eta^E = \eta^0 g(\sigma) \tag{4.32}$$

$$\eta^0 = \tfrac{4}{15}n^2\sigma^4\sqrt{\pi mkT} = \frac{64\pi}{75}n^2\sigma^6\eta^* \tag{4.33}$$

η_s^H is also a contribution of $g_{2s}^{(1)}$ to $\mathbf{J}_V^H$.

$$\eta_s^H = \eta^0 S \tag{4.34}$$

where

$$S = -\tfrac{1}{2}G(\sigma) - \tfrac{1}{16}\int_1^\infty dx\, G(\sigma x)\frac{d}{dx}(P_2 - P_4) \tag{4.35}$$

The term η_p is the sum of η_p^H, a contribution of $g_{2p}^{(1)}$ to $\mathbf{J}_V^H$,

$$\eta_p^H = \frac{\eta^0}{8}\int_1^\infty dx\, G(\sigma x)x[3P_4 - P_2] \tag{4.36}$$

and $\eta_p{}^S$, a contribution of $g_{2p}^{(1)}$ to $\mathbf{J}_V{}^S$,

$$\eta_p{}^S = \frac{128\pi^5 n^2 \sigma^2 \eta^*}{74kT}(3W_2 + W_3) \tag{4.37}$$

Finally, the quantity Φ_p is composed of the terms

$$\Phi_p{}^H = -\tfrac{5}{16}\eta^0 \int_1^\infty dx\, xP_2\left(XT\frac{\partial}{\partial T} + \tfrac{1}{2}X'\right)G(\sigma x) \tag{4.38}$$

and

$$\Phi_p{}^S = \frac{128\pi^5 n^2 \sigma^2 \eta^*}{75kT}\left[\frac{1}{3}(4W_2 + W_3) + \left(\tfrac{1}{2}X' - XT\frac{\partial}{\partial T}\right)(3W_1 + W_2)\right] \tag{4.39}$$

where

$$X \equiv \frac{1}{n}\left(\frac{\partial P}{\partial \bar{e}}\right)_n \equiv \frac{2}{3} + X' \tag{4.40}$$

$\bar{e}$ is the thermodynamic energy per particle.

Misguich[14] has also obtained viscosity and thermal conductivity formulas for binary mixtures.

C. Comparison of the RA and PNM Theories with Experiment

Misguich's correction to the PNM thermal conductivity will be ignored in the following calculations since no numerical calculations of the correction are available presently.

In Tables II and III shear viscosities and thermal conductivities predicted by the PNM and the RA theories are compared with experiment[29–31]

TABLE II

Thermal Conductivity of Liquid Argon[a]

State	90°K 1.3 atm	128°K 50 atm	133.5°K 100 atm	185°K 500 atm
κ(obs.)[b]	2.96	1.89	1.86	1.87
κ(calc.)[c]	3.42	...	...	...
κ(calc.)[d]	1.05	1.35	1.29	1.51
κ(calc.)[e]	1.65	1.70	1.59	1.70

[a] Units are 10^{-4} cal/(sec)(cm)(°K).
[b] Ikenberry and Rice's experimental data.[31]
[c] PNM theory.
[d] RA theory.
[e] RA theory with scale factors.[12]

TABLE III

Shear Viscosity of Liquid Argon[a]

State	90°K 1.3 atm	128°K 50 atm	133.5°K 100 atm	185.5°K 500 atm
η(obs.)[b]	2.39	0.84	0.84	0.87
η(calc.)[c]	1.31	0.82	0.83	0.86
η(calc.)[d]	1.41	0.69	0.70	0.86
η(calc.)[e]	1.74	0.73	0.73	0.87

[a] Units are in millipoise.
[b] After Lowry, Rice, and Gray.[29,12]
[c] PNM theory.
[d] RA theory.
[e] RA theory with scale factors.[12]

for liquid argon. The lowest temperature (90°K) is the only point below the critical pressure. At this point, all three theories seriously underestimate the viscosity: the PNM result is slightly better than the original RA result.

Above the critical pressure, the PNM predictions always remain in the experimental range of error (2%); the average absolute errors of the RA predictions are about 12%.

All these predictions are based on the modified Lennard-Jones potential and Kirkwood, Lewinson, and Alder's computation of the equilibrium pair-correlation function for the truncated Lennard-Jones potential model. The RA results use an experimental value of the friction coefficient, ζ. However, Palyvos and Davis[32] have shown that the RA results do not change greatly when theoretical values of the friction coefficient are used.

By introducing scale factors determined by requiring the equation of state to give the correct pressures at the given temperature and density, Rice and Gray improved the RA predictions considerably. Their "scaled" values of η are shown in the last two rows of Table II. The scaled values are improvements over the unscaled values of the viscosity. However, above critical pressure the PNM model still gives the better predictions.

The thermal conductivities are compared in Table III. Below the critical pressure the PNM model gives somewhat better agreement with experiment than the RA model. Above critical pressure there are no available PNM results.

In Table IV and Fig. 1 we compare the PNM and RA predictions of the bulk viscosity. Unfortunately, the available computations of ϕ for liquid argon based on the RA theory are for conditions under which there are no experimental values of the bulk viscosity. The single experimental

TABLE IV

Bulk Viscosity of Liquid Argon[a]

State	90°K 1.3 atm	128°K 50 atm	133.5°K 100 atm	185.5°K 500 atm
ϕ(obs.)[b]	1.9–2.1			
ϕ(calc.)[c]	1.81			
ϕ(calc.)[d]	<0	0.90	0.93	1.12

[a] Units are in millipoise.
[b] Inferred from Naugle, Lunsford, and Singer's data.[33]
[c] PNM theory.
[d] RA theory.

value in the first row of the table was obtained by extrapolation of the data of Naugle, Lunsford, and Singer.[33] The PNM result agrees fairly well with the extrapolated value while the RA theory predicted a negative value for Φ at that temperature and pressure. It should be pointed out, however, that the RA predictions of bulk viscosity agree very well with experiment for gaseous argon over rather wide ranges of temperature and density. This agreement is shown in Fig. 2.

In Table V the PNM theory is compared to experiments on argon along the vapor-pressure curve. Figs. 2 and 3 give similar comparisons of the PNM theoretical thermal conductivities and experiment for liquids krypton and xenon.

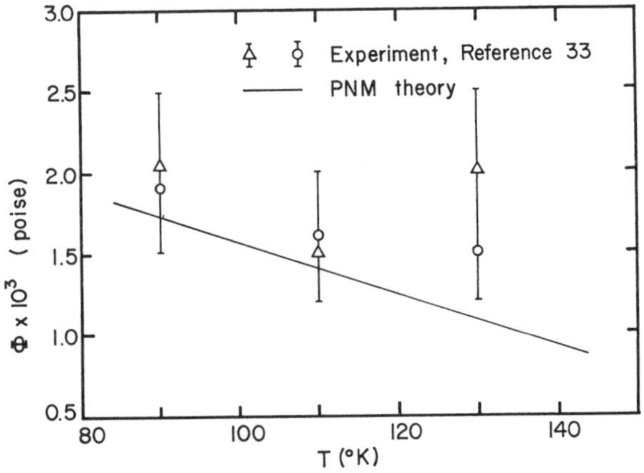

Fig. 1. Bulk viscosity versus temperature for argon along vapor-pressure curve. △ and ○ are experimental points estimated from data on Φ/η (using experimental values[29] of η) and Φ, respectively.

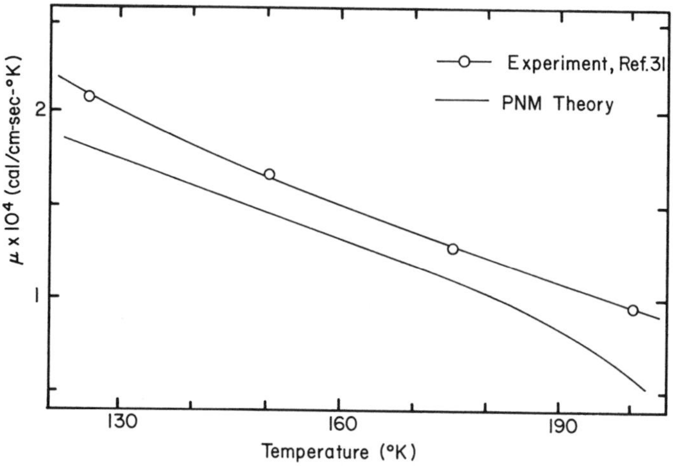

Fig. 2. Thermal conductivity versus temperature for liquid krypton along the vapor-pressure curve.

The RA and PNM theories both seem to predict transport coefficients generally with no more than about 30–40% error. The kinetic contributions neglected in the PNM theory but retained in the effective potential theory of Section II and in the generalized RA theory to be discussed in Section V would give about a 5–15% correction to the PNM results. This increase would be an overall improvement in the comparisons made here except in the case of the thermal conductivity of xenon. In this case the PNM predictions are too high.

TABLE V

PNM Results for Argon Along Vapor-Pressure Curve, with Percentage of the Experimental Values (A, 110°K; B, 130°K; C, 150°K)

	$\eta \times 10^{+3}$ (poise)			$\zeta \times 10^{+3}$ (poise)			$\kappa \times 10^{+4}$ [cal/(sec)(cm)(°K)]		
	A	B	C	A	B	C	A	B	C
PNM	1.04	0.73	0.44	1.44	1.06	0.67	2.16	1.28	0.70
	77%	83%	86%	90–94%	54–71%	...	93%	76%	54%
Exptl.[a]	1.36	0.88	0.51	1.54	1.50	...	2.32	1.69	1.31
				1.60	1.96				
				(±25%)	(±25%)				

[a] Ref. 29 for shear viscosity, Ref. 31 for thermal conductivity, Ref. 33 for bulk viscosity.

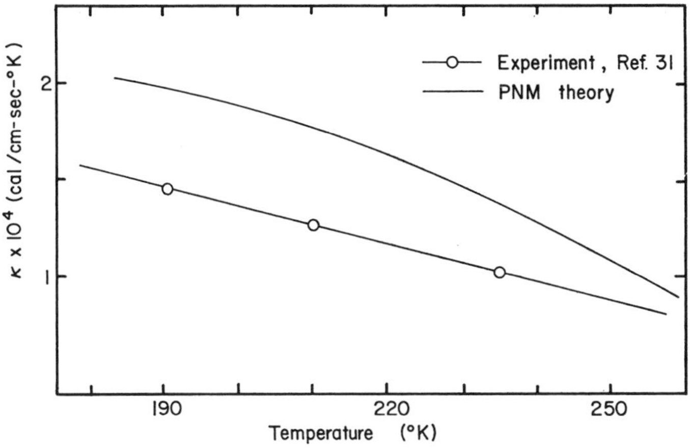

Fig. 3. Thermal conductivity versus temperature for liquid krypton along the vapor-pressure curve.

V. A GENERALIZED RICE-ALLNATT THEORY

A. The Kinetic Equation

Recently, the author[34] presented a theory which utilizes the RA idea of treating separately the strongly repulsive and "soft" collisional contributions to the singlet kinetic equation and incorporates the PNM approximation to the local equilibrium part of the doublet distribution function. A very similar theory was published independently by Misguich and Nicolis[35,36] under the name "generalized Rice-Allnatt theory," or the GRA theory. In what follows, we shall present the author's version of the theory, not to imply a judgment of the merits of one version over the other but rather to make the author's task easier. The spirit of the GRA theory, like that of the effective potential theory of Section II, is to try to combine the desirable features of the RA and PNM theories in constructing a kinetic equation valid for diffusional as well as other transport coefficients and in low-density as well as high-density regions.

As our starting point, we again take Severne's[2] exact long-time limit for g_2. Since we are interested only in linear transport processes, we write

$$\prod_{i=1}^{s} f(\mathbf{x}_i, \mathbf{v}_i, t) = \prod_{i=1}^{s} f^0(\mathbf{x}_i, \mathbf{v}_i, t) + \prod_{i=1}^{s} f^0(\mathbf{x}_i, \mathbf{v}_i, t) \sum_{\alpha=1}^{s} \phi(\mathbf{x}_\alpha, \mathbf{v}_\alpha, t) \quad (5.1)$$

where f^0 is the local equilibrium value of f and $f^0\phi \equiv f - f^0$ represents the deviation of f from local equilibrium. Inserting (5.1) into the defining expression for g_2, (4.1), and approximating by g_2^{PNM} the contribution to

g_2 from the first member of the right-hand side of (5.1), we obtain the following *linearized* expression for g_2:

$$g_2(\mathbf{l}, \mathbf{v}_1, \mathbf{v}_2; \mathbf{x}_1, t) = g_2^{PNM} + \sum_{s=2}^{N} \int \prod_{j=3}^{s} d\mathbf{v}_j \; c^{(s)}(\mathbf{l}, \{0\}; +io) \prod_{i=1}^{s} f^0(i) \sum_{\alpha=1}^{s} \phi(\alpha) \tag{5.2}$$

where $f^0(i) \equiv f^0(\mathbf{x}_1, \mathbf{v}_i, t)$, $\phi(\alpha) \equiv \phi(\mathbf{x}, \mathbf{v}_\alpha, t)$, and

$$C^{(s)}(\mathbf{l}, \{0\}; +io) = -\frac{\Omega}{8\pi^3} c^{-s+2} \sum_{n=1}^{\infty} \langle \mathbf{l}, -\mathbf{l} | \, [(L_0 - io)^{-1}(-\delta L)]^n \, | 0 \rangle_s^{ir}$$

$$\equiv \frac{1}{\mathbf{l} \cdot \mathbf{v}_{12} - io} \langle \mathbf{l}, -\mathbf{l} | \, (\delta L) A \, | 0 \rangle_s^{ir} \tag{5.3}$$

The second equality in (5.3) is the defining equation for the operator A. At this point, the second member of the right-hand side of (5.2) is exact to the linear approximation. Thus (5.2) is based on two of the three approximations of the PNM theory (the third PNM approximation, that the $\phi(\alpha) = 0$, is not used here). Now let us introduce the *third approximation* of the GRA theory: it is assumed that the quantity $\sum_{\alpha=1}^{s} \phi(\alpha)$ commutes with the operator A in (5.3) that is, the commutator of A and $\Sigma\phi(\alpha)$ is neglected. Under this assumption the author has shown that the resulting expression for g_2 is

$$g_2(\mathbf{l}, \mathbf{v}_1, \mathbf{v}_2; \mathbf{x}_1, t) = g_2^{PNM} + g_2^{(1)} \tag{5.4}$$

where $g_2^{(1)}$ is identical to the corresponding term [defined by (2.21)] obtained in the effective potential theory. The commutation approximation used here is similar to the PNM approximation in the sense that if we read the time axis from right to left in the operator product of (5.3), then by commuting the quantity $\Sigma\phi(\alpha)$ to the left of A we get the contribution to g_2 arising from the most recent (nontrivial) effect of the "inhomogeneity" $\Sigma\phi(\alpha)$ on the collisional process. (If the quantity $\Sigma\phi(\alpha)$ were commuted to the left of $C^{(s)}$ the entire term would vanish. This is what we mean by most recent nontrivial effect.) The error involved in the commutation approximation could perhaps be estimated by examining the first few terms in the expansion of $C^{(s)}$. This has not been done up to the present.

The *fourth approximation* of the GRA theory is that the contribution of the short-range strongly repulsive collisions to the singlet kinetic equation arise from binary hard sphere collisions. Thus the linearized GRA singlet kinetic equation is

$$\partial_t^0 f^0 + \mathbf{v}_1 \cdot \nabla_1 f^0 + m^{-1}\mathbf{F}^{(S)0} \cdot \partial_1 f^0 + \mathscr{B}(g^{PNM})$$

$$= \int_{\mathbf{v}_{21}\cdot\mathbf{k}>0} [f_2(\mathbf{x}_1, \mathbf{x}_1, +\sigma\mathbf{k}, \mathbf{v}_1', \mathbf{v}_2', t)$$

$$- f_2(\mathbf{x}_1, \mathbf{x}_1, -\sigma\mathbf{k}, \mathbf{v}_1, \mathbf{v}_2, t)]\mathbf{v}_{21} \cdot \mathbf{k}\, d\mathbf{k}\, d\mathbf{v}_2 + \mathscr{A}_S\phi \qquad (5.5)$$

where the operators $\mathscr{B}$ and $\mathscr{A}_S$ are defined by (2.29), and (2.34) respectively, and f_2 is to be linearized as follows:

$$f_2(\mathbf{x}_1, \mathbf{x}_1 \pm \sigma\mathbf{k}) = f^0(1)f^0(2)(1 + \phi(1) + \phi(2))$$
$$\pm f^0(1)\sigma\mathbf{k} \cdot \nabla_1 f^0(2) + g_2(\mathbf{x}_1, \mathbf{x}_1 \pm \sigma\mathbf{k}) \qquad (5.6)$$

where g_2 is given by the inverse Fourier transform of (5.4). As in the effective potential theory, we have not used Enskog's chaos assumption. This unnecessary restriction was used for evaluating the Enskog operator in the original GRA kinetic equation that the author proposed. J. Misguich kindly pointed out that the restriction was unnecessary.

B. The Transport Coefficients

Taking the forms of ϕ and g_2 to be those given by (2.41) and (5.4), respectively, and using the defining expressions given in Section I for the various parts of the heat flux, momentum flux, and mass flux, we obtain the following transport coefficients for the GRA model.

Thermal Conductivity

$$\kappa = \kappa_K + \kappa_{KV} + \kappa^{PNM} \qquad (5.7)$$

$$\kappa_K = \tfrac{5}{2}nk^2Ta_1 \qquad (5.8)$$

$$\kappa_{KV} = \kappa_K\left[\frac{2\pi}{5}n\sigma^3 g(\sigma) + \frac{\pi}{15}n\sigma^3 G(\sigma)(1 + 2\chi_3) - \frac{e^S + 9P^S}{15nkT}\right] \qquad (5.9)$$

and κ^{PNM} is the PNM thermal conductivity given by (4.20).

Shear Viscosity

$$\eta = \eta_K + \eta_{KV} + \eta^{PNM} \qquad (5.10)$$

$$\eta_K = -\frac{b_0}{2}nmkT \qquad (5.11)$$

$$\eta_{KV} = \eta_K\left[\frac{4\pi}{15}n\sigma^3 g(\sigma) + \frac{2\pi}{15}n\sigma^3 G(\sigma)(1 - 3\chi_2) + \frac{2(e^S - P^S)}{5nkT}\right] \qquad (5.12)$$

and η^{PNM} is given by (4.30).

Bulk Viscosity

$$\Phi = \Phi_{KV} + \Phi^{PNM} \tag{5.13}$$

$$\Phi_{KV} = b_2\left[\frac{5\pi}{2} n\sigma^3 G(\sigma) nkT - \tfrac{9}{2}P^S\right] \tag{5.14}$$

and Φ^{PNM} is given by (4.31).

Self-Diffusion

$$D = \frac{kT}{\zeta^H_{(1)} + \zeta^H_{(3)} + \zeta^S} \tag{5.15}$$

The self-diffusion coefficient is the same as the effective potential result because the quantity $g_2^{(1)}$, giving $\zeta^H_{(3)}$, appears in both approximations to g_2. The quantities e^S, P^S, χ_n, $\zeta^H_{(1)}$, etc., are defined in various of the (2.47)–(2.59). The quantities κ_{KV}, η_{KV}, and Φ_{KV} are potential contributions arising from the deviations of the singlet distribution from local equilibrium.

As mentioned above, in the original work the author obtained the quantities a_1, b_0, and b_2 from the GRA equation using Enskog's chaos assumption for f_2 in the hard core collision operator. Since the chaos assumption is no longer used, we shall not give the original results but rather suggest the effective potential values, (2.49), (2.47), and (2.48) as the best available values for a_1, b_0, and b_2. The author is presently solving (5.5) for these quantities.

VI. SUMMATION OF DIAGRAMS

A. Some Numerical Studies

In certain cases, in both equilibrium and nonequilibrium theories, diagramic analysis has allowed the systematic summation of dominant contributions in many-body expansions such as Mayer cluster expansions, Feynmann diagrams, the Brussels school's expansions of the type illustrated by (4.2). For example, owing to the long range of the Coulomb interaction, in low-density plasmas and dilute electrolytes a summable class of many-body diagrams may be shown to be the dominant class of terms in many-body expansions. Summing this class of diagrams gives Debye-Huckel theory[37] for equilibrium systems and the Balescu-Guernsey-Lenard kinetic equation[38–40] for plasmas. As another example, low-density diagrams may be summed to give Boltzmann's equation for dilute gases. In a recent series of papers, using the Prigogine-Balescu diagram theory, Allen and Cole[41,42] have presented a kinetic theory of simple liquids based on suming certain summable classes of diagrams which they hope are dominant in dense fluids. In what follows in this section, we present briefly some numerical studies by Dowling and the author[34,43] which indicate that is is unlikely that a term by term analysis of diagrams will reveal dominant contributions for simple nonpolar liquids. This is perhaps not surprising since even in equilibrium theory no class of diagrams has been shown to be dominant in liquids. The hypernetted chain class is the only summable class of diagrams which gives good qualitative equilibrium results for liquids; and these diagrams cannot be shown to dominate other terms in a cluster expansion. Thus in equilibrium theory and nonequilibrium theory it appears that the choice of diagrams to sum will have to be those which one can sum explicitly and the validity of the result tested by comparison with experiment.

Also included in what follows is a brief discussion of Allen and Cole's original kinetic theory. In the following section, an improved version of the theory, given by Foster and Cole,[44] will be discussed.

A few comments about the Prigogine-Balescu diagram theory[38,45] are useful before discussing Dowling and the author's calculations. In the diagram theory, one represents terms in operators such as the collision operator $\psi(\mathbf{l}, \{\nabla\}, io + i\partial_t)$, given by (7.9), by a diagram which tells how many particles are involved in the interaction and what order in interaction the term is. The matrix element of the free particle resolvent, $(L_0 - io)^{-1}$, is represented by a number of superposed lines equal to the number of nonzero Fourier vectors $\{\mathbf{k}\}$ involved in the matrix element and each line is labeled with the index of the particle associated with that nonzero Fourier vector. The matrix element of an interaction term, say, δL_{ij}, is represented as a number of superposed lines equal to the number of nonzero vectors associated with particles other than i and j (and therefore conserved in the interaction) and a vertex representing the possible transitions $\mathbf{k}_i, \mathbf{k}_j \rightarrow \mathbf{k}_i', \mathbf{k}_j'$ under δL_{ij} consistent with the conservation law, (4.7), imposed by the assumption of pairwise additive forces. A few examples, typical of those we shall discuss in this section, are

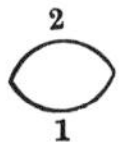

$$= \frac{8\pi^3}{m^2} c \int d\mathbf{l}\, V_l \mathbf{l} \cdot \partial_{12} \frac{1}{\mathbf{l} \cdot \mathbf{v}_{12} - io} V_l \mathbf{l} \cdot \boldsymbol{\partial}_{12}$$

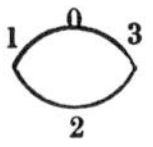

$$= \frac{(8\pi^3 c)^2}{m^3} \int d\mathbf{l}\, V_l \mathbf{l} \cdot \boldsymbol{\partial}_{12} \frac{1}{\mathbf{l} \cdot \mathbf{v}_{12} - io} V_l \mathbf{l}$$

$$\cdot\, \boldsymbol{\partial}_{13} \frac{1}{\mathbf{l} \cdot \mathbf{v}_{12} - io} V_l \mathbf{l} \cdot \boldsymbol{\partial}_{23} \tag{6.1}$$

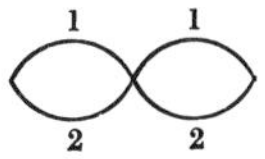

$$= \frac{8\pi^3 c}{m^3} \int d\mathbf{l}\, d\mathbf{l}'\, V_l \mathbf{l} \cdot \boldsymbol{\partial}_{12} \frac{1}{\mathbf{l} \cdot v_{12} - io} V_{(\mathbf{l}-\mathbf{l}')}(\mathbf{l} - \mathbf{l}')$$

$$\cdot\, \boldsymbol{\partial}_{12} \frac{1}{l' \cdot v_{12} - io} V_{l'} \mathbf{l}' \cdot \boldsymbol{\partial}_{12}$$

These diagrams represent the first few terms in the homogeneous collision operator, $\psi(\mathbf{l}, \{0\}; io)$. It is the homogeneous operator that acts on the unknown ϕ after the exact transport equation, (7.8), has been linearized. In Table VI the diagrams contributing to the singlet kinetic equation are depicted through fourth order in the interaction.

In order to estimate the order of magnitude of the diagrams shown in Table VI we shall replace the velocity and velocity operators by thermal velocity, that is, $\mathbf{v}_i = (kT/m)^{1/2}$ and $\boldsymbol{\partial}_i = (m/kT)^{1/2}$, and introduce the reduced variables

$$\mathbf{l}_r = \sigma \mathbf{l} \tag{6.2}$$

$$V_r(l_r) = \frac{V_l}{\epsilon\sigma^3} \tag{6.3}$$

where σ and ϵ are the characteristic molecular diameter and interaction energy, respectively. With this reduction the diagrams in (6.1) become

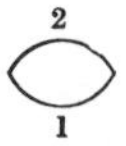

$$\approx (c\sigma^3)\,\frac{(kT/m)^{1/2}}{\sigma}\left(\frac{\epsilon}{kT}\right)^2 8\pi^3 \int d\mathbf{l}_r\, V_r^2(l_r) l_r$$

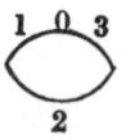

$$\approx (c\sigma^3)^2\,\frac{(kT/m)^{1/2}}{\sigma}\left(\frac{\epsilon}{kT}\right)^3 (8\pi^3)^2 \int d\mathbf{l}_r\, V_r^3(l_r) l_r \tag{6.4}$$

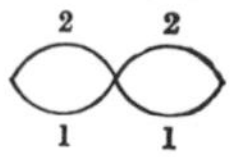

$$\approx (c\sigma^3)\,\frac{(kT/m)^{1/2}}{\sigma}\left(\frac{\epsilon}{kT}\right)^3 8\pi^3$$

$$\times \int d\mathbf{l}_r\, d\mathbf{l}_r'\, V_r(l_r) V_r(|\mathbf{l}_r - \mathbf{l}_r'|)\, |\mathbf{l}_r - \mathbf{l}_r'|\, V(l_r')$$

Similar reductions may be performed for the other diagrams depicted in Table VI.

Allen and Cole argue that the collision operator is dominated by the chain diagrams

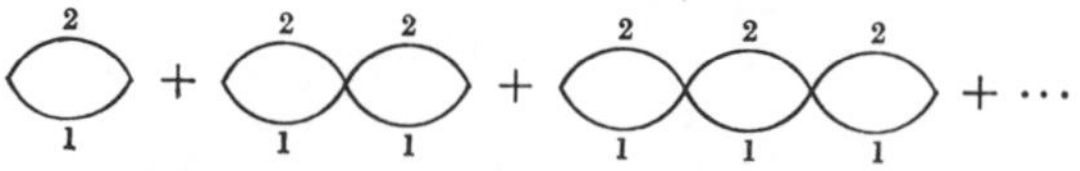

(6.5)

known to lead to Boltzmann's binary collision operator,[45,46] and the ring diagrams

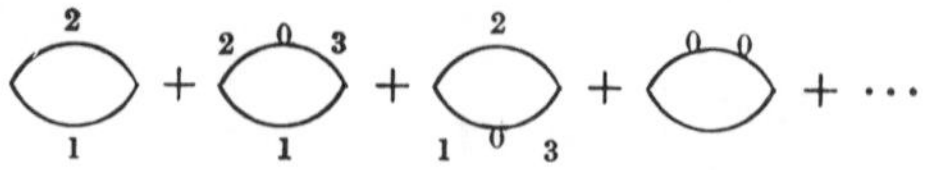

(6.6)

known to dominate for low-density plasmas. Summing these two classes of diagrams, they obtain for the collision operator Boltzmann's collision operator plus the plasma operator minus the cycle operator  since it was counted both in (6.5) and (6.6). Moreover, unlike Rice and Allnatt, they argue that it is appropriate to use the total potential energy in the Boltzmann and plasma operators rather than decompose it into a strongly repulsive part contributing to the Boltzmann operator and a "soft" part contributing to the plasma operator.

In order to utilize their theory Allen and Cole have to require that the pair potential have a Fourier transform. In their numerical work[42] they use the truncated Lennard-Jones potential

$$V(r) = 4\epsilon\left[\left(\frac{\sigma}{r}\right)^{12} - \left(\frac{\sigma}{r}\right)^{6}\right] \qquad r \geqslant 0.7\sigma \tag{6.7}$$
$$= 0 \qquad r \leqslant 0.7\sigma$$

the justification of the cutoff being that not many particles have the energy to achieve an interparticle separation $r < 0.7\sigma$. An alternative to (6.7), used by Dowling and the author, is to use a potential of the form

$$V(r) = \epsilon b[Ae^{-\alpha(r/\sigma)^2} - e^{-\gamma(r/\sigma)^2}] \tag{6.8}$$

where the constants ϵ and σ are the 6–12 Lennard-Jones constants and b, A, α, and γ are determined so as to ensure that (6.8) have the same position and depth of the minimum as (6.7) and that (6.8) be 0 at $r = \sigma$, where (6.7) is zero. For argon these parameters are $\epsilon/k = 121°\text{K}$, $\sigma = 3.418 \times 10^{-8}$ cm, $b = 6.07$, $A = 1400$, $\alpha = 8.54$, and $\gamma = 1.3$. In Fig. 4, (6.7) and (6.8) are compared, and in Fig. 5 the second virial coefficient is compared for the two potentials. Clearly, one expects on the basis of these figures that (6.7) and (6.8) should yield similar predicted values for equilibrium and transport properties. The motivation, of course, to choosing (6.8) over (6.7) is that its Fourier transform can be obtained explicitly with a resulting simplification of the numerical work.

Using the potential function given by (6.8), Dowling and the author have estimated the orders of magnitudes of the diagrams ranging through fourth order in the interaction potential. The column headed "exact" corresponds to using the estimates in (6.4) and a similar estimate for . In the other columns upper and lower bounds are placed on the terms. The details of these bounds are given elsewhere.[43] Examination of the entries in Table VI leads to the conclusion that all the diagrams are comparable to the ring diagram of the same order in the interaction. This conclusion is made even stronger by recognizing that $c\sigma^3 \leqslant \frac{1}{2}$ rather than

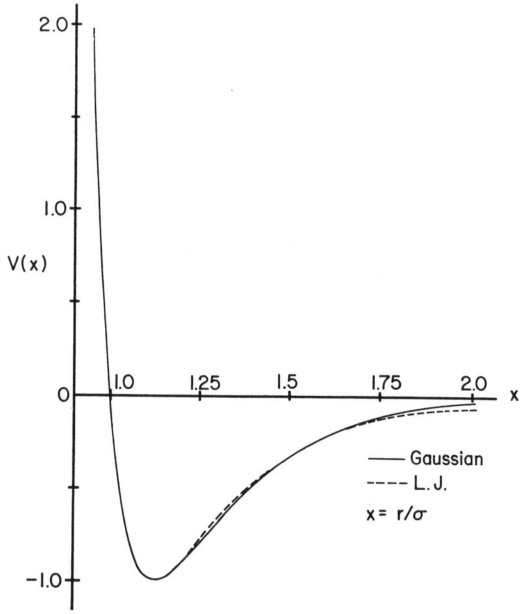

Fig. 4. Comparison of the Lennard-Jones and the Gaussian (6.8) potential functions, respectively. The potential energies are plotted in units of ϵ versus the reduced distance r/σ. The parameters of the Gaussian potential are those for argon given in the text.

the value of unity used for Table VI. Thus there seems to be no justification for keeping the ring diagrams over any others except for the appealing fact that the rings are a summable class.

As a further examination of the Allen-Cole theory, Dowling and the author computed the equilibrium radial distribution function from the expression Allen[42] used for the doublet distribution function in obtaining thermal conductivity from their theory. In the equilibrium limit the expression given is simply the ring sum which can be done explicitly. The result is

$$g(r) - 1 = -\beta \int d\mathbf{l} \, \frac{V_l e^{i\mathbf{l}\cdot\mathbf{r}}}{1 + 8\pi^3 c\beta V_l} \tag{6.9}$$

As seen in Fig. 6, this expression drastically underestimates the first peak in the correlation function. Another test of (6.9) may be obtained by recalling the grand canonical ensemble expression for the isothermal compressibility κ_T:

$$ckT\kappa_T - 1 = c \int d\mathbf{r}[g(r) - 1] \tag{6.10}$$

TABLE VI

Estimated Contributions of Various Diagrams for Gaussian Potential[a]

Diagram	Lower bound	"Exact"	Upper bound
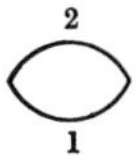		1.63×10^5	
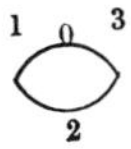		9.20×10^9	
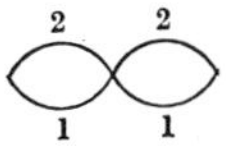	5.36×10^9	5.74×10^{10}	5.16×10^{11}
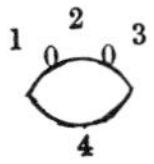		1.06×10^{14}	
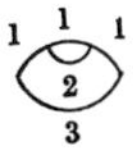	3.68×10^{14}		
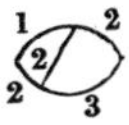	3.11×10^{14}		
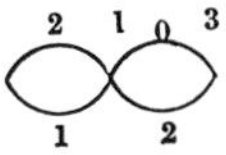	6.85×10^{13}		
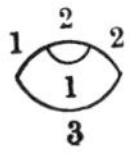	4.02×10^{13}		6.43×10^{16}
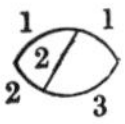	2.83×10^{13}		
	9.00×10^{12}		5.10×10^{17}
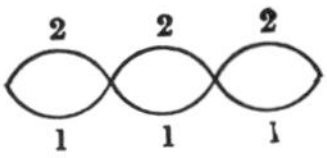	1.04×10^{14}		6.02×10^{17}

[a] Diagrams reduced by $32\pi^4[b/8(\gamma\pi)^{3/2}]^s(kT/m)^{1/2}\sigma^{-1}$ and $c\sigma^3$ and ϵ/kT have been set to unity. s denotes the order of interaction of a given diagram.

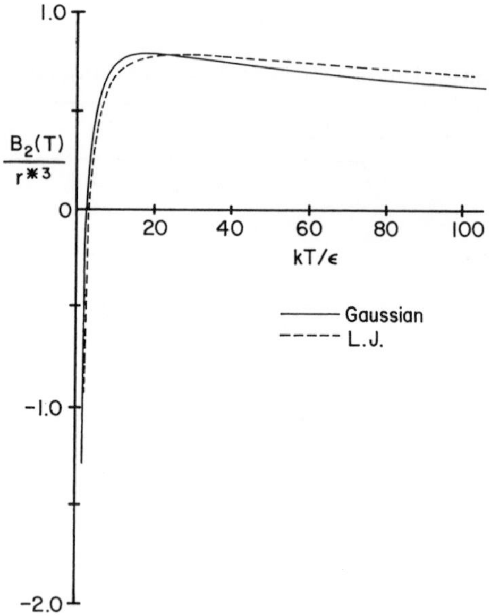

Fig. 5. Comparison of reduced second virial coefficients predicted by the Lennard-Jones and Gaussian potentials.

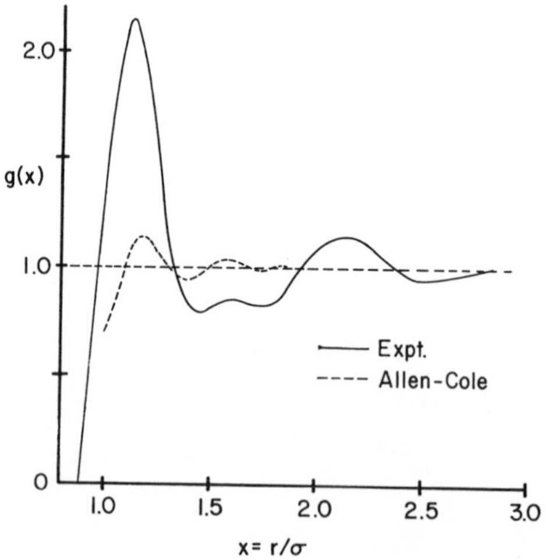

Fig. 6. Comparison of the experimental radial distribution function (C. J. Pings, private communication) with that predicted by (6.9) for the Gaussian potential. The experimental data are for argon at 138.15°K and a density of 0.982 g/cm^3.

Inserting (6.9) into the integrand of (6.10), we obtain

$$ckT\kappa_T - 1 = -\frac{8\pi^3 cV_0}{1 + 8\pi^3 c\beta V_0} \tag{6.11}$$

where

$$V_0 = \frac{1}{8\pi^3}\int d\mathbf{r} V(r) \tag{6.12}$$

For argon at 87°K and under its vapor pressure, the Gaussian potential yields $\kappa_T = 1.82 \times 10^{-6}$ atm^{-1} compared to the experimental value of $\kappa_T = 2.18 \times 10^{-4}$ atm^{-1}. For the truncated Lennard-Jones model (6.12) has the unacceptable feature of depending completely on the truncation value of r.

As another test of the theory, Dowling and the author computed self-diffusion coefficients for potential models (6.7) and (6.8) and obtained negative diffusion coefficients in both cases, the reason being that the contribution of the subtracted cycle diagram simply swamps the other contributions. Thus the original Allen-Cole theory does not seem to describe liquid behavior.

B. Foster and Cole's Theory

Foster and Cole[44] have modified the Allen-Cole theory to obtain for the singlet distribution function an equation identical to the RA equation given by (3.1), except that the term denoted by $\mathscr{B}(g_2^{(1)})$ in the RA equation is replaced in Foster and Cole's work by a modified plasma operator of the form

$$M_1 = \frac{8\pi^4}{m^2}\int d\mathbf{l}\, d\mathbf{v}_2\, V_l^S \mathbf{l}\cdot\boldsymbol{\partial}_1 \frac{\delta(\mathbf{l}\cdot\mathbf{v}_{12})}{|\varepsilon|^2} V_l^S \mathbf{l}\cdot\boldsymbol{\partial}_{12} f(\mathbf{x}_1, \mathbf{v}_1, t) f(\mathbf{x}_1, \mathbf{v}_2, t) g(\mathbf{x}_{12}) \tag{6.13}$$

where

$$\varepsilon = 1 + 8\pi^3 m^{-1} V_l^S \int d\mathbf{v}_3 \frac{1}{\mathbf{l}\cdot\mathbf{v}_{13} - io}\mathbf{l}\cdot\boldsymbol{\partial}_3 f(\mathbf{x}_1, \mathbf{v}_3, t) \tag{6.14}$$

and $g(\mathbf{x}_{12})$ is the equilibrium radial distribution function evaluated at the local thermodynamic conditions at $\mathbf{x}_1$. They also derive a similar equation for the doublet distribution function.

Equation (6.13) is not the result of a systematic summation of diagrams: it results from summing the ring diagrams of dilute plasma theory and then arbitrarily inserting the quantity $g(x_{12})$ into the integrand. In fact, the appearance of $g(x_{12})$ in M_1 poses a serious problem, because M_1 can depend only on $\mathbf{x}_2$ since it appears in the singlet equation. Thus (6.13) as given cannot be a collision operator for the singlet kinetic equation.

Moreover, even if $g(x_{12})$ is set to unity, the resulting plasma operator neglects the temporal delocalization and memory contributions (accounted for by the quantities $g_{2p}^{(1)}(st.)$ and $g_{2p}^{(1)}(n.st.)$ in the PNM and GRA theories and by $g_2^{(3)}$ and $g_2^{(4)}$ in the effective potential model). Neglect of these terms is a good approximation for plasmas, but, as was discussed in Section II, not for dense fluids.

As far as the author knows, Foster and Cole have not determined the transport coefficients from their theory, so, for this reason, and because of the inconsistency mentioned above, we shall not consider the theory further in this article.

VII. SOME GENERAL ASPECTS OF KINETIC EQUATIONS

A. Master Equations

As mentioned earlier one of the important developments in kinetic theory in the last decade or so was the derivation of "master" equations which are exact reductions of the Liouville equation (or von Neumann's equation in the case of quantum mechanics) and which under certain conditions give all the reduced distribution functions as functionals of the singlet distribution function. Van Hove[47] derived the first such equation in 1957 for quantum mechanical fluids, and during the next five to ten years several other derivations of master equations were presented. Zwanzig[1] reviewed the subject in 1964 and demonstrated the identity of all the master equations existing at the time except that of Van Hove. An example of such master equations, derived by Severne[2] and based on the infinite-order perturbation theory developed by Prigogine and co-workers,[2,38,45,46] has been taken as the starting point of the theories discussed in Sections II, IV, V, and VI. Severne's exact singlet equation is of the form

$$\begin{aligned} \partial_t f(\mathbf{x}_1, \mathbf{v}_1, t) + \mathbf{v}_1 \cdot \nabla_1 f(\mathbf{x}_1, \mathbf{v}_1, t) + m^{-1}\mathbf{F}(\mathbf{x}_1, t) \cdot \boldsymbol{\partial}_1 f(\mathbf{x}_1, \mathbf{v}_1, t) \\ = \mathscr{D}(\mathbf{x}_1, \mathbf{v}_1, t) + i\int_0^t dt' \sum_{s=2}^{N} \int (d\mathbf{v}\, d\mathbf{x})^{s-1} \\ \prod_{i=2}^{s} \delta(\mathbf{x}_1 - \mathbf{x}_i)\psi^{\{s\}}(\{-i\nabla\}; t - t') \prod_{j=1}^{s} f(\mathbf{x}_j, \mathbf{v}_j, t') \end{aligned} \tag{7.1}$$

where the abbreviation $(d\mathbf{v}\, d\mathbf{x})^{s-1} = d\mathbf{v}_2 \cdots d\mathbf{v}_s\, d\mathbf{x}_2 \cdots d\mathbf{x}_s$ has been used and where

$$\mathbf{F}(\mathbf{x}_1, t) = -\nabla_1 \int d\mathbf{x}_2\, d\mathbf{v}_2\, V(\mathbf{x}_{12}) f(\mathbf{x}_2, \mathbf{v}_2, t) \tag{7.2}$$

$$\mathscr{D}(\mathbf{x}_1, \mathbf{v}_1, t) = c\int d\mathbf{v}^{N-1} \sum_{\mathbf{k}_1} e^{i\mathbf{k}_1\cdot\mathbf{x}_1}\, \mathscr{D}_{\mathbf{k}_1}(\mathbf{v}^N, t; f_N(0)) \tag{7.3}$$

with

$$\mathscr{D}_{\mathbf{k}_1}(\mathbf{v}^N,t) = -\frac{1}{2\pi}\oint dz\, e^{-izt} \sum_{\{\mathbf{k}\}}^{ir} \sum_{n=1}^{\infty} (-1)^n \langle \mathbf{k}_1 | [\delta L(L_0 - z)^{-1}]^n | \{\mathbf{k}\}\rangle\, \Gamma_{\{\mathbf{k}\}}(t=0) \tag{7.4}$$

and

$$\psi^{(s)}(\{-i\nabla\}; t) = -\frac{1}{2\pi}\oint dz\, e^{-izt} c^{-s+1} \times \sum_{n=2}^{\infty} (-1)^n \langle \mathbf{q}_1 + \mathbf{q}_2 + \cdots + \mathbf{q}_N | \times \delta L[(L_0 - z)^{-1}\delta L]^{n-1} | \mathbf{q}_1, \ldots, \mathbf{q}_N\rangle_s^{ir} \big|^{\mathbf{q}_1 \to -i\Delta_j} \tag{7.5}$$

The quantity $\Gamma_{\{\mathbf{k}\}}\ (t = 0)$ is the Fourier component of $f_N(\Gamma, t = 0) - \prod_{i=1}^{N} f(\mathbf{x}_i, \mathbf{v}_i, t = 0)$ and, therefore, arises from the initial correlation among the particles of the system. The meanings of the superscript "*ir*" and the subscript "*s*" on the ket in (7.5) have been discussed in Section IV.A. The ket notation $\langle \mathbf{q}_1 + \cdots + \mathbf{q}_N|$ in (7.5) means that $\mathbf{k}_1' = \sum_{i=1}^{N} \mathbf{q}_i$ $\mathbf{k}_i' = 0$, $j = 2, \ldots, N$. As in the definition of $C^{(s)}$—(4.2)—after the matrix elements of $\psi^{(s)}$ have been fully expressed the vectors $\mathbf{q}_j$ are replaced by the operators $-i\nabla_j$.

The so-called destruction operator $\mathscr{D}(\mathbf{x}_1, \mathbf{v}_1, t)$ accounts for the effects of initial intermolecular correlations on the time evolution of the singlet distribution function. Assuming the initial correlations to be of finite extension in space and considering the first few terms in the perturbation expansion defining $\mathscr{D}_{\mathbf{k}_1}(\mathbf{v}^N, t)$, Prigogine and Resibois[2,45,46] for homogeneous systems and Severne[2] for inhomogeneous systems concluded that for potential models qualitatively correct for fluids

$$\mathscr{D}(\mathbf{x}_1, \mathbf{v}_1, t) \to 0 \qquad \text{for} \quad t \gg t_c \tag{7.6}$$

where the collision time $t_c \sim r^0/\bar{v}$, r^0 being the range of the interaction potential and $\bar{v}$ the average molecular speed. Under similar considerations, they also concluded that the non-Markovian collision operator $\psi^{(s)}$ has the property that

$$\psi^{(s)}(\{-i\nabla\}, t \to 0 \qquad \text{for} \quad t \gg t_c \tag{7.7}$$

Assuming (7.6) and (7.7) to be generally true for fluids, we obtain the following long-time or "transport" limit of (7.1):

$$\partial_t f(\mathbf{x}_1, \mathbf{v}_1, t) + \mathbf{v}_1 \cdot \nabla_1 f(\mathbf{x}_1, \mathbf{v}_1, t) + m^{-1}\mathbf{F}(\mathbf{x}_1, t) \cdot \boldsymbol{\partial}_1 f(\mathbf{x}_1, \mathbf{v}_1, t) = i \sum_{s=2}^{N} \int (d\mathbf{v}\, d\mathbf{x})^{s-1} \prod_{i=2}^{s} \delta(\mathbf{x}_1 - \mathbf{x}_i)\tilde{\psi}^{(s)}(\{-i\nabla\}; i\partial_t + io) \prod_{j=1}^{s} f(\mathbf{x}_i, \mathbf{v}_j, t) \tag{7.8}$$

where

$$\tilde{\psi}^{(s)}(\{-i\nabla\}; i\partial_t + io) = c^{-s+1}\sum_{n=2}^{\infty}(-1)^n \langle \mathbf{q}_1 + \cdots + \mathbf{q}_N| \times \delta L[(L_0 - z)^{-1}\,\delta L]^{n-1}\,|\mathbf{q}_1, \ldots, \mathbf{q}_N\rangle_s^{ir}\Big|_{\substack{\mathbf{q}_j \to -i\nabla_j \\ z \to i\partial t + io}} \tag{7.9}$$

Equation (7.8) is the general transport equation for the class of fluids obeying (7.6) and (7.7). Such a class of fluids must include at least non-turbulent Newtonian fluids. For Newtonian fluids undergoing turbulent motion, the full kinetic equation, (7.1) must be considered since, owing to the long-range correlations, the destruction term $\mathscr{D}(\mathbf{x}_1, \mathbf{v}_1, t)$ may decay too slowly to be neglected. Also, the full equation should be retained in dealing with viscoelastic fluids since the response and relaxation of these systems are known to depend sensitively on their preparation or initial state.

In addition to deriving (7.1), Severne, generalizing to inhomogeneous systems the work of the Brussels school on homogeneous systems, was able to express the higher-order distribution functions in terms of functionals of the singlet distribution function plus "transient" functions of the initial correlations of the type represented by $\mathscr{D}(\mathbf{x}_1, \mathbf{v}_1, t)$. Under the same conditions as those underlying (7.6) and (7.7), the transient parts of the higher-order distribution functions vanishes for $t \gg t_c$ giving the functions as functional only of the singlet distribution function. Equation (4.1) is Severne's long-time result for the doublet distribution function.

An especially simple technique of deriving master equations is the projection operator method introduced by Zwanzig.[48,49,1] We shall outline this method here. Consider the projection operator P which separates f_N into its "relevant" part $f_N{}^R$ and its "irrevelant" part $f_N{}^I$. Thus

$$f_N{}^R = Pf_N \tag{7.10}$$

$$f_N{}^I = f_N - f_N{}^R = (1 - P)f_N \tag{7.11}$$

where P has the projection operator property

$$P^2 = P \tag{7.12}$$

and will be chosen to be time independent. Since P is chosen to be time independent it can be applied directly to the Liouville equation to yield

$$Pi\frac{\partial f_N}{\partial t} = i\frac{\partial f_N{}^R}{\partial t} = PL(f_N{}^R + f_N{}^I) \tag{7.13}$$

Also,

$$(1 - P)i\frac{\partial f_N}{\partial t} = i\frac{\partial f_N{}^I}{\partial t} = (1 - P)L(f_N{}^R + f_N{}^I) \tag{7.14}$$

The formal solution of (7.14) is

$$f_N{}^I(t) = e^{-it(1-P)L} f_N{}^I(0) - i\int_0^t e^{-is(1-P)L}(1 - P)Lf_N{}^R(t - s)\, ds \quad (7.15)$$

Inserting this result into (7.13), we obtain

$$i\frac{\partial f_N{}^R(t)}{\partial t} - PLf_N{}^R(t)$$

$$= PLe^{-it(1-P)L} f_N{}^I(0) - i\int_0^t PLe^{-is(1-P)L}(1 - P)Lf_N{}^R(t - s)\, ds \quad (7.16)$$

This is a formal master equation for any projection of f_N.

To obtain a particular kinetic equation the projection operator P must be specified at this point. For example, Zwanzig has shown[1] that if

$$f_N{}^R = Pf_N \equiv \frac{1}{\Omega^N}\int dx^N f_N = \frac{\varphi^N}{\Omega^N} \quad (7.17)$$

then (7.16) yields a master equation for the velocity distribution function φ_N; and if $\exp[-is(1 - P)L]$ is expanded in an infinite perturbation series about the free particle propagator $\exp[-is(1 - P)L_0]$, then the equation is identical to the Prigogine-Resibois[2] master equation for φ_N. Another projection operator must be considered in treating inhomogeneous systems. For example, if P is defined to be

$$PG = \frac{1}{\Omega^{N-1}}\int d\mathbf{x}_2 \cdots d\mathbf{x}_N G(\mathbf{x}^N, \mathbf{v}^N, t) \quad (7.18)$$

then

$$f_N{}^R = \frac{1}{\Omega^{N-1}}\int d\mathbf{x}^{N-1} f_N \equiv \frac{1}{N\Omega^{N-1}} f_{1,N}(\mathbf{x}_1, \mathbf{v}_1, \ldots, \mathbf{v}_N, t) \quad (7.19)$$

For the classical systems with finite range interactions, Balescu[38] has shown that $f_{1,N}$ factors as follows

$$f_{1,N} = f(\mathbf{x}_1, \mathbf{v}_1, t)\varphi_{N-1}(\mathbf{v}^{N-1}, t) = f(\mathbf{x}_1, \mathbf{v}_1, t)\prod_{i=2}^{N} \varphi(\mathbf{v}_i, t) \quad (7.20)$$

where

$$\varphi(\mathbf{v}_i, t) = \int f(\mathbf{x}_i, \mathbf{v}_i, t)\, d\mathbf{x}_i$$

is the singlet velocity distribution function. Inserting the result of (7.20) into (7.16) and integrating over the velocities $d\mathbf{v}^{N-1} = d\mathbf{v}_2 \cdots d\mathbf{v}_N$, we obtain

$$\frac{\partial f(\mathbf{x}_1, \mathbf{v}_1, t)}{\partial t} + \mathbf{v}_1 \cdot \nabla_1 f(\mathbf{x}_1, \mathbf{v}_1, t)$$

$$= -i \int d\mathbf{v}^{N-1} PLe^{-it(1-P)L} f_N{}^I(0)$$

$$- \int_0^t ds \int d\mathbf{v}^{N-1} PLe^{-is(1-P)L}(1 - P)L \prod_{j=2}^{N} \varphi(\mathbf{v}_j, t - s) f(\mathbf{x}_1, \mathbf{v}_1, t - s) \tag{7.21}$$

Under conditions such that the term in (7.21) involving the initial value of the irrevelant part, $f_V{}^I(0)$, vanishes for long time, (7.21) reduces to a kinetic equation depending only on the history of the singlet distribution function. This is similar to Severne's result.

The advantage over the projection operator technique over the infinite-order perturbation theory is the simplicity and generality with which master equations of the type of (7.16) can be obtained. However, the operators appearing in (7.16) are so complicated that to obtain explicit results one generally is forced to the kind of infinite-order perturbation expansion and Fourier representation typical of that used by the Brussels school in deriving master kinetic equations.

B. Homogeneous Solutions to General Linear Collision Operator of Singlet Kinetic Equation

In fluids composed of monatomic particles interacting with pairwise additive centrally symmetric forces, Severne's exact transport equation[2] for the singlet distribution function f may be linearized by the Chapman-Enskog[2,16] scheme to give[34]

$$d_t{}^0 f^{(0)} + \mathscr{B}(f^{(0)}) = \mathscr{A}(\phi) \tag{7.22}$$

where

$$d_t{}^0 \equiv \partial_t + \mathbf{v}_1 \cdot \nabla_1 + \mathbf{F}^{(0)}(\mathbf{x}_1, t) \cdot \boldsymbol{\partial}_1 \tag{7.23}$$

$$\mathbf{F}^{(0)}(\mathbf{x}_1, t) = -\nabla_1 \int V(\mathbf{x}_1 - \mathbf{x}_2) f^{(0)}(\mathbf{x}_2, \mathbf{v}_2, t)\, d\mathbf{x}_2\, d\mathbf{v}_2 \tag{7.24}$$

$$\mathscr{B}(f^{(0)}) = \sum_{s=2}^{N} i \int d\Gamma^{s-1} \delta^{s-1} \tilde{\psi}^{(s)}(\{-i\nabla\}; io + i\partial_t) \prod_{j=1}^{s} f^{(0)}(\mathbf{x}_j, \mathbf{v}_j, t) \tag{7.25}$$

$$\mathscr{A}(\phi) = \sum_{s=2}^{N} i \int d\mathbf{v}^{s-1} \tilde{\psi}^{(s)}(\{0\}; io) \prod_{j=1}^{s} f^{(0)}(\mathbf{x}_1, \mathbf{v}_j, t) \sum_{\alpha=1}^{s} \phi(\mathbf{x}_1, \mathbf{v}_\alpha, t) \tag{7.26}$$

$$\Gamma^{s-1} = (\mathbf{x}_2, \ldots, \mathbf{x}_s, \mathbf{v}_2, \ldots, \mathbf{v}_s) \quad \text{and} \quad \delta^{s-1} = \prod_{i=2}^{s} \delta(\mathbf{x}_1 - \mathbf{x}_i) \tag{7.27}$$

In these expressions $f^{(0)}$ is the solution to the zeroth-order Chapman-Enskog equation and is of the form

$$f^{(0)}(\mathbf{x}, \mathbf{v}, t) = n\left(\frac{m}{2\pi kT}\right)^{3/2} \exp\left[-\frac{m}{2kT}(\mathbf{v}-\mathbf{u})^2\right] \tag{7.28}$$

n, T, and $\mathbf{u}$ are velocity-independent parameters which depend on $\mathbf{x}$ and t and can be identified with the local hydrodynamic density, temperature, and mass average velocity at a later stage in the theory. The quantity ϕ, defined as $\phi = f/f^{(0)} - 1$, accounts for the deviation of f from its "local equilibrium" value $f^{(0)}$. The collision operator $\tilde{\psi}^{(s)}$ appearing in $\mathscr{B}(f^{(0)})$ was defined by (7.9). It is the same operator, but with $i\nabla_j$ and $i\partial_t$ set equal to zero, that appears in $\mathscr{A}(\phi)$.

What we wish to do in this section is to study the homogeneous solutions of $\mathscr{A}(\phi)$. The main points we wish to make can be summarized as the following two theorems valid for fluids composed of particles interacting with pairwise additive centrally symmetric forces:

Theorem 1. $\phi_k = 1, \mathbf{v}$, *and* v^2 *are solutions of the homogeneous equation*

$$\mathscr{A}(\phi_h) = 0 \tag{7.29}$$

Theorem 2. *If calculated from the Chapman-Enskog theory, the kinetic contribution to the bulk viscosity is identically zero.*

Theorems 1 and 2 are well known for gases sufficiently dilute that Boltzmann's equation holds. However, the proof given here (due to Dowling and the author[50]) is true for a fluid of arbitrary density.

An important consequence of Theorem 1 is that it enables one to identify the constants n, $\mathbf{u}$, and T appearing in $f^{(0)}$ with the local hydrodynamic density, mass average velocity, and temperature. Thus the local hydrodynamic variables are determined completely by $f^{(0)}$ and the linear kinetic constitutive relations (i.e., the kinetic fluxes of energy and momentum linear in gradients of the temperature and the hydrodynamic velocity) are determined uniquely by $f^{(0)}\phi$.

To prove Theorem 1, we shall use the following result:

$$\begin{aligned} A &\equiv \sum_{s=2}^{N} i \int d\mathbf{v}^{s-1}\, d\mathbf{x}^{s-1} \psi^{(s)}(\{0\}; io) \prod_{j=1}^{s} f^{(0)}(\mathbf{x}_1, \mathbf{v}_i, t) \\ &= \int d\mathbf{v}_2 i\boldsymbol{\partial}_1 \cdot \sum_{l} \langle 0|\, \mathbf{F}_{12}\, |\mathbf{l}, -\mathbf{l}'\rangle \sum_{s=2}^{N} iNc^{-s+1} \sum_{n=2}^{\infty} \int d\mathbf{v}^{s-2} \\ &\quad \times \langle \mathbf{l}, -\mathbf{l}|\, [(L_0 - io)^{-1}(-\delta L)]^{n-1}\, |0\rangle \prod_{j=1}^{s} f^{(0)}(\mathbf{x}_1, \mathbf{v}_j, t) = 0 \end{aligned} \tag{7.30}$$

Equation (7.30) is, in fact, the zeroth-order Chapman-Enskog equation for $f^{(0)}$. The solution has already been given above by (7.28). That (7.30) is zero for $f^{(0)}$ given by (7.28) follows from the fact[46] that the quantity appearing to the right of the matrix element of the force between Particles 1 and 2, (i.e., to the right of $\langle 0|\, \mathbf{F}_{12}\, |\mathbf{l}, -\mathbf{l}\rangle$) is equal to the Fourier transform of

$$[g_2^{(0)}(|\mathbf{x}_1 - \mathbf{x}_2|; \mathbf{x}_1, t) - 1]f^{(0)}(\mathbf{x}_1, \mathbf{v}_1, t)f^{(0)}(\mathbf{x}_1, \mathbf{v}_2, t)$$

where $g_2^{(0)}$ is the equilibrium radial distribution function evaluated at the density $n(\mathbf{x}_1, t)$ and temperature $T(\mathbf{x}_1, t)$. Thus the quantity appearing to the right of $\boldsymbol{\partial}_1$ in (7.30) is proportional to the average of $\mathbf{F}_{12}$ over an equilibrium ensemble. Since this force is zero, (7.30) is identically zero.

We can now prove Theorem 1 by noting that $\tilde{\psi}^{(s)}_{(\{0\};io)}$ is independent of $n(\mathbf{x}_1, t)$, $\beta(\mathbf{x}_1, t) = 1/kT(\mathbf{x}_1)\, t)$, and $\mathbf{u}(\mathbf{x}_1, t)$. Therefore, we have the identities

$$\mathscr{A}(\phi_h = 1) = n(\mathbf{x}_1, t)\frac{\partial A}{\partial n(\mathbf{x}_1, t)} \tag{7.31}$$

$$\mathscr{A}(\phi_h = \mathbf{v} - \mathbf{u}) = \frac{1}{m\beta(\mathbf{x}_1, t)}\frac{\partial A}{\partial \mathbf{u}(\mathbf{x}_1, t)} \tag{7.32}$$

$$\mathscr{A}(\phi_h = \tfrac{1}{2}m(\mathbf{v} - \mathbf{u})^2) = -\frac{\partial A}{\partial \beta(\mathbf{x}_1, t)} + \frac{3}{2}\frac{n(\mathbf{x}_1, t)}{\beta(\mathbf{x}_1, t)}\frac{\partial A}{\partial n(\mathbf{x}_1, t)} \tag{7.33}$$

Since A was shown to be identically zero for arbitrary $n(\mathbf{x}_1, t)$, $T(\mathbf{x}_1, t)$ and $\mathbf{u}(\mathbf{x}_1, t)$, the right-hand sides of (7.31)–(7.33) are zero, thus proving the theorem.

Theorem 2 now follows directly from Theorem 1. Since 1, $\mathbf{v}$, and v^2 are homogeneous solutions to (7.22), we are free to require the auxiliary conditions on ϕ:

$$\int d\mathbf{v} f^{(0)}\phi\phi_h = 0 \tag{7.34}$$

for $\phi_h = 1$, $\mathbf{v} - \mathbf{u}$, and $(\mathbf{v} - \mathbf{u})^2$. These conditions allow, as already mentioned, the identification of the n, $\mathbf{u}$, and T appearing in $f^{(0)}$ with the local density, hydrodynamic velocity, and temperature defined by the moments

$$\int f\alpha\, d\mathbf{v} \tag{7.35}$$

where $\alpha = 1$, $\mathbf{v}/n$, and $(m/3kn)(\mathbf{v} - \mathbf{u})^2$, respectively.

The kinetic part of the bulk viscosity is proportional to the trace of the quantity

$$\int f^{(0)}\phi m(\mathbf{v} - \mathbf{u})(\mathbf{v} - \mathbf{u})\, d\mathbf{v} \tag{7.36}$$

The trace of (7.36) is identically zero by the auxiliary condition for $\phi_h = (\mathbf{v} - \mathbf{u})^2$. Thus Theorem 2 is proved. It is easy to prove Theorem 2 for mixtures.

In the dilute gas limit, the operator $\mathscr{A}$ becomes the linearized Boltzmann operator, and the fact that mass, momentum, and kinetic energy are solutions to the homogeneous equation, $\mathscr{A}(\phi) = 0$, follows because these quantities are invariants of the binary collisional process defining $\mathscr{A}$. Thus it is interesting that in the general case considered in Theorem 1 the homogeneous equation still *has the same solutions but not for the same reason*, since the general operator describes a complicated many-body collisional process.

In a paper in which they establish (in extension of Résibois' work[51] on shear viscosity) the equivalence of transport coefficients calculated from Severne's transport equation by the Chapman-Enskog method with those calculated from autocorrelation (or time correlation) function theory, Nicolis and Severne[52] have pointed out that the nonequilibrium temperature defined by (7.35) with conditions (7.36) is not identical to the nonequilibrium temperature that one deduces from the assumptions of autocorrelation function theory. In autocorrelation function theory the auxiliary condition (7.34) is replaced by the following condition[52,53] on the total nonequilibrium energy $\bar{e}$

$$\begin{aligned}\bar{e} &\equiv \frac{1}{2}\int d\mathbf{v}_1\, d\mathbf{v}_2\, d\mathbf{x}_{12}\{\tfrac{1}{2}mv_1^{\,2} + \tfrac{1}{2}mv_2^{\,2} + V(\mathbf{x}_{12})\}f_2 \\ &= \frac{1}{2}\int d\mathbf{v}_1\, d\mathbf{v}_2\, d\mathbf{x}_{12}\{\tfrac{1}{2}mv_1^{\,2} + \tfrac{1}{2}mv_2^{\,2} + V(\mathbf{x}_{12})\}f_2^{\,eq} \equiv \bar{e}^{eq} \qquad (7.37)\end{aligned}$$

where $f_2^{\,eq}$ is the equilibrium pair correlation function. $\bar{e}^{eq}$ a is known functional of the equilibrium temperature T so that (7.37) defines a relationship between the nonequilibrium energy and the equilibrium temperature T, that is, $\bar{e} = \bar{e}^{eq}(T)$, whereas the Chapman-Enskog method is to define, by (7.34), a relationship between the T and nonequilibrium kinetic energy $e_k = n^{-1}\int \tfrac{1}{2}mv_1^{\,2}f(\mathbf{x}_1, \mathbf{v}_1, t)\, d\mathbf{v}_1$. With the autocorrelation function definition of temperature the kinetic part of the bulk viscosity will not be zero. However, the total of the kinetic and potential parts of the bulk viscosity must be the same for either the Chapman-Enskog or the autocorrelation approach.

In the same paper Nicolis and Severne showed that the nonstationary contributions (which we called memory effects in Sections II–V) to the collision operator must be retained in the Chapman-Enskog solution of the transport equation in order to obtain equivalence between the kinetic theory and autocorrelation theory of the transport coefficients.

VIII. TIME CORRELATION FUNCTIONS AND MEMORY FUNCTIONS

A. Time Correlation Formulas for the Transport Coefficients

The methods used for the derivation of correlation formulas for the transport coefficients are of two types. One deals with the approach to equilibrium of an isolated system and the other deals with the response of a system to the external constraints, thermodynamical or mechanical. Examples of the first type are Green's method[54] of Brownian motion approach, Kubo, Yokota, and Nakajima's[55] method based on Onsager's fluctuation–regression assumption, and Mori's[56,57] method wherein it is assumed that the deviation from local equilibrium is due to the coupling among different parts of the system. Examples of the second type are Kubo's method[58,59] of treating conduction and magnetic problems, and McLennan's extension[60,61] of Kubo's method to include the effects of external heat and mass reservoir and the nonconservative mechanical forces. In spite of the variety of the methods used, they all lead to equivalent formulas for the transport coefficients. A quite complete review of the various methods of deriving time correlation formulas was given by Zwanzig,[3] so we shall not present any details of these methods except one example for the purpose of illustration. We shall consider the method of Kubo, Yokota, and Nakajima[55] in the following. The method is based on Onsager's[62] assumption that the average behavior of the fluctuation of a physical quantity in an aged system is governed by the macroscopic physical law which governs the macroscopic change of the corresponding macroscopic variable. Let α_1, $i = 1, 2, \ldots$, be a set of macroscopic variables which have the values $\alpha_i(t)$ at time t. According to Onsager's assumption, the values $\alpha_i(t + \tau)$ of these variables at time $t + \tau$ (τ is short in the macroscopic sense, but long on a microscopic scale) is given by the relation

$$\alpha_i(t + \tau) - \alpha_i = \sum_k G_{ik} \frac{\partial S}{\partial \alpha_k} \tau \tag{8.1}$$

where S is the entropy of the system; $\partial S/\partial \alpha_i$ multiplied by temperature gives a generalized force; G_{ik} is a macroscopic transport coefficient. Multiplying (8.1) by $\alpha_j(t)$ and averaging over an appropriate equilibrium distribution, one obtains

$$\langle \alpha_i(t + \tau)\alpha_j(t)\rangle - \langle \alpha_i(t)\alpha_j(t)\rangle = -\tau k G_{ij} \tag{8.2}$$

where $\langle \quad \rangle$ means an average over an equilibrium distribution and k is the Boltzmann constant. From (8.2) and the principle of microscopic reversibility Onsager's reciprocal law was deduced. Equation (8.2) also provides a method for calculating the coefficients G_{ij}. Equation (8.2) can be rearranged in the following way.

$$\begin{aligned} G_{ij} &= -\frac{1}{k\tau}\langle \alpha_i(t+\tau)\alpha_j(t) - \alpha_i(t)\alpha_j(t)\rangle \\ &= \frac{1}{k\tau}\int_0^{\tau}\int_0^{t'}\langle \dot{\alpha}_i(t+s)\dot{\alpha}_j(t)\rangle\, ds\, dt' \\ &= \frac{1}{k}\int_0^{\tau}\left(1 - \frac{s}{\tau}\right)\langle \dot{\alpha}_i(t+s)\dot{\alpha}_j(t)\rangle\, ds \end{aligned} \tag{8.3}$$

where the property that the correlation $\langle \alpha_i(t+s)\alpha_j(t)\rangle$ is independent of t and the assumption that

$$\langle \dot{\alpha}_i(t)\alpha_j(t)\rangle = -\langle \alpha_i(t)\dot{\alpha}_j(t)\rangle = 0 \tag{8.4}$$

have been used. If the correlation time of α_i and α_j is much shorter than τ, one may write (8.3) as

$$G_{ij} = \frac{1}{k}\int_0^{\infty}\langle \dot{\alpha}_i(0)\dot{\alpha}_j(s)\rangle\, ds \tag{8.5}$$

which is the general form of time correlation formulas for the transport coefficients. An extension of this method to non-Markovian cases has been done by Zwanzig.[63]

The derivation of correlation formulas in the phase space description is, notationwise, much more complicated, so we shall not consider any derivation of this sort, but simply summarize the results of Green,[54] Mori,[56,57] and others in the following:

1. *Diffusion Coefficients*

$$D_{\alpha\beta} = D_{\beta\alpha} = \frac{1}{3kT\Omega}\int_0^{\infty}\langle \mathbf{J}^{(\alpha)}(0)\cdot\mathbf{J}^{(\beta)}(t)\rangle\, dt \tag{8.6}$$

$$\mathbf{J}^{(\alpha)} = \sum_{i(\alpha)} \mathbf{p}_i \tag{8.7}$$

where Ω is the volume of the system; the Greek letters as superscript or subscript refer to the molecular species; the Roman letters as subscripts refer to specific molecules; the symbol $\sum_{i(\alpha)}$ means a summation over all particles of species α.

2. *Self-Diffusivity*

$$D_1 = \lim_{\epsilon\to 0}\frac{1}{3m_1^{\,2}}\int_0^{\infty} e^{-\epsilon t}\langle \mathbf{p}_1(0)\cdot\mathbf{p}_1(t)\rangle\, dt \tag{8.8}$$

3. *Shear Viscosity*

$$\eta = \lim_{\epsilon\to 0} \frac{1}{kT\Omega} \int_0^\infty dt\, e^{-\epsilon t}\langle J_\eta(0) J_\eta(t)\rangle \tag{8.9}$$

$$J_\eta = \sum_j \left(\frac{p_j^x p_j^y}{m} + F_j^x x_j^y\right) \tag{8.10}$$

and $\sum_i$ denotes a summation over all particles of the system. p_i^x and p_i^y denote the xth and yth Cartesian components of the momentum of the jth particle, while F_i^x and x_j^y are the corresponding components of the force on and the position of the jth particle.

4. *Bulk Viscosity*

$$\Phi = \lim_{\epsilon\to 0} \frac{1}{9kT\Omega} \int_0^\infty dt\, e^{-\epsilon t}\langle J_\Phi(0) J_\Phi(t)\rangle \tag{8.11}$$

$$J_\Phi = \text{trace}\left\{\sum_j \left(\frac{\mathbf{p}_i\mathbf{p}_j}{m} + \mathbf{F}_j\mathbf{x}_j\right) - \mathbf{1}\left[P\Omega + \left(\frac{\partial P\Omega}{\partial\langle E\rangle}\right)_T (E - \langle E\rangle)\right]\right\} \tag{8.12}$$

where $\mathbf{1}$ is the unit second-order tensor, P the pressure, $E = \sum \frac{1}{2}mv_j^2 + \frac{1}{2}\sum_{i,j} V(x_{ij})$, and $\langle E\rangle \equiv U$ is the thermodynamic energy function.

5. *Thermal Conductivity*

$$\kappa = \lim_{\epsilon\to 0} \frac{1}{3kT^2\Omega} \int_0^\infty e^{-\epsilon t}\langle \mathbf{J}_\kappa(0) \cdot \mathbf{J}_\kappa(t)\rangle\, dt \tag{8.13}$$

$$\mathbf{J}_\kappa = \sum_i \frac{p_i^2}{2m_i}\frac{\mathbf{p}_i}{m_i} + \frac{1}{2}\sum_{i,j} (V_{ij}\mathbf{1} + \mathbf{R}_{ij}\mathbf{F}_{ij}) \cdot \frac{p_i}{m_i} - \sum_i h_0^{(i)} \frac{\mathbf{p}_i}{m_i} \tag{8.14}$$

where $h_0^{(i)}$ is the enthalpy of particle i at equilibrium; V_{ij} is the pair potential function between particle i and particle j.

6. *Thermal Diffusivity*

$$D_T^{(\alpha)} = \lim_{\epsilon\to 0} \frac{1}{3kT\Omega} \int_0^\infty e^{-\epsilon t}\langle \mathbf{J}_\kappa(0) \cdot \mathbf{J}^{(\alpha)}(t)\rangle\, dt \tag{8.15}$$

The formulas given above are valid if the average denoted by $\langle\ \rangle$ is taken over a canonical ensemble. When other equilibrium ensembles are used, a proper modification of the flux $\mathbf{J}$ is necessary. If the time average of $\mathbf{J}$,

$$\hat{\mathbf{J}} = \lim_{T\to\infty} \frac{1}{T} \int_0^T \mathbf{J}(t)\, dt \tag{8.16}$$

does not vanish for some ergodic part of finite measure, we have to replace **J** by

$$\mathbf{J} - \hat{\mathbf{J}} \tag{8.17}$$

in the correlation formulas given above. This has been pointed out by McLennan.[64] Discussion concerning the use of different ensembles can be found in the articles of McLennan[64] and Green.[65]

The diffusivity and the thermal diffusion coefficient given by (8.6) and (8.15) correspond to the following phenomenological relation

$$\langle \mathbf{J}^{(\alpha)} \rangle = -D_T^{(\alpha)} \nabla \ln T - \frac{\sum_\beta D_{\alpha\beta}(\nabla \mu_\beta + s_\beta \nabla T)}{m_\beta} \tag{8.18}$$

where μ_β is chemical potential of a particle of species β, and s_β is its entropy. Another problem concerns the boundary of the system. One may assert that the correlation time is usually very short so that the boundary effect may be neglected. However, such a statement needs more careful consideration, because in these formulas we are interested in the correlation of fluctuations. To be more cautious about this, one may follow Green[54] by assuming that the system is contained in a three-dimensional toroidal space. This is equivalent to applying periodic boundary conditions to the system. The boundary problem thus can be avoided.

The discovery of the time correlation function expressions for the transport coefficients was probably the most important advance in transport theory since the Boltzmann equation: these expressions provide a general starting point for computing transport coefficients just as the partition function offers a general starting point for statistical thermodynamics. On the other hand, one is still faced with the N-body problem when trying to compute transport properties from the time correlation functions. Thus, as in dealing with the transport limit of the general kinetic equation, special models, and/or approximations must be made to obtain tractable expressions. The next section will be concerned with one simple model and the section after that will be concerned with another such model.

B. Gaussian Approximation of Time Correlation Functions

Perhaps the simplest procedure for estimating time correlation functions is to assume they decay in time according to some known function involving unknown parameters. Then, assuming the same function describes the short- as well as the long-time behavior of the time correlation function, one can determine the unknown parameters from the time derivatives of the correlation function evaluated at $t = 0$. The choice of exponential decay is ruled out (for smooth interaction potentials) since time correlation functions of the form

$$\psi(t) = \langle A(0)A(t)\rangle \tag{8.19}$$

are even functions of time and have no term linear in t in their series expansion about $t = 0$. (In this section we shall restrict our attention to the self-diffusion coefficient and to the transport coefficients of one-component systems so that the time correlation functions are of the form of (8.19).)

A Gaussian approximation to time correlation functions of the form of (8.19) will be discussed now. This approximation was first introduced by Rice,[66] motivated by a dimensional analysis by Kirkwood,[67] in treating self-diffusion. Later Wei[68] and the author[69] used the same approximation to obtain formulas for the shear and bulk viscosities and, independently, Forster, Martin, and Yip[70] used the equivalent approximation to compute shear viscosity. The procedure is to write (8.19) in the form

$$\psi(t) = \nu e^{-t^2/\tau^2} \tag{8.20}$$

and to note that if (8.20) is valid for all times then

$$\psi(0) = \nu \tag{8.21}$$

and

$$\frac{d^2\psi}{dt^2}(t = 0) = -\frac{2\nu}{\tau^2} \tag{8.22}$$

Thus the values of zeroth and second derivative of $\psi(t)$ at $t = 0$ yield the parameters ν and τ.

Since $A(t)$ can be written in the form

$$A(t) = e^{iLt}A(0) \tag{8.23}$$

where L is the Liouville operator, the nth derivative of $\psi(t)$ at $t = 0$ is given by

$$\begin{aligned}\frac{d^n\psi}{dt^n}(t = 0) &= i^n\langle A(0)L^nA(0)\rangle \\ &= i^n\langle (L^mA(0))(L^{n-m}A(0))\rangle\end{aligned} \tag{8.24}$$

with m an arbitrary positive integer less than or equal to n. The second form of the right-hand side of (8.24) is the result of the symmetry of the

Liouville operator. By parity

$$\frac{d^{2n+1}\psi}{dt^{2n+1}}(t=0)=0 \tag{8.25}$$

so that the only nonzero derivatives of ψ at $t = 0$ may be written in the form

$$\frac{d^{2n}\psi}{dt^{2n}}(t=0)=(-1)^n\langle(L^nA(0))^2\rangle \tag{8.26}$$

with $n = 0, 1, 2, \ldots$. For the Gaussian model, we have

$$\nu=\langle(A(0))^2\rangle \tag{8.27}$$

and

$$\frac{2\nu}{\tau^2}=\langle(LA(0))^2\rangle \tag{8.28}$$

The general form of the transport coefficients considered here is

$$\theta_A=\lim_{\epsilon\to 0}\alpha_A\int_0^\infty dt\, e^{-\epsilon t}\psi_A(t) \tag{8.29}$$

so that to the Gaussian approximation we obtain

$$\theta_A=\alpha_A\nu_A\tau_A\frac{\sqrt{\pi}}{2} \tag{8.30}$$

For a given transport coefficient α_A is known and ν_A and τ_A will be determined from (8.27) and (8.28), respectively. The results for self-diffusion, viscosity, and thermal conductivity are summarized below:

Self-Diffusion

$$A=\mathbf{p} \tag{8.31}$$

$$\alpha_A=\frac{1}{3m^2} \tag{8.32}$$

$$\nu_A=3mkT \tag{8.33}$$

$$\frac{2\nu_A}{\tau_A{}^2}=nkT\int d\mathbf{r}\, g(r)\left[V''(r)+\frac{2}{r}V'(r)\right] \tag{8.34}$$

where the primes on $V(r)$ in the integrand of (8.34) denote derivatives with respect to r. Putting the quantities of (8.32)–(8.34) into (8.30) yields Rice's[66] original approximation for the self-diffusion coefficient.

Shear Viscosity

$$A = J_\eta \quad \text{of (8.10)}$$

$$\alpha_A = \frac{1}{kT\Omega} \tag{8.35}$$

$$\nu_A = \langle (J_\eta(0))^2 \rangle$$

$$= N(kT)^2 \left\{ 1 + \frac{2\pi n}{15kT} \int_0^\infty dr\, r^2 [4rV'(r) + r^2 V''(r)] \right\} \tag{8.36}$$

$$\frac{2\nu_A}{\tau_A{}^2} = \langle (LJ_\eta(0))^2 \rangle$$

$$= \frac{4\pi}{15m} nN(kT)^2 \left\{ \int_0^\infty dr\, g(r) \left[58rV'(r) + 47r^2 V''(r) \right.\right.$$

$$\left. + 6r^3 V'''(r) + \frac{1}{kT} (4(rV')^2 + (r^2 V'')^2) \right]$$

$$+ \frac{\pi n}{kT} \int_0^\infty \int_0^\infty dr\, dr'\, r^2 r'^2 \int_{-1}^1 d\mu\, \mu g_3(r, r', \sqrt{r^2 + r'^2 - 2rr'\mu})$$

$$\times [5V'(r)V'(r') + 6rV''(r)V'(r') - rr'V''(r)V''(r')$$

$$+ 3\mu^2 (rV''(r) - V'(r))(r'V''(r') - V'(r'))] \tag{8.37}$$

Bulk Viscosity

$$A = J_\Phi \quad \text{of (8.12)}$$

$$\alpha_A = \frac{1}{9kT\Omega} \tag{8.38}$$

$$\nu_A = \langle (J_\Phi(0))^2 \rangle$$

$$= N(kT)^2 \left\{ 6 + \frac{2\pi n}{kT} \int_0^\infty dr\, g(r) r^4 \left[V''(r) + \frac{1}{r} V'(r) \right] \right\} - 9\Omega kTK_0 \tag{8.39}$$

$$\frac{2\nu_A}{\tau_A{}^2} = \langle (LJ_\Phi(0))^2 \rangle$$

$$= \frac{2\pi}{5} \frac{nN}{m} (kT)^2 \left\{ \int_0^\infty dr\, g(r) \left[38rV'(r) + 37r^2 V''(r) \right.\right.$$

$$\left. + 6r^3 V'''(r) + \frac{1}{3kT} (2(rV'(r))^2 + 3(r^2 V''(r)^2) \right]$$

$$+ \frac{8\pi}{3} \frac{n}{kT} \int_0^\infty \int_0^\infty dr\, dr'\, r^2 r'^2 \int_{-1}^1 d\mu\, \mu g_3(r, r', \sqrt{r^2 + r'^2 - 2rr'\mu})$$

$$\times [rr'V''(r)V''(r') + 4rV''(r)V'(r')$$

$$\left. + 2\mu^2 (rV''(r) - V'(r)(r'V''(r') - V'(r'))] \right\} \tag{8.40}$$

The quantity K_0 in (8.39) is the isentropic bulk modulus, that is,

$$K_0 = -\Omega\left(\frac{\partial P}{\partial \Omega}\right)_S \tag{8.41}$$

Expressions (8.36) and (8.39) were first obtained by Zwanzig and Mountain,[71] who showed that these expressions are (aside from a factor of $(kT\Omega)^{-1}$) the high-frequency shear and bulk moduli, respectively. Equations (8.37) and (8.40) are expressed in the forms presented by Forster, Martin, and Yip.

Thermal Conductivity

$$A = \mathbf{J}_\kappa \quad \text{of (8.14)}$$

$$\alpha_A = \frac{1}{3kT^2\Omega} \tag{8.42}$$

$$\begin{aligned}
\nu_A &= \langle \mathbf{J}_\kappa(0)\cdot\mathbf{J}_\kappa(0)\rangle \\
&= \frac{15N}{2m}(kT)^3 + \frac{2\pi}{m} Nn(kT)^2\int_0^\infty dr\, g(r)\Big\{12r^2V(r) + \frac{13}{3}r^3V'(r) \\
&\quad + \frac{1}{3}r^4V''(r) - \frac{2r^2}{kT}V^2(r) + \frac{r^3}{3kT}V(r)V'(r)\Big\} \\
&\quad + \frac{2\pi^2}{m}Nn^2(kT)\int_0^\infty dr\int_0^\infty dr'\int_{-1}^1 d\mu\, g_3(r, r', \sqrt{r^2+r'^2-2rr'\mu}) \\
&\quad \times \Big\{-\frac{4}{3}r^3r'^2V'(r)V(r) - \frac{m2r^4r'^3}{3\sqrt{r^2+r'^2-2rr'\mu}} \\
&\quad \times V'(r')V'(\sqrt{r^2+r'^2-2rr'\mu}) + \frac{3r^5r'^2\mu}{\sqrt{r^2+r'^2-2rr'\mu}} \\
&\quad \times V'(\sqrt{r^2+r'^2-2rr'\mu})V'(r)\Big\}
\end{aligned} \tag{8.43}$$

$$\frac{2\nu_A}{\tau_A^2} = \langle L\mathbf{J}_\kappa(0)\cdot L\mathbf{J}_\kappa(0)\rangle$$

= About 50 integrals involving two-, three-, and four-particle radial distribution functions (8.44)

It is well known that the Gaussian approximation to the velocity autocorrelation function does not give the negative values observed (in molecular dynamic studies) for times on the order of 5×10^{-13} sec in dense low-temperature liquids. In Fig. 7, the Gaussian approximation to the (normalized) velocity autocorrelation function is compared with that observed in molecular dynamic simulations of liquid argon. Although the

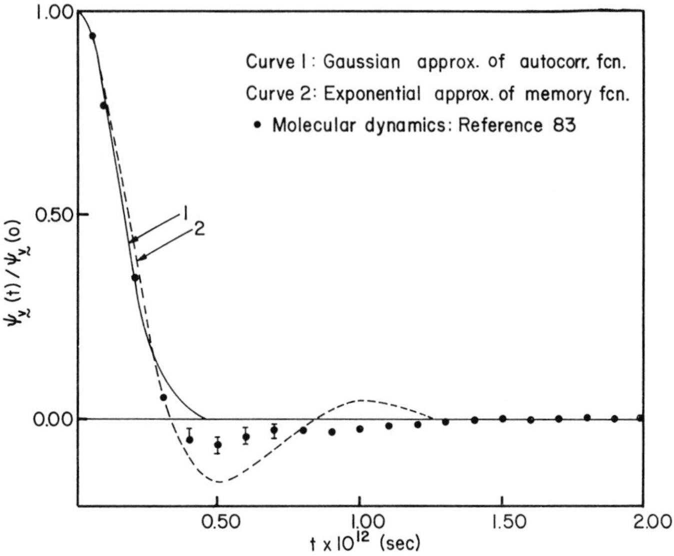

Fig. 7. Calculated normalized velocity autocorrelation function for argon molecules in the liquid at $\rho_m = 1.374\ g/cm^3$ and $T = 90°K$. The data used in calculating these curves were; $D = 2.72 \times 10^{-5}\ cm^2/sec$, Ref. 24; $\langle \nabla^2 V \rangle = 11.25 \times 10^3\ erg/cm^3$, Boato et al., *J. Chem. Phys.*, **40,** 2419 (1964). Figure taken from Ref. 12.

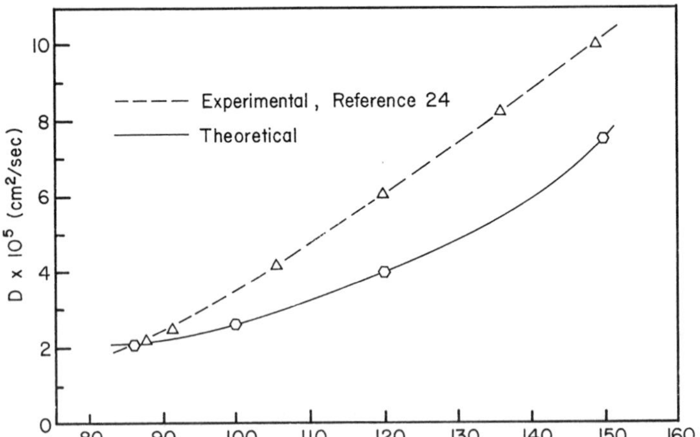

Fig. 8. Self-diffusivity of liquid argon along the vapor-pressure curve. Theoretical values obtained from Gaussian model for autocorrelation function. Values of $\langle F_1^2 \rangle$ used in theory are experimental values determined by G. Boato, G. Casanova, and A. Levi, *J. Chem. Phys.*, **40,** 2419 (1964).

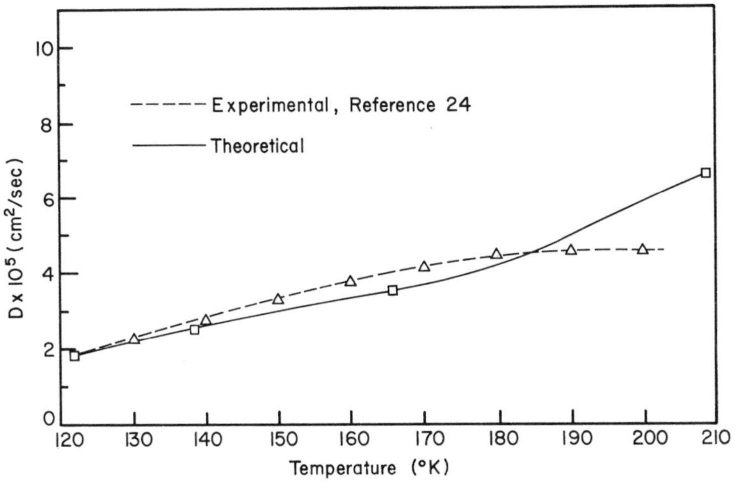

Fig. 9. Self-diffusivity of liquid krypton along the vapor-pressure curve. Gaussian model for autocorrelation function. Values of $\langle F_1^2 \rangle$ determined from argon data by law of corresponding states, scaling with critical parameters.

two curves in Fig. 7 differ for times greater than about 4×10^{-13} sec, the area under the two curves differs by only about 20%, indicating that the self-diffusion coefficient predicted by the Gaussian approximation would differ from experiment by the same amount. In Figs. 8–10 the self-diffusion coefficients measured in liquids argon, krypton, and xenon are compared with those predicted by the Gaussian approximation.

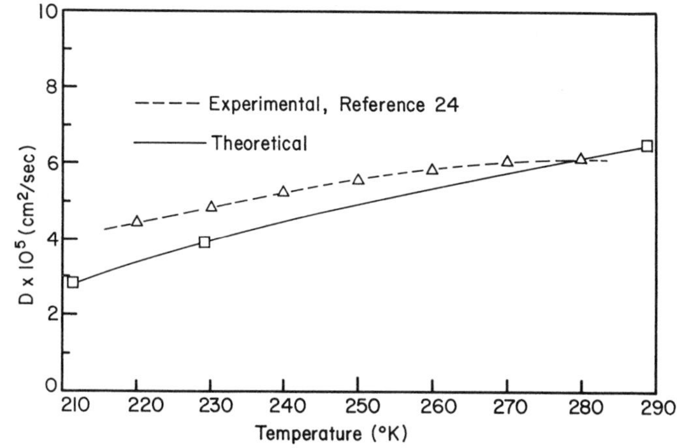

Fig. 10. Self-diffusivity of liquid xenon along the vapor-pressure curve. Gaussian model for autocorrelation function. Values of $\langle F_1^2 \rangle$ determined from argon data by law of corresponding states, scaling with critical parameters.

TABLE VII

Comparison of Theoretical and Observed Values of Shear Viscosity (in units of 10^{-3} g/cm-sec) of Liquid Argon. Theoretical Values are Computed Using the Gaussian Approximation Results of (8.36) and (8.37), the Superposition Approximation for the Triplet Correlation Function, and the Radial Distribution Functions Determined by Computer Molecular Dynamics Experiments.[a] Measured Values are those Reported by Zhdanova[b] and Those Quoted by Rice and Gray[c] at 1.12 g/cm³. Table Taken from Ref. 70

mn (g/cm³)	T (°K)	η (calc.)	η (obs.)
1.37	98.5	1.67	2.23
1.25	99.5	1.24	1.50
	128.2	1.24	1.43
	157	1.24	1.37
1.18	108	1.00	1.14
	124	0.99	1.17
			0.84[c]
	190	1.00	1.22
			0.87[c]
	219	0.94	1.24
			0.89[c]
0.915	317	0.61	0.71
0.834	163	0.45	0.60
0.75	186.5	0.36	0.54
	205	0.37	0.55
	352	0.38	0.58

[a] L. Verlet, *Phys. Rev.*, **165,** 201 (1968).

[b] N. F. Zhdanova, *Zh. Eksperim. i Teor. Fiz.*, **31,** 724 (1956) [English transl.: *Soviet Phys. JETP*, **4,** 749 (1957)].

[c] Ref. 12.

Although there are no data available on the time correlation functions of the momentum and energy fluxes, studies on memory functions (to be discussed in the next section) indicate that the Gaussian approximation is no worse for the time correlation functions of these fluxes than for the velocity time correlation function. The shear viscosity computations of Forster, Martin, and Yip[70] based on the Gaussian approximation are compared to experiment in Table VII.

C. Memory Functions

About twelve years ago, Zwanzig showed in an important work[49] that the time correlation functions, $\psi_A(t) \equiv \langle A(0)A(t)\rangle$, obey Volterra equations of the form

$$\frac{d\psi_A(t)}{dt} = -\int_0^t ds\, K_A(s)\psi_A(t-s) \tag{8.45}$$

with the initial condition

$$\psi_A(t=0) = \langle (A(0))^2 \rangle \tag{8.46}$$

The kernel $K_A(s)$ is the so-called "memory function" which relates the rate of change of $\psi_A(t)$ at time t to its past history. The form Zwanzig derived for the memory function is

$$K_A(s) = \langle LA(0)e^{is(1-P_A)L}LA(0)\rangle \tag{8.47}$$

where P_A is a projection operator defined by

$$P_A G = \frac{A(\Gamma)\int d\Gamma'\, A(\Gamma')G(\Gamma')f^{eq}(\Gamma')}{\langle (A)^2 \rangle} \tag{8.48}$$

$\Gamma \equiv (\mathbf{x}_1, \ldots, \mathbf{x}_N, \mathbf{v}_1, \ldots, \mathbf{v}_N)$ denotes a point in phase space (in the initial time ensemble, that is, $A(\Gamma) = A(0)$) and $d\Gamma \equiv d\mathbf{x}_1 \cdots d\mathbf{v}_N$ denotes a volume element in phase space.

The value of $K_A(s)$ and all its derivatives at $s = 0$ may be related to the values $\psi(t)$ and its derivatives at $t = 0$ through (8.19). Differentiating (8.19) n times with respect to time, we obtain

$$\frac{d^n\psi_A(t)}{dt^n} = -\sum_{m=0}^{n-2} \frac{d^m K_A(t)}{dt^m}\frac{d^{n-m-2}\psi_A(0)}{dt^{n-m-2}} - \int_0^t ds\, K_A(s)\frac{d^{n-1}\psi_A(t-s)}{dt^{n-1}} \tag{8.49}$$

for $n \geqslant 2$. Setting $t = 0$, we find the set of equations

$$\frac{d^n\psi_A(0)}{dt^n} = -\sum_{m=0}^{n-2} \frac{d^m K_A(0)}{dt^m}\frac{d^{n-m-2}\psi_A(0)}{dt^{n-m-2}} \tag{8.50}$$

of which the first few equations are

$$\frac{d^2\psi_A(0)}{dt^2} = -K_A(0)\psi_A(0) \tag{8.51}$$

$$\frac{d^3\psi_A(0)}{dt^3} = -K_A(0)\frac{d\psi_A(0)}{dt} + \frac{dK_A(0)}{dt}\psi_A(0) \tag{8.52}$$

$$\frac{d^4\psi_A(0)}{dt^4} = -K_A(0)\frac{d^2\psi_A(0)}{dt^2} - \frac{dK_A(0)}{dt}\frac{d\psi_A(0)}{dt} - \frac{d^2K_A(0)}{dt^2}\psi_A(0) \tag{8.53}$$

Since the odd derivatives of ψ_A are zero at $t = 0$, (8.52) implies

$$\frac{dK_A(0)}{dt} = 0 \tag{8.54}$$

Rearranging (8.51) and (8.53), we obtain

$$K(0) = -\frac{1}{\psi_A(0)}\frac{d^2\psi_A(0)}{dt^2} \tag{8.55}$$

and

$$\frac{d^2K_A(0)}{dt^2} = -\frac{1}{\psi_A(0)}\frac{d^4\psi_A(0)}{dt^4} - \left[\frac{1}{\psi_A(0)}\frac{d^2\psi_A(0)}{dt^2}\right]^2 \tag{8.56}$$

where it may be recalled from Section VIIIB that

$$\frac{d^{2n}\psi_A(0)}{dt^{2n}} = (-1)^n\langle(L^n A(0))^2\rangle \tag{8.57}$$

Let us define the Laplace transform of ψ_A by

$$\tilde{\psi}_A(\omega) = \int_0^\infty dt\, e^{-\omega t}\psi_A(t) \tag{8.58}$$

The transform of (8.45) yields the result

$$\tilde{\psi}_A(\omega) = \frac{\psi_A(t=0)}{\omega + \tilde{\mathbf{K}}_A(\omega)} \tag{8.59}$$

so that a transport coefficient θ_A of the form given by (8.29) becomes

$$\theta_A = \lim_{\epsilon\to 0}\frac{\alpha_A\psi_A(0)}{\varepsilon + \int_0^\infty dt\, e^{-\epsilon t}K_A(t)} \tag{8.60}$$

Thus, the initial value of ψ_A and the value of the Laplace transform $\tilde{K}_A(\omega)$ at $\omega = 0$ determine the transport coefficient θ_A.

Evaluation of the memory function is no easier than evaluation of the time correlation function itself. However, mathematically modeling K_A is a different starting point from modeling ψ_A and, therefore, might yield new insights or better numerical results. The first attempt to model the memory function was that of Berne, Boon, and Rice[72] for the velocity autocorrelation function. They assumed the form of the memory function to be

$$K_v(t) = \gamma e^{-\alpha|t|} \tag{8.61}$$

The parameter γ was determined by the exact condition (8.25) on K_v at $t = 0$, that is,

$$K_v(0) = \gamma = \frac{1}{m^2\langle \mathbf{v}^2 \rangle} \langle (\mathbf{F}_1(0))^2 \rangle$$
$$= \frac{n}{3m} \int d\mathbf{r}\, g(r)\, \nabla^2 V(r) \tag{8.62}$$

The Laplace transform of (8.62) is

$$\tilde{\mathbf{K}}_v(\omega) = \frac{\gamma}{\omega + \alpha} \tag{8.63}$$

so that the self-diffusion coefficient becomes

$$D = \frac{(kT/m)}{\gamma/\alpha} = \frac{\alpha kT}{m\gamma} \tag{8.64}$$

Thus, for the exponential model given by (8.61), the initial value of $d^2\psi_v/dt^2$ plus the value of the diffusion coefficient can be used to determine the unknown constants in the memory function. Then by solving (8.45) the autocorrelation function can be obtained as a function of time. Berne, Boon, and Rice's predicted values for the normalized velocity autocorrelation function are compared to computer experiments in Fig. 7. The qualitative features are an improvement over the simple Gaussian approximation for $\psi_v(t)$—the time at which the autocorrelation function is predicted to first go negative is in pretty good agreement with experiment. Of course, since the diffusion coefficient was used in fitting the parameters α and γ, the model cannot be used to predict the diffusion coefficient.

In the exponential model one parameter, γ, is determined by the short-time behavior of $K_v(t)$ while the decay rate α is determined from the long-time (actually the time integral) behavior of $K_v(t)$. Such a procedure is an attempt to interpolate the entire time behavior of the memory function by building in accurate limiting behavior.

Writing the self-diffusion coefficient in terms of the time integral of the autocorrelation function and solving (8.64) for α/γ, Berne[4] has expressed the exponential memory function in the rather nice form

$$K_v(t) = \frac{\langle (\mathbf{F}_1)^2 \rangle}{m^2\langle v^2 \rangle} \exp\left[- \frac{\langle (\mathbf{F}_1)^2 \rangle t}{m^2\langle \mathbf{v}^2 \rangle} \int_0^\infty ds\, \langle \mathbf{v}(0) \cdot \mathbf{v}(s) \rangle \right] \tag{8.65}$$

Actually the detailed behavior of $K_{v(t)}$ near $t = 0$ is rather poorly represented by the exponential approximation. In particular, we showed earlier that all the odd derivatives of $K_v(t)$ vanish at $t = 0$ for sufficiently smooth (Lennard-Jones, for example) interaction potentials. From (8.61), on the other hand, we predict that all the odd derivatives are finite if the derivatives at $t = 0$ are defined as limits from $t > 0$. Singwi and Tosi[73] eliminated the problem with the odd derivatives by assuming the following Gaussian memory function

$$K_{\mathbf{v}}(t) = \frac{\langle(\mathbf{F}_1)^2\rangle}{m^2\langle v^2\rangle} \exp - \left[\left(\frac{\pi}{4}\right)^{1/2} \frac{\langle(\mathbf{F}_1)^2\rangle}{m^2\langle v^2\rangle} t \int_0^\infty ds\langle\mathbf{v}(0)\cdot\mathbf{v}(s)\rangle\right]^2 \quad (8.66)$$

As in the case of the exponential approximation, the form of (8.66) is chosen so that $K_{\mathbf{v}}(t = 0)$ is exact and the diffusion coefficient is given exactly. In Fig. 11 the normalized memory function deduced from computer experiments is compared with the above exponential and Gaussian models. The Gaussian model appears to be an improvement over the exponential model and is quite accurate up to times of the order of 5×10^{-13} sec for the fluid considered. In Fig. 12 computer experimental and theoretical autocorrelation functions are compared for the exponential and Gaussian models. Predictions of the Gaussian model are significantly better than those of the exponential model.

As was done for the autocorrelation function, calculations in Section VIIIB, the parameters in the Gaussian model

$$K_v(t) = \gamma e^{-(t/\tau)^2} \quad (8.67)$$

if this form of (8.67) is assumed to be valid for all times, can be determined by equating for $t = 0$ the zeroth and second derivatives of (8.67)

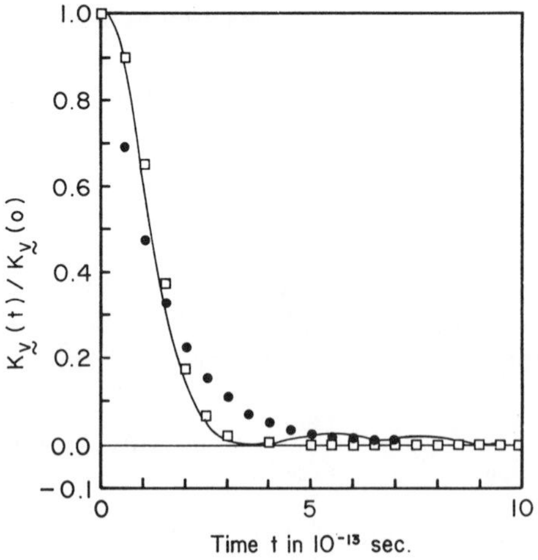

Fig. 11. Normalized memory function deduced from computer experiments (solid curve), the exponential [(8.61) (solid dots)], and the Gaussian [(8.66) (squares)] models. Figure taken from Ref. 4.

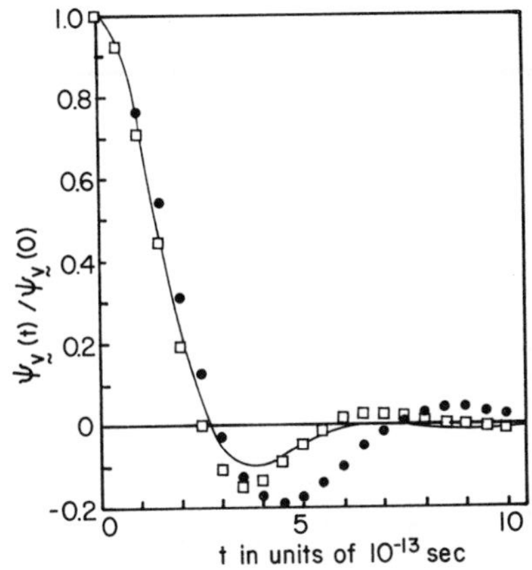

Fig. 12. Normalized velocity autocorrelation function deduced from computer experiments (solid curve) and the exponential [(8.61) (solid dots)] and Gaussian [(8.66) (squares)] models for the memory function. Figure taken from Ref. 4.

to the exact values given by (8.55) and (8.56). Thus

$$K_{\mathbf{v}}(0) = \gamma = \frac{\langle (\mathbf{F}_1)^2 \rangle}{m^2 \langle v^2 \rangle} \tag{8.68}$$

$$\frac{d^2 K_{\mathbf{v}}(0)}{dt^2} = \frac{\gamma}{\tau^2} = +\frac{1}{m^2 \langle v^2 \rangle} \langle (L\mathbf{F}_1)^2 \rangle - \left(\frac{\langle (\mathbf{F}_1)^2 \rangle}{m^2 \langle v^2 \rangle} \right)^2 \tag{8.69}$$

In terms of γ and τ of the Gaussian model the self-diffusion coefficient is

$$D = \frac{kT}{m\gamma\tau} \tag{8.70}$$

An advantage of this Gaussian model over (8.66) is that the diffusion coefficient is predicted by the model rather than being used in predicting the behavior of $K_{\mathbf{v}}$. In Figs, 13 and 14, the velocity memory function and autocorrelation function predicted by determining γ and τ according to (8.68) and (8.69) are compared with computer experiments. The agreement is quite good and indicates that the model should yield good predictions of diffusion coefficients. Martin and Yip, who studied this Gaussian model,[74] using computer-generated data to calculate γ and τ from (8.68) and (8.69), predict $D = 2.95 \times 10^{-5}$ cm^2/sec compared with the computer-generated value of $D = 1.88 \times 10^{-5}$ cm^2/sec. The computer experiment was for a Lennard-Jones potential model and for conditions appropriate for argon at 85.5°K and 1.407 g/cm^3. Surprisingly the agreement is no better than that obtained from the lower-order Gaussian approximation in Section VIIIB.

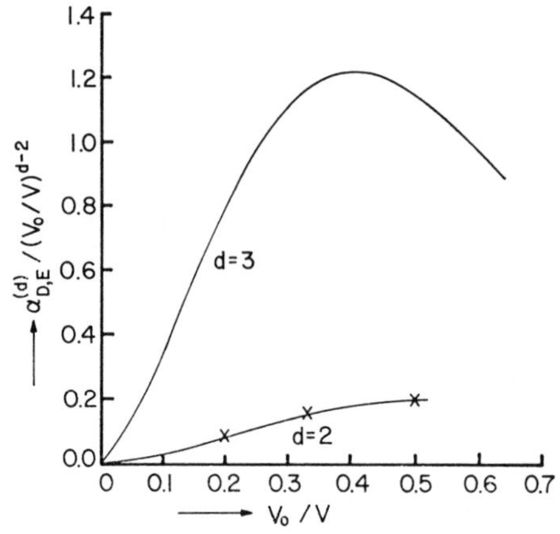

Fig. 13. Normalized velocity memory function predicted (squares) by the Gaussian memory model of (8.67). Solid curve deduced from computer experiments on velocity autocorrelation function. Figure taken from Ref. 4.

Extension of the Gaussian memory model to the other correlation functions is immediately obvious, although the quantity $d^4\psi_A(0)/dt^4$ is so complicated for the heat and momentum fluxes that it is doubtful that the extension would prove fruitful for actually computing $\psi_A(t)$ or the transport coefficients.

For an excellent discussion of theoretical and computer experimental momentum autocorrelation functions and memory functions the reader

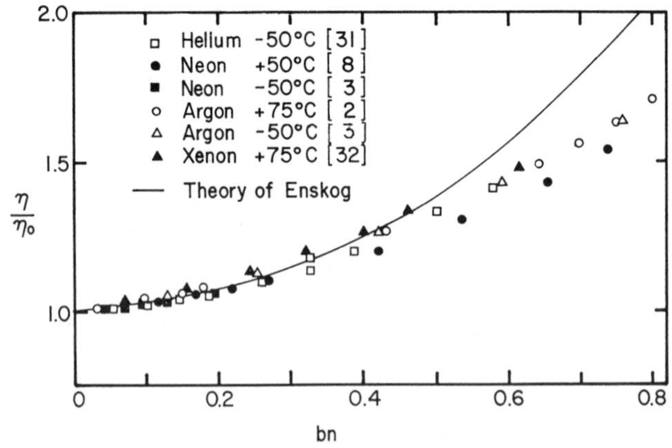

Fig. 14. Normalized velocity autocorrelation function predicted (solid dots) by the Gaussian memory model of (8.67). Solid curve is from computer experiments. Figure taken from Ref. 4.

is referred to Berne's work in Ref. 4. Angular momentum and spin autocorrelation function studies are discussed also in this reference.

D. Long-Time Decay of Time Correlation Functions

In computer studies of hard sphere systems Alder and Wainwright[75,76] observed a positive slowly varying long-time tail on the velocity correlation function. They observed that for times longer than about 10 mean collision times the asymptotic form of the normalized velocity correlation function, $\rho(t)$, is

$$\rho(t) \sim \frac{\alpha}{t^{d/2}} \tag{8.71}$$

where $d(=2, 3)$ is the dimension of the system and α is a constant for a given density. By studying the velocity correlation between a molecule and its neighboring molecules, Alder and Wainwright[75] discovered a vortex flow pattern which could explain the observed persistence of velocity, that is, the slowly decaying tail of $\rho(t)$. They showed that a transient solution of the Navier-Stokes equation reproduced rather well the observed velocity field and asymptotic value of $\rho(t)$ for times longer than 10 mean collision times. Thus the persistence of velocity appears to be associated with the shear damping of vortex rings.

In the two-dimensional case the asymptotic forms of $\rho(t)$ and of the kinetic parts of the autocorrelation functions of J_η^{xy} and $\mathbf{J}_K$, which are expected from the hydrodynamical model to behave similarly to $\rho(t)$, have the special significance of implying that the transport coefficients diverge in time and that hydrodynamics in the usual sense does not exist for two dimensions. This surprising result is the subject of a great deal of investigation and speculation currently.

Alder and Wainwright's conclusions have been reproduced by a variety of techniques[77–81] including one kinetic theoretical model for rigid disks and rigid spheres developed by Dorfman and Cohen.[79] These investigators expand the Laplace transform $\tilde{\rho}(\omega)$ of $\rho(t)$ in the following series ordered according to the nature of the collisional terms (i.e., diagrams) retained:

$$\tilde{\rho}(\omega) = \tilde{\rho}_1(\omega) + n\tilde{\rho}_2(\omega) + \cdots \tag{8.72}$$

They evaluate the term $\tilde{\rho}_1(\omega)$ in terms of uncorrelated binary collisions, that is, from the Boltzmann equation. They then approximate $\tilde{\rho}_2(\omega)$ by summing sequences of correlated binary collisions. The sequences of correlated binary collisions represent, to each order in density, the most divergent terms in a density expansion of $\tilde{\rho}(\omega)$. The term $\tilde{\rho}_1(\omega)$ gives a contribution to $\rho(t)$ of the form e^{-t/t_0}, where t_0 is the collision frequency.

For $t \gg t_0$, Dorfman and Cohen obtain from $\tilde{\rho}_2(\omega)$ the following contribution to $\rho(t)$:

$$\rho^{(d)}(t) \approx \alpha_0^{(d)} \left(\frac{t_0}{t}\right)^{d/2} \tag{8.73}$$

where d denotes the dimension of the system, the subscript zero on $\alpha_0^{(d)}$ means the lowest-order density approximation, and

$$\alpha_0^{(2)} = \left[8\pi n^2 \sigma^d \left(D_0 + \frac{\eta_0}{nm}\right) t_0\right]^{-1} = \frac{1}{4} \tag{8.74}$$

$$\alpha_0^{(3)} = \left[\pi (n\sigma^d)^{4/3} \left(D_0 + \frac{\eta_0}{nm}\right) t_0\right]^{-3/2} \Big/ 12n$$
$$= (\tfrac{64}{11})^{3/2}/12 \tag{8.75}$$

where D_0 and η_0 are the first Chapman-Enskog approximation to the dilute gas self-diffusion and viscosity coefficients, respectively.

In order to generalize their results to dense systems, Dorfman and Cohen use Enskog's dense hard sphere theory to account for the dense fluid spatial correlations and obtain the following dense fluid version of (8.74) and (8.75):

$$\alpha_{D,E}^{(2)} = \left[8\pi n \left(D_E + \frac{\eta_E}{nm}\right) t_0\right]^{-1} \tag{8.76}$$

$$\alpha_{D,E}^{(3)} = \left[\pi \left(D_E + \frac{\eta_E}{nm}\right) t_0\right]^{-3/2} \Big/ 12n \tag{8.77}$$

where D_E and η_E are Enskog's dense gas self-diffusion and viscosity coefficients. In Fig. 15 these values for $\alpha_{D,E}^{(d)}$ are plotted versus reduced volume and the two-dimensional result is compared with the computer results of Alder and Wainwright. The agreement is very good, indicating that consecutive correlated binary collisions (between molecules whose spatial correlations are given by Enskog's local equilibrium radial distribution function approximation) are primarily responsible for setting up the correlated motions observed as vortex patterns. Dorfman and Cohen, in agreement with other investigators, also show that the kinetic parts of the autocorrelation functions of J_η^{xy} and $\mathbf{J}_\kappa$ decay as $\alpha(t_0/t)^{d/2}$, *and* they obtain expressions for the characteristic constant α.

With regard to the question of existence of transport coefficients in two dimensions, it should be mentioned the Dorfman and Cohen state that their results may be valid only up to times $t \simeq 30t_0$. Beyond this other collisional classes may play a dominant role.

In closing this section, let us discuss briefly why the approximate kinetic theories discussed in Sections II–VI and the correlation function

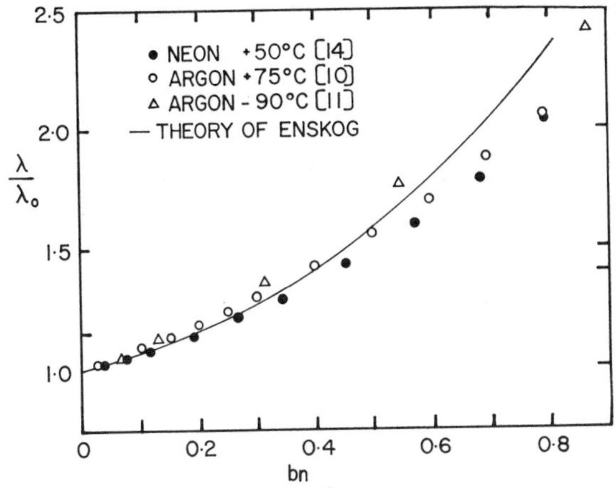

Fig. 15. $\alpha_{D,E}^{(d)}$ as a function of reduced volume for hard disks ($d = 2$) and hard spheres ($d = 3$). Solid curves are from Dorfman and Cohen's theory.[79] Crosses are from Alder and Wainwright's computer experiments.

models discussed in Sections VIIIB and VIIIC are not rejected in the three-dimensional cases even though they have neglected the long-time behavior discussed here. In the two-dimensional case, they must of course be discarded if the conclusion is eventually reached that transport coefficients and the usual hydrodynamics do not exist in two dimensions. On the basis of Alder and Wainwright's[82,75] computer results on hard spheres and Rahman's[83,78] results on Lennard-Jones particles the long-time part of $\rho(t)$ contributes about 10–15% of the self-diffusion coefficient for rigid spheres at liquidlike densities. The state of the art of calculating liquid state transport coefficients is not good enough at this point in time to yield results to better than 85–90% accuracy, as would be required to test the seriousness of omitting the long-time correction. For one thing the potential models and the theoretical pair correlation functions will have to be refined a great deal before a 10% correction can be tested. In the case of the viscosities and the thermal conductivity, neglecting the long-time contribution should be a good approximation for dense fluids since the kinetic parts (which are the only parts thought to have the asymptotic time dependence) of these transport coefficients are no more than about 30% of the total for dense fluids. A 15% correction of a 30% contribution will be less than 5%. Finally, even if the state of the art of transport calculations develops to the point of worrying about 5–15% corrections, the short-time model kinetic equations will provide useful starting points for developing for real fluids a kinetic theory including the long-time contributions. In fact, this was done by Dorfman and Cohen, who took

Enskog's kinetic equation as the starting point and then corrected it for the long-time contributions.

IX. ENSKOG THEORY OF FLUIDS COMPOSED OF MOLECULES INTERACTING WITH IMPULSIVE FORCES

A. Hard Sphere Fluids

Because of its influence on current kinetic theories of dense fluids, its success in correlating dense gas transport properties, and its relevance to computer experiments on hard sphere systems, Enskog's original theory[16,20] of smooth hard sphere fluids and extensions of the original theory will be reviewed briefly in concluding this article. The impulsive models, though crude, have the advantage of simplicity over more general models and it may well be that if an accuracy of only a factor of 2 is desired these models are the answer.

Under the assumption that the rate of collisional change of the singlet distribution function is determined entirely by uncorrelated binary collisions, the kinetic equation for a pure hard sphere fluid is

$$\partial_t f + \mathbf{v}_1 \cdot \nabla_1 f = \sigma^2 \int_{\mathbf{v}_{21}\cdot\mathbf{k}>0} [f_2(\mathbf{x}_1, \mathbf{x}_1 + \sigma\mathbf{k}, \mathbf{v}_1', \mathbf{v}_2', t) - f_2(\mathbf{x}_1, \mathbf{x}_1 - \sigma\mathbf{k}, \mathbf{v}_1, \mathbf{v}_2, t)]\mathbf{v}_{21} \cdot \mathbf{k}\, d\mathbf{v}\, d\mathbf{v}_2 \quad (9.1)$$

If we then approximate the doublet distribution function by

$$f_2(\mathbf{x}_1, \mathbf{x}_1 \pm \sigma\mathbf{k}, \mathbf{v}_1, \mathbf{v}_2, t) = g\left(\mathbf{x}_{12} = \pm\sigma\mathbf{k};\ T\left(\frac{\mathbf{x}_1 + \mathbf{x}_2, t}{2}\right), n\left(\frac{\mathbf{x}_1 + \mathbf{x}_2, t}{2}\right)\right) \times f(\mathbf{x}_1, \mathbf{v}_1, t) f(\mathbf{x}_1 \pm \sigma\mathbf{k}, \mathbf{v}_2, t) \quad (9.2)$$

we obtain Enskog's kinetic equation. The g in (9.2) is the contact value of the equilibrium radial distribution function evaluated at the local thermodynamic state midway between the centers of the spheres.

To the first approximation (i.e., keeping, as outlined in Section II, the first terms in the Sonine polynomial expansion for the singlet perturbation ϕ) the following transport coefficients are predicted by the Enskog theory:

Self-Diffusion

$$D = \frac{3}{8n\sigma^2 g(\sigma)}\left(\frac{kT}{\pi m}\right)^{1/2} \quad (9.3)$$

Thermal Conductivity

$$\kappa = \kappa^*\left[(1 + \tfrac{3}{5}bng(\sigma))^{-1} + \frac{32}{25\pi}(bn)^2 g(\sigma)\right) \tag{9.4}$$

where κ^* is defined by (2.82) and

$$b = \frac{2\pi}{3}\sigma^3 \tag{9.5}$$

Shear Viscosity

$$\eta = \eta^*\left[(1 + \tfrac{2}{5}bng(\sigma))^2(g(\sigma))^{-1} + \frac{48}{25\pi}(bn)^2 g(\sigma)\right] \tag{9.6}$$

where η^* is defined by (2.77).

Bulk Viscosity

$$\Phi = \eta^*\left[\frac{16}{5\pi}(bn)^2 g(\sigma)\right] \tag{9.7}$$

Collision rates predicted by Enskog's theory agree with computer results[84] for hard spheres. Predicted[82] self-diffusion coefficients differ from computer experimental by about the 10–15% known to arise from the persistence of velocity (vortex formation) effect discussed in Section VIIID. This is to be expected since Enskog's equation also neglects correlated binary collisions.

If the hard sphere diameter σ is determined for a given temperature by equating the theoretical transport coefficients to experimental in the zero density limit, Enskog's theory predicts fairly well the density dependence of the transport coefficients of simple fluids. Figures 16 and 17 give comparisons of experimental shear viscosities and thermal conductivities with computations Sengers[85] has made using the zero density fits for σ. Agreement is rather good up to reduced densities of $n\sigma^3 \approx 0.5$ and even at the highest densities shown theory is only about 20% high. For the calculated curves in these figures Sengers used the following known virial expansion for $g(\sigma)$:

$$g(\sigma) = 1 + 0.6250bn + 0.2869(bn)^2 + 0.115(bn)^3 + 0.109(bn)^4 \tag{9.8}$$

One could equally well use the form of $g(\sigma)$ found by the scaled particle theory[17] of hard spheres, that is,

$$(g\sigma) = \frac{2 - \frac{1}{2}bn + \dfrac{(bn)^2}{32}}{2\left(1 - \dfrac{bn}{4}\right)^3} \tag{9.9}$$

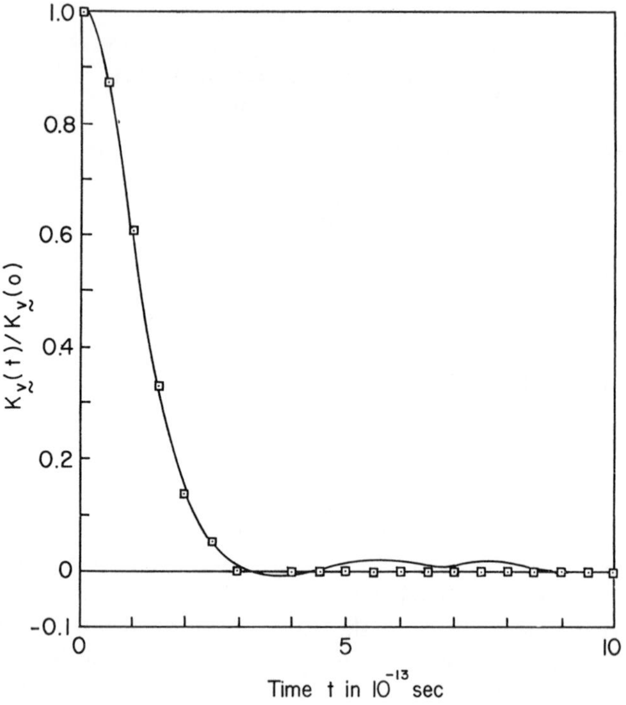

Fig. 16. Density dependence of the viscosity of helium, neon, argon, and xenon compared with Enskog's theory. Figure taken from Ref. 85.

This form of $g(\sigma)$ leads to similar predictions as the exact virial expansion for the range of densities in Figs. 16 and 17 and is a good approximation to $g(\sigma)$ over the whole density range.

In order to improve the agreement between Enskog's theory and dense gas experiments, Michels and Gibson[86] introduced the procedure of determining $g(\sigma)$ from the measured thermal pressure according to the equation

$$bng(\sigma) = \frac{1}{nk}\left(\frac{\partial P}{\partial T}\right)_n - 1 \tag{9.10}$$

whereas to ensure that $\lim_{n\to 0} g(\sigma) = 1$ the quantity b is determined by the expression

$$b = \frac{d}{dT}(BT) \tag{9.11}$$

where B is the second virial coefficient. Sengers has used this procedure to determine b and $g(\sigma)$ for the inert gases and has used these in Enskog's

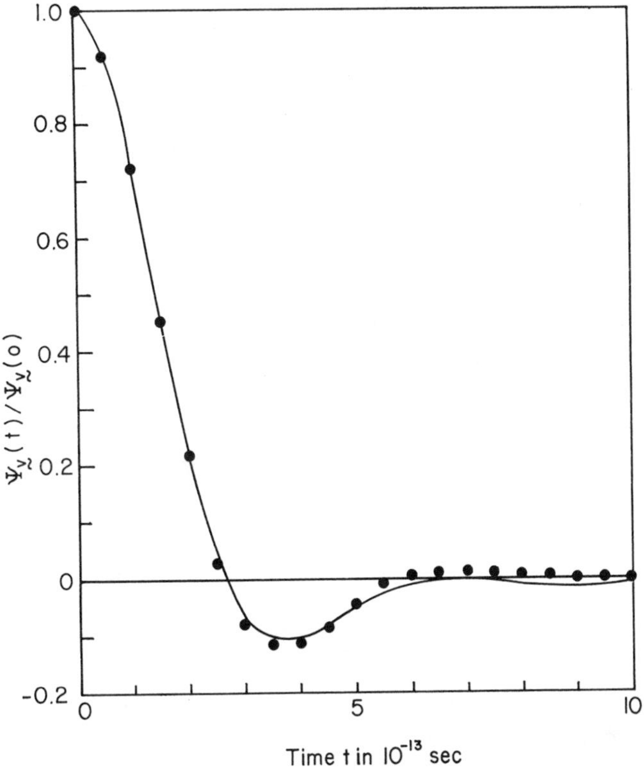

Fig. 17. Density dependence of the thermal conductivity of neon and argon compared with Enskog's theory. Figure taken from Ref. 85.

equations to predict shear viscosities and thermal conductivities for the inert gases. His results are compared with experiment in Table VIII. The agreement is very good. Sengers used the same procedure with equal success in predicting the shear viscosity of dense gaseous hydrogen (at 25°C) and nitrogen (at 75°C). Moreover, by adding the Eucken polyatomic correction suggested by Hirschfelder, Curtiss, and Bird,[20] namely,

$$\kappa_{COR} = \frac{\kappa_0 - \kappa_0'}{g(\sigma)} \tag{9.12}$$

where $\kappa_0' = \lim_{n\to 0} \kappa_{\text{exp}}$ and $\kappa_0 = \lim_{n\to 0} \kappa_{\text{Enskog}} = \kappa^*$, Sengers obtained good predictions of thermal conductivities of gaseous nitrogen (at 75°C) over the density range 0–600 amagat. One amagat is the density of the substance at 0°C and 1 atm.

TABLE VIII

Comparison of the Viscosity and the Thermal Conductivity of the Noble Gases with the Empirical Modification of the Theory of Enskog. Table taken from Ref. 85

	$\frac{\eta_{calc}-\eta_{exp}}{\eta_{exp}}$ (%)						
ρ (amagat)	100	200	300	400	500	600	700
Helium 0°C	0%	0%	−1%	−2%	−4%		
Neon +75°C	+1%	+1%	+1%	+1%	+1%	+1%	+1%
Neon +25°C	+1%	+1%	+1%	+1%	+1%	+2%	+2%
Argon +75°C	0%	−3%	−5%	−8%	−10%	−12%	
Argon −50°C	+5%	+5%	0%				
Xenon +75°C	+14%	+10%	0%	−7%			

	$\frac{\lambda_{calc}-\lambda_{exptl}}{\lambda_{exptl}}$ (%)						
ρ (amagat)	100	200	300	400	500	600	700
Neon +75°C	0%	0%	0%	−1%	−2%	−2%	−2%
Neon +25°C	0%	−1%	−2%	−2%	−3%	−4%	−4%
Argon +75°C	−3%	−7%	−11%	−12%	−13%	−15%	
Argon −90°C	+1%	−4%	−8%	−6%	−4%		

It is interesting that the viscosity of small polyatomic molecules does not seem to be strongly affected by the polyatomic degrees of freedom.

Thorne[16] extended Enskog's theory to binary hard sphere mixtures, and Tham and Gubbins[87] worked out the multicomponent theory.

B. Square-Well Fluids

In an attempt to extend Enskog's theory to a model system capable of having a liquid state, Rice, Sengers, and the author[88] considered a fluid composed of molecules interacting according to a pairwise additive square-well potential defined as follows:

$$\begin{aligned} V(r) &= 0 \qquad && r > R\sigma \\ &= -\epsilon && \sigma < r < R\sigma \\ &= +\infty && r \leqslant \sigma \end{aligned} \tag{9.13}$$

Following Enskog, they assumed that only uncorrelated binary collisions contribute to the collision integral and that the double distribution function factors in the radial distribution function times the singles distribution functions as illustrated by (9.2). Thus their kinetic equation it similar to (9.1) except that the collision integral is composed of a sum of four parts arising from the four possible distinct types of binary collisions between square-well molecules. The four types of collisions may be summarized as follows:

Type 1. When two particles collide at a distance equal to the repulsive diameter σ_1, they will exchange energy and momentum in the same manner as rigid spheres.

Type 2. When two particles approach each other from a distance greater than the attractive diameter σ_2, they will experience a sudden increase in velocity and kinetic energy due to an amount of potential energy ϵ being converted into kinetic energy.

Type 3. When two particles, initially moving apart with relative kinetic energy greater than ϵ, reach the separation σ_2, relative kinetic energy of amount ϵ will instantaneously be converted to potential energy.

Type 4. When two particles moving apart have a relative kinetic energy less than ϵ, the relative velocity is simply reversed with the conservation of kinetic energy and momentum.

Davis, Rice, and Sengers' square-well equation of state and transport coefficients are summarized here.

Pressure

$$P = nkT\{1 + bn[g(\sigma) + R^3 g(R\sigma)(1 - e^{\epsilon/kT})]\} \qquad (9.14)$$

where $g(R\sigma)$ is the value of the equilibrium radial distribution function for a pair of molecules separated infinitesimally more than $R\sigma$; that is, it is the limit of $g(r)$ as the molecules approach $R\sigma$ from a greater separation.

Self-Diffusion

$$D = \frac{3}{8n\sigma^2}\left(\frac{kT}{\pi m}\right)^{1/2}[g(\sigma) + R^2 g(R\sigma)\Xi]^{-1} \qquad (9.15)$$

where

$$\Xi = \exp\left(\frac{\epsilon}{kT}\right) - \frac{\epsilon}{2kT} - 2\int_0^\infty x^2\left(x^2 + \frac{\epsilon}{kT}\right)\exp(-x^2)\,dx \qquad (9.16)$$

Values are given for Σ in Table IX.

Thermal Conductivity

$$\kappa = \kappa^*\left\{\frac{[1 + \frac{3}{5}bn[g(\sigma) + R^3 g(R\sigma)H]^2}{g(\sigma) + R^2 g(R\sigma)[\Xi + \frac{11}{16}(\epsilon/kT)]^2} + \frac{32}{25\pi}(bn)^2[g(\sigma) + R^4 g(R\sigma)\Xi]\right. \qquad (9.17)$$

TABLE IX

Values of the Functions H and Ξ Defined by (9.18) and (9.16), Respectively, Versus the Inverse Reduced Temperature, ϵ/kT

$\frac{\epsilon}{kT}$	H	Ξ
0	0.0000	0.0000
0.1	−0.0012	0.0086
0.2	−0.0066	0.0324
0.3	−0.0184	0.0710
0.4	−0.0387	0.1252
0.5	−0.0685	0.1959
0.6	−0.1105	0.2846
0.7	−0.1665	0.3927
0.8	−0.2385	0.5221
0.9	−0.3290	0.6747
1.0	−0.4403	0.8528
1.1	−0.5752	1.0589
1.2	−0.7368	1.2958
1.3	−0.9282	1.5666
1.4	−1.1530	1.8747
1.5	−1.4153	2.2239

where

$$H = 1 - \exp\left(\frac{\epsilon}{kT}\right) + \frac{\epsilon}{2kT}\left[1 + \frac{4}{\sqrt{\pi}}\exp\left(\frac{\epsilon}{kT}\right)\int_{\sqrt{\epsilon/kT}}^{\infty}\exp(-x^2)x^2\,dx\right] \tag{9.18}$$

Values of H are also given in Table IX. κ^* is defined by (2.82).

Shear Viscosity

$$\eta = \eta^*\left\{\frac{[1 + \frac{2}{5}bn[g(\sigma) + R^3g(R\sigma)H]^2}{g(\sigma) + R^2g(R\sigma)\left[\Xi + \frac{1}{6}\left(\frac{\epsilon}{kT}\right)^2\right]} + \frac{48}{25\pi}(bn)(g(\sigma) + R^4g(R\sigma)\Xi)\right\} \tag{9.19}$$

where η^* is defined by (2.87).

Bulk Viscosity

$$\Phi = \eta^*(bn)^2 \times \left\{\frac{[R^3g(R\sigma)(\epsilon/kT)^2(4\sqrt{\epsilon/kT} + \sqrt{\pi})]^2}{80g(\sigma) + 80R^2g(R\sigma)\{\Xi + \frac{1}{32}(\epsilon/kT)[(\epsilon/kT)^3 - 2(\epsilon/kT)^2 + 31]\}} + \frac{16}{5\pi}(g(\sigma) + R^4g(R\sigma)\Xi)\right. \tag{9.20}$$

The final formulas as originally given by Davis, Rice, and Sengers (DRS) for κ, η, and Φ were more complicated than the ones above. Sengers[89] obtained the above simpler but equivalent expressions for κ and η and the above expression for Φ, which includes the cross kinetic-potential contribution arising from the distortion

$$b_2\left\{\frac{15}{8} - \frac{5m\mathfrak{C}^2}{4kT} + \frac{m^2\mathfrak{C}^4}{8(kT)^2}\right\}\nabla_1 \cdot \mathbf{u}$$

to ϕ [see (2.41)]. This contribution, the first member in the curly brackets of (9.20) was neglected by DRS—at liquid densities and temperatures the term amounts to no more than about 5% of the term kept by DRS.

Perhaps the best test of the square-well theory is a computation done by Luks and Davis[90] in which they used experimental equation-of-state and thermal conductivity data to obtain $g(\sigma)$ and $g(R\sigma)$ from the theoretical expressions for P and κ. Using these values of $g(\sigma)$ and $g(R\sigma)$, they predicted shear viscosities and self-diffusion coefficients for liquid argon. The results are shown in Table X. The predicted viscosities agree rather

TABLE X

Calculated and Observed Values of the Viscosity Coefficient η and the Self-Diffusion Coefficient D for Liquid Argon at Various Temperatures T and Pressures P. Experimental Values for the Thermal Conductivity Were Used to Determine the Pair Correlation Functions. Table Taken from Ref. 90

T (°K)	P (atm)	$\eta \times 10^2$ (calc.) (poise)	$\eta \times 10^2$ (obs.) (poise)	$D \times 10^5$ (calc.) (cm²/sec)	$D \times 10^5$ (obs.) (cm²/sec)
87.5	1.0	0.1999	0.246	1.70	2.27
91.0	1.5	0.194	0.220	1.81	2.50
105.6	5.0	0.159	0.148	2.54	4.20
120.3	12.3	0.113	0.113	3.46	6.03
136.3	26.8	0.0927	0.077	4.96	8.26
149.4	45.5	0.0599	0.051	6.57	10.00
91.04	23.9	0.196		1.80	2.32
91.27	100.0	0.203		1.76	1.72
91.06	498.8	0.356		0.955	
105.57	25.0	0.163		2.45	4.13
105.47	100.3	0.173		2.36	3.07
105.35	498.1	0.217		1.76	
120.25	24.7	0.127		3.33	5.96
120.46	100.1	0.142		3.15	4.96
120.48	500.8	0.203		2.32	
136.26	27.9	0.0942		5.09	8.35
135.78	100.0	0.122		4.01	6.37
135.87	500.5	0.176		3.00	
149.63	100.0	0.109		4.33	8.16
149.60	500.8	0.143		3.68	

well with experiment while the predicted self-diffusion coefficients are 35% too low. The predicted temperature dependence of the self-diffusion coefficient is good, however. The values used for σ, $R\sigma$, and ϵ/k in constructing Table X were taken from virial coefficient data[20] to be 3.16 Å, 5.86 Å, and 69.4°K, respectively.

Since $g(R\sigma)$ is usually close to unity a useful approximate scheme for calculating transport coefficients from the square-well formulas is to set $g(R\sigma) \equiv 1$ and solve the equation of state for $g(\sigma)$, that is,

$$g(\sigma) = \left[\frac{P}{nkT} - bnR^3(1 - e^{\epsilon/kT})\right](bn)^{-1} \tag{9.21}$$

With this approximation and experimental equation of state data one obtains agreement between experiment and theory similar to that reported in Table X.

Recently, Schrodt and Luks[91] have computed square-well radial distribution functions using the superposition approximation. They found good agreement between theory and experiment for K and η for liquid Ar. The author would be inclined to use their computed values for $g(R\sigma)$ plus equation-of-state data to get the best estimate of $g(\sigma)$.

The square-well theory has been extended to binary mixtures by McLaughlin and the author.[92,93]

C. Loaded Sphere Fluids

One of the simplest molecular models allowing for internal energy transfer is the loaded sphere model. A loaded sphere is here defined to be a hard sphere of diameter σ whose center of mass is displaced from its center of geometry by a distance ξ. Also it is required that a straight line passing through the centers of mass and geometry is a symmetry axis of the internal distribution of molecular mass. The dynamical state of a loaded sphere can be characterized by the position $\mathbf{x}$ and velocity $\mathbf{v}$ of its center of mass; by θ and ζ, the polar spherical coordinates of $\boldsymbol{\xi} = \xi\mathbf{e}$, the vector lying between the center of mass and center of geometry; and by the angular velocity $\boldsymbol{\omega}[= \mathbf{e}x(d\mathbf{e}/dt)]$ of the molecular axis as measured in the laboratory frame of reference. Sandler and Dahler[94] have developed Enskog's theory for a fluid of loaded sphere molecules. Their starting equation is

$$\left(\partial_t + \mathbf{v}_1 \cdot \nabla_1 + \mathbf{e}_1 \cdot \frac{\partial}{\partial \mathbf{e}_1}\right) f(\mathbf{x}_1, \nu_1; t)$$

$$= \sigma^2 \iint_{\mathbf{g}_{21}\cdot\mathbf{k}>0} [f_2(\mathbf{x}_1, \nu_1'; \mathbf{x}_1 + \boldsymbol{\delta}_1 - \boldsymbol{\delta}_2, v_2'; t)$$

$$- f_2(\mathbf{x}_1, \nu_1; \mathbf{x}_1 + \boldsymbol{\delta}_1^* - \boldsymbol{\delta}_2^*; t)]\, d\mathbf{k}\, d\nu_2 \tag{9.22}$$

with Enskog's chaos assumption in the form

$$f_2(\mathbf{x}_1, \nu_1'; \mathbf{x}_1 + \boldsymbol{\delta}_1 - \boldsymbol{\delta}_2, \nu_2'; t) \\ = g(|\mathbf{x}_{12}| = \sigma; T(\mathbf{x}_1 + \boldsymbol{\delta}_1; t), n(\mathbf{x}_1 + \boldsymbol{\delta}_1; t)) \\ \times f(\mathbf{x}_1, \nu_1'; t) f(\mathbf{x}_1 + \boldsymbol{\delta}_1 - \boldsymbol{\delta}_2, \nu_2'; t) \quad (9.23)$$

The symbol ν_i denotes the collection of variables $(\mathbf{v}_i, \boldsymbol{\omega}_i, \mathbf{e}_i)$; $\boldsymbol{\delta}_i^*$ and $\boldsymbol{\delta}^i$ are the vectors extending from the centers of mass to the point of contact of colliding spheres for direct and inverse encounters, respectively. In particular, $\boldsymbol{\delta}_1^* - \boldsymbol{\delta}_2^* = \xi(\mathbf{e}_1 - \mathbf{e}_2) - \sigma\mathbf{k}$ and $\boldsymbol{\delta}_1 - \boldsymbol{\delta}_2 = \xi(\mathbf{e}_1 - \mathbf{e}_2) + \sigma\mathbf{k} \cdot \mathbf{g}_{21}$ is the relative velocity of a pair of points lying along the line of geometric centers of colliding spheres 1 and 2.

Sandler and Dahler determined from (9.22) the linear laws of heat and momentum transfer and the coefficients of viscosity and thermal conductivity. Fourier's law is unchanged for polyatomic fluids, but Newton's for the pressure tensor is no longer symmetric, being of the form

$$\mathbf{J} = [P - \Phi\nabla_1 \cdot \mathbf{u}]\mathbf{1} - \eta[\nabla_1\mathbf{u} + \nabla_1\mathbf{u}^T - \tfrac{2}{3}\nabla_1 \cdot \mathbf{u}\mathbf{1}] + \Phi_s\mathbf{1} \wedge \nabla \wedge \mathbf{u} \quad (9.24)$$

where the new quantity Φ_s represents the spin or vortex viscosity.

Based on extensive numerical computation Sandler and Dahler concluded that the rotational contribution to the thermal conductivity of loaded spheres affects only the kinetic term and that in this term the effect is essentially that obtained by replacing in Enskog's hard sphere theory the monatomic specific heat ($\tilde{C}_v = \frac{3}{2}k$) by the polyatomic value ($\tilde{C}_v = \frac{5}{2}k$) appropriate to loaded spheres. This is known as the Encken correction for applying the monatomic theory to polyatomic systems, and it is essentially what Sengers did empirically in applying Enskog's theory to dense gaseous nitrogen. Sandler and Dahler also found that the loaded sphere shear viscosity is essentially equal (within 5% over the gas to close packing density range) to Enskog's hard sphere viscosity, again in agreement with Sengers' application of Enskog's theory of shear viscosity to gaseous hydrogen and nitrogen. Thus on the basis of the loaded sphere model one is led to anticipate that for fluids of small nonpolar polyatomic molecules theories of monatomic fluids can be used directly for shear viscosity and, with only the Encken correction, for thermal conductivity.

The bulk viscosity, on the other hand, is sensitive to the polyatomic nature of the molecules. This is not surprising since the kinetic part of Φ is zero monatomic fluids and is of the order of the kinetic part of η for polyatomic fluids. Even at high densities, the loaded sphere bulk viscosity is substantially different from the smooth quantity; for example, at $bn = 1.48$ a typical value of $\Phi_{\text{loaded sphere}}\Phi_{\text{smooth sphere}}$ is about 1.5. Baleiko[95] has found a similar sensitivity of the coefficient of thermal diffusion to the rotational degree of freedom of loaded spheres. The self-diffusion coefficient of loaded spheres, on the other hand, will be seen in a later section to be roughly the same as the smooth diffusion coefficient.

The vortex viscosity Φ_s is predicted to be of the same order of magnitude as η and Φ. Thus the antisymmetric part of the pressure tensor can be large under rotational flow conditions. Condiff and Dahler have considered several situations[96] in which Φ_s would be important.

Dahler and co-workers have examined other loaded molecules, such as spherocylinders, but they have restricted these studies mostly to dilute gases. Baleiko and the author[95,97] have developed the Enskog theory of loaded sphere mixtures.

D. Rough Sphere Fluids

Another simple "polyatomic" model allowing rotational energy transfer is the rough sphere. A rough sphere is defined to be a hard sphere with a rough surface such that when two such spheres collide the relative velocity of the points of contact of the spheres is reversed. In the case of smooth hard spheres only the component of the relative velocity lying along the line of centers of the colliding spheres is reversed, whereas for rough spheres this component as well as the component tangent to the colliding spheres is reversed. The dynamical state of a rough sphere is characterized by the position **x** and velocity **v** of its center of mass and by the angular velocity of the sphere with respect to its center of mass. The Enskog equation for a rough sphere fluid is

$$(\partial_t + \mathbf{v}_1 \cdot \nabla_1) f(\mathbf{x}_1, \tau_1; t) = \sigma^2 \int_{\mathbf{v}_{21}\cdot\mathbf{k}>0} d\mathbf{k}\, d\tau_2 [f_2(\mathbf{x}_1, \tau_1'; \mathbf{x}_1 + \sigma\mathbf{k}, \tau_2'; t) - f_2(\mathbf{x}_1, \tau_1; \mathbf{x}_1 - \sigma\mathbf{k}, \tau_2; t)] \quad (9.25)$$

with the chaos assumption

$$f_2(\mathbf{x}_1, \tau_1; \mathbf{x}_1 \pm \sigma\mathbf{k}, \tau_2; t) = g\left(x_{12} = \sigma; T\left(\frac{\mathbf{x}_1 + \mathbf{x}_2}{2}, t\right), n\left(\frac{\mathbf{x}_1 + \mathbf{x}_2}{2}, t\right)\right) \times f(\mathbf{x}_1, \tau_1; t) f(\mathbf{x}_1 \pm \sigma\mathbf{k}, \tau_2; t) \quad (9.26)$$

The symbol τ_i denotes the collection of variables $(\mathbf{v}_i, \boldsymbol{\omega}_i)$.

McCoy, Sandler, and Dahler (MSD)[98] have obtained from (9.25) expressions for the coefficients of thermal conductivity and shear, bulk, and vortex viscosities for rough sphere fluids. As was the case with loaded sphere fluids the vortex viscosity is of the same order of magnitude as the shear and bulk viscosities. The range of moment of inertia I theoretically possible for rough spheres is $0 \leqslant (4I/m\sigma^2) \leqslant \frac{2}{3}$. In the moment of inertia range MSD found that rough sphere thermal conductivities are about 5–25 higher than Encken-corrected hard sphere thermal conductivities. The thermal conductivities increase by about 10–30% as I varies from 0 to $\frac{1}{6}m\sigma^2$. MSD found also that the shear viscosity increases with increasing I ranging from Enskog's hard sphere value when $I = 0$, a value about 30% higher than Enskog's value at the upper limit of I. The bulk viscosity varies rather strongly with I while the vortex viscosity is independent of I in MSD's calculations. As we shall see in a later section the self-diffusion coefficient decreases by about 30% as I ranges from 0 to $\frac{1}{6}m\sigma^2$. Thus for the rough sphere model the Encken-corrected version of Enskog's hard sphere results involves a systematic error of as much as 30% for maximum allowed values of I.

It is likely that conclusions based on the loaded sphere model are more applicable to real fluids of small polyatomic molecules, since the rough sphere interaction represents the maximum coupling (for an impulsive potential) of rotational and translational motions.

Stenzel, Baleiko, and the author[95,99,100] have obtained the transport coefficients for mixtures of rough spheres.

E. Self-Diffusion Coefficients for Smooth, Loaded, and Rough Spheres Interacting with a Square-Well Potential

Brown and Davis[101] have compared the first Chapman-Enskog approximation to the Enskog self-diffusion coefficient for smooth, loaded, and rough spheres interacting with a square-well pair potential. For *smooth spheres* they use the known result

$$D = D^*[g(\sigma) + R^2 g(R\sigma)\Xi]^{-1} \tag{9.27}$$

for *rough spheres* they derive

$$D = D^*\left[\frac{(8I/m\sigma^2) + 1}{(4I/m\sigma^2) + 1}\, g(\sigma) + R^2 g(R\sigma)\Xi\right]^{-1} \tag{9.28}$$

and for *loaded spheres* they derive

$$D = \frac{4}{U(\alpha)}\,[g(\sigma) + R^2 g(R\sigma)\Xi]^{-1} \tag{9.29}$$

The quantity α is defined by

$$\alpha = \frac{m\xi^2}{2\Gamma} \tag{9.30}$$

TABLE XI

Values of $U(\alpha)$ as a Function of α

α	$U(\alpha)$
0.0	4.00000
0.0001	3.99973
0.001	3.99734
0.005	3.98674
0.01	3.97362
0.02	3.94781
0.03	3.92254
0.04	3.89780
0.05	3.87356
0.06	3.84980
0.625	3.84394
0.07	3.82652
0.08	3.80369
0.09	3.78130
0.167	3.62288

where Γ is the moment of inertia of the loaded sphere. Ξ is defined by (9.16) and is given versus ϵ/kT in Table IX. The quantity $U(\alpha)$ is defined by

$$U(\alpha) = \int_{-1}^{1}\int_{-1}^{1} [1 + \alpha(2 - \mu_1{}^2 - \mu_2{}^2]^{-1/2}\, d\mu_1\, d\mu_2 \qquad (9.31)$$

and is given in Table XI as a function of α over a physically realistic range of α ($\alpha \approx 0.167$ for the molecule *HT*).

A glance at Table XI reveals that the loaded sphere and smooth sphere result would differ by at most about 10% at the highest values of α. In the high-temperature limit ($\Xi \simeq 0$), the rough sphere diffusion coefficients can be as much as 30% lower (when $4I/m\sigma^2 = \frac{2}{3}$) than smooth sphere results. On the other hand, at temperatures characteristic of a liquid where $\Xi \simeq 1$ the rough sphere results differ from the smooth sphere results by only about 10%. Thus in the liquid range the models studied here indicate one can ignore the polyatomic contributions of the self-diffusion in small-molecule liquids.

References

1. For a short review of work on master equations up to 1964, see R. Zwanzig, *Physica*, **30**, 1109 (1964).
2. Of special importance to several of the kinetic models we shall discuss is the master equation developed by the Brussels School. The papers in which the formalism is fully developed are I. Prigogine and P. Résibois, Physica, **27**, 629 (1961); and G. Severne, *Physica*, **31**, 877 (1965).

3. Time correlation functions and their relation to transport coefficients were reviewed by R. Zwanzig, *Ann. Rev. Phys. Chem.*, **16,** 67 (1965).
4. A good source for recent developments in time correlation function theory and the related generalized hydrodynamics is a contribution by B. Berne to *Physical Chemistry, An Advanced Treatise*, Vol. VIII B, H. Eyring, D. Henderson, and W. Jost, Eds., Academic Press, New York, 1971.
5. See Ref. 4 for a description of recent molecular dynamics studies of diatomic molecular models.
6. L. Kadanoff and J. Swift, *Phys. Rev.*, **166,** 89 (1968).
7. P. Résibois and M. De Leener, *Phys. Rev.*, **152,** 305, 318 (1966); **178,** 806, 819 (1969).
8. P. Résibois, *Lecture Notes in Physics*, Vol. 7, J. Ehlers, K. Hepp, and H. A. Weidenmuller, Eds., Springer-Verlag, Berlin, 1971, p. 76.
9. S. A. Rice, J. P. Boon, and H. T. Davis, *Simple Dense Fluids*, H. L. Frisch and Z. W. Salsburg, Eds., Academic Press, New York, 1968, p. 251.
10. H. E. Stanley, *Introduction to Phase Transitions and Critical Phenomena*, Oxford University Press, Oxford, 1971.
11. *Critical Phenomena, Proceeding of a Conference Held in Washington, D.C., April, 1965*, M. S. Green and J. V. Sengers, Eds., *Natl. Bur. Std. Misc. Publ.*, **273** (Dec. 1, 1966).
12. S. A. Rice and P. Gray, *The Statistical Mechanics of Simple Liquids*, Wiley, New York, 1966.
13. I. Prigogine, G. Nicolis, and J. Misguich, *J. Chem. Phys.*, **43,** 4516 (1965).
14. J. Misguich, Ph.D. Thesis, Universite Libre de Bruxelles, 1968; *J. Phys. Radium*, **30,** 221 (1969).
15. J. Palyvos, H. T. Davis, J. Misguich, and G. Nicolis, *J. Chem. Phys.*, **49,** 4088 (1968).
16. S. Chapman and T. G. Cowling, *The Mathematical Theory of Non-uniform Gases*, Cambridge University Press, Cambridge, 1939.
17. H. Reiss, *Advances in Chemical Physics*, Vol. IX, I. Prigogine and S. A. Rice, Eds., Wiley-Interscience, New York, 1965, p. 1.
18. G. Dowling, Ph.D. Thesis, University of Minnesota, 1971.
19. G. Dowling and H. T. Davis, *J. Chem. Phys.* (submitted for publication).
20. J. O. Hirschfelder, C. F. Curtiss, and R. B. Bird, *Molecular Theory of Gases and Liquids*, Wiley, New York, 1954.
21. J. H. Irving and J. G. Kirkwood, *J. Chem. Phys.*, **18,** 17 (1950).
22. N. G. van Kampen, *Phys. Rev. A*, **135,** 362 (1964).
23. J. G. Kirkwood, V. A. Lewinson, and B. J. Alder, *J. Chem. Phys.*, **20,** 929 (1952).
24. J. Naghizadeh and S. A. Rice, *J. Chem. Phys.*, **36,** 2710 (1962).
25. M. Baleiko and H. T. Davis, *J. Chem. Phys.*, **52,** 2427 (1970).
26. I. B. Schrodt, J. S. Ku, and K. D. Luks, *Phys. Chem. Liq.*, **2,** 147 (1971).
27. C. C. Wei and H. T. Davis, *J. Chem. Phys.*, **45,** 2533 (1966); **46,** 3456 (1967).
28. J. Misguich, private communication, 1972.
29. B. A. Lowry, S. A. Rice, and P. Gray, *J. Chem. Phys.*, **40,** 3673 (1964).
30. C. A. Cook, *Argon, Helium and the Rare Gases*, Interscience, New York, 1961.
31. L. Ikenberry and S. A. Rice, *J. Chem. Phys.*, **39,** 1561 (1963).
32. J. Palyvos and H. T. Davis, *J. Phys. Chem.*, **71,** 439 (1967).
33. D. G. Naugle, J. H. Lunsford, and J. R. Singer, *J. Chem. Phys.*, **45,** 4669 (1966); **44,** 741 (1966).

34. H. T. Davis, "Kinetic Theory of Liquids," Proceedings of the Conference on Statistical Mechanics held by the International Union of Applied Physics, April, 1971, Chicago, Ill.
35. J. Misguich, *Report EUR-CEA-FC-573* (November 1970).
36. J. Misguich and G. Nicolis, *J. Mol. Phys.* (submitted for publication).
37. J. Mayer, *J. Chem. Phys.*, **18,** 1426 (1950).
38. R. Balescu, *Statistical Mechanics of Charged Particles*, Wiley, New York, 1964.
39. A. Lenard, *Ann. Phys.*, **3,** 390 (1960).
40. R. Guernsey, *Phys. Fluids*, **5,** 322 (1962).
41. P. M. Allen and G. H. A. Cole, *Mol. Phys.*, **14,** 413 (1968); **15,** 549, 557 (1968).
42. P. M. Allen, *Mol. Phys.*, **18,** 349 (1970).
43. G. Dowling and H. T. Davis, *Can. J. Phys.*, **50,** 317 (1972).
44. M. J. Foster and G. H. A. Cole, *Mol. Phys.*, **20,** 417 (1971).
45. I. Prigogine, *Non-Equilibrium Statistical Mechanics*, Wiley-Interscience, New-York, 1963.
46. P. Résibois, "Irreversible Processes in Classic Gases," in *Physics of Many-Particle Systems*, E. Meeron, Ed., Gordon and Breach, 1966.
47. L. van Hove, *Physica*, **23,** 441 (1957).
48. R. W. Zwanzig, *Phys. Rev.*, **124,** 983 (1961).
49. R. W. Zwanzig, *Lectures in Theoretical Physics*, Vol. 3, W. E. Brittin, W. B. Downs, and J. Downs, Eds., Wiley-Interscience, New York, 1962, p. 106.
50. H. T. Davis and G. Dowling, *J. Stat. Phys.*, **4,** 193 (1972).
51. P. Résibois, *J. Chem. Phys.*, **41,** 2979 (1964).
52. G. Nicolis and G. Severne, *J. Chem. Phys.*, **44,** 1477 (1966).
53. M. H. Ernest, Thesis, University of Amsterdam, 1965.
54. M. S. Green, *J. Chem. Phys.*, **20,** 1281 (1952); **22,** 398 (1954).
55. R. Kubo, M. Yokota, and S. Nakajima, *J. Phys. Soc. Japan*, **11,** 1029 (1957).
56. H. Mori, *Phys. Rev.*, **111,** 694 (1958); **112,** 1829 (1958).
57. H. Mori, *Progr. Theoret. Phys.* (*Kyoto*), **28,** 763 (1962).
58. R. Kubo, *J. Phys. Soc. Japan*, **12,** 570 (1957).
59. R. Kubo, *Lectures Theoret. Phys.* (*Boulder*), **1,** 120 (1958).
60. J. A. McLennan, *Phys. Rev.*, **115,** 1405 (1959).
61. J. A. McLennan, *Phys. Fluids*, **3,** 493 (1960).
62. L. Onsager, *Phys. Rev.*, **37,** 405 (1931); **38,** 2265 (1931).
63. R. Zwanzig, *J. Chem. Phys.*, **40,** 2527 (1964).
64. J. A. McLennan, *Progr. Theoret. Phys.* (*Kyoto*), **30,** 408 (1963).
65. M. S. Green, *Phys. Rev.*, **119,** 829 (1960).
66. S. A. Rice, *J. Chem. Phys.*, **33,** 1376 (1960).
67. J. G. Kirkwood, *J. Chem. Phys.*, **14,** 180 (1946).
68. C. C. Wei, Ph.D. Thesis, University of Minnesota, 1967.
69. C. C. Wei and H. T. Davis, Midwest Theoretical Conference, University of Wisconsin, 1968.
70. D. Forster, P. Martin, and S. Yip, *Phys. Rev.*, **170,** 160 (1968).
71. R. Zwanzig and R. D. Mountain, *J. Chem. Phys.*, **44,** 2777 (1966).
72. B. Berne, J. P. Boon, and S. A. Rice, *J. Chem. Phys.*, **45,** 1086 (1966).
73. K. Singwi and S. Tosi, *Phys. Rev.*, **157,** 153 (1967).
74. P. Martin and S. Yip, *Phys. Rev.*, **170,** 151 (1968).
75. B. J. Alder and T. E. Wainwright, *Phys. Rev. A*, **1,** 18 (1970).
76. T. E. Wainwright, B. J. Alder and D. M. Gass, *Phys. Rev. A*, **4,** 233 (1971).
77. Y. Pomeau, *Phys. Lett.* **27A,** 601 (1968).

78. R. Zwanzig and M. Bixon, *Phys. Rev.*, **2,** 2005 (1970).
79. J. R. Dorfmann and E. G. D. Cohen, *Phys. Rev. Lett.*, **25,** 1257 (1970).
80. M. H. Ernst, F. H. Hauge, and J. M. J. van Leeuwan, *Phys. Rev. Lett.*, **25,** 1254 (1970).
81. A. Widom, *Phys. Rev.* (to be published).
82. B. J. Alder and T. E. Wainwright, *Phys. Rev. Lett.*, **18,** 988 (1967).
83. A. Rahman, *Phys. Rev. A*, **136,** 405 (1964); *J. Chem. Phys.*, **45,** 2585 (1966).
84. B. J. Alder and T. E. Wainwright, in *Transport Processes in Statistical Mechanics*, I. Prigogine Ed., Interscience, New York, 1958.
85. J. V. Sengers, *Intern. J. Heat Mass Transfer*, **8,** 1103 (1965).
86. A. Michels and R. O. Gibson, *Proc. Roy. Soc. (London) Ser. A*, **134,** 288 (1931).
87. M. K. Tham and K. E. Gubbins, *J. Chem. Phys.*, **55,** 268 (1971).
88. H. T. Davis, S. A. Rice, and J. V. Sengers, *J. Chem. Phys.*, **35,** 2210 (1961); H. T. Davis, Ph.D. Thesis, University of Chicago, 1962.
89. J. V. Sengers, private communication, 1962.
90. H. T. Davis and K. D. Luks, *J. Phys. Chem.*, **69,** 869 (1965).
91. I. B. Schrodt and K. D. Luks, *J. Chem. Phys.*, to be published.
92. I. L. McLaughlin and H. T. Davis, *J. Chem. Phys.*, **45,** 2020 (1966).
93. J. Palyvos, K. D. Luks, I. L. McLaughlin, and H. T. Davis, *J. Chem. Phys.*, **47,** 2082 (1967).
94. S. I. Sandler and J. S. Dahler, *J. Chem. Phys.*, **46,** 3520 (1967).
95. M. Baleiko, Ph.D. Thesis, University of Minnesota, 1971.
96. See for example, D. W. Condiff, Ph.D. Thesis, University of Minnesota, 1965.
97. M. Baleiko and H. T. Davis, *J. Phys. Chem. Liq.* (to be published).
98. B. J. McCoy, S. I. Sandler, and J. S. Dahler, *J. Chem. Phys.*, **45,** 3485 (1966).
99. R. Stenzel, Master's Thesis, University of Minnesota, 1967.
100. M. Baleiko, R. Stenzel, and H. T. Davis, *J. Phys. Chem. Liq.* (to be published).
101. R. Brown and H. T. Davis, *J. Phys. Chem.*, **75,** 1970 (1971).

AUTHOR INDEX

Numbers in parentheses are reference numbers and show that an author's work is referred to although his name is not mentioned in the text. Numbers in *italics* indicate the pages on which the full references appear.

Abrikosova, I. I., *87*
Abrikosova, I. L., *87*
Adair, T. W., 404(506), *93*
Adamenko, I. I., *79*
Adams, J., 57(273), *73, 78*
Adey, C. O., *80, 84*
Ailawadi, N., *78*
Ailawadi, N. K., 10, 12(511), 46(218), *74, 93*
Aizawa, Y., 156(9), 183(9), *184*
Alcock, A. J., *87*
Alder, B. J., 274, 325, 326, 327, 329(82, 84), *341–343*
Alekhin, A. D., *63, 83*
Alexhin, A. D., *84*
Allen, P. M., 292–296, *342*
Alpert, S. S., *79, 84*
Andersen, C. M., 239, 241, 242(73), *255*
Andersen, H. C., 46(219), *74*
Angus, J. C., 59(313), *80*
Anisimov, M. A., *78, 79, 84*
Anosov, D. V., 156(13), 159, 161(13), *184*
Arefev, I. M., 43, *62, 65, 78, 79, 84, 87*
Arnold, V. I., 155(3), 156(3), 159(20), 160(21), 161, 163(26), 164(28), 171(3), 177(3,43), *184*
Arnowitt, R., 248(80), *255*
Aslaksen, E. W., *62, 65*
Atakhodzhaev, A. K., *62, 75, 77*
Aval, G. M., *76*
Avez, A., 155(3), 156(3), 159(20), 160(21), 161, 163(26), 164(28), 171(3), 177(3, 43), *184*

Bak, C. S., *79, 84*
Baleiko, M., 275, 338, 339, *341, 343*
Balescu, R., 292(38), 293(38), 300(38), 303, *342*
Balta, Y., *79, 84*
Baranov, V. G., *79*
Bardeen, J., 233, 234(65), 238(65), *255*
Baret, J. F., *68, 74*
Barger, R. L., *69*
Barker, J. A., 231, *255*
Barocchi, F., 41(37,38), *62, 80, 87*
Bartoli, F. J., *77*
Batra, I. P., *87, 88*
Belland, P., *62*
Benedek, G. B., 2, 8(57), 27(90), 34, 39, 42(90), 55(213), 56(213), 57(213,307), 58(362,363), *60–69, 74, 80, 82–86*
Benjamin, R., 52(198), *73*
Ben-Reuben, A., 46(22), *74, 75, 77*
Berezin, F. A., 198(20,22), *254*
Berge, P., 22(310), 24(28,310), *62, 79, 80, 83*
Berne, B., 258(4,5), *320,* 321, 322(4), 323(4), 324(4), 325, *341, 342*
Berne, B. J., 17(505), 46(218), 59(286), *74, 78–80, 93*
Bersohn, R., *70*
Bertolotti, M., 25, *65, 78*
Bespalov, V. I., *87*
Bhatia, A. B., 19(30), *62*
Billingsley, P., 155(4), *184*
Bird, R. B., 268(20), 328(20), 331, *341*
Birkhoff, G. D., 163(24), 168, 169, *184*
Birnbaum, G., 50, 52, *71, 72*
Biryukov, V. N., *79, 84, 87*
Bixon, M., 325(78), *343*
Blend, H., *62*
Bloch, C., 201(24), 202(24), *254*
Bloembergen, N., *63, 67, 81, 87*
Bloom, G. H., *90*
Blum, L., 59(287–289), *73, 79*
Boato, G., 316
Bochkova, O. M., *87*
Bogachev, A. P., *76*

Bogoliubov, N. N., 190, 208, 234, 235(67), 238, 240, 242(70), 244(70), *254, 255*
Boley, C. D., *67, 70, 83*
Boon, J. P., 1, 2, 8(32), 9(32), 13(32), 17(32,505), 18(32), 20(102), 22(102), 24, 25(297,512), 40, 41(159), *62, 66, 67, 70, 80, 83, 93*, 259, 320, 321, *341, 342*
Born, M., 26(509), *93*
Botherol, P., *75*
Bouchiat, M. A., 55(203,214), 57, *73, 74, 78, 86*
Bourke, P. J., *82*
Boyer, L., *61, 66*
Bozhkov, A. I., *88*
Bradford, E., 58(363), *82, 83*
Brady, G. W., *80, 84, 85*
Brahic, A., 177, 180, *185*
Braun, P., 39(40), *63, 84*
Bret, G., *88*
Brewer, R. G., *88*
Brillouin, L., *66*
Broida, H. P., *77*
Brossel, J., *73*
Brown, W. B., 190, 191, *254*
Brown, R., 339, *343*
Broz, A., *63, 66*
Bucaro, J. A., 50(187), 52(186), *63, 70–72, 77*
Bunkin, F. V., *88*
Burlefinger, E., *88*
Byrns, F. L., 248–250, *255*

Calmettes, P., 24(28), *62, 79, 83, 86*
Caloin, M., 41(158), *66, 70*
Camerini-Otero, R. D., 22(357), *83*
Candau, S. J., 6, *61, 66, 69, 76*
Candau, S. U., 41(158), *70*
Canfield, F. B., 231(60,61), *255*
Cannell, D. S., 27, 28, 42, *65, 66, 84, 86, 87*
Careri, G., 225, *255*
Carlson, F. D., 22(356), 58(356), *83*
Carlson, P. D., 58(361), *83*
Caroli, C., *80*
Carome, E. F., 16, 19(134), *66, 68*
Carroll, M. M., 143, 147, *154*
Casanova, G., 316
Cazabat-Longequeue, A. M., *65, 66, 70*
Cecchi, L., *61, 66*
Chabbal, R., 27(25), *62*
Chabrat, J. P., *80*
Chaerat, J. P., *74*
Chalyi, A. V., *63, 84*
Champion, J. V., 52(163), *71*
Chandrasekhar, S., 24(510), *93*
Chandrasekharan, V., *66*
Chang, R. F., *80, 84*
Chapman, S., 261(16), 266(16), 268(16), 271(16), 304(16), 328(16), 332(16), *341*
Chase, L., 30(20), *61*
Chen, S. H., 22(359), 43, 58(359), 59, *70, 80, 83, 84, 87*
Chernets, A. N., *66*
Chernyauska, I. O., *79*
Chernysheva, E. O., *75*
Chiao, R. Y., 41, *66, 67, 88*
Chirikov, B. V., 156(8), 171, 177, 178, 179(8), 183(8), *184, 185*
Cho, C. W., *63, 88, 89, 91, 92*
Chopan, A. A., *88*
Chu, B., 2, 40, *60, 80, 83, 84, 86*
Chung, C. H., 45(44), 46, *63, 66*
Clark, N. A., 48(280), 49, 50(280), 57(307), *78, 80*
Cohen, E. G. D., 325–327, 343
Cohen, M. G., *66, 67, 70*
Cohen-Solal, G. W., *88*
Cole, G. H. A., 292–295, 299, 300, *342*
Coleman, A. J., 216(36), 217(36), *254*
Coleman, B. D., 95, 97(17), 98(1, 2, 9, 18–20, 22,23), 99, 100, 102, 103, 108, 109, 110(22,23), 111(23), 112(23), 113, 114, 115(22,23), 116, 117(1,2,9), 124(1,2,9, 18–20,22), 128, 130, 132, 141, 142(26), 143, 144, 146, 147, 148(1,2,26), 149–152, *153, 154*
Colles, M. J., *88*
Condiff, D. W., 338, *343*
Contopoulos, G., 156(6), 183(6), *184*
Cook, C. A., *341*
Cooper, L. N., 233, 234(65), 238(65), *255*
Cooper, V. G., *64, 71, 75*
Cornall, W. S., *69*
Coumou, D. J., *75*
Courtens, E., *77*
Cowling, T. G., 261(16), 266(16), 268(16), 271(16), 304(16), 328(16), 332(16), *341*
Craddock, H. C., *71, 75*
Cruchon, D., *74*
Cummins, H. Z., 2, 32, 33(3), 34, 39(58, 81,82), 42(111), 43(111), 58(361), *60, 61, 63, 65–67, 83, 85, 86*

Curtiss, C. F., 268(20), 328(20), 331, *341*

Dahler, J. S., 336–338, *343*
Daino, B., 25(272), *78*
Danby, J. M. A., 172(35), *184*
Danielmeyer, H. G., *61, 66*
Daniels, A., *66*
Daniels, W. B., 52(166), 53(166), 54(166), *71*
Danileiko, Y. U. K., *92*
Daree, K., *88*
Davis, B., *78*
Davis, H. T., 259, 262, 275(25), 278(27), 286, 289(34), 293(34,43), 295(43), 304(34), 312(69), 332(88), 333, 335, 336(92,93), 338(97), 339(100,101), *341–343*
Davis, J. L., 30(20), *61*
Day, W. A., 95, 149, *153*
Debye, P., 43, *80, 85*
Dechow, F. J., *68, 74*
De Genner, P. G., 228, *255*
Degennes, P. G., 57(205), *73, 80, 85*
Degiorgio, V., 32, 34, 35(19), 36, 37(19), *61*
Deguent, P., 2, 8(32), 9(32), 13(32), 17(32), 18(32), 20(102), 22(102), *62, 66, 93*
DeLeener, M., 258, *341*
De Micheils, C., *62, 87*
Demicheli, R., *66*
Denariez, M., *88*
De Pasquale, D., *65*
Desai, R. C., *65, 70, 83*
Deutch, 22
Deutsch, C., *78*
Devlin, G. E., 30, *61*
Devonshire, A. F., 225, *255*
Diel, J. F., *81*
Dietz, D. R., *63, 92*
Dill, J. F., *67*
Dill, E. H., 95, 99, 100, 108, 128, 141–144, 146, 147, 148(26), 149, *153*
DiPorto, F., *78*
DiPorto, P., 25(272), *78*
Domb, C., 228, *255*
deDominicis, C., 201(24), 202(24), *254*
Dorfmann, J. R., 325–327, *343*
Dowling, G., 262, 293, 295, 296, *341, 342*
Dransfeld, K., *67, 89*
Dubin, S. B., 58(362), *80, 83*
Dubois, M., 22(310), 24(28,310), *62, 80, 83*
Du Martin, S., *88*
Dunning, J. W., 59(313), *80*
Dupois, M., *79*
Du Pre, D. B., 50(180), 51(180), 52(180), *72*
Durand, S., *67, 78*
Dyakov, Y. U. E., *88*

Eastman, D. P., 41(105), *67, 76*
Edwards, R. V., 59(313), *80*
Edwards, S. F., 228, *255*
Ellenson, W., *73*
Emmett, J. L., *89*
Englisch, W., *64*
Enns, R. H., 63, *87–89*
Enright, G., 44(269), *78*
Ericksen, J. L., 108, *153*
Ernest, M. H., 307(53), *342*
Ernst, M. H., 325(80), *343*
Evans, D. E., *60*

Fabelinskii, I. L., 44(224, 256), *60–62, 68, 75, 77, 87, 89–92*
Faizulloyev, S. F., *75*
Falk, H., 222, *254*
Feborov, M. V., *88*
Feke, G. T., 39(51), *63, 85*
Fenby, D. V., 40(354), *83, 86*
Fermi, E., 178, *185*
Figgins, R., *61*
Filatova, L. S., *82*
Fink, U., *69*
Finnigan, J. A., *75*
Fisher, M. E., 220–222, 226, *254*
Fixman, M., *85*
Fleury, P. A., 1, 2, 6(9), 26(9), 27(9), 40, 41, 50(180), 51(180), 52–54, *61, 66, 67, 70–73*
Foltz, N. D., *88, 89, 91, 92*
Ford, J., 155, 156(7,12), 167(30), 169(30), 172, 174(36), 175(36), 183(7,12), *184, 185*
Ford, N. C., 27, 42, 43(109,127), *61, 63, 67, 68, 81, 85, 87*
Forster, D., 46(218), *74,* 312, 315, 318, *342*
Fosdick, R. L., 143, 147, *154*
Foster, M. J., 293, 299, 300, *342*
Franklin, R. M., 22(357), *83*
French, M. J., 59(313), *80*

Friedricks, K. O., 198(21), *254*
Frisch, H. L., 17(110), 59(286), *67, 70, 73, 79, 85*
Fritsch, K., *65, 68, 81*
Frorvin, A. I., *72*
Fujime, S., 58(316), *81*

Gammon, R. W., 42, 43(111), *63, 66, 67, 85*
Ganguly, B. N., *63, 67, 81*
Garmire, E., *89*
Garrod, C., 216(37), 218(37,40), 220(37), *254*
Gartenhaus, S., 239–242, *255*
Gass, D. M., 325(76), *342*
Gerjouy, E., 188(5), *253*
Gershon, N. D., 46(220), *74, 75, 77*
Gersten, J. I., 50(168), *71*
Geschwind, S., 30(20), *61*
Gewurtz, S., *67*
Gibbs, J. W., 190, 191, 194, *254*
Gibson, R. O., 330, *343*
Giglio, M., *63, 65, 74, 85, 86*
Giradeau, M. D., 187, 190, 196(17), 208(27), 234, 244(75), 247(79), 248(80), *254, 255*
Giuliano, C. R., *89*
Giulotto, L., *66*
Goldblatt, N., *89*
Goldburg, W. I., 79, *82–84, 86*
Golden, S., 194, 195, 204(16), 209, *254*
Goldstein, H., 173(37), *185*
Golger, A. L., *89*
Gordon, E. I., *66, 67*
Gornall, W. L., *65, 70, 83*
Gornall, W. S., 24, 41(115,121), 52(169), *63, 67, 68, 71, 81*
Grasyuk, A. Z., *89*
Gravatt, C. C., *63, 67, 79, 84*
Gray, C. G., *71*
Gray, M. A., *89*
Gray, P., 259(12), 271(12), 275(12), 276(12), 278(12), 286, 287(29), 288(29), 316(12), 318(12), *341*
Green, M. S., 259(11), 308, 309, 311, *341, 342*
Greene, J. M., 169(33), 171(33), *184*
Greytak, T. J., 8(57), 51, 52, *63, 65, 67, 69, 73*
Griffin, A., *63, 81*
Griffiths, R. B., 212(33), *254*
Grimm, H., *67, 89*
Groe, K., *89*
Gross, E. F., *75*
Gubbins, K. E., 332, *343*
Guberman, B. S., *89*
Gudimenko, G. L., *64, 67, 69*
Guenther, A. H., *92*
Guernsey, R., 292(40), *342*
Guo, Zhong-Heng, 147, *154*
Gurtin, M. E., 95, 128, 132, 152, *153*

ter Haar, D., 156(14), 157(14,16), 158(17), *184*
Haar, W., 57(273), *73, 78*
Haering, R. R., *63*
Hagenlocker, E. E., *89, 91*
Hall, L. H., *73*
Haller, I., 49(284), *79*
Halley, J. W., 51, *73*
Halmos, P. R., 155(1), 157(1), 159(1), *183*
Hammer, D., 39(40), *63, 84*
Hammerstein, A., 239, 240(71), 241(71), *255*
Hanes, G. R., *67, 69*
Hansen, R. S., *68, 74*
Hara, E. H., *63, 65, 67, 69–71, 81*
Hardy, G. H., 213(35), 251, *254*
Harrigan, M., *63, 66*
Hauge, F. H., 325(80), *343*
Hawkins, G., *65*
Hawkins, G. A., 39(51), 55(213), 56(213), 57(213), *63, 74, 85, 86*
Heiles, C., 164, 167(27), *184*
Heinicke, W., *89, 92*
Heller, J. P., 158, *184*
Heller, W., *81*
Henderson, D., 231, *255*
Henon, M., 164, 167(27), *184*
Henry, D. L., 39(58), *63, 85, 86*
Herbert, J. J., 58(361), *83*
Herbert, T., 22(356), 58(356), *83*
Herman, R. M., *89*
Hess, S., *75*
Hirooka, H., 156(9), 183(9), *184*
Hirschfelder, J. O., 268(20), 328(20), 331, *341*
Hollinger, A., 41(105), *67, 76*
Hoover, W. G., 227, *225*
Hopf, E., 158, *184*
van Hove, L., 300, *342*

Howard-Lock, H. E., 52(169), *71*
Huang, J. S., *73, 85*
Huber, A., 189, 190, 195, *253*
Hunt, J. T., 244(75), *255*
Hunter, J. L., *67, 81*

Ieda, M., *81*
Ikenberry, L., 285, 288(31), *341*
Ingard, K. U., 55(207,208), *73*
Inoue, K., *89, 90*
Irving, J. H., 271(21), *341*
Isihara, A., 229, *255*
Ito, S., *82*
Ito, T., *90*
Iwamoto, F., *71*
Izrailev, F. M., 171(34), *184*

Jackson, D. A., *64, 67, 68, 71, 72, 75–77, 81, 82*
Jacobs, D. J., *75*
Jakeman, E., 32, 36, *61*
Jakub, M. T., 147, *154*
Jansen, M., *64, 71, 75*
Jaseja, T. S., *90*
Jennings, D. A., *69, 92*
Jones, R., *67*
Jordan, P., 198(19), *254*
Jordan, P. C., 194, 195, 204(16), *254*
Jorma, S., *90*

Kachnowski, T. A., 39(63), *64, 85*
Kadonoff, L., 258, *341*
Kaiser, W., *64, 85, 86, 88, 90, 91*
van Kampen, N. G., 274, *341*
Karasz, F. E., *81*
Kasten, R., *63, 66*
Kato, E., *68*
Katsura, S., 245(77), *255*
Katyl, R. H., 55(207,208), *73*
Kaufman, A. N., 183(47), *185*
Kawasaki, K., 39, *81, 85, 93*
Keating, P. N., *78*
Keeler, R. N., 2, 22(347), 59(347), *82, 90*
Keijser, R. A. J., *64, 71, 75*
Kelley, H. C., *81*
Kellog, O. D., 188(1), *253*
Kenemuth, J., 41(105), *67*
Kern, S., *85*
Keyes, P. H., *80, 84*
Keyes, T., *71, 75, 78*

Khazov, L. D., *91*
Khromov, A. S., *77*
Kielich, S., *71, 72, 75, 77, 81, 92, 93*
Kiess, E. M., *69*
Kijewski, L. J., 216(38), 218(38–40), 219(38), 220(38), *254*
Kilpatrick, J. E., 224(45), *254*
Kirkwood, J. G., 225, 226, *255,* 271(21), 274, 312, *341, 342*
Kivelson, D., *71, 75, 78*
Kolpakov, Y. U. K., *85*
Koncvalov, E. V., *82*
Koppe, H., 191, 192, *254*
Koppel, D. E., 22(357), *83*
Kopylovskii, B. D., *62*
Korcbkin, V. V., *90*
Kosaki, M., *81*
Kozierowski, M., *93*
Knaap, H. E. P., 41(121), *63, 64, 67, 68, 71, 75, 81*
Knable, N., *63*
Knirk, D. L., 59(324), *81*
Knowles, J. K., 143, 147, *154*
Kruiskii, N. P., *83*
Ku, J. S., 276, *341*
Kubarev, A. M., *87*
Kubo, R., 308, *342*
Kumar, B., *90*
Kunsitis-Swyt, C. R., *66, 68*
Kyzylasov, Y. U. I., *90, 92*

Lagues, I., *83, 86*
Laj, C., *83, 86*
Lalanne, C. R., *93*
Lalanne, J. R., *72, 75–77, 93*
Lallemand, P., 50(200), *65, 68, 70, 73, 75, 87*
Lambropoulos, P., *85*
La Macchia, J. T., *62, 69*
Landau, L., 155(2), *183*
Langevin, D., 57, *74, 78*
Langley, K. H., 27, 42(109,127), 43(109, 127), *61, 67, 68, 70, 85, 86*
Larson, E. V., 40(506), *93*
Lastovka, J. B., 32, 34, 35(19), 36, 37(19), 39(51), *61–65, 68, 85*
Laubereau, A., *64*
Laugley, K. E., 39(63), *64*
Lavrinovich, N. N., *90*
Lebowitz, J. L., 156(11), 177, 178(39), 183(11), *184, 185*

Lee, C. S., *68, 81*
van Leeuwan, J. M. J., 325(80), *343*
Leff, H., 224, *254*
Leidecker, H. W., *62*
Leite, R. C. C., *64, 76*
Lekkerkerker, H. N., 24, 25(512), *93*
Leland, T. W., 232(64), *255*
Leland, T. W., Jr., 231, *255*
Lenard, A., 292(39), *342*
Lennard-Jones, J. E., 225, *255*
Levenson, M. D., *71, 76*
Levi, A., 316
Levine, H. B., 50, *71*
Lewinson, V. A., 274, *341*
Lichtenberg, A. J., 177, 178, 180, 182(42), *185*
Lieb, E., 193, 245(76), *255*
Lieberman, M. A., 177(42), 178, 180, 182(42), *185*
Lifshitz, E., 155(2), *183*
Lim, T. K., 39(63), *64, 85*
Lipworth, E., *79*
Litan, A., *64*
Litovitz, T. A., 19(132), 50(187), 52(186), *68–72, 76, 77, 89, 90*
Litster, J. D., 47–49, 50(280), *78, 79*
Littlewood, J. E., 213(35), 251, *254*
Lo, S. M., *81, 85*
Loj, C., 24(28), *62, 79, 80*
Longequeue, A. M., *68*
Low, W., *70*
Lowdermilk, W. H., *87*
Lowry, B. A., 286, 287(29), 288(29), *341*
Luban, M., 248, *255*
Lucas, H. C., *72, 76, 81*
Luks, K. D., 276, 335, 336, *341, 343*
Lunacek, J. H., 57(307), 58(362), *65, 80, 83, 86, 87*
Lunsford, G. H., 167(30), 169(30), 172, 174(36), 175(36), *184, 185*
Lunsford, J. H., 287, 288(33), *341*
Luxemburg, W. A. J., 114, *154*

Mack, J. E., 27, *62*
Mack, M. E., *90*
Madigosky, W. K., *90*
Maeda, H., *81*
Maier, M., *86, 90*
Makarenko, S. P., *82*
Maker, P. D., *93*
Mancini, M., 41(38), *62, 80*
Mann, J. A., *68, 74*
Mansoori, G. A., 231, *255*
Marrucci, G., 143, 147, *154*
Marshall, A. G., *76, 81*
Martin, F. B., *72, 76, 77*
Martin, P., 312, 315, 318, 323, *342*
Martin, P. E., *93*
Martynov, A. D., *92*
Mash, D. I., *62, 68, 89–91*
Massey, H. S. W., 188(3), *253*
Matsubara, T., 201, 202(23), *254*
Matsuoka, M., *87*
Mattioli, M., *62*
Mattis, D., 245(76), *255*
May, A. D., 15(66), *63–65, 67, 69–71, 81*
Mayer, J., 292(37), *342*
Mayer, J. E., 225, *255*
Mazo, R. M., 224(46), 230, 248–250, *254, 255*
McClintock, M., *69*
McCoy, B. J., 338, *343*
McIntyre, D., 2, *61, 80, 84*
McLaughlin, I. L., 336, *343*
McLennan, J. A., 308, 311, *342*
McNutt, D. P., 27(25), *62*
McTague, J. P., 50–52, *71–73, 75*
Meeten, G. H., 52(163), *71*
Meiikyan, A. O., *69*
Melikyan, A. O., *64*
Mercher, M., *89*
Mermin, N. D., 241, *255*
Meunier, J., 55(203,214), 57(214), *73, 74, 86*
Meyer, H., *91*
Mezutani, J., *81*
Michels, A., 330, *343*
Miller, B. N., *81*
Miller, G. A., *68, 81*
Minck, R. W., *89, 91*
Misguich, J., 259, 267(14), 280–282, 285, 289, 291, *341, 342*
Mitchell, A. C., *90*
Mizel, V. J., 95, 98, 100, 102, 108–110, 111(23), 112(23), 113–118, 124, 130, 132, 144, 149, *153*
Mohr, R., 42, 43(127), *68, 70, 85, 86*
Monkewicz, A., *63, 66, 90*
Montrose, C. J., *67, 68, 81*
Moore, R. S., *64, 76*

Mori, H., 308, 309, *342*
Morozov, V. V., *79, 84, 89–91*
Moser, J., 163(25), 167, *184*
Mott, N. F., 188(3), *253*
Mountain, R. D., 2, 14(131), 15(131), 19(132), 20(131), 22, 23(330), 41, *61, 68, 73, 81, 86,* 315, *342*
Mueller, R. K., *85*
Murch, R. E., 143, 147, *154*
Myer, M., 80, *84*
Mynchenko, Y. U. B., *83*

Naghizadeh, J., 270(24), 316(24), *341*
Nakagaki, M., *81*
Nakajima, S., *72,* 308, *342*
Naugle, D. G., 40(506), *93,* 287, 288(33), *341*
Nelkin, M., 8(133), *68*
Nelson, E., 210, *254*
von Neumann, J., 198(18), *254*
Nichols, W. H., 16, 19(134), *66, 68*
Nicolis, G., 259, 280, 289, 307, *341, 342*
Nikitin, I. K., *91*
Nikolaenko, P. T., *72, 76*
Noll, W., 97, 98(18–20), 103, 109, 124, 141, 143, 144, 147, 149, *153, 154*
Nomoto, O., *68*
Nossal, R., 22(359), 58(359), 59, *83*

O'Connor, C. L., 41(137), *68*
Oksemgorm, B., *88*
Oksengorn, B., *72, 73*
Okubo, S., 189(6), *253*
Olivei, A., *64*
Onsager, L., 308, *342*
Ooyama, N., 156(9), 183(9), *184*
Orsay Liquid Crystal Corp., 49(277), *78*
Overdeck, J. T. H., 57(212), *74*
Owen, D. R., 95, 99, 116, 132, 149–152, *153*
Owen, J. E. M., *81*

Page, D. I., *82*
Palin, G. J., *64, 69, 82*
Palyvos, J., 259(15), 286, 336(93), *341, 343*
Pande, P. C., *90*
Pao, Y. H., *70*
Papazian, L., *81*
Papoular, M., *74, 85*
Parkash, V., *90*
Parodi, C., *80*
Pasmanik, G. A., *87*
Paul, D. M., *64, 68*
Pecora, R., 2, 46(219), 58, 59(14), *61, 74, 76, 80–82*
Peierls, R. E., 193, *254*
Pen-Reuven, A., *77*
Penrose, O., 177, *185*
Percus, J. K., 216(37,38), 218(37–40), 219(38), 220(37,38), *254*
Perzyna, P., 109, *153*
Pesin, M. S., *69, 76*
Peterson, L. M., *91*
Piercy, H. E., *67*
Piercy, J. E., *69*
Pike, E. R., 32, 36, 41, *61, 62, 64, 65, 67, 69, 70, 82*
Pine, A. S., *91*
Pinnow, D. A., *69, 76*
Platzman, P. M., *66, 70*
Pohl, D., *85, 86, 91*
Pohyakova, A. L., *91*
Polonsky, N., 43, *70, 80, 83, 84, 87*
Polya, G., 213(35), 251, *254*
Pomeau, Y., 325(77), *342*
Popovichev, V. I., *89*
Porto, S. P. S., *64, 76*
Pound, R. V., 103, *153*
Powles, J. G., *71, 72, 75–77*
Prigogine, I., 258(2), 259, 261(2), 262, 278(2), 280, 289(2), 293(45), 294(45), 300, 301, 303, 304(2), *340, 342*
Prorbin, A. I., *76*
Puell, H., *88*
Puglielli, V. G., 42(109), 43(109), *67, 85, 87*
Purcell, E. M., 103, *153*
Pusey, P. N., 22(357), *79, 82–84, 86*

Quate, C. E., *61*

Rado, W. G., *89, 91*
Ragulskii, V. V., *89*
Rahman, A., 10, 12(511), *93,* 327, *343*
Rahn, O., *90*
Rakhimov, Z. K. H., *75*
Ralph, H. I., *71*
Ramsey, N. F., 103, *153*
Rangnekar, S. S., *89*
Rank, D. H., 41(105), *67, 69, 76, 88, 89, 91, 92*

Rao, Dvgln, *67, 78, 91*
Rawson, E. G., 15(66), *64, 69*
Reinhold, L., *91*
Reiss, H., 228, *255,* 261(17), 329(17), *341*
Renner, G., *90*
Résibois, P., 258, 261(2), 278(2), 289(2), 294(46), 300(2,46), 301, 303, 304(2), 307, *340, 342*
Rice, S. A., 17(505), *93,* 259, 270(24), 271(12), 275(12), 276(12), 278(12), 285, 286, 287(29), 288(29,31), 312, 313, 316(12,24), 318(12), 320, 321, 332, 333, 335, *341, 343*
Rieckhoff, B., *88*
Ripper, J. E., *64*
Rippler, J. E., *76*
Romanov, V. P., *75, 82*
Romanova, L. M., *76*
Roseler, F. L., 27(25), *62*
Roshchina, G. P., *64, 67, 69, 82*
Rosina, M., 218(40), *254*
Rother, W., *90, 91*
Rouch, J., *74, 80*
Rowell, R. L., *76*
Rowlinson, J. S., 232, *255*
Rozhdestvenskaya, N. B., *76, 77*
Ruelle, D., 212, 222, *254*
Ruvalds, J., *72*
Rytov, S. M., *64, 69, 76*

Sabirov, I. M., 44(224), *69, 75*
Sacchi, C. A., *91*
Sadownik-Wodzinska, A., *93*
Sagitova, E. V., *77*
Saito, N., *82,* 156(9), 183(9), *184*
Saito, T. T., *91*
Saji, Y., *68*
Sakamoto, S., *82*
Salsburg, Z. W., 59(288,289,324), *79, 81*
Salto, N., *81*
Sandercock, J. R., 28, 29, *62*
Sandler, S. I., 336–338, *343*
Sather, G. A., 232(64), *255*
Sato, T., *64, 82*
Saxon, D. S., 188(4,5), *253*
Schaefer, D. W., 22(357), 58, *82, 83*
Schawlow, A. L., *69, 71, 76, 89*
Schlupf, J. P., 41(137), *68*
Schoenes, F. J., 40(305,306), *80, 84*
Schofield, P., 58(363), *82, 83*
Schrieffer, J. R., 233, 234(65), 238(65), *255*
Schrodt, I. B., 276, 336, *341, 343*
Schultz, T., 245(76), *255*
Schuster, T. M., 22(358), *83*
Schwinger, J., 188(4), *253*
Scott, R. L., 230, *255*
Scudier, F., *78*
Searby, G. M., *69*
Sechkarev, A. V., *72*
Sengers, J. V., 2, *61, 80, 84,* 259(11), 329, 330(85), 331–333, 335, *341, 343*
Series, G. W., *69*
Sette, D., *65, 78*
Severne, G., 258(2), 261, 278, 289, 300–302, 304, 307, *340, 342*
Shaham, Y. J., *91*
Shapero, D. C., *88*
Shapiro, S. L., *69, 77*
Shchelev, M. Y. A., *90*
Shen, Y. R., *91*
Shilin, H. V., *62*
Shirkov, D. N., 238(70), 240(70), 242(70), 244(70), *255*
Shpak, M. T., *92*
Silin, V. V., *66*
Simic-Glavaski, B., *71, 72, 75–77*
Sinai, Ja. G., *184*
Sinai, Ya. G., 156, 159, 161(13), *184*
Singal, S. P., *66, 68*
Singer, J. R., 287, 288(33), *341*
Singwi, K., 322, *342*
Skripov, V. P., *85*
Skrypnik, N. G., *87*
Slusher, R. E., 50(168), 55, *71, 72*
Smith, I. W., *65, 74, 86*
Smith, R. A., *61*
Sokerov, S., *82*
Solana, J., *72*
Solecki, R., 147, *154*
Solovev, V. A., *68, 75, 82*
Soren, M. S., *69*
Stanley, H. E., 259, *341*
Starunov, V. S., 44(224,256), *68, 69, 75, 77, 87, 89–92*
Steele, W. A., *76, 77*
Stegeman, G. I. A., 41(115), 44(259,269), 45(259), *67, 69, 77, 78*
Stenzel, R., 339, *343*
Stepanova, T. S., *91*

Stephen, M. J., 51, 58, *73, 79, 82*
Stevens, B., 59
Stinson, T. W., 47–49, 50(280), *78*
Stoicheff, B. P., 41(115), 44, 45(259), 52(169), *66, 67, 71, 77, 78*
Stoicheff, P. P., 41(121), *68*
Stolen, R. H., 41(115), *67*
Stolyov, S. P., *82*
Stoner, E. C., 245, *255*
St. Peters, R. L., *69*
Stranahan, G., 239–242, *255*
Sugawara, A., 8(72), *64, 69*
Surko, C. M., 50(168), 55, *71, 72*
Swift, J., 258, *341*
Swinney, H. L., 2, 32, 33(3), 34, 39(58,63, 81,82), 40, 42(111), 43(111), *60, 61, 63–67, 85, 86, 93*
Szoke, A., *77*

Tabibian, R., *81*
Tabisz, G. C., *72*
Takuma, H., *90, 92*
Tanaka, M., *64, 69*
Tang, C. L., *92*
Ter-Mikablyan, M. L., *69*
Ter-Mikaelyan, M. L., *64*
Tham, M. K., *332, 343*
Thibeau, M., *72*
Thiebeau, M., *73*
Thiel, D., 40(354), *83, 86*
Thompson, C. J., 209, *254*
Tiganov, E. V., 44(256), *77, 90, 91*
Tikhonov, E. A., *92*
Timchenko, A. I., *66*
Tolmachev, V. V., 190, 238(70), 240(70), 242(70), 244(70), *254, 255*
Tong, E., 19(30), *62*
Tosi, S., 322, *342*
Townes, C. H., *89*
Trotter, H., 210, *254*
Truesdell, C., 109, 143, 147, *154*
Tscharnuter, W., 39(40), 40(354), *63, 83, 84, 86*
Tserkovnikov, Iu. A., 208, 234, *254*
Tsintsadze, N. L., *92*
Tukhvatuilin, E. K. H., *62*
Tukhvatullin, F. K. M., *77*
Turner, R., *67*

Uchida, N., *70*
Uhlenbeck, G. E., 157, 178(44), 182(44), *184, 185*
Ulam, S. M., 178, *185*

Valatin, J. G., 235(68), *255*
Vallauri, R., 41(38), *62, 80*
Vaughan, J. M., *69, 82*
Vaughn, J. M., *64, 69, 82*
Venkateswaran, C. S., *70*
Verlet, L., 318
Vinen, W. F., *64, 69, 70, 82*
Vodar, B., *72, 88*
Volterra, V., 41(115), 46, 50(187), 152(186), *67, 69, 72, 77*
Voronel, A. V., *78, 79, 84*
Voucamps, C., *74, 80*
Vrij, V., 57(212), *74*
Vuks, M. F., *82*

Wainwright, T. E., 325–327, 329(82,84), *342, 343*
Walda, G., *92*
Walder, J., *92*
Walker, G. H., 156(7), 172, 183(7), *184*
Wang, C.-C., 98(21), 109, 116, *153, 154*
Wang, C. H., *72, 77*
Wang, C. S., *63, 65, 67, 70, 81, 83, 87, 92*
Wang, M. C., 178(44), 182(44), *185*
Webb, J. P., *86*
Webb, W. W., 55(215), 57(215), *73, 74, 85, 86*
Wei, C. C., 278, 312, *341, 342*
Weinberg, D. L., *93*
Weinzierl, P., 39(40), *63, 84*
Welsh, H. L., 15(66), *64, 69*
Wendl, G., *90*
Wentzel, G., 208, 237, 248, *254, 255*
Whittaker, E. T., 98, *154*, 163(23), 164(23), *184*
Whittle, C. D., 52(163), *71*
Wick, R. V., *92*
Widom, A., 325(81), *343*
Wiggins, T. A., *63, 67, 88, 89, 91, 92*
Wightman, A. S., 156(10), 157(10), 158(10), 162(10), 183, *184*
Wigner, E. P., 198(19), *254*
Wilcox, R. M., 192(14), *254*
Wilkinson, C. D. W., *61*
Wims, A., *80, 84*
Winslow, D. K., *61*

Winterling, G., *89, 92*
Woerner, R., 52(198), *73*
Wolf, E., 26(509), *93*
Wolinski, I. M., *93*
Woods, G., 58(361), *83*
Woolf, M. A., *66, 70*
Worlock, J. M., 52(166), 53(166), 54(166), *71*
Wright, D., 247(79), *255*
Wright, R. B., *72, 77*
Wu, E. S., 55(215), 57(215), *74, 86*
Wysocki, J., 57(273), *73, 78*

Yajima, T., *90*
Yamakawa, H., 229, *255*
Yan, J., 51, 52, *73*
Yang, C. C., *63, 67, 81*
Yeh, R., 229, *255*
Yeh, R. H. T., 211, 223, *254*
Yeh, Y., 2, 22(346,347,358), 59(346,347), *63, 65, 79, 82, 83, 86*
Yip, S., 8(72,76,133), 45(44), 46, *63–66, 68–70, 83,* 312, 315, 318, 323, *342*
Yokota, M., 308, *342*
Yphantis, D. A., 22(358), *83*
Yulmetev, R. M., *65*

Zaanen, A. C., 114, *154*
Zaitsev, G. I., *92*
Zaitsev, G. L., *77*
Zamir, E., *75, 77*
Zaslavskii, G. M., 177(40), 178, *185*
Zawadowski, A., *72*
Zhdanova, N. F., 318
Zhong-Heng, Guo, 147, *154*
Zollweg, J., 55(213), 56, 57(213), *74, 86*
Zollwegg, J., *65*
Zoppi, M., *87*
Zubarev, D. B., 208, 234, *254*
Zubkov, L. A., *76, 77*
Zverev, G. Z., *92*
Zwanzig, R., 10, 12(511), *93,* 258(1,3), 300, 302, 303, 308, 309, 315, 318, 319, 325(78), *340–343*

SUBJECT INDEX

Adiabatic process, 107
Admissible process, 96, 98, 106, 120
Allen and Cole kinetic theory, 292
Anisotropic molecules, 43
Applied forces, 96, 103
Area-preserving mapping, 160, 164, 174, 177, 179
Argon, 40, 50, 52, 53

Baker's transformation, 177
Benard instability, 24
Bifurcation, 238, 242, 244, 246
Boltzmann's collision operator, 295
Bounds on derivative, 220
Bragg-Williams approximation, 243, 244
Brillouin, 19
  modes, 23
  peaks, 42
  scattering, 8, 40, 48
  spectra, 20, 43
  width, 23
Bulk viscosity, 273, 277, 284, 287, 305, 310, 314, 329, 334

Canonical form of the dynamical equations, 137
Canonical free energy, 108, 137
Chain rule, 116
Chemical reaction, 2, 22, 59
Clipped correlation, 19, 36
CO_2, 42, 57
Coherence area, 36
Collision induced, 50
Compound interferometers, 27
Concentration, 24
  fluctuations, 4, 43
  mode, 25
Condensation, 239
Configurational coordinate, 96, 106
Constitutive assumption, 96, 97, 109
Constitutive equations, 106, 109
Constitutive functionals, 119–120
  smoothness of pages, 119–120
  restrictions on, 123–124 ff
Convection, 24
Convexity, 193, 212, 221, 222, 251
Correlation function, 32, 33, 34, 43
Correlation lengths, 38, 40, 47
Creation operator, 279
Critical exponents, 39
Critical mixing, 40
Critical phenomenon, 2, 3, 8
Critical transport phenomena, 258
  microscopic model of spin systems of resibois and De Leener, 258
  semi-microscopic theory of fluids and spin systems of Kadanoff and Swift, 258
Cross relations, 127, 128, 132
C-systems, 159, 160, 171
Cylindrical tubes, 139–145
  Hamiltonian functional, 144
  inertia of, 141
  Lagrangian functional, 144

Density correlation function, 51, 55
Density fluctuations, 43
Depolarized Rayleigh wing, 50
Depolarized scattering, 43
Depolarized spectra, 44, 46
Derivatives, 116
  instantaneous, 116
  past-history, 116
Diagrams, 292
Diagram theory, 257
Diffusion coefficients, 58, 309
Digital correlators, 3, 34
Dispersion, 20, 22
  effects, 20
Dominated convergence property, 114
Doublet correlation function, 261
Doublet distribution function, 259, 270
Dynamic liquid structure, 2
Dynamic shear waves, 44

Effective potential kinetic theory, 259, 265, 263, 264

doublet distribution function, 263
kinetic equation, 264
nondissipative hydrodynamic equations, 267
spatial and time delocalization effects, 265
Effective potential theory, 271
numerical predictions, 274
transport coefficients, 271
Electronic spectroscopy, 2, 31, 32
Enskog's theory of hard sphere fluids, 328
agreement between Enskog's theory and dense gas experiments, 330
Entropy, 96, 98, 103
Ergodicity, 156, 157
Excluded volume, 227, 228
Exponential behavior, 171
Exponential instability, 159, 160, 171

Fabry-Perot interferometer, 26, 35
free spectral range, 26
finesse, 26
resolving power, 26
Fading memory, 108–122
generalizations of, 148–152
principle of, 120
Fokker-Planck equation, 182
Frechet derivative, 115
Free volume theory, 225, 226
Full correlation, 36
Functional differential equations, 99
Function norm, 110

Gas mixtures, 24
Gaussian fields, 33
Generalized hydrodynamics, 2, 3, 9
Generalized hydrodynamics equations, 11
Generalized Rice-Allnatt theory, 289
Grating spectrometer, 25

Hammerstein theorems, 239, 240
Hamiltonian functional, 137
Hartree-Fock method, 195, 197, 203, 237, 239, 245
Helium, 41, 51
Hellman-Feynman theorem, 199
Helmholtz free energy, 97, 98, 129
High frequency effects, 16
Histories, 110
History, 112
of a function, 106
past, 112
smooth, 117
tame, 114
Homogeneous solutions of exact linearized singlet kinetic equation, 304
Hydrodynamic instabilities, 4
Hydrodynamic limit, 12, 19
Hydrodynamic modes, 13
Brillouin modes, 13
thermal diffusivity mode, 13
thermal relaxation mode, 13
viscous relaxation mode, 13
Hypersonic velocity, 40
Hypersound dispersion, 9
Hypersound velocity, 21

I_2 absorption cell, 30
Inertia function, 97, 103
Inertial force, 97, 104
Inertia of tensor, 97, 104, 141, 146
Influence measure, 110
Instantaneous Poisson bracket, 137, 138
Instantaneous derivative, 116
Integrated intensity, 47
Intensity correlations, 31, 32
Interference, 46
Intermolecular light scattering, 44, 50, 55
Intermolecular spectra, 52
Internal energy, 105
Internal force relation, 97, 98, 104, 105, 126
Irreversible rate equation, 176, 178
Isothermal process, 107
Isotope mixtures, 248

Kinetic and transport properties of many-body systems, 257
Kinetic theory of polyatomic systems, 258
Krypton, 50

Lagrangian functional, 135
Landon-Placzek ratio, 15, 19
Lasers, 25
Lattice vibration, 163, 172
Law of balance of energy, 96, 103
Light beating spectroscopy, 56
Liquid crystals, 3, 46, 49, 57
Liquid mixtures, 22
Loaded sphere fluids, 336
self diffusion coefficient, 339
Lower bounds, 209

Golden-Thompson inequality, 209
reduced density matrices, 213

Macromolecules, 57, 58
Master equations, 300
from perturbation theory, 300
from projection operator methods, 302
Matsubara's theorem, 201, 202, 204, 208, 236
Memory functions, 318
derivatives at 5 = 0, 319
exponential model, 320
Gaussian models, 322
relation to transport coefficient, 320
Mermin model, 241
Michelson interferometer, 27
Mixing, 157, 158
Molecular field model, 242
Molecular field theory, 232
Monotomic liquid, 40

Nematic liquid, 49
Neon, 40, 52
Nonequilibrium phenomenon, 24
Nonlinear oscillator systems, 163
Nonlinear resonances, 191
Numerical studies of nonequilibrium diagrams, 292
Gaussian potential energy function, 295
Lennard Jones potential energy function, 295

Orientation fluctuation, 45, 50

Past history derivative, 116
Perturbation theory, 207, 231
Phase transitions, 3
Phase velocity, 20
Photocurrent spectrum, 33
Poincaré surface of section, 164
Prigogine-Nicolas-Misguich kinetic theory, 259, 278, 280, 281, 283
comparison with experiment, 285
doublet correlation function, 281
Principle of fading memory, 124

Quasi-particles, 197, 198, 202, 204, 207, 208, 235, 237

Rayleigh-Brillouin spectrum, 19, 38, 41
Rayleigh number, 24
Rayleigh peak, 23, 25
Rayleigh width, 39, 40
Regularity, 115
preservation of, 115
Relaxation, 16
mountain peak, 16
Relaxation processes, 14, 20, 41, 43, 46
Relaxation property, 113, 128
Representability problems, 213, 216
Resonant reabsorption contrast, 29
Rice-Allnatt theory, 285
comparison with experiment, 285
doublet distribution function, 275
doublet kinetic equation, 275
singlet kinetic equation, 275
transport coefficients, 276
Ripplons, 55, 57
Rotational diffusion, 43
Roton pairs, 51
Roton spectra, 52
Rough sphere fluids, 338
self diffusion coefficient, 339

Scaling laws, 38
Scattering intensity spectrum, 6
σ-Section, 111
Second law of thermodynamics, 106–108, 122, 123
Second order Raman, 51, 55
Second virial coefficient, 223
Self-diffusion, 274, 277, 309, 313, 321, 323, 328, 333, 339, 392
Sets of histories, 109
SF_6, 42, 47
Shear viscosity, 272, 277, 284, 286, 291, 310, 329, 334
Shear waves, 50
Singlet distribution function, 259
Smooth thermodynamic process, 103
Solutions, 22, 57, 59
Solution theory, 230
Sound velocity, 22, 41
Spectrum, 18, 22, 23
Spectrum analyzers, 36
Spherical shells, 145–148
Hamiltonian functional, 148
inertia of, 146
Lagrangian functional, 147–148
Spider, dangling, 100–102, 138, 139

Square-well fluids, 332
  equation of state and transport coefficients, 333
Static continuation, 111
Statistical mechanical variational principles, 188, 189, 193
  Gibbs-Bogoliubov inequality, 189
  Jordan-Golden inequality, 194
  Peierls inequality, 193
Stoner theory, 245
Superconductivity, 233
Surface scattering, 55
Surface tension, 55, 56
Systems with memory, 95

Tandem-Fabry-Perot, 3
Tandem spherical (FP) interferometer, 27
Temperature, 96, 103
Tensor, 141, 146
  inertia of, 141, 146
Thermal conductivity, 272, 277, 283, 291, 310, 315, 329, 333
Thermal diffusivity, 38, 310
Thermal relaxation, 9, 20, 21
Time correlation formulas for the transport coefficients, 308
Time correlation functions, 311
  a Gaussian approximation, 312
  derivative at t = 0, 312
  long time decay, 325
Transport coefficients, 257, 283
Transport fluxes, 257
Transport function, 17
  diffusion mode, 18
  "high frequency" effects, 18
  second order effects, 18

Ultrasonics, 20
Ultrasonic velocity, 21

Velocity correlation functions, long time tail, 325

Xenon, 42, 56